D1693716

Flow Cytometry with Plant Cells

Edited by
Jaroslav Doležel, Johann Greilhuber, and Jan Suda

1807–2007 Knowledge for Generations

Each generation has its unique needs and aspirations. When Charles Wiley first opened his small printing shop in lower Manhattan in 1807, it was a generation of boundless potential searching for an identity. And we were there, helping to define a new American literary tradition. Over half a century later, in the midst of the Second Industrial Revolution, it was a generation focused on building the future. Once again, we were there, supplying the critical scientific, technical, and engineering knowledge that helped frame the world. Throughout the 20th Century, and into the new millennium, nations began to reach out beyond their own borders and a new international community was born. Wiley was there, expanding its operations around the world to enable a global exchange of ideas, opinions, and know-how.

For 200 years, Wiley has been an integral part of each generation's journey, enabling the flow of information and understanding necessary to meet their needs and fulfill their aspirations. Today, bold new technologies are changing the way we live and learn. Wiley will be there, providing you the must-have knowledge you need to imagine new worlds, new possibilities, and new opportunities.

Generations come and go, but you can always count on Wiley to provide you the knowledge you need, when and where you need it!

William J. Pesce
President and Chief Executive Officer

Peter Booth Wiley
Chairman of the Board

Flow Cytometry with Plant Cells

Analysis of Genes, Chromosomes and Genomes

Edited by
Jaroslav Doležel, Johann Greilhuber, and Jan Suda

WILEY-VCH Verlag GmbH & Co. KGaA

The Editors

Prof. Dr. Jaroslav Doležel
Inst. of Experimental Botany
Lab. Mol. Cytogenetics & Cytometry
Sokolovská 6
77200 Olomouc
Czech Republic

Prof. Dr. Johann Greilhuber
University of Vienna
Department of Systematic and
Evolutionary Botany
Rennweg 14
1030 Vienna
Austria

Dr. Jan Suda
Charles Univ., Fac. of Science
Dept. of Botany
Benátská 2
12801 Prague
Czech Republic

and

Institute of Botany
Academy of Sciences
Průhonice 1
25243 Průhonice
Czech Republic

■ All books published by Wiley-VCH are carefully produced. Nevertheless, authors, editors, and publisher do not warrant the information contained in these books, including this book, to be free of errors. Readers are advised to keep in mind that statements, data, illustrations, procedural details or other items may inadvertently be inaccurate.

Library of Congress Card No.: applied for
British Library Cataloguing-in-Publication Data
A catalogue record for this book is available from the British Library.

Bibliographic information published by the Deutsche Nationalbibliothek
Die Deutsche Nationalbibliothek lists this publication in the Deutsche Nationalbibliografie; detailed bibliographic data are available in the Internet at ⟨http://dnb.d-nb.de⟩.

© 2007 WILEY-VCH Verlag GmbH & Co. KGaA, Weinheim

All rights reserved (including those of translation into other languages). No part of this book may be reproduced in any form – by photoprinting, microfilm, or any other means – nor transmitted or translated into a machine language without written permission from the publishers. Registered names, trademarks, etc. used in this book, even when not specifically marked as such, are not to be considered unprotected by law.

Printed in the Federal Republic of Germany
Printed on acid-free paper

Typesetting Asco Typesetters, Hong Kong
Printing betz-Druck GmbH, Darmstadt
Binding Litges & Dopf Buchbinderei GmbH, Heppenheim
Wiley Bicentennial Logo Richard J. Pacifico

ISBN 978-3-527-31487-4

Contents

Preface *XVII*

List of Contributors *XXI*

1   **Cytometry and Cytometers: Development and Growth** *1*
Howard M. Shapiro

         Overview *1*
1.1   Origins *1*
1.2   From Absorption to Fluorescence, from Imaging to Flow *2*
1.2.1   Early Microspectrophotometry and Image Cytometry *3*
1.2.2   Fluorescence Microscopy and the Fluorescent Antibody Technique *3*
1.2.3   Computers Meet Cytometers: The Birth of Analytical Flow Cytometry *4*
1.2.4   The Development of Cell Sorting *7*
1.3   The Growth of Multiparameter Flow Cytometry *8*
1.4   Bench-tops and Behemoths: Convergent Evolution *11*
1.5   Image Cytometry: New Beginnings? *14*
         References *15*

2   **Principles of Flow Cytometry** *19*
J. Paul Robinson and Gérald Grégori

         Overview *19*
2.1   Introduction *19*
2.2   A Brief History of Flow Cytometry *20*
2.3   Components of a Flow Cytometer *21*
2.3.1   Fluidics *22*
2.3.2   Optics *25*
2.3.3   Electronic Systems *27*
2.4   Flow Cytometric Informatics *30*
2.5   Spectral Compensation *33*
2.6   Cell Sorting *34*
2.7   Calibration Issues *37*

Flow Cytometry with Plant Cells. Edited by Jaroslav Doležel, Johann Greilhuber, and Jan Suda
Copyright © 2007 WILEY-VCH Verlag GmbH & Co. KGaA, Weinheim
ISBN: 978-3-527-31487-4

2.8 Conclusions 37
References 39

3 Flow Cytometry with Plants: an Overview 41
Jaroslav Doležel, Johann Greilhuber, and Jan Suda

Overview 41
3.1 Introduction 42
3.2 Fluorescence is a Fundamental Parameter 43
3.3 Pushing Plants through the Flow Cytometer 44
3.3.1 Difficulties with Plants and their Cells 44
3.3.2 Protoplasts are somewhat "Easier" than Intact Cells 45
3.3.3 Going for Organelles 46
3.4 Application of Flow Cytometry in Plants 47
3.4.1 Microspores and Pollen 47
3.4.2 Protoplasts 47
3.4.2.1 Physiological Processes 48
3.4.2.2 Secondary Metabolites 48
3.4.2.3 Gene Expression 48
3.4.2.4 Somatic Hybrids 49
3.4.2.5 DNA Transfection 49
3.4.3 Cell Nuclei 49
3.4.3.1 Ploidy Levels 50
3.4.3.2 Aneuploidy 51
3.4.3.3 B Chromosomes 52
3.4.3.4 Sex Chromosomes 52
3.4.3.5 Cell Cycle and Endopolyploidy 52
3.4.3.6 Reproductive Pathways 53
3.4.3.7 Nuclear Genome Size 54
3.4.3.8 DNA Base Content 55
3.4.3.9 Chromatin Composition 56
3.4.3.10 Sorting of Nuclei 56
3.4.4 Mitotic Chromosomes 57
3.4.5 Chloroplasts 57
3.4.6 Mitochondria 58
3.4.7 Plant Pathogens 58
3.4.8 Aquatic Flow Cytometry 59
3.5 A Flow Cytometer in Every Laboratory? 59
3.6 Conclusions and Future Trends 60
References 61

4 Nuclear DNA Content Measurement 67
Johann Greilhuber, Eva M. Temsch, and João C. M. Loureiro

Overview 67
4.1 Introduction 67
4.2 Nuclear DNA Content: Words, Concepts and Symbols 69

4.2.1	Replication–Division Phases	69
4.2.2	Alternation of Nuclear Phases	70
4.2.3	Generative Polyploidy Levels	70
4.2.4	Somatic Polyploidy	71
4.3	Units for Presenting DNA Amounts and their Conversion Factors	72
4.4	Sample Preparation for Flow Cytometric DNA Measurement	74
4.4.1	Selection of the Tissue	74
4.4.2	Reagents and Solutions	75
4.4.2.1	Isolation Buffers and DNA Staining	76
4.5	Standardization	80
4.5.1	Types of Standardization	80
4.5.2	Requirement of Internal Standardization – a Practical Test	82
4.5.3	Choice of the Appropriate Standard Species	83
4.5.3.1	Biological Similarity	83
4.5.3.2	Genome Size	84
4.5.3.3	Nature of the Standard	84
4.5.3.4	Availability	84
4.5.3.5	Cytological Homogeneity	85
4.5.3.6	Accessibility	85
4.5.3.7	Reliability of C-Values	85
4.5.4	Studies on Plant Standards	86
4.5.5	Suggested Standards	88
4.6	Fluorescence Inhibitors and Coatings of Debris	89
4.6.1	What are Fluorescence Inhibitors and Coatings of Debris?	89
4.6.2	Experiments with Tannic Acid	92
4.6.3	A Flow-cytometric Test for Inhibitors	95
4.7	Quality Control and Data Presentation	95
4.8	Future Directions	98
	References	99

5	**Flow Cytometry and Ploidy: Applications in Plant Systematics, Ecology and Evolutionary Biology**	**103**
	Jan Suda, Paul Kron, Brian C. Husband, and Pavel Trávníček	
	Overview	103
5.1	Introduction	103
5.2	Practical Considerations	104
5.2.1	Relative DNA Content, Ploidy and Flow Cytometry	104
5.2.2	General Guidelines for Ploidy-level Studies	105
5.2.3	Use of Alternative Tissues	108
5.2.3.1	Preserved or Dormant Tissue	108
5.2.3.2	Pollen	111
5.2.4	Other Considerations/Pitfalls	113
5.2.4.1	Holokinetic Chromosomes (Agmatoploidy)	113
5.2.4.2	DNA Content Variation within Individuals	113

5.3	Applications in Plant Systematics *114*
5.3.1	Systematics of Heteroploid Taxa *114*
5.3.1.1	Detecting Rare Cytotypes *117*
5.3.1.2	Phylogenetic Inference *117*
5.3.2	Systematics of Homoploid Taxa *118*
5.4	Applications in Plant Ecology and Evolutionary Biology *119*
5.4.1	Spatial Patterns of Ploidy Variation *119*
5.4.1.1	Invasion Biology *119*
5.4.2	Evolutionary Dynamics of Populations with Ploidy Variation *120*
5.4.3	Ploidy Level Frequencies at Different Life Stages (Temporal Variation) *121*
5.4.4	Reproductive Pathways *122*
5.4.4.1	Unreduced Gametes and Polyploidy *122*
5.4.4.2	Asexual Seed Production *124*
5.4.4.3	Hybridization *124*
5.4.5	Trophic Level Interactions and Polyploidy *125*
5.5	Future Directions *126*
	References *128*

6 Reproduction Mode Screening *131*
Fritz Matzk

	Overview *131*
6.1	Introduction *131*
6.2	Analyses of the Mode of Reproduction *134*
6.2.1	Traditional Techniques *134*
6.2.2	Ploidy Analyses of Progenies Originating from Selfing or Crossing *139*
6.2.2.1	Identification of B_{III}, B_{IV} and M_I Individuals after Selfing or Intraploidy Pollinations *139*
6.2.2.2	Crossing of Parents with Different Ploidy or with Dominant Markers *140*
6.2.3	Flow Cytometric Analyses of the Relative DNA Content of Microspores or Male Gametes *141*
6.2.4	The Ploidy Variation of Embryo and Endosperm Depending on the Reproductive Mode *142*
6.3	A Recent Innovative Method: the Flow Cytometric Seed Screen *142*
6.3.1	Advantages and Limitations of the FCSS *143*
6.3.2	Applications of the FCSS *146*
6.3.2.1	Botanical Studies *146*
6.3.2.2	Evolutionary Studies *147*
6.3.2.3	Genetical Analyses of Apomixis *147*
6.3.3	Methodological Implications *147*
6.4	Flow Cytometry with Mature Seeds for other Purposes *149*
6.5	Conclusions *150*
	References *151*

7	**Genome Size and its Uses: the Impact of Flow Cytometry** *153*	
	Ilia J. Leitch and Michael D. Bennett	
	Overview *153*	
7.1	Introduction *153*	
7.2	Why is Genome Size Important? *154*	
7.3	What is Known about Genome Size in Plants? *155*	
7.3.1	Angiosperms *156*	
7.3.2	Gymnosperms *157*	
7.3.3	Pteridophytes *158*	
7.3.4	Bryophytes *158*	
7.3.5	Algae *158*	
7.4	The Extent of Genome Size Variation across Plant Taxa *159*	
7.5	Understanding the Consequences of Genome Size Variation: Ecological and Evolutionary Implications *160*	
7.5.1	Influence of Genome Size on Developmental Lifestyle and Life Strategy *161*	
7.5.2	Ecological Implications of Genome Size Variation *163*	
7.5.3	Implications of Genome Size Variation on Plants' Responses to Environmental Change *166*	
7.5.3.1	Genome Size and Plant Response to Pollution *166*	
7.5.3.2	Genome Size and Threat of Extinction *166*	
7.5.4	Consequences of Genome Size Variation for Survival in a Changing World *167*	
7.6	Methods of Estimating Genome Size in Plants and the Impact of Flow Cytometry *168*	
7.6.1	The Development of Flow Cytometry for Genome Size Estimation in Angiosperms *169*	
7.6.1.1	Choice of Fluorochromes *169*	
7.6.1.2	Internal Standardization *169*	
7.6.1.3	The Need for Cytological Data *170*	
7.6.1.4	Awareness of the Possible Interference of DNA Staining *170*	
7.6.2	Potential for the Application of Flow Cytometry to Other Plant Groups *171*	
7.6.2.1	Gymnosperms *171*	
7.6.2.2	Pteridophytes *172*	
7.6.2.3	Bryophytes *172*	
7.7	Recent Developments and the Future of Flow Cytometry in Genome Size Research *172*	
	References *174*	
8	**DNA Base Composition of Plant Genomes** *177*	
	Armin Meister and Martin Barow	
	Overview *177*	
8.1	Introduction *177*	
8.2	Analysis of Base Composition by Flow Cytometry *178*	

8.2.1	Fluorescence of Base-Specific Dyes: Theoretical Considerations	180
8.2.2	Base Composition of Plant Species Determined by Flow Cytometry and its Relation to Genome Size and Taxonomy	185
8.2.3	Comparison of Flow Cytometric Results with Base Composition Determined by other Physico-Chemical Methods	204
8.2.4	Possible Sources of Error in Determination of Base Composition by Flow Cytometry	205
8.3	Conclusions	211
	References	213

9 Detection and Viability Assessment of Plant Pathogenic Microorganisms using Flow Cytometry 217
Jan H. W. Bergervoet, Jan M. van der Wolf, and Jeroen Peters

	Overview	217
9.1	Introduction	217
9.2	Viability Assessment	218
9.2.1	Viability Tests for Spores and Bacteria	219
9.3	Immunodetection	222
9.3.1	Microsphere Immuno Assay	224
9.3.1.1	Detection of Plant Pathogenic Bacteria and Viruses	225
9.3.1.2	Paramagnetic Microsphere Immuno Assay	226
9.4	Conclusions and Future Prospects	227
	References	229

10 Protoplast Analysis using Flow Cytometry and Sorting 231
David W. Galbraith

	Overview	231
10.1	Introduction	231
10.1.1	Protoplast Preparation	231
10.1.2	Adaptation of Flow Cytometric Instrumentation for Analysis of Protoplasts	233
10.1.3	Parametric Analyses Available for Protoplasts using Flow Cytometry	234
10.2	Results of Protoplast Analyses using Flow Cytometry and Sorting	237
10.2.1	Protoplast Size	237
10.2.2	Protoplast Light Scattering Properties	238
10.2.3	Protoplast Protein Content	239
10.2.4	Protoplast Viability and Physiology	239
10.2.5	Protoplast Cell Biology	243
10.2.6	Construction of Somatic Hybrids	244
10.2.7	The Cell Cycle	244
10.3	Walled Plant Cells: Special Cases for Flow Analysis and Sorting	246
10.4	Prospects	247
	References	248

11	**Flow Cytometry of Chloroplasts** *251*
	Erhard Pfündel and Armin Meister
	Overview *251*
11.1	Introduction *251*
11.1.1	The Chloroplast *252*
11.2	Chloroplast Signals in Flow Cytometry *255*
11.2.1	Autofluorescence *255*
11.2.2	Light Scattering *259*
11.3	Progress of Research *259*
11.3.1	Chloroplasts from C_3 Plants *260*
11.3.2	Chloroplasts from C_4 Plants *261*
11.4	Conclusion *263*
	References 264

12	**DNA Flow Cytometry in Non-vascular Plants** *267*
	Hermann Voglmayr
	Overview *267*
12.1	Introduction *267*
12.2	Nuclear DNA Content and Genome Size Analysis *271*
12.2.1	General Methodological Considerations *272*
12.2.1.1	Isolation and Fixation of Nuclei *272*
12.2.1.2	Standardization *274*
12.2.1.3	Fluorochromes for Estimation of Nuclear DNA Content *275*
12.2.1.4	Secondary Metabolites as DNA Staining Inhibitors *276*
12.2.2	DNA Content and Genome Size Studies *276*
12.2.2.1	Algae *277*
12.2.2.2	Bryophytes *280*
12.3	Future Perspectives *283*
12.4	Conclusion *284*
	References 285

13	**Phytoplankton and their Analysis by Flow Cytometry** *287*
	George B. J. Dubelaar, Raffaella Casotti, Glen A. Tarran, and Isabelle C. Biegala
	Overview *287*
13.1	Introduction *288*
13.2	Plankton and their Importance *288*
13.2.1	Particles in Surface Water *288*
13.2.2	Phytoplankton *289*
13.2.3	Distributions in the Aquatic Environment *289*
13.3	Considerations for using Flow Cytometry *291*
13.3.1	Analytical Approach *291*
13.3.2	Limitations and Pitfalls of using Biomedical Instruments *292*
13.3.3	Instrument Modification and Specialized Cytometers *293*

13.3.4	Sizing and Discrimination of Cells 295
13.3.5	More Information per Particle: From Single Properties to (Silico-) Imaging 297
13.4	Sampling: How, Where and When 301
13.4.1	Sample Preparation 301
13.4.2	Critical Scales and Sampling Frequency 302
13.4.3	Platforms for Aquatic Flow Cytometry 303
13.5	Monitoring Applications 305
13.5.1	Species Screening: Cultures 305
13.5.2	Phytoplankton Species Biodiversity 307
13.5.3	Harmful Algal Blooms 308
13.6	Ecological Applications 308
13.6.1	Population-related Processes 308
13.6.2	Cell-related Processes and Functioning 311
13.6.3	Plankton Abundance Patterns in the Sea: Indicators of Change 314
13.7	Marine Optics and Flow Cytometry 314
13.8	Future Perspectives 315
	References 319

14 Cell Cycle Analysis in Plants 323
Martin Pfosser, Zoltan Magyar, and Laszlo Bögre

Overview 323
- 14.1 Introduction 323
- 14.2 Univariate Cell Cycle Analysis in Plant Cells 325
- 14.3 BrdUrd Incorporation to Determine Cycling Populations 326
- 14.4 Cell Cycle Synchronization Methods: Analysis of Cell Cycle Transitions in Cultured Plant Cells 327
- 14.5 Plant Protoplasts to Study the Cell Cycle 335
- 14.6 Root Meristems for Cell Cycle Synchronization 335
- 14.7 Study of Cell Cycle Regulation by using Synchronized Cell Cultures and Flow Cytometry 336
- 14.8 Cell Cycle and Plant Development 338
- 14.9 Flow Cytometry of Dissected Tissues in Developmental Time Series 339
- 14.10 Cell Type-specific Characterization of Nuclear DNA Content by Flow Cytometry 339
- 14.11 Other Methods and Imaging Technologies to Monitor Cell Cycle Parameters and Cell Division Kinetics in Developing Organs 340
- 14.12 Concluding Remarks 342
  References 343

15 Endopolyploidy in Plants and its Analysis by Flow Cytometry 349
Martin Barow and Gabriele Jovtchev

Overview 349
- 15.1 Introduction 349

15.2	Methods to Analyze Endopolyploidy	*351*
15.2.1	Microscopy	*351*
15.2.1.1	Chromosome Counts	*351*
15.2.1.2	Feulgen Microdensitometry, Fluorescence Microscopy, Image Analysis	*352*
15.2.2	Flow Cytometry	*352*
15.2.2.1	Evaluation of Histograms	*353*
15.2.2.2	Quantification of the Degree of Endopolyploidy	*354*
15.3	Occurrence of Endopolyploidy	*355*
15.3.1	Endopolyploidy in Species	*356*
15.3.2	Endopolyploidy in Ecotypes and Varieties	*356*
15.3.3	Endopolyploidy in Different Life Strategies	*357*
15.3.4	Endopolyploidy in Organs	*359*
15.4	Factors Modifying the Degree of Endopolyploidization	*362*
15.4.1	Genome Size and Endopolyploidy	*362*
15.4.2	Environmental Factors	*363*
15.4.3	Symbionts and Parasites	*364*
15.4.4	Phytohormones	*365*
15.5	Dynamics of Endopolyploidization	*366*
15.6	Endopolyploidy and Plant Breeding	*367*
15.6.1	Endopolyploidy in Crop Plants	*367*
15.6.2	*In vitro* Culture and Plant Regeneration	*368*
15.7	Conclusions	*369*
	References	*370*
16	**Chromosome Analysis and Sorting**	*373*
	Jaroslav Doležel, Marie Kubaláková, Pavla Suchánková, Pavlína Kovářová, Jan Bartoš, and Hana Šimková	
	Overview	*373*
16.1	Introduction	*374*
16.2	How Does it Work?	*375*
16.3	How it All Began	*377*
16.4	Development of Flow Cytogenetics in Plants	*379*
16.4.1	Preparation of Suspensions of Intact Chromosomes	*379*
16.4.1.1	Biological Systems for Chromosome Isolation	*379*
16.4.1.2	Cell Cycle Synchronization and Metaphase Accumulation	*383*
16.4.1.3	Preparation of Chromosome Suspensions	*383*
16.4.2	Chromosome Analysis	*385*
16.4.2.1	Bivariate Analysis of AT and GC Content	*385*
16.4.2.2	Fluorescent Labeling of Repetitive DNA	*386*
16.4.2.3	The Use of Cytogenetic Stocks	*386*
16.4.2.4	Assignment of Chromosomes to Peaks on Flow Karyotypes	*386*
16.4.3	Chromosome Sorting	*387*
16.4.3.1	Estimating the Purity in Sorted Fractions	*389*
16.4.3.2	Improving the Sort Purity	*389*

16.4.3.3 Two-step Sorting *389*
16.4.3.4 Purities and Sort Rates Achieved *390*
16.5 Applications of Flow Cytogenetics *390*
16.5.1 Flow Karyotyping *390*
16.5.2 Chromosome Sorting *392*
16.5.2.1 Physical Mapping and Integration of Genetic and Physical Maps *392*
16.5.2.2 Cytogenetic Mapping *392*
16.5.2.3 Analysis of Chromosome Structure *396*
16.5.2.4 Targeted Isolation of Molecular Markers *396*
16.5.2.5 Recombinant DNA Libraries *396*
16.6 Conclusions and Future Prospects *398*
References *400*

17 Analysis of Plant Gene Expression Using Flow Cytometry and Sorting *405*
David W. Galbraith

Overview *405*
17.1 Introduction *405*
17.2 Methods, Technologies, and Results *406*
17.2.1 Current Methods for Global Analysis of Gene Expression *406*
17.2.1.1 Methods Based on Hybridization *407*
17.2.1.2 Methods Based on Sequencing *408*
17.2.1.3 Emerging Sequencing Technologies *409*
17.2.1.4 Other -omics Disciplines and Technologies *410*
17.2.2 Using Flow Cytometry to Monitor Gene Expression and Cellular States *411*
17.2.2.1 Transgenic Markers Suitable for Flow Cytometry and Sorting *411*
17.2.2.2 Subcellular Targeting as a Means for Transgenic Analysis *412*
17.2.3 Using Flow Sorting to Measure Gene Expression and Define Cellular States *414*
17.2.3.1 Protoplast and Cell Sorting Based on Endogenous Properties *414*
17.2.3.2 Protoplast Sorting Based on Transgenic Markers *416*
17.2.3.3 Sorting of Nuclei Based on Transgenic Markers *417*
17.3 Prospects *418*
17.3.1 Combining Flow and Image Cytometry *418*
17.3.2 Use of Protoplasts for Confirmatory Studies *418*
17.3.3 Analysing Noise in Gene Expression *419*
References *421*

18 FLOWer: A Plant DNA Flow Cytometry Database *423*
João Loureiro, Jan Suda, Jaroslav Doležel, and Conceição Santos

Overview *423*
18.1 Introduction *423*
18.2 Taxonomic Representation in DNA Content Studies *425*
18.3 Nuclear Isolation and Staining Buffers *427*

18.4	Standardization and Standards 430
18.5	Fluorochromes 433
18.6	Quality Measures of Nuclear DNA Content Analyses 434
18.7	The Uses of DNA Flow Cytometry in Plants 435
18.8	Instrumentation 435
18.9	Where Are the Results Published? 436
18.10	Conclusion 437
	References 438

Index *439*

Preface

This book is being published more than three decades after the publication of the first formal report on flow cytometric analysis of plant material by F. O. Heller in 1973. This pioneering work did not find immediate favor with researchers and it was only after a considerable period of time that the usefulness of the technique was recognized with numerous applications of flow cytometry being developed and applied in plant science and industry. The reason for the growing popularity of flow cytometry is not hard to guess, as the method provides a unique means with which to analyze and manipulate plant cells and subcellular particles. Several optical parameters of particles can be analyzed simultaneously, quantitatively and at high speed. Statistically relevant data are quickly provided and the detection of subpopulations is possible. The ability to purify specific subpopulations of particles by flow sorting then provides a tool for their manipulation and analysis using other methods. As a result, current flow cytometry is now able to provide answers to the once utopian suite of challenging questions on plant growth, development, function and evolution at subcellular, cellular, organismal, and population levels.

Despite significant progress in the development of instrumentation, and the growing number of reported applications, researchers continue to be frustrated when searching for first-hand information on plant flow cytometry. Such information is currently scattered in a number of books and various journals. Due to some fundamental differences between plant, human and animal cells and tissues, and the fact that the scientific targets of those working with these different cell and tissue types only partially overlap, the plethora of biomedical publications cannot provide a substitute. One of the gurus of flow cytometry, Howard Shapiro, pertinently characterized the state of affairs in his fourth edition of *Practical Flow Cytometry* (2003, p. 512): "There are now enough references to justify a book on applications of flow cytometry to plants, but, as far as I know, nobody has written one."

Sharing the same opinion, and stimulated by our long-term experience with plant flow cytometry, we arrived at the conclusion in late 2003 that the time was ripe for the publication of such a treatise. Our intention was to prepare a comprehensive, instructive and stimulating title which would cover virtually all fields of current plant flow cytometric research and offer an easily accessible source of information. We trust that we have succeeded and look forward to the comments from the readers.

So what is on the menu? We start by describing the origin and evolution of flow cytometry and explaining the principles of flow cytometry and sorting (Chapters 1 and 2). Chapter 3 provides a general overview of plant flow cytometry, setting the stage for the more specialized topics discussed in Chapters 4–17. The first three of which cover the analysis of nuclear DNA content and its applications in the determination of genome size (Chapter 4), ploidy level (Chapter 5) and mode of reproduction (Chapter 6). Chapter 7 then explains the importance of research on nuclear genome size and Chapter 8 discusses the use of flow cytometry to estimate base composition in plant genomes. Focusing on microorganisms, Chapter 9 describes the use of flow cytometry in plant pathology, while Chapter 10 brings us back to plants and explains the analysis and sorting of naked plant cells, or protoplasts. We then move on to Chapter 11, the analysis of chloroplasts. Entering more exotic worlds, non-vascular plants and their DNA content is considered in Chapter 12 and the characterization of phytoplankton provides the subject of Chapter 13. Chapter 14 deals with the analysis of the cell cycle and is logically followed by a discussion on endopolyploidy (Chapter 15). Moving on to genome analysis, Chapter 16 describes the analysis and sorting of mitotic chromosomes and Chapter 17 introduces flow cytometry as a powerful tool for analyzing gene expression. The book closes with Chapter 18, which presents the FLOWer, a plant DNA flow cytometry database, and offers interesting quantitative data retrieved from publications in this area of flow cytometry.

Although the book was written by leading authorities and includes the most recent information, every effort has been made to avoid jargon and to explain all specific terms. Thus the book should be appreciated by users at every level of experience. Indeed, we are very happy with the final outcome and we hope that we have not only filled the gap in the current literature but created a reference volume for plant flow cytometry.

This book would never have materialized without the hard work, encouragement and support of many people. We are particularly indebted to the authors of the individual chapters who joined us on our formidable journey and provided excellent contributions. It is their hard work which makes this book a valuable reference text. We extend our gratitude to them all.

We greatly appreciate the highly professional, efficient, and conscientious work of the team at Wiley-VCH, who made publication of this book possible and who guided us carefully through the whole process. All three of us are excited by the graphical design of the book and the attractive front cover featuring the flower of lotos (*Nelumbo nucifera*). We felt it appropriate to include the nuclear genome size of the cover plant, and our original estimates are 1010 Mbp/1C (Prague) and 1017 Mbp/1C (Vienna).

We thank our colleagues for their assistance with editing the manuscripts and their preparation for submission. In particular we appreciate the assistance of Eva M. Temsch and Hermann Voglmayr with the revision of some of the figures.

We hope that the investment of our time, which was to a significant extent at the expense of our private lives, was justified and that it will promote the use of flow cytometry in plant science and production. We sincerely hope that the readers will enjoy exploring the fascinating world of plant flow cytometry as much as we enjoyed writing and editing this book.

January 2007

Jaroslav Doležel, Olomouc (Czech Republic)
Johann Greilhuber, Vienna (Austria)
Jan Suda, Prague (Czech Republic)

List of Contributors

Martin Barow
Leibniz Institute for Plant
Genetics and Crop Plant
Research (IPK)
Department of Cytogenetics
Corrensstraße 3
06466 Gatersleben
Germany

Jan Bartoš
Institute of Experimental Botany
Laboratory of Molecular
Cytogenetics and Cytometry
Sokolovská 6
772 00 Olomouc
Czech Republic

Michael D. Bennett
Jodrell Laboratory
Royal Botanic Gardens
Kew, Richmond
Surrey TW9 3AB
UK

Jan H. W. Bergervoet
Plant Research International
Business Unit BioScience
GreenFlow
Bornsesteeg 65
6708 PD Wageningen
The Netherlands

Isabelle C. Biegala
Institute de Recherche pour le
Développement (IRD)
Centre d'Océanologie de Marseille
Rue de la Batterie des Lions
13007 Marseille
France

Laszlo Bögre
School of Biological Sciences
Royal Holloway
University of London
Egham TW20 0EX
UK

Raffaella Casotti
Stazione Zoologica A. Dohrn di Napoli
Laboratory of Ecophysiology
Villa Comunale
80121 Naples
Italy

Jaroslav Doležel
Institute of Experimental Botany
Laboratory of Molecular Cytogenetics
and Cytometry
Sokolovská 6
772 00 Olomouc
Czech Republic

List of Contributors

George B. J. Dubelaar
CytoBuoy b.v.
Weiland 70C
Nieuwerbrug 2451BD
The Netherlands

David W. Galbraith
University of Arizona
Department of Plant Sciences
303 Forbes Building
Tucson, AZ 85721
USA

Gérald Grégori
Laboratoire de Microbiologie,
Géochimie et Ecologie Marines
CNRS UMR 6117
163 Avenue de Luminy, Case 901,
Bâtiment TPR1
13288 Marseille cedex 9
France

Johann Greilhuber
Department of Systematic and
Evolutionary Botany
University of Vienna
Rennweg 14
1030 Vienna
Austria

Brian C. Husband
Department of Integrative Biology
University of Guelph
Ontario
N1G 2W1
Canada

Gabriele Jovtchev
Leibniz Institute for Plant
Genetics and Crop Plant
Research (IPK)
Department of Cytogenetics
Corrensstraße 3
06466 Gatersleben
Germany

Pavlína Kovářová
Institute of Experimental Botany
Laboratory of Molecular Cytogenetics
and Cytometry
Sokolovská 6
772 00 Olomouc
Czech Republic

Paul Kron
Department of Integrative Biology
University of Guelph
Ontario
N1G 2W1
Canada

Marie Kubaláková
Institute of Experimental Botany
Laboratory of Molecular Cytogenetics
and Cytometry
Sokolovská 6
772 00 Olomouc
Czech Republic

Ilia J. Leitch
Jodrell Laboratory,
Royal Botanic Gardens
Kew, Richmond
Surrey TW9 3AB
UK

João Carlos Mano Loureiro
Laboratory of Biotechnology
and Cytomics
Department of Biology
University of Aveiro
Campus Universitário de Santiago
3810-193 Aveiro
Portugal

Fritz Matzk
Leibniz Institute for Plant Genetics
and Crop Plant Research (IPK)
Corrensstraße 3
06466 Gatersleben
Germany

Armin Meister
Leibniz Institute for Plant
Genetics and Crop Plant
Research (IPK)
Department of Cytogenetics
Corrensstraße 3
06466 Gatersleben
Germany

Jeroen Peters
Plant Research International
Business Unit BioScience
GreenFlow
Bornsesteeg 65
6708 PD Wageningen
The Netherlands

Martin Pfosser
Biology Centre of the Upper
Austrian Museums
J.-W.-Klein-Strasse 73
4040 Linz
Austria

Erhard Pfündel
Universität Würzburg
Julius-von-Sachs-Platz
97082 Würzburg
Germany

Zoltan Magyar
School of Biological Sciences
Royal Holloway
University of London
Egham TW20 0EX
UK

J. Paul Robinson
Purdue University Cytometry
Laboratories
Bindley Bioscience Center
1203 W. State Street
West Lafayette, IN 47907
USA

Conceição Santos
Laboratory of Biotechnology
and Cytomics
Department of Biology
University of Aveiro
Campus Universitário de Santiago
3810-193 Aveiro
Portugal

Howard M. Shapiro
The Center for Microbial Cytometry
283 Highland Avenue
West Newton, MA 02465-2513
USA

Hana Šimková
Institute of Experimental Botany
Laboratory of Molecular Cytogenetics
and Cytometry
Sokolovská 6
772 00 Olomouc
Czech Republic

Pavla Suchánková
Institute of Experimental Botany
Laboratory of Molecular Cytogenetics
and Cytometry
Sokolovská 6
772 00 Olomouc
Czech Republic

Jan Suda
Department of Botany
Faculty of Science
Charles University in Prague
Benátská 2
128 01 Prague
Czech Republic

and

Institute of Botany
Academy of Sciences of the
Czech Republic
Průhonice 1
252 43 Prague
Czech Republic

Glen A. Tarran
Plymouth Marine Laboratory
Prospect Place
West Hoe
Plymouth PL1 3DH
UK

Eva Maria Temsch
Department of Systematic and
Evolutionary Botany
University of Vienna
Rennweg 14
1030 Vienna
Austria

Pavel Trávníček
Department of Botany
Faculty of Science
Charles University in Prague
Benátská 2
128 01 Prague
Czech Republic

and

Institute of Botany
Academy of Sciences of the
Czech Republic
Průhonice 1
252 43 Prague
Czech Republic

Jan M. van der Wolf
Plant Research International
Business Unit BioScience
GreenFlow
Bornsesteeg 65
6708 PD Wageningen
The Netherlands

Hermann Voglmayr
Department of Systematic and
Evolutionary Botany
University of Vienna
Rennweg 14
1030 Vienna
Austria

1
Cytometry and Cytometers: Development and Growth

Howard M. Shapiro

Overview

It took almost 200 years of microscopy, from the mid-1600s until the mid-1800s, before objective data could be derived from specimens under the microscope by photography. The subsequent development of both image and flow cytometry for use by biologists followed the development of photometry, spectrometry, and fluorometry by physicists and chemists. Early cytometers measured cellular characteristics, such as nucleic acid content at the whole cell level; since few reagents were available that could specifically identify different types of cells, higher resolution imaging systems were developed for this task, but were too slow to be practical for many applications. The development of flow cytometry and cell sorting facilitated the development of more specific reagents, such as monoclonal antibodies and nucleic acid probes, which now allow cells to be precisely identified and characterized using simpler, low-resolution imaging systems. Although the most complex cytometers remain expensive, these newer instruments may bring the benefits of cytometry to a much wider community of users, including botanists in the field.

1.1
Origins

If the microscopic structures in cork to which Robert Hooke gave the name "cells" in the mid-17th century may be compared to the surviving stone walls of an ancient city, to what are we to compare the vistas available to 21st-century microscopists, who can follow the movements of individual molecules through living cells?

Between the time Hooke named them and the time that Schleiden, Schwann, and Virchow established cells as fundamental entities in plant and animal structure, function, and pathology, almost two centuries had elapsed. During most of that period, the only record of what could be seen under the microscope was an observer's drawing, and, even with the aid of a camera lucida, it was difficult if not impossible to eliminate subjective influences on the research product. The development of photography in the 1830s was quickly followed by the marriage of the Daguerrotype camera and the microscope, but it was only in the 1880s that photomicrography became accepted as the definitive objective method in microscopy, due in large measure to Robert Koch's advocacy (Breidbach 2002).

Even by that time, what we would today properly call cytometry, that is, the measurement of cells, was restricted to the quantification of morphologic characteristics, such as the sizes and numbers of cells and their organelles. The visualization of organelles themselves was greatly facilitated by differential staining methods, the development of which accelerated in the late 1800s with the availability of newly synthesized aniline dyes (Baker 1958; Clark and Kasten 1983); Paul Ehrlich's initial researches in this area were to lead directly to the transformation of pharmacology from alchemy to science, and his appreciation of the specificity of antigen–antibody reactions provided an early milestone on the path toward modern immunochemical reagents.

Spectroscopy, a tool of physics adapted to chemistry and astronomy in the 19th century, became a mainstay of cytometry shortly thereafter. Microspectrophotometric measurement, either of intrinsic optical characteristics of cellular constituents or of optical properties of dyes or reagents added to cells, provided objective, quantitative information about cells' chemistry that could be correlated with their functional states.

The subsequent development of both cytometry and cytometers has been characterized by the use of such information, wherever possible in place of the inherently subjective and less quantitative results obtained by human observers.

In the remainder of this chapter, I will consider the history of cytometry from the 20th century onwards. Although much of the material has been covered, sometimes in greater detail, in several of my earlier publications (Shapiro 2003, 2004a, 2004b), this version of the story will pay special attention to one of the principal uses of cytometry in botany, namely, the determination of the genome sizes of plants by measurement of nuclear DNA content (Bennett and Leitch 2005; Doležel and Bartoš 2005; Greilhuber et al. 2005).

1.2
From Absorption to Fluorescence, from Imaging to Flow

It is easy, and probably easier for younger than for older readers, to forget that both Feulgen's staining procedure (Feulgen and Rossenbeck 1924) and Caspersson's ultraviolet (UV) absorption microspectrophotometric method for quantification of nuclear DNA content (Caspersson and Schultz 1938) were developed years before it was established that DNA was the genetic material. The evolution of cytometers from microscopes began in earnest in the 1930s in Torbjörn Caspersson's laboratory at the Karolinska Institute in Stockholm. He developed a series of progressively more sophisticated microspectrophotometers, and confirmed that, as had been suggested by conventional histologic staining techniques of light microscopy, tumor cells were likely to have abnormalities in DNA and RNA content (Caspersson 1950). In a memoir, which in itself provides useful insights on the development of cytometry, Leonard Ornstein (1987) documents the influence of Caspersson's work in establishing the genetic role of DNA. The first report that DNA contents of haploid, diploid, and tetraploid plant cells were in the ratio of 1:2:4 was published in 1950 by Swift, who made measurements using the Feulgen technique; his paper also introduced the terms C, 2C, and 4C to describe the respective DNA contents for cells of a particular species.

1.2.1
Early Microspectrophotometry and Image Cytometry

Microspectrophotometers were first made by putting a small "pinhole" aperture, more properly called a field stop, in the image plane of a microscope, restricting the field of view to the area of a single cell, and placing a photodetector behind the field stop. Using progressively smaller field stops permits measurement of light transmission through correspondingly smaller areas of the specimen, and, by moving the stage in precise incremental steps in the plane of the slide, and recording the information, it becomes possible to measure the integrated absorption of a cell, and/or to make an image of the cell with each pixel corresponding in intensity to the transmission or absorption value. This was the first, and, until the 1950s, the only approach to scanning cytometry, and, even when measurements were made at the whole cell level, the process was extremely time-consuming, especially since there was no practical way to store data other than by writing down measured values as one went along. Publications were unlikely to contain data from more than a few hundred cells. By the 1960s, Zeiss had commercialized a current version of Caspersson's apparatus, and others had begun to build high-resolution scanning microscopes incorporating a variety of technologies. During the 1950s, what we now call "cytometry" was known as "analytical cytology". The first and second editions of a book with the latter title appeared in 1955 and 1959 (Mellors 1959). The book included chapters on histochemistry, on absorption measurement, on phase, interference, and polarizing microscopy, and on Coons's fluorescent antibody method (Coons et al. 1941).

1.2.2
Fluorescence Microscopy and the Fluorescent Antibody Technique

Fluorescence microscopy was developed around the turn of the 20th century. The earliest instruments used UV light for excitation; later systems could employ excitation at blue and longer wavelengths, but the requirement for relatively high power at relatively short wavelengths made it necessary to use arc lamps, rather than filament lamps, as light sources. Fluorescence microscopy, in principle, allows visualization of bright objects against a dark background. Earlier systems, however, were likely to fall short of achieving this goal because they were essentially transmitted-light microscopes with colored glass filters in both the excitation path, that is, between the light source and the condenser, and the emission path, that is, between the objective and the eyepiece. The combination of stray light transmission through both excitation and emission filters and fluorescence excited in the emission filter often resulted in the background being too bright to permit observation of weakly fluorescent material.

An extremely important application of fluorescence microscopy developed during the 1940s was the fluorescent antibody technique introduced by Coons et al. (1941). Other workers had demonstrated that azo dye-conjugated antisera to bacteria retained their reactivity with the organisms and would agglutinate them to form faintly colored precipitates; however, the absorption of the dye-conjugated sera was not strong enough to permit visual detection of bacterial antigens in tissue preparations.

Albert Coons surmised that it might be easier to detect small concentrations of antibody labeled with fluorescent material against a dark background using fluorescence microscopy. He and his coworkers labelled anti-pneumococcal antibodies with anthracene and could detect both isolated organisms and, more importantly, antibody bound to antigen in tissue specimens, by the UV-excited blue fluorescence of this label, as long as tissue autofluorescence and background were not excessive.

In 1950, Coons and Kaplan reported that fluorescein gave better results as an antibody label than did anthracene, because the blue-excited yellow–green fluorescence of fluorescein was easier to discriminate from autofluorescence. Thereafter, fluorescein became and has remained the most widely used immunofluorescent label.

A significant advance in fluorescence microscopy, epiillumination, was made in 1967 by Ploem (1967), who substituted dichroic mirrors for the half-silvered mirror normally used in an incident light microscope, and added excitation and emission filters to the optical path. Even when color glass filters were still used for excitation and emission wavelength selection, this configuration greatly reduced both stray light transmission and filter fluorescence, yielding much lower backgrounds. Within a short time, it had been reported that, when an epiilluminated apparatus was employed, measurements of nuclei stained by a fluorescent Feulgen procedure using acriflavine yielded results equivalent to those obtained by the standard absorption method (Böhm and Sprenger 1968).

1.2.3
Computers Meet Cytometers: The Birth of Analytical Flow Cytometry

By the mid-1950s, it had become clear that malignant cells often contained more nucleic acid than normal cells, and Mellors and Silver (1951) proposed construction of an automatic scanning instrument for screening cervical cytology (Papanicolaou or "Pap" smears). Their prototype measured fluorescence rather than absorption, and anticipated Ploem (1967) in introducing UV epiillumination. Tolles (1955) described the "Cytoanalyzer" built for cervical cytology. A disc containing a series of apertures rotated in the image plane of a transmitted light microscope, producing a raster scan of a specimen with approximately 5-µm resolution. A hardwired analyzer extracted nuclear size and density information; cells were then classified as normal or malignant using these parameters. The Cytoanalyzer proved unsuitable for clinical use, but its performance was encouraging enough for the American Cancer Society and the US National Cancer Institute to continue funding research on cytology automation in the United States.

Recording and storing cell images was a nontrivial task in the 1960s, when mainframe computers occupied entire rooms, required kilowatts of power for both the computer and the mandatory air-conditioning, and cost millions of dollars, for which the buyer received a computer with speed and storage capacity exceeded a thousand-fold by a 2005 model laptop costing under US$1000. Nonetheless, when minicomputers became available in the middle of the decade, there were at least a few groups of analytical cytologists ready to use them. The TICAS system, assembled at the University of Chicago in the late 1960s, interfaced Zeiss's (Oberkochen, Germany) commercial version of the Caspersson microspectrophotometer to a minicomputer, with the aim of automating interpretation of Pap smears (Wied and Bahr 1970).

The use of stage motion for scanning made operation extremely slow; it could take many minutes to produce a high-resolution scanned image of a single cell, even when there were computers available to capture the data. Somewhat higher speed could be achieved by using discs or galvanometer-driven moving mirrors for image scanning, and limiting the tasks of the motorized stage to bringing a new field of the specimen into view and into focus; this required some electronic storage capability, and made measurements susceptible to errors due to uneven illumination across the field, although this could be compensated for. My colleagues and I at the US National Institutes of Health (NIH; Bethesda, MD, USA) built "Spectre II" (Stein et al. 1969), which incorporated a galvanometer mirror scanning system (Ingram and Preston 1970) and a Digital Equipment Corporation LINC-8 minicomputer. While this system had sufficient computer power to capture high-resolution cell images (0.2 µm pixels), data had to be recorded on 9-track tape and transported to a mainframe elsewhere on the NIH campus for analysis (Shapiro et al. 1971).

Although imaging cytometers of the 1960s were not based on video cameras, for a number of reasons, not least of which was the variable light sensitivity of different regions of a camera tube, which made quantitative measurements difficult, it was recognized that the raster scan mechanism of a cathode ray tube could be used on the illumination side of an image analysis system, with the "flying spot" illuminating only a small segment of the specimen plane at any given time (Young 1951). The CYDAC system, a flying spot scanner built at Airborne Instruments Laboratory (Long Island, NY, USA), was used in studies of the automation of differential leukocyte counting (Prewitt and Mendelsohn 1966) and chromosome analysis (Mendelsohn 1976).

During World War II, the US Army became interested in developing devices for rapid detection of bacterial biowarfare agents in aerosols; this would require processing a relatively large volume of sample in substantially less time than would have been possible using even a low-resolution scanning system. The apparatus built by Gucker et al. (1947) in support of this project achieved the necessary rapid specimen transport by injecting the air stream containing the sample into the center of a larger sheath stream of flowing air that passed through the focal point of a dark-field microscope. Particles passing through the system scattered light into a collection lens, eventually producing electrical signals from a photodetector. The instrument could detect objects in the order of 0.5 µm in diameter, and is generally recognized as having been the first flow cytometer used for observation of biological cells. Although Moldavan had suggested counting cells in a fluid stream a decade earlier (Moldavan 1934), his account suggests that he failed to build a working apparatus.

By the late 1940s and early 1950s, the principles of the Gucker apparatus, including the use of sheath flow, were applied to the detection and counting of red blood cells in saline solutions (Crosland-Taylor 1953), providing effective automation for a diagnostic test notorious for its imprecision when performed by a human observer using a hemocytometer and a microscope. Neither the bacterial counter nor the early red cell counters had any significant capacity either for discriminating different types of cells or for making quantitative measurements. Both types of instrument were measuring what users of flow cytometers now call side-scatter signals; although larger particles, in general, produced larger signals than smaller ones, correlations between particle sizes and signal amplitudes were not particularly strong.

An alternative flow-based method for cell counting was developed in the 1950s by Wallace Coulter (Coulter 1956). Recognizing that cells, which are surrounded by a lipid membrane, are relatively poor conductors of electricity as compared to saline, he devised an apparatus in which cells passed one by one through a small (<100 µm) orifice between two chambers filled with saline. When a cell passed through, the electrical impedance of the orifice increased in proportion to the volume of the cell, producing a voltage pulse. The Coulter counter (Coulter Electronics, now Beckman Coulter, Hialeah, FL, USA) was widely adopted in clinical laboratories for blood cell counting; it was soon established that it could provide more accurate measurements of cell size than had previously been available (Brecher et al. 1956; Mattern et al. 1957).

In the early 1960s, investigators working with Leitz (Wetzlar, Germany) conceived a hematology counter that added a fluorescence measurement to the light scattering measurement used in red cell counting (Hallermann et al. 1964). If a fluorescent dye such as acridine orange was added to the blood sample, white cells would be stained much more brightly than red cells; the white cell count could then be derived from the fluorescence signal, and the red cell count from the scatter signal. It was also noted that acridine orange fluorescence could be used to discriminate mononuclear cells from granulocytes. It is not clear whether the device, which would have represented a new level of sophistication in flow cytometry, was actually built.

Around the same time, the promising results obtained with the Cytoanalyzer in attempts to automate reading of Pap smears (Tolles 1955) encouraged executives at the International Business Machines Corporation (IBM, Armonk, NY, USA) to look into producing an improved instrument. Assuming this would be some kind of image analyzer, IBM gave technical responsibility for the program to Louis Kamentsky, who had developed a successful optical character reader. He did some calculations of what would be required in the way of light sources, scanning rates, and computer storage and processing speeds to solve the problem using image analysis, and concluded that a different approach would be required.

Having learned from pathologists in New York that cell size and nucleic acid content could provide a good indicator of whether cervical cells were normal or abnormal, Kamentsky traveled to Caspersson's laboratory in Stockholm and learned microspectrophotometry. He then built a microscope-based flow cytometer that used a transmission measurement at visible wavelengths to estimate cell size and a 260-nm UV absorption measurement to estimate nucleic acid content (Kamentsky 1973; Kamentsky et al. 1965). Subsequent versions of this instrument, which incorporated a dedicated computer system, could measure as many as four cellular parameters (Kamentsky and Melamed 1969). A brief trial on cervical cytology specimens indicated that the system had some ability to discriminate normal from abnormal cells (Koenig et al. 1968); it could also produce distinguishable signals from different types of cells in blood samples stained with a combination of acidic and basic dyes.

The first commercial flow cytometric differential leukocyte counter, introduced in the early 1970s, was the Hemalog D (Technicon Corporation, now Bayer, Tarrytown, NY, USA); Ornstein was a prime mover in its development, having interacted with Kamentsky's group along the way (Ornstein 1987). The Hemalog D analyzed three separate aliquots of sample, making light scattering and absorption measurements at different wavelengths in three different flow cytometers to classify leukocytes based on the relatively specific cytochemical staining procedures used by hematologists for such purposes as determination of lineage of leukemic cells. Although the apparatus performed well, it was initially regarded with a great deal of suspicion, at least in part due to the novelty of flow cytometry. The developers and manufacturers of then-contemporary image analyzing differential counters, which certainly did not perform much better than did the Hemalog D, did what they could to keep potential users suspicious of flow cytometry for as long as possible; the technology would eventually be legitimized by its dramatic impact on immunology, which was facilitated by the introduction of cell sorting and immunofluorescence measurements.

1.2.4
The Development of Cell Sorting

Although impedance (Coulter) counters and optical flow cytometers could analyze hundreds of cells/second, providing a high enough data acquisition rate to be useful for clinical use, microscope-based static cytometers offered a significant advantage. A system with computer-controlled stage motion could be programmed to reposition a cell on a slide within the field of view of the objective (Stein et al. 1969), allowing the cell to be identified or otherwise characterized by visual observation; it was, initially, not possible to extract cells with known measured characteristics from a flow cytometer. Until this could be done, it would be difficult to verify any cell classification arrived at using a flow cytometer, especially where the diagnosis of cervical cancer or leukemia might be involved.

This problem was solved in the mid-1960s, when both Mack Fulwyler (Fulwyler 1965), working at the Los Alamos National Laboratory (Los Alamos, NM, USA), and Kamentsky, at IBM (Kamentsky and Melamed 1967), demonstrated cell sorters built as adjuncts to their flow cytometers. Kamentsky's system used a syringe pump to extract selected cells from its relatively slow-flowing sample stream. Fulwyler's was based on ink jet printer technology then recently developed by Richard Sweet (Sweet 1965) at Stanford University (Stanford, CA, USA); following passage through the cytometer's measurement system (originally a Coulter orifice), the saline sample stream was broken into droplets, and those droplets that contained cells with selected measurement values were electrically charged at the droplet break-off point. The selected charged droplets were then deflected into a collection vessel by an electric field; uncharged droplets went, as it were, down the drain.

1.3
The Growth of Multiparameter Flow Cytometry

In the early 1970s, the group at Los Alamos led the way in implementation of practical multiparameter flow cytometers; their larger instruments, with droplet sorting capability, combined two-color fluorescence measurements with measurements of Coulter volume and (thanks to the contributions of Paul Mullaney, Gary Salzman, and others) light scattering at several angles (Mullaney et al. 1969; Salzman et al. 1975a, 1975b; Steinkamp et al. 1973). The cytometers were interfaced to Digital Equipment Corporation (Maynard, MA, USA) minicomputers. Several instruments made at Los Alamos were delivered to the National Institutes of Health; other institutions copied most or all of the Los Alamos design in their own laboratory-built apparatus.

Fluorescence measurement had been introduced to flow cytometry in the late 1960s to improve both quantitative and qualitative analyses. By that time, Van Dilla et al. (1969) at Los Alamos and Dittrich and Göhde (1969) in Germany had built fluorescence flow cytometers to measure cellular DNA content; the Los Alamos investigators used a fluorescent Feulgen staining procedure, whereas Dittrich and Göhde's publication introduced the use of ethidium bromide as a rapid DNA stain, facilitating analysis of abnormalities in tumor cells and of cell cycle kinetics in both neoplastic and normal cells. The Los Alamos instrument incorporated the orthogonal "body plan" now standard in laser-source instruments, with

Fig. 1.1 Three historic flow cytometers and a post-historic image cytometer. (a) Phywe Impulscytophotometer ICP 22, which was licensed to Phywe by Partec, was one of the first compact bench-top flow cytometers (1975). (b) The BD Biosciences FACSAria, a modern high-speed cell sorter reduced to the size of a typical bench-top instrument (BD Biosciences). (c) A 1974 NIH photograph of Leonard Herzenberg with the B-D FACS-1, the first commercial cell sorter. (d) The author with a simple imaging cytometer, minus the laptop computer used for data collection and analysis.

the optical axes of illumination and light collection at right angles to each other and to the direction of sample flow. Kamentsky, who had left IBM to found Bio/Physics Systems (Mahopac, NY, USA), produced the Cytofluorograf, an orthogonal geometry fluorescence flow cytometer that was the first commercial product to incorporate an argon ion laser; Wolfgang Göhde's Partec (Münster, Germany) Impulscytophotometer (ICP) instrument built around a fluorescence microscope with arc lamp illumination, was distributed commercially by Phywe (Göttingen, Germany). Using this instrument, Heller (1973) was the first to describe flow cytometric measurements of DNA content in plant cell nuclei (see Chapter 3). Partec instruments (Fig. 1.1a), which are relatively compact and, in their simpler configurations, can be run on batteries, remain popular for use in botanical applications.

Leonard Herzenberg and his colleagues (Herzenberg et al. 1976) at Stanford, realizing that fluorescence flow cytometry and subsequent cell sorting could provide a useful and novel method for purifying living cells for further study, developed a series of instruments after exposure to a Kamentsky prototype lent by IBM (Saunders and Hulett 1969). Although their original apparatus (Hulett et al. 1969), with arc lamp illumination, was not sufficiently sensitive to permit them to achieve their objective of sorting cells from the immune system, based on the presence and intensity of staining by fluorescently labelled antibodies, the second version (Bonner et al. 1972), which used a water-cooled argon laser, was more than adequate. This was commercialized as the Fluorescence-Activated Cell Sorter (FACS) in 1974 by Becton-Dickinson (B-D, now BD Biosciences, San Jose, CA, USA) (Fig. 1.1b).

Coulter Electronics (now Beckman Coulter), which by 1970 had become a very large and successful manufacturer of laboratory hematology counters, pursued the development of fluorescence flow cytometers through a subsidiary, Particle Technology, under Mack Fulwyler's direction in Los Alamos. The TPS-1 (Two Parameter Sorter), Coulter's first product in this area, reached the market in 1975. It used an air-cooled 35-mW argon ion laser source and could measure forward scatter and fluorescence.

Multiple wavelength fluorescence excitation was introduced to flow cytometry in apparatus built at Block Engineering (Cambridge, MA, USA) during an abortive attempt to develop a hematology instrument. The first instrument (Curbelo et al. 1976) derived five illuminating beams from a single arc lamp; the second (Shapiro et al. 1977) used three laser beams; both could analyze over 30 000 cells per second and, using hardwired pre-processors and integral minicomputers, identify cells comprising less than 1/100 000 of the total sample. The laser source system incorporated forward and side scatter measurements, which permitted lymphocyte gating (Shapiro 1977), influenced by work carried out at Los Alamos (Salzman et al. 1975a, 1975b). Block also built a slow flow system intended for detection of hepatitis B virus and antigen in serum; it could discriminate scatter singles from large viruses (Hercher et al. 1979) and could theoretically detect a few dozen fluorescein molecules above background. The Block cytometers were never sold commercially, but influenced the optical, electronic, and systems design of later instruments.

By the time the Society for Analytical Cytology (now ISAC) came into being in 1978, B-D, Coulter, and Ortho (a division of Johnson & Johnson that bought Bio/Physics Systems and was ultimately acquired by B-D) were producing flow cytometers that could measure small- (forward scatter) and large- (side scatter) angle light scattering and fluorescence in at least two wavelength regions, analyzing several thousand cells per second, and with droplet deflection cell sorting capability. Ortho was also distributing the ICP, which, by virtue of its optical design, could make higher precision measurements of DNA content than could laser-based flow cytometers. DNA content analysis was receiving considerable attention as a means of characterizing the aggressiveness of breast cancer and other malignancies, and, at least in part due to the results of a Herzenberg sabbatical in

Cesar Milstein's laboratory in Cambridge (UK), monoclonal antibodies had begun to emerge as practical reagents for dissecting the stages of development of cells of the blood and immune system. Loken, Parks, and Herzenberg had successfully performed a two-color immunofluorescence experiment, introducing fluorescence compensation in the process (Loken et al. 1977), although it was clear that a great deal needed to be done in the area of fluorescent-label development to realize the potential of monoclonal antibodies.

Although the instruments Kamentsky built at IBM were computer controlled, computers were expensive options for most flow cytometers until the early 1980s, by which time microprocessor-based systems could do the work of an earlier generation of minicomputers. Without computers, instruments might be able to measure four or more parameters per cell, but did not have the processing power to implement true multiparameter methodology for gating, that is, selection of subsets of cells from a heterogeneous population using combinations of several measurement values, or for sort control. Once dedicated microcomputers became "standard equipment" on commercial flow cytometers, multiparameter measurement techniques became practical for many more researchers than had previously been able to use them.

1.4
Bench-tops and Behemoths: Convergent Evolution

From the early 1970s on, commercial production of instruments has allowed researchers who cannot develop and build their own apparatus to pursue increasingly sophisticated applications of fluorescence flow cytometry and sorting. Advances in the technology itself have continued to occur primarily in a relatively small community of academic, government, and industrial laboratories.

Los Alamos provided the inoculum for the subsequent growth of another major center for flow cytometer development at Lawrence Livermore Laboratory (Livermore, CA, USA), where high-speed flow sorting was perfected as a means for separating human chromosomes stained with a combination of A-T selective (Hoechst 33342) and G-C selective (chromomycin A_3) DNA dyes (Gray et al. 1987; Peters et al. 1985). The MoFlo high-speed sorter developed by Ger van den Engh and others at Livermore was subsequently refined by Cytomation (now DakoCytomation, Fort Collins, CO, USA), and has been produced commercially by them since 1994.

Chromosome sorting (also of major interest to botanists (Doležel et al. 2001; Chapter 16)) and high-speed sorting in general initially required high-powered, water-cooled ion lasers, in part because the first generation of high-speed instruments from B-D and Cytomation made measurements of cells in a jet in air, necessitating the use of relatively inefficient light collection optics. Systems built around fluorescence microscopes, such as the original Partec Impulscytophotometer and the system originally described by Tore Lindmo and Harald Steen in 1979 (Steen 1980; Steen and Lindmo 1979), can make optimal use of the relatively low excitation power available from mercury or xenon arc lamps by using high numerical aperture microscope lenses or their equivalent for both illumination and light collection. The Lindmo–Steen apparatus, sensitive enough to make precise DNA content measurements of bacteria (Steen et al. 1982), was originally commercialized by Leitz; later versions were sold by Skatron (Lier, Norway) and Bio-Rad (Milano, Italy), and the latest, now being produced by Apogee (Hemel Hempstead, Herts, UK) can use a low-power laser source to measure scatter signals from viruses (Steen 2004). Another bench-top arc source flow cytometer, the Quanta, developed by NPE Systems (Miami, FL, USA), makes both fluorescence and impedance measurements, and is now being sold by Beckman Coulter.

In the mid and late 1970s, Kamentsky's Bio/Physics Systems and its successor, Ortho Diagnostics Systems, introduced laser source flow cytometers and sorters in which measurements were made in flat-sided quartz flow cuvettes, and in which "high-dry" microscope objectives were used to increase light collection. This made it possible to use air-cooled rather than water-cooled lasers for immunofluorescence measurements, decreasing the size, cost, and power consumption of instruments. In the early 1980s, B-D introduced its FACS analyzer, a small but sensitive bench-top system employing an arc lamp source; within a few years, it was supplanted by the FACScan, a three-color bench-top analyzer using a rectangular cuvette with a gel-coupled lens for highly efficient light collection, allowing more sensitive immunofluorescence measurements to be made using a 15-mW air-cooled argon laser source than were possible using 10 times more laser power in jet-in-air sorters. The FACScan was followed by the FACSort, which included a relatively slow fluidic sorter; both were succeeded by the FACSCalibur, which offers both a fluidic sorting option and a fourth fluorescence channel with excitation from a red (635–640 nm) diode laser.

The emphasis in the Herzenbergs' laboratory at Stanford has remained on sorting cells on the basis of immunofluorescence signals with the aim of isolating morphologically indistinguishable viable lymphocytes with differences in functional characteristics. This required development of a large armamentarium of monoclonal antibodies, of labels with diverse spectral characteristics, and of the hardware and software necessary to achieve multiparameter fluorescence compensation (Bagwell and Adams 1993) and gating rapidly enough to implement complex sorting strategies. Until the algal photosynthetic pigment phycoerythrin was introduced as an antibody label in 1982 (Oi et al. 1982), two large lasers were required for two-colour immunofluorescence measurements; the combination of fluorescein- and phycoerythrin-labeled antibodies could be excited effectively at 488 nm, while providing sufficient separation of emission maxima to discriminate the fluorescence of the two labels. It is now possible, using a combination of organic dye labels, phycobiliproteins, tandem conjugates of both, and semiconductor nanocrystals ("quantum dots") as labels, to carry out 17-color immunofluorescence experiments on flow cytometers with three laser beams (Perfetto et al. 2004). Large lasers are no longer required.

From the mid-1990s on, there has been a proliferation of diode and solid-state lasers, and these small, energy-efficient, and (usually) relatively inexpensive sources have increasingly been incorporated into flow cytometers. The use of violet (395–415 nm) diode lasers in cytometry was first described at a meeting in 2000 (Shapiro and Perlmutter 2001); by 2002, several manufacturers had incorporated such sources into their instruments. They can be used to excite DNA dyes such as DAPI and Hoechst 33342, which are normally used with UV excitation. Frequency-doubled diode-pumped YAG lasers, emitting green light at 532 nm, have also come into use, and are available in bench-top analyzers from BD Biosciences, Guava Technologies (Hayward, CA, USA), and Luminex (Austin, TX); all but the last of these companies also offer doubled semiconductor lasers emitting at 488–492 nm in lieu of argon lasers.

An increasing amount of the internal electronics of flow cytometers has become computer-based, with the latest systems incorporating special-purpose large-scale integrated circuits, microprocessors, microcontrollers, and digital signal-processing chips.

The development of digital audio, telephony, and video has resulted in large increases in the performance, and decreases in the price, of analog-to-digital converters (ADCs), which are critical elements in data acquisition systems for any type of instrumentation, flow cytometers included. The ADCs originally used with flow cytometers had only 8- or 10-bit resolution, making it necessary to use logarithmic amplifiers to process signals with a large dynamic range. This necessitated the use of hardware for fluorescence compensation. While this approach is feasible when three or four colors are measured, it is essentially impossible to implement for modern multibeam instruments in which measurements of 12 or more colors may be made. The alternative is software compensation (Bagwell and Adams 1993), which is best applied to linear data digitized to at least 16-bit resolution. In the early 1990s, software compensation was implemented in the Beckman Coulter EPICS XL analyzer, which captures 20-bit linear data, eliminating the need for logarithmic amplifiers. BD Biosciences's DiVa electronics use high-speed digitization to permit digital computation of pulse height, width, and area, while DakoCytomation and Partec have developed their own approaches to high-resolution digital data analysis. As has been the case for audio and video, digital techniques can be expected to become predominant in cytometry.

With the introduction of the FACSAria sorter (Fig. 1.1c) in late 2002, BD Biosciences successfully hybridized the behemoth high-speed cell sorter and the bench-top analyzer; this bench-top apparatus incorporates digital electronics and measures cells in a cuvette using as many as three beams, typically at 407, 488, and 633 nm, all derived from low-power, air-cooled lasers. The InFlux, a high-speed sorter recently introduced by Ger van den Engh's new company, Cytopeia (Seattle, WA, USA), also features a small footprint and the ability to operate using only air-cooled lasers.

Multilaser bench-top analyzers are now available from BD Biosciences, Beckman Coulter, DakoCytomation, Luminex, and Partec; the latter also offers a combination of arc lamp and laser sources. Users not completely satisfied with the multiparameter software available from their cytometer manufacturers can choose from among the offerings of a number of third-party providers.

With the aid of a continually expanding repertoire of reagents, many of which are exquisitely specific, an ever more colorful palette of fluorescent labels, and increasingly sophisticated data analysis procedures, multiparameter fluorescence flow cytometers can identify many more lineages and sublineages of both prokaryotic and eukaryotic cells than were contemplated a few decades ago. Remarkably, all of the work is carried out without the benefit of morphologic information. Although the instruments also typically measure small-angle or forward light scatter, often erroneously referred to as a "size" measurement, and large-angle or side scatter, which provides information about internal granularity and surface roughness but does not resolve cellular detail, most of the information needed for cell identification comes from the intensity of fluorescence measurements made at various excitation and emission wavelengths.

Flow cytometry remains a highly effective, accurate, and precise way to obtain objective and quantitative information from single cells; its principal disadvantages lie in the complexity and cost of the apparatus. It is thus logical to ask whether the advances made in electronics and electro-optics in recent years might provide a simpler and more affordable alternative. Recent work in my laboratory (Shapiro 2004b; Shapiro and Perlmutter 2006) and elsewhere (Jelinek et al. 2001; Mazzini et al. 2005; Rodriquez et al. 2005; Stothard et al. 2005; Varga et al. 2004; Wittrup et al. 1994) provides grounds for optimism.

1.5
Image Cytometry: New Beginnings?

In 1994, Wittrup et al. described a cytometric apparatus called the Fluorescence Array Detector (FAD), in which camera lenses were used to form a 1:1 image of a 1×1 cm field of view on a cooled 512×512 pixel charge-coupled device (CCD) detector with 20 µm square pixels. Since each pixel collected light from an area larger than the area of a typical cell, no morphologic information was available. The instrument had only one moving part, a focusing stage; low-intensity (1 mW/cm^2) illumination of the field came from the expanded beam of a 488-nm air-cooled argon ion laser. Although a software shading correction was used in an attempt to compensate for the uneven illumination obtained from the laser beam, the coefficient of variation (CV) of the fluorescence intensity distribution of 6-µm polystyrene beads was reported to be 12.9%; this relatively large variance, which would be unacceptable in the context of DNA content measurement, was attributed to imperfect shading correction. Sensitivity was impressive; noise due to dark current and stray light was equivalent to only a few hundred fluorescein MESF (molecules of equivalent soluble fluorochrome)/pixel, although fluorescence from conventional glass microscope slides increased the background fluorescence by approximately 10-fold.

At the time the FAD was constructed, a cooled CCD camera cost tens of thousands of dollars; and a 50-mW air-cooled 488-nm argon laser cost at least $8000, eliminating the instrument from consideration as a low-cost replacement for a flow cytometer. At present, CCD and complementary metal oxide-silicon (CMOS) cameras with the requisite sensitivity are available for no more than a few hundred dollars; however, argon ion lasers and solid-state lasers emitting at the same wavelength still cost thousands of dollars. While in 1994, an arc lamp, also costing thousands of dollars, would probably have been the only feasible alternative to a laser as the light source for an instrument similar to the FAD, it has recently been shown (Mazzini et al. 2005) that high-intensity light-emitting diodes (LEDs), available for only a few dollars, can provide sufficient illumination over the required area to be usable as light sources for fluorescence microscopy as well as for an FAD-like device. We have established (Shapiro and Perlmutter 2006) that an apparatus using an LED for excitation, relatively inexpensive 35-mm camera lenses or low power microscope lenses for light collection, and a CCD or CMOS camera as a detector (Fig. 1.1d) can detect low-level fluorescence signals in the range that would be expected from cells stained with fluorescent antibodies, as well as the substantially stronger signals associated with cells stained for DNA. Although we have not definitively established a range of measurement precision, we note that Varga et al. (2004) achieved CVs of less than 4% by applying a correction for uneven illumination to fluorescence measurements made using a CCD camera on a conventional fluorescence microscope, and believe that at least equivalent performance can be achieved in a large-field imaging system analogous to the FAD.

Rodriguez et al. (2005) have demonstrated that a prototype low-magnification imaging system can be as effective as a flow cytometer for immunofluorescence-based counting of CD4+ T lymphocytes in the blood of patients infected with HIV, and there is evidence that well-designed, extremely inexpensive "toy" microscopes are sufficiently rugged and of sufficient optical quality for use in tropical environments (Jelinek et al. 2001; Stothard et al. 2005). Thus, a new generation of image cytometers, under development to facilitate infectious disease diagnosis in resource-poor countries, may soon provide botanists with an effective and economical means of determining genome sizes of plants in the field, a welcome and not completely unintended consequence.

Acknowledgment

The author thanks Dr. Roland Göhde (Partec GmbH, Münster, Germany) for the photo of the Phywe Impulscytophotometer ICP 22 in Fig. 1.1a.

References

Bagwell, C. B., Adams, E. G. **1993**, *Ann. NY Acad. Sci.* 677, 167–184.

Baker, J. R. **1958**, *Principles of Biological Microtechnique*, Methuen, UK.

Bennett, M. D., Leitch, I. J. **2005**, *Ann. Bot.* 95, 1–6.

Böhm, N., Sprenger, E. **1968**, *Histochemie* 16, 100–118.

Bonner, W. A., Hulett, H. R., Sweet, R. G., Herzenberg, L. A. **1972**, *Rev. Sci. Instr.* 43, 404–409.

Brecher, G., Schneiderman, M. G., Williams, Z. **1956**, *Am. J. Clin. Pathol.* 26, 1439–1449.

Breidbach, O. **2002**, *J. History Biol.* 35, 221–250.

Caspersson, T. O. **1950**, *Cell Growth and Cell Function*, Norton, USA.

Caspersson, T. O., Schultz, J. **1938**, *Nature* 142, 294–295.

Clark, G., Kasten, F. H. **1983**, *History of Staining*, 2nd edn, Williams & Wilkins, USA.

Coons, A. H., Kaplan, M. H. **1950**, *J. Exp. Med.* 91, 1–13.

Coons, A. H., Creech, H. J., Jones, R. N. **1941**, *Proc. Soc. Exp. Biol. Med.* 47, 200–202.

Coulter, W. H. **1956**, *Proc. Natl Elect. Conf.* 12, 1034–1042.

Crosland-Taylor, P. J. **1953**, *Nature* 171, 37–38.

Curbelo, R., Schildkraut, E. R., Hirschfeld, T., Webb, R. H., Block, M. J., Shapiro, H. M. **1976**, *J. Histochem. Cytochem.* 24, 388–395.

Dittrich, W., Göhde, W. **1969**, *Zeitschrift Naturforsch.* 24b, 360–361.

Doležel, J., Bartoš, J. **2005**, *Ann. Bot.* 95, 99–110.

Doležel, J., Lysák, M. A., Kubaláková, M., Šimková, H., Macas, J., Lucretti, S. **2001**, *Methods Cell Biol.* 64, 3–31.

Feulgen, R., Rossenbeck, H. **1924**, Hoppe-Seyler's *zeitschr. physiol. Chemie* 135, 203–248.

Fulwyler, M. J. **1965**, *Science* 150, 910–911.

Gray, J. W., Dean, P. N., Fuscoe, J. C., Peters, D. C., Trask, B. J., van den Engh, G. J., Van Dilla, M. A. **1987**, *Science* 238, 323–329.

Greilhuber, J., Doležel, J., Lysák, M. A., Bennett, M. D. **2005**, *Ann. Bot.* 95, 255–260.

Gucker Jr., F. T., O'Konski, C. T., Pickard, H. B., Pitts Jr., J. N. **1947**, *J. Am. Chem. Soc.* 69, 2422–2431.

Hallermann, L., Thom, R., Gerhartz, H. **1964**, *Verhand. Deutsch. Gesellschaft Inn. Med.* 70, 217–219.

Heller, F. O. **1973**, *Ber. Deutsch. Bot. Gesellschaft* 86, 437–441.

Hercher, M., Mueller, W., Shapiro, H. M. **1979**, *J. Histochem. Cytochem.* 27, 350–352.

Herzenberg, L. A., Sweet, R. G., Herzenberg, L. A. **1976**, *Scientific American* 234, 108–117.

Hulett, H. R., Bonner, W. A., Barrett, J., Herzenberg, L. A. **1969**, *Science* 166, 747–749.

Ingram, M., Preston Jr., K. **1970**, *Scientific American* 223, 72–78.

Jelinek, L., Peters, G., Okuley, J., McGowan, S. **2001**, *Intel Technol. J.* 5, 1–10.

Kamentsky, L. A. **1973**, *Adv. Biol. Med. Phys.* 14, 93–161.

Kamentsky, L. A., Melamed, M. R. **1967**, *Science* 156, 1364–1365.

Kamentsky, L. A., Melamed, M. R. **1969**, *Proc. IEEE* 57, 2007–2016.

Kamentsky, L. A., Melamed, M. R., Derman, H. **1965**, *Science* 150, 630–631.

Koenig, S. H., Brown, R. D., Kamentsky, L. A., Sedlis, A., Melamed, M. R. **1968**, *Cancer* 21, 1019–1026.

Loken, M. R., Parks, D. R., Herzenberg, L. A. **1977**, *J. Histochem. Cytochem.* 25, 899–907.

Mattern, C. F. T., Brackett, F. S., Olson, B. J. **1957**, *J. Appl. Physiol.* 10, 56–70.

Mazzini, G., Ferrari, C., Baraldo, N., Mazzini, M., Angelini, M. **2005**, Improvements in fluorescence microscopy allowed by high power light emitting diodes, in *Current Issues on Multidisciplinary Microscopy Research and Education*, (vol. 2), ed. A. Méndez-Vilas, L. Labajos Broncano, Formatex, Spain, pp. 181–188.

Mellors, R. C. (ed.) **1959**, *Analytical Cytology*, 2nd edn, McGraw-Hill, USA.

Mellors, R. C., Silver, R. **1951**, *Science* 114, 356–360.

Mendelsohn, M. L. (ed.) **1976**, *Automation of Cytogenetics*, CONF-751158, Lawrence Livermore Laboratory, USA.

Moldavan, A. **1934**, *Science* 80, 188.

Mullaney, P. F., Van Dilla, M. A., Coulter, J. R., Dean, P. N. **1969**, *Rev. Scient. Instr.* 40, 1029–1032.

Oi, V. T., Glazer, A. N., Stryer, L. **1982**, *J. Cell Biol.* 93, 981–986.

Ornstein, L. **1987**, *Electrophoresis* 8, 3–13.

Perfetto, S. P., Chattopadhyay, P. K., Roederer, M. **2004**, *Nature Rev. Immunol.* 4, 648–655.

Peters, D. C., Branscomb, E., Dean, P. N., Merrill, T., Pinkel, D., Van Dilla, M. A., Gray, J. W. **1985**, *Cytometry* 6, 290–301.

Ploem, J. S. **1967**, *Zeitschrift Wissenschaft. Mikrosk.* 68, 129–142.

Prewitt, J. M. S., Mendelsohn, M. L. **1966**, *Ann. NY Acad. Sci.* 128, 1035–1053.

Rodriguez, W. R., Christodoulides, N., Floriano, P. N., Graham, S., Mohanty, S., Dixon, M., Hsiang, M., Peter, T., Zavahir, S., Thior, I., Romanovicz, D., Bernard, B., Goodey, A. P., Walker, B. D., McDevitt, J. T. **2005**, *PLoS Med*. 2, 663–672.

Salzman, G. C., Crowell, J. M., Goad, C. A., Hansen, K. M., Hiebert, R. D., LaBauve, P. M., Martin, J. C., Ingram, M., Mullaney, P. F. **1975a**, *Clin. Chem*. 21, 1297–1304.

Salzman, G. C., Crowell, J. M., Martin, J. C., Trujillo, T. T., Romero, A., Mullaney, P. F., LaBauve, P. M. **1975b**, *Acta Cytol*. 19, 374–377.

Saunders, A. M., Hulett, H. R. **1969**, *J. Histochem. Cytochem*. 17, 188.

Shapiro, H. M. **1977**, *J. Histochem. Cytochem*. 25, 976–989.

Shapiro, H. M. **2003**, *Practical Flow Cytometry*, 4th edn, Wiley-Liss, USA.

Shapiro, H. M. **2004a**, *Cytometry* 58A, 13–20.

Shapiro, H. M. **2004b**, *Cytometry* 60A, 115–124.

Shapiro, H. M., Perlmutter, N. G. **2001**, *Cytometry* 44, 133–136.

Shapiro, H. M., Perlmutter, N. G. **2006**, *Cytometry* 69A, 620–630.

Shapiro, H. M., Bryan, S. D., Lipkin, L. E., Stein, P. G., Lemkin, P. F. **1971**, *Exp. Cell Res*. 67, 81–85.

Shapiro, H. M., Schildkraut, E. R., Curbelo, R., Turner, R. B., Webb, R. H., Brown, D. C., Block, M. J. **1977**, *J. Histochem. Cytochem*. 25, 836–844.

Steen, H. B. **1980**, *Cytometry* 1, 26–31.

Steen, H. B. *Cytometry* **2004**, 57A, 94–99.

Steen, H. B., Lindmo, T. **1979**, *Science* 204, 403–404.

Steen, H. B., Boye, E., Skarstad, K., Bloom, B., Godal, T., Mustafa, S. **1982**, *Cytometry* 2, 249–257.

Stein, P. G., Lipkin, L. E., Shapiro, H. M. **1969**, *Science* 166, 328–333.

Steinkamp, J. A., Fulwyler, M. J., Coulter, J. R., Hiebert, R. D., Horney, J. L., Mullaney, P. F. **1973**, *Rev. Scient. Instr*. 44, 1301–1310.

Stothard, J. R., Kabatereine, N. B., Tukahebwa, E. M., Kazibwe, F., Mathieson, W., Webster, J. P., Fenwick, A. **2005**, *Am. J. Trop. Med. Hyg*. 73, 949–955.

Sweet, R. G. **1965**, *Rev. Scient. Instr*. 36, 131–136.

Swift, H. **1950**, *Proc. Natl Acad. Sci., Washington* 36, 643–654.

Tolles, W. E. **1955**, *Trans. NY Acad. Sci*. 17, 250–256.

Van Dilla, M. A., Trujillo, T. T., Mullaney, P. F., Coulter, J. R. **1969**, *Science* 163, 1213–1214.

Varga, V. S., Bocsi, J., Sipos, F., Csendes, G., Tulassay, Z., Molnár, B. **2004**, *Cytometry* 60A, 53–62.

Wied, G. L., Bahr, G. F. (eds.) **1970**, *Automated Cell Identification and Cell Sorting*, Academic Press, USA.

Wittrup, K. D., Westerman, R. J., Desai, R. **1994**, *Cytometry* 16, 206–213.

Young, J. Z. **1951**, *Nature* 167, 231.

2
Principles of Flow Cytometry

J. Paul Robinson and Gérald Grégori

Overview

Flow cytometry holds a unique place among biomedical tools as it is the only technology whereby single cells can be evaluated in high content, classified as required, and then sorted into single-cell units or a homogeneous population. This ability of single-cell analysis provides a powerful opportunity for cytomic analysis whereby any obtainable cytome can be interrogated at the highest possible level for detection systems. The second aspect of flow cytometry that is unique is the ability to collect multiple variables on the order of 10–20 actual measurements. While many high-content systems claim to produce a high number of parameters, they frequently base this claim on a few variables with each variable generating multiple parametric determinations. Flow cytometry can collect simultaneously multiple angles of scatter, and multiple spectral components, while at the same time classifying each cell into a cluster of similar cells. This multiparametric capacity separates flow cytometry from all other technologies.

Given this potential, it is clear that flow cytometry is of crucial importance in the biological armament. Regardless of the application field, this capability of basing analyses on population statistics using every single particle or cell as an individual multiparametric unit is unrivaled in biological measurement systems. In addition to all of this, the ability to physically separate cells or particles based on any available parameter, under sterile conditions if necessary, makes flow cytometry a truly amazing technology.

2.1
Introduction

During the last decade or more, flow cytometry (FCM) has become a very powerful technology that has impacted a wide range of fields from basic cell biology to genetics, as well as immunology, molecular biology, and environmental science (particularly aquatic microbiology). The principles of FCM can be easily understood from its definition: measurement of the properties of isolated cells flowing in single file within a liquid sheath as they are intercepted by a high-intensity light source focused in a very small region. Cells are interrogated in a very short time (a few microseconds) during which multiple signals are collected, mainly light scatter and fluorescence emissions in the visible spectrum. Thanks to the progress in electronics and informatics, flow cytometers can readily analyze single particles/cells at rates of up to 100 000 per second. It is thus possible to discriminate particles/cells into clusters based on statistical analyses of the set of parameters collected for each particle. Using these statistical analyses, it is possible to electronically separate these populations and identify them using multivariate analytical techniques.

Flow cytometry has two key advantages. The first is that a large number of particles can be evaluated in a very short time, which makes the results statistically strong and representative of the whole population. Even at rates up to 100 000 per second, approximately 20 parameters from each particle can be collected and analyzed. The second key advantage is the ability to physically separate single particles/cells from mixed populations at rates up to 70 000 cells per second, by a process known as cell sorting. Each sorted particle can be physically placed into a defined vessel (a tube, a slide, any well from a 96-well plate) for further analysis, culture, or chemical decomposition. Indeed, sorting preserves the viability of most cells. If necessary, this process can be performed under sterile conditions. Even if the cell of interest is a rare event (i.e. 1:100 000 or more), it is still possible to identify it and physically isolate a rare population.

In 1983, Howard Shapiro noted that multiparameter FCM was now a "reality in the field" because of the ready availability of commercial instruments. Since then, the field has expanded well beyond anything that was considered possible at that time. Today's instruments have the capacity to measure many spectral bands simultaneously together with a variety of scatter signals. With modern computers, faster electronics, and advanced digital signal processing, it is possible to perform complex multiparametric analyses in real time, creating the opportunity to make complex sorting decisions within a few microseconds after measurements are made. The result of this technology is that FCM together with the concomitant development of fluorescent dyes provides new approaches for single-cell analysis to better characterize cell systems. Moreover, this is performed in real time (Robinson et al. 1991), something that is difficult, if not impossible, by other analytical techniques.

2.2
A Brief History of Flow Cytometry

The basic principles of FCM are based on some old ideas generated early in the 20th century and, of course, on the principles of laminar flow defined by Osborne Reynolds in the late 19th century. Some 50 years after Reynolds, Andrew Moldavan (1934) designed an instrument that could have identified single cells using a microscope and a photodetector. In the 1940s, George Papanicolaou demonstrated that he could identify cancerous cells from cervical smears by observing the staining patterns obtained using specifically designed stains (Papanicolaou and Traut 1941; Chapter 1). This suggested several opportunities for the identification of abnormal cells, primarily using image-analysis techniques. It proved to be quite difficult to create analytical technology based on the capability of computers and imaging technology at that time, and the result was a movement toward single-cell analysis, as opposed to image processing and recognition.

It was in the 1960s that Louis Kamentsky began the drive to design and build single-cell analyzers. Working at IBM's Watson laboratories, Kamentsky was interested in using optical character-recognition techniques to identify cancer cells. Because of the lack of computational power, this became a difficult goal. Kamentsky shifted his focus from image-based technology to single-cell analysis and the design of a cytometer that measured light absorption and scatter (Kamentsky et al. 1965), and shortly thereafter added the ability to sort cells using fluidic switching (Kamentsky and Melamed 1967). At the same time, Mack J. Fulwyler was trying to solve a problem generated by the study of red blood cells using a Coulter volume analysis system. A bimodal distribution of red blood cells observed using a Coulter volume detector suggested the possibility of two different types of red blood cells, contrary to accepted medical understanding. Fulwyler recognized that physically separating these "different" cells was necessary to determine if the two populations were in fact different. He became aware of Richard Sweet's development of high-speed chart recorders using electrostatic drop generation (Sweet 1965), and after visiting Sweet's laboratory he utilized this technology to design and build the first electrostatic cell sorter to separate red blood cells (Fulwyler 1965). Ironically, upon completion of the instrument, it took only a short time to recognize that the supposed bimodal distribution was related to spatial orientation rather than to inherent red blood cell variability (M. J. Fulwyler, personal communication). Amazingly, this finding of great significance has never been formally published. It was immediately obvious to Fulwyler that sorting of white blood cells was an opportunity not to be missed. The history of the development of cell sorting is well covered by Shapiro (2003). The cell sorter became a reality in 1965 and has proved to be one of the most important technologies available for cell analysis.

2.3
Components of a Flow Cytometer

Although flow cytometers appear to be very complex instruments they can be broken down into three essential parts – fluidics, optics, and electronics – and a vital additional component, informatics.

2.3.1
Fluidics

The fluidic system brings the sample into the instrument, then separates, aligns, and carries the particles to the interrogation point where they are intercepted individually by a light source (frequently supplied by one or more lasers). Within the fluidic system, the keystone is the flow chamber, in which a fluid sheath (particle free) meets the sample and results in the separation and alignment of sample particles into a stream of single particles (Fig. 2.1a). This phenomenon is referred to as hydrodynamic focusing. To achieve this, the flow within the flow chamber must be kept laminar.

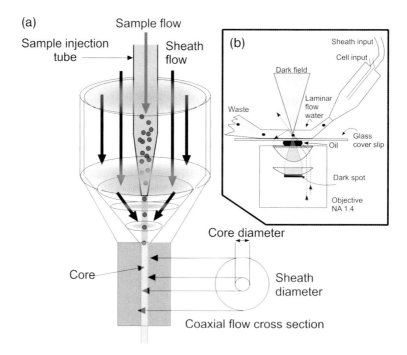

Fig. 2.1 (a) A functional view of a flow chamber in a typical flow cytometer. The fluid sheath flowing through a large area is forced under pressure into a much smaller orifice. Placed at the very center of the flow cell, an injection tube injects the sample (i.e. cells and other particles) into the centre of the flowing stream, thereby forcing the cells to undergo hydrodynamic focusing. A coaxial cross-section of the sheath and core is also shown. (b) An alternative flow cell based on an axial flow system typically used in microscope-based flow cytometers. In these instruments the laminar stream flows across a coverglass to a waste collector on the opposite side. Of note is the dark field objective. The central obscuration in the real focal plane of the objective produces a dark cone in the illumination field and since it extends above the object plane contains only the fluorescence and scattered light signal in the absence of the illumination signal. NA = numerical aperture.

Reynolds worked out the equation expressing fluid flow:

$$\mathrm{Re} = vd\rho/\eta,$$

where Re is the Reynolds number (a dimensionless number), v the average velocity (m s^{-1}), d the tube diameter (m), ρ the fluid density (kg m^{-3}), and η dynamic fluid viscosity (kg m^{-1} s^{-1}). For a Reynolds number $<$ 2300, flow will be laminar, a mandatory factor for quality optical measurements in sheath flow-based flow cytometers. Careful design of the fluidic system, particularly the flow chamber components, is crucial to maintain non-turbulent flow. Cells are thus hydrodynamically focused in a so-called core stream encased within the liquid sheath (Fig. 2.1a). This sheath-flow principle was derived from the work of Moldavan and subsequently Crosland-Taylor (1953), who designed a system similar to that in common use today, in which an insertion rod (needle) deposits cells within a flowing stream of fluid sheath (usually water or saline) forming a coaxial flow which moves from a larger to a smaller orifice, creating a parabolic velocity profile with a maximum at the center of the profile. Because of the hydrodynamic focusing effect, cells that are injected by the needle remain in the centre of the "core" fluid, thus allowing very accurate excitation with subsequent measurement having excellent sensitivity and precision within the flowing stream.

The fluid sheath generally flows at a rate of a few millilitres per minute, whereas the sample flows at a rate lower than 100 μl per minute. In air-pressured systems there is a pressure differential between the sheath and the sample (which is within the core), whereby the sample is 1 to 2 psi (0.069 to 0.138 bars; 1 pound per square inch (psi) converts to 6.89476 kilopascal (kPa)) above the sheath, forcing alignment of cells in single file throughout the core. Assuming that the flow is perfectly laminar, it is possible to calculate the diameter of the sample core and of the jet (core + fluid sheath) using the following equations:

$$d_{core} = 2\sqrt{\frac{f_{sample}}{\pi v_{jet}}}$$

$$\text{and because } v_{jet} = \frac{F_{sheath}}{\pi \left(\frac{D_{jet}}{2}\right)^2} \text{ then } d_{core} = \sqrt{\frac{f_{sample}}{F_{sheath}}} D_{jet}$$

where d_{core} is the sample core diameter coming out from the nozzle (in cm); D_{jet} the diameter of the jet (sample + fluid sheath) coming out from the nozzle (in cm); v_{jet} the jet velocity coming out from the nozzle (in cm s^{-1}); f_{sample} the sample flow rate (in cm^3 s^{-1}); and F_{sheath} the fluid sheath flow rate (in cm^3 s^{-1}). This equation shows that increasing the sample flow rate increases the diameter of the sample core. The consequence is an increase in the number of particles flowing per unit of time and thus the likelihood of coincidence – two or more particles within the focused light source (called the interrogation volume) at the same time. If the sample flow is too high, the increased core diameter destabilizes the flowing cells and reduces the accuracy and precision of the measurement. To prevent this, a simple solution consists of a concomitant increase in the sheath fluid flow rate as well. The sample and sheath flow rates depend on the instrument design as well as on the kind of analysis being performed (e.g. cell cycle analyses must be run at a low flow rate). Thus, it is important to be able to control both flow velocities on an instrument. If a highly accurate system is desired, multiple sheath inlets can be used to create more stable flow streams, but this is generally not provided in commercial systems.

The sample is injected into the instrument through a nozzle at a constant and controllable flow rate. Most commercial flow cytometers employ one of two main ways to inject the sample:

(i) On a majority of flow cytometers a pressure differential between the sample and the liquid sheath regulates the number of particles flowing through the interrogation volume;

(ii) On some syringe-based systems there is a volumetric sample injection using a syringe to inject the sample into the flow chamber at a very accurate flow rate. The piston of the syringe is controlled by a step motor so that the volume analyzed can be exactly measured, thus allowing the cell abundance to be calculated, a key parameter in many fields (e.g. microbiology, aquatic sciences, biotechnologies).

The design of the flow chamber is a crucial component. The fluid flows from a very large area to a very constrained channel (core) and its velocity increases significantly as the square of the ratio of the larger to smaller diameter. Within this channel, the velocity profile is parabolic, with the velocity at a maximum at the center of the stream and almost zero at the walls. This becomes a critical issue for biological specimens because they may contain proteins and a variety of released molecules; surface binding may eventually increase turbulence and destroy the hydrodynamic nature of the flow. In addition, such molecules may stick to the surface of the chamber where the velocity approaches zero, causing build up of turbulence-producing surfaces.

The acceleration at the core of the flow chamber is an important aspect in flow cytometers because particles are injected into the very centre of the flowing sheath stream. The fluid sheath highly accelerates the central core and thus induces spatial separation of particles within this rather long core stream. This hydrodynamic focusing creates the ability to more accurately analyze the signals from single cells. Once particles are spatially separated within the core and accurately identified by their optical properties, it is then possible to physically separate them in a process known as particle sorting, which will be addressed in more detail later in this chapter.

An alternative system to that described above uses axial flow, in which cells are injected with a regular nozzle onto the surface of a microscope objective to obtain laminar flow. The particles flow across the objective and are extracted from the system on the other side using a low-pressure region, as shown in Fig. 2.1b. This is similar to systems designed by Harald Steen and others (Petersen 1983; Steen 1983; Steen and Lindmo 1979) and has several advantages: (i) the ability to use high-numerical aperture microscope objectives providing high photon collection efficiency; (ii) excellent resolution and signal-to-noise ratio; and (iii) the ability to use a regular arc lamp as the light source. This system also has extraordinary sensitivity for forward scatter and is one of the most sensitive systems available for scatter. It was initially designed to be optimized for very small particles such as bacteria.

2.3.2 Optics

Most modern flow cytometers are provided with lasers (monochromatic light sources) as excitation sources, whereas the earliest systems used mercury lamps. In the late 1960s, relatively large water-cooled ion lasers were identified as the most desirable source of coherent light at 488 nm, which is the best excitation wavelength for fluorescein. These high-cost, large (and inefficient) light sources functioned to shape the design of the instruments themselves, making them enormous constructs often taking 60 to 80 square feet of floor space and requiring high cooling-water volumes and high current levels. Thanks to the advent of more recent solid-state lasers, the footprint of flow cytometers has been significantly improved and reduced. Further, in the mid 1980s, there was an emerging market in flow cytometers that did not require sorting. First known as "analyzers," these instruments are now commonly referred to as "bench-top" analyzers. This is somewhat of a misnomer, as the third generation of sorters is almost indistinguishable from the bench-top analyzers of the past.

As already mentioned, the key to the efficiency and sensitivity of flow cytometers is the laser-based coherent light sources. The excitation wavelength available is the chief criterion for selection of a laser. The beam should be segmented in a transverse emission mode (TEM) of TEM 00, although in some circumstances a mixed TEM 00 and TEM 01 mode does not exclude the usefulness of such a beam mode (Fig. 2.2b) (A simple explanation of TEM is available at http://en.wikipedia.org/wiki/Transverse_mode). Of course, the excitation light wavelength must match the absorption spectrum of the fluorochromes of interest. One reason that early systems used large water-cooled argon-ion lasers was that multiple lines could be obtained from these lasers by rotating the laser mirrors. They could produce lines in the UV (350 nm), deep blue (457 nm), blue (488 nm), and blue-green (514 nm) regions of the visible spectrum, making them a very useful light source. In addition, the argon laser was the only satisfactory coherent source of excitation for the most-used fluorochrome in the field: fluorescein.

The light beam emitted from a laser must then be focused to a spot in the desired shape, also referred to as the interrogation volume, where the flowing cells will be intercepted one by one. This is accomplished by using a beam-shaping optic to obtain the desired crossed-cylindrical beam shape. As displayed in Fig. 2.2,

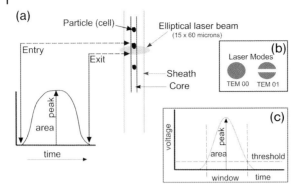

Fig. 2.2 (a) As cells pass through the interrogation point, they generate a pulse that can be characterized as shown here. At the point of entry into the laser beam, the pulse rises to a peak and holds for as long as the particle (cell) is in the stream. Once the particle begins to leave the laser beam profile, the signal intensity decreases to zero. The maximum signal is the peak, and the time taken between entering and leaving the beam is the time of flight (TOF). It is common to measure the total area under the curve (integral signal) for total fluorescence. Also shown is the excitation light beam profile most commonly used in flow cytometry. (b) Laser mode TEM 00 is the most desirable in flow cytometry. However, it is possible to mix the TEM 00 and the TEM 01 modes. (c) Definitions of each component of the signal from a cell passing through an elliptical beam.

the most desirable beam shape is an elliptical spot about 15 by 60 μm. This beam has a large, relatively flat cross-section, which reduces the variation in intensity of the excitation spot should the particle move around within the excitation area. Such a beam is optimal for particles smaller than the beam width. Reducing the beam even further would have the effect of "slit-scanning" the travelling particle.

In flow cytometers equipped with lasers (monochromatic light sources), there is no need to filter the light in order to obtain the desired excitation wavelength. But in flow cytometers equipped with arc lamps, which emit in a broad spectrum, excitation filters are required to select the proper excitation wavelength and eliminate the others. Similarly, light scattered by or fluorescence emitted by particles as they are intercepted by the excitation light must be filtered in order to separate the different wavelength ranges and direct them to their respective photodetectors. Two main groups of filters are used for these purposes in flow cytometers: colored glass filters and interference filters.

Colored glass or absorptive filters are generally composed of glass or plastic colored by dyes that absorb unwanted wavelength regions and transmit most of the light in the desired region. They can be long-pass or short-pass, meaning that they transmit wavelengths longer or shorter than a certain value, respectively. The filtering efficiency, that is, the quantity of light absorbed by the filter, can be enhanced by changing the angle of the filter with respect to the light-source direction in order to increase the pathway of the light within the filter. Absorptive filters must be used with caution however, as they tend to produce fluorescence from the dye used in their manufacture.

Interference filters, also called dichroic, dielectric, or reflective filters, are more complex filters composed of transparent glass or quartz with several layers of dielectric materials deposited on the surface. Compared to absorptive filters, interference filters usually transmit less light, but permit greater selectivity and allow sharper transitions between rejected and transmitted regions. The light entering the filter produces a series of constructive or destructive interferences as it passes through the different layers. Only specific wavelength ranges are eventually transmitted by the filter.

Within interference filters, there are two principal types: band-pass filters and dichroic beam-splitting mirrors. Band-pass filters transmit only a particular wavelength range; in other words, they block wavelengths above and below the transmitted region. These filters are characterized by (i) the wavelength range corresponding to the maximal transmittance; (ii) the width (in nm) of the transmitted spectrum at 50% of transmittance; and (iii) the percentage of light transmitted at the maximal transmittance.

Dichroic mirrors (or edge filters) may be short-pass or long-pass filters. Short-pass filters transmit only wavelengths shorter than a particular value, whereas long-pass filters transmit only wavelengths longer than a certain value. Both are commonly used in flow cytometers as dichroic mirrors in order to split the incident light into two directions according to the color ranges. They are generally placed at an angle of 45° with respect to the incident beam. On most flow cytometers these filters are used to separate the different fluorescence emissions and direct them toward specific photodetectors.

Dichroic mirrors are characterized by (i) the transmittance maximum and (ii) the wavelength at 50% of transmittance, also called cut-off for short-pass filters or cut-on for long-pass filters.

2.3.3
Electronic Systems

Thanks to the filters described above, photons which have been scattered or emitted by each particle are separated according to their energy level (i.e. wavelength) and directed to appropriate photodetectors. These detectors take in photons and put out electrons. The current produced is then converted into a digital value recorded on a computer.

In most flow cytometers, two main types of photodetectors are used: photodiodes and photomultiplier tubes (PMTs).

Photodiodes are made of photosensitive materials. They are fast, relatively cheap devices, and do not require a power supply, but their gain is much lower than that of PMTs and often limits their use for fluorescence in flow cytometry, as opposed to the collection of forward-angle light scatter and/or light absorbance.

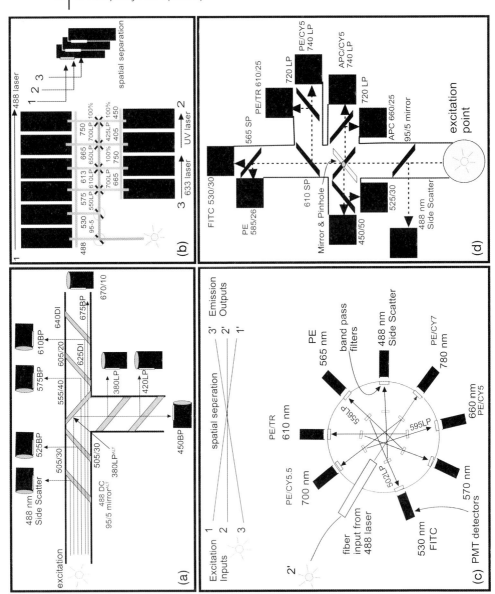

PMTs are more complex devices. They present a photocathode placed behind a glass or quartz window. The window can be on the end or the side of the PMT. Each photon impinging on the photocathode on the external side generates a photoelectron on the internal side; this photoelectron is accelerated toward the first dynode as a result of the difference of potential (DOP) between these two electrodes. The photoelectron gains energy during this acceleration, and when it hits the dynode it dislodges more electrons. These electrons are then accelerated and driven to the next dynode, and so on, as a result of the DOP between the electrodes. The higher the DOP, the higher the energy provided to the electrons at each step and the higher the signal amplification. The signal reaching a PMT can thus be enhanced by a factor of up to 1 million. This is the reason that PMTs are more sensitive and thus more suitable than regular photodiodes for the collection of low-intensity signals such as those produced by small and/or dimly fluorescent particles.

It has become standard design to utilize a PMT for each spectral wavelength of interest. In most instruments designed up to about 1990, a maximum of four or five spectral bands was collected. However, in the last few years, five to 10 spectral detectors became more common. Each spectral band is collected by a PMT strategically placed within an optical system, of which there are many current designs. Figure 2.3 shows several different optical layouts currently used in commercial systems. It is now evident that many biological requirements are in the range of 10 to 15 spectral bands, far more than the three to five available for the first 20 years in the field. Next-generation systems will include a vast number of PMTs, avalanche photodiodes, or multichannel PMTs (Robinson 2004) in addition to high-speed cameras. The disadvantage of the multichannel PMT is that detection sensitivity is reduced and it is not currently possible to adjust the sensitivity of each channel as can be achieved with individual PMTs. The advantage is that the complexity and number of optical components is reduced, and the opportunity arises for advanced automated classification.

Most cytometers use PMTs as detectors for both fluorescence and side scatter. The pulse of a particle crossing the excitation beam will depend upon the beam shape, width, and intensity, and on the particle size and velocity (Fig. 2.2a). Systems running at 10 m s^{-1} will cross a 10-μm beam in 1 μs, or a 5-μm beam in

Fig. 2.3 Different optical pathways from several commercial flow cytometers. (a) Beckman-Coulter ALTRA. (b) Dako-Cytomation CYAN instrument with photodetectors placed on three beams with slightly different trajectories. (c) Becton-Dickinson ARIA system based on an innovative photomultiplier tube (PMT) array in which PMTs are placed on a ring that allows the emission signal to bounce around the ring. There are six additional detectors on this system (not shown), which come from the first and third lasers (see diagram). (d) Becton-Dickinson Vantage system with a typical configuration of photodetectors. In all cases (a–d) above, a narrow bandpass filter is placed immediately in front of each PMT in addition to the dichroic mirrors that are used to direct the various emission spectra.

only 500 ns. The majority of instruments prior to publication of this chapter were designed around analog detection, rather than digital electronics. Essentially, once the threshold voltage is met (based on the discriminator circuit described below), the signal (usually 0–10 V) is fed into an analog-to-digital converter (ADC) circuit called a comparator circuit, whose purpose is to identify the presence of a measurable signal and send a signal to trigger the rest of the detection system (Fig. 2.2c). This is a binary decision only. Once a decision to collect is reached, several measurements for each variable are made, such as peak (maximum), integral (area), and time of flight (length of the signal). Several complications can cause problems in the detection electronics. For instance, if two particles pass the interrogation point in a very short interval, both signals must be aborted if this time is shorter than the reset time for the electronics. Another circuit is required to make this decision.

To further complicate matters, many systems use several lasers (two or more beams) delayed by a few microseconds only. Each particle must be analyzed at the perfect time by each laser, so data generated from the first beam must be stored while the system waits for the particle to reach the second laser beam, and so on. If the beam separation is large enough, several cells might be analyzed by the first beam before the first cell passes the second beam. This rather complex system is not necessary on simpler analysis instruments, but it is absolutely mandatory on more advanced multi-laser cell sorters. In addition, the time taken for all the analysis components sets the maximum analysis rate of a flow cytometer. The faster the system, the shorter the dead time must be; for example, a dead time of less than 10 μs would be necessary to analyze 100 000 cells per second. Actually, depending on how many events must be analyzed to achieve a speed of 100 000 cells per second, the dead time would need to be considerably shorter.

2.4
Flow Cytometric Informatics

After signals have been collected, filtered, and converted to digital values, data are stored on a computer. Most commercial flow cytometers store their data in a particular format called Flow Cytometry Standard (FCS).

The first accepted FCS format was published in 1984 (Murphy and Chused 1984). It was then revised in 1990 by the Data File Standards Committee of the Society for Analytical Cytology (now the International Society for Analytical Cytology, or ISAC) to give birth to the FCS 2.0 version (Anonymous 1990). In 1997, this version was further revised to handle data files > 100 MB and to support UNICODE text for keyword values (Seamer et al. 1997).

Within FCS files, storage is made in real time and most often in listmode. The structure of an FCS data file is based on three to four segments:
  (i) The Header identifies the file as an FCS file and specifies the version of FCS used. It also contains numerical values identifying the position of the following Text segment.

(ii) The Text segment contains several Keywords and numerical values used to describe the sample and the experimental conditions.
(iii) The Data are stored in the form of numerical values in a format specified in the Text segment. They are stored in sequential fashion as they were generated by each cell as it passed the interrogation point.
(iv) The Analysis segment is optional (example: results from cell cycle analysis).

Flow cytometry provides a multiparametric analysis of each single particle analyzed by the instrument. The data are displayed in one (histograms) or two (cytograms) dimensions. Histograms show the frequency distribution of particles for any particular parameter. Cytograms display correlated data from any two parameters. Data can also be displayed in 3-D plots, the third axis being either the abundance (number of events) or a third parameter (Fig. 2.4).

Flow cytometers collect vast amounts of data very quickly. In fact, they are in a class of instruments that push the limits of data collection. For example, it is currently possible to collect at least 11 or more fluorescent spectral bands simultaneously with at least two scatter signals on thousands of cells per second, creating a multivariate analysis challenge (De Rosa and Roederer 2001).

The key principle of FCM is that every particle is identified individually and classified into a category or population according to multivariate analysis solutions. Advanced statistics of the data are crucial to the establishment of separation criteria for analysis and real-time sorting.

Every particle that passes the interrogation point will produce a signal that can be collected on every detector, which would cause a data overload problem. To prevent this, a circuit called a discriminator is included and set to exclude signals below a preset voltage (Fig. 2.2c). On many current instruments it is possible to use discriminators on any or all detectors. That is to say, multiple detectors must register a preset signal level or nothing is collected by the data collection system. Once a threshold setting is satisfied, the discriminator triggers the entire data collection system, and all identified detectors will collect the signal. Very often, the forward-angle light-scatter signal is used as discriminator to detect the presence of a measurable particle. However, it is also useful to use a fluorescence detector when recording only those particles detectable above a certain level of fluorescence.

The most frequently recognizable detection system in flow cytometers nowadays is that of fluorescence. The initial detection system used in the earliest instruments was Coulter volume (Coulter 1956), based on the original patent of Wallace Coulter, whereby the principle of impedance changes was transferred from cell-counting instruments to flow cytometers. This technology was also the basis for the first cell sorter. In addition to impedance, light scatter was also measured. Current systems have taken a rather complex pathway for the measurement of fluorescence. There are, however, some complications in measuring fluorescence signals.

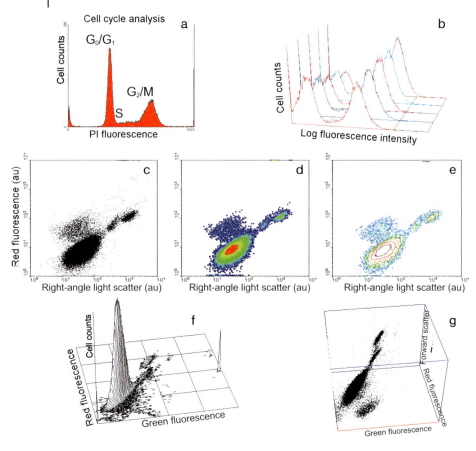

Fig. 2.4 Ways to display flow cytometric data. (a, b) Histograms display one parameter as a frequency distribution. (c) Dotplots (cytograms) display correlated data from any two parameters. Each dot corresponds to an event analyzed by the flow cytometer. Of course, several events can occupy the same dot if they have the same parameter intensities. (d) Density plots display two parameters as a frequency distribution. Color is used to code the different frequencies of events. (e) Contour plots display correlated data from any two parameters, with contour lines joining points of equal elevation (frequency distribution). (f, g) Data can also be displayed in three dimensions, the third dimension being either the frequency distribution (f) or any parameter (g).

Linear amplifiers produce signals that are proportional to their inputs, but require very expensive hardware that is only just becoming available at the level of commercial systems. While it is possible to amplify these signals, most immunofluorescence applications have huge dynamic ranges that are beyond amplification in the linear domain of most current systems. For this reason, logarithmic amplifiers with scales of three to five decades are required. Log amplifiers are particularly useful for samples in which some particles exhibit very small amounts of signal, while others have signals that are four orders of magnitude larger.

2.5
Spectral Compensation

When a particle or cell contains several fluorophores (fluorescent molecules) with signals in multiple spectral bands, the identification and analysis becomes considerably more complex because of the likely spectral overlap among the fluorophore emissions (Fig. 2.5). For example, a system containing a detector with a band-pass filter designed to collect fluorescence from FITC (fluorescein isothiocyanate, 525 nm) and another detector designed to collect signals at 550 nm will register photons in both detectors. When a single fluorophore is being collected, this presents no problem, but it is a different story when two or more fluorophores with close emission bands are simultaneously present. It is then necessary to identify which fluorophore was the real emitter of the photons collected on each detector. To achieve this, a process known as spectral compensation must be carried out, whereby a percentage of the signal collected by one detector is subtracted from the signal collected by the other. Of course, the complexity of spectral compensation increases with the number of fluorophores. A special set of circuits must be designed that allows for a varying percentage of each signal to be subtracted from every other detector. While this can be performed perfectly well off-line in software (Bagwell and Adams 1993), if the very goal of the analysis is to sort a certain population of cells, the compensation must be performed in real time between the time the cell passes the excitation beam (interrogation point) and the time it reaches the last point by which a sort or abort decision must be made. Because compensation in FCM is very sophisticated, it requires a large number of controls to establish appropriate compensation settings and

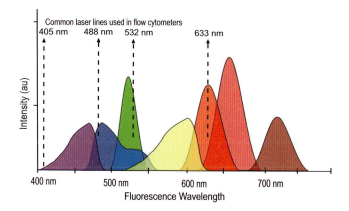

Fig. 2.5 Example of spectral overlap between different fluorophore emissions. Shown in this figure are the excitation lines of several common lasers frequently used in flow cytometers (e.g. violet diode laser (405 nm), argon (488 nm), diode-pumped YAG (532 nm), and helium/neon (HeNe; 633 nm)). Because of the overlap of many fluorochrome emissions, it is necessary to identify and make allowance for the spectral overlap.

photomultiplier setup. As fluorescent dyes increase in number and spectral proximity, the need for spectral compensation circuitry becomes more urgent. This is far more involved than anything currently available in image-analysis systems.

2.6
Cell Sorting

Among current flow cytometers, analyzers (also referred to as bench-top flow cytometers) and more complex instruments called sorters can be distinguished. Sorters are able to both analyze and physically separate (sort) particles of interest based on their optical properties.

There are two basic types of cell sorters. The first, the fluid-switching cell sorter (Fig. 2.6c), uses a mechanical device to deflect the particles into a tube. This system is very simple to use but is rather slow (maximum 300 cells sorted per second) compared to the second type, the droplet cell sorter, which is based on the generation and deflection of droplets containing the particle of interest and is able to sort up to several thousand particles per second. The majority of current sorters are equipped with this electrostatic technology.

Actually, there is also a third kind of cell sorter, based on a photodamage technique. Cells are not really sorted, but all particles except the cells of interest are rendered non-viable by illumination with a high-energy laser pulse, which induces photodamage to DNA (Herweijer et al. 1988). This system is not widely used and presents several drawbacks, such as the use of very expensive pulsed-laser systems, the high abundance of cellular debris in the sorted fraction, and the need to calibrate the pulse.

The principle of cell sorting was included in instruments designed by Fulwyler (1965), Kamentsky and Melamed (1967), and also Dittrich and Göhde (1969) in order to definitively analyze a cell of interest. It was Fulwyler, however, who identified the technique developed by Sweet (1965) for electrostatic droplet separation for use in high-speed inkjet printers as the ideal technology for cell sorting. This evolved into the technique of choice for virtually all current commercial cell sorters (Fig. 2.6a). The implementation of this idea into a commercial system was carried out by Hertzenberg's group in the early 1970s (Bonner et al. 1972).

The principle of electrostatic sorting is based on the ability to identify a cell of interest by its optical properties (light scatter and/or fluorescence), determine its physical position in the jet with a high degree of accuracy, break off the jet into droplets using a piezoelectric crystal, place a charge on the stream at exactly the right time (when the cell reaches the last attached droplet), and then physically deflect and collect only the droplet containing the sorted cell (Fig. 2.6a).

The technology of high-speed sorting has already been well defined by van den Engh (2000), who discusses in detail the complex issues involved. In brief, the speed and accuracy of a cell sorter are based on several factors. Firstly, despite the initial discussion pointing out that fully stable laminar flow is required for accurate analysis, for cell sorting the stream must be vibrated using a piezoelectric

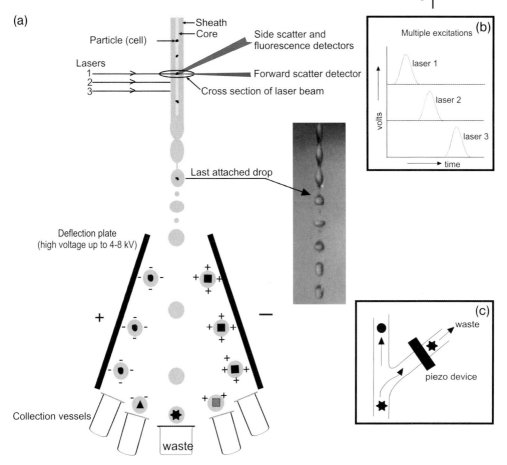

Fig. 2.6 (a) The principle of electrostatic cell sorting based on Sweet's inkjet printer technology. In this figure a stream of liquid intersects a laser beam (or multiple laser beams 1, 2, 3). The stream is vibrated by a piezoelectric crystal oscillator at frequencies from 10 to 300 kHz, depending upon the orifice size, stream velocity, nature of the stream, and particle size. Typically 30 000–50 000 Hz is used to create droplets at the same frequency. Once a cell/particle is identified as desirable, a charge is placed on the stream which remains with the last drop (last attached drop) that leaves the stream. Using a computation method, this drop is sorted by being attracted toward a plate almost parallel with the stream and containing opposite charges in the vicinity of 4–8 kV. Each droplet containing a desirable particle can be placed into one of several containers (shown is a four-way sorting system). In the center of the figure is a video image of the droplets strobed at the same frequency as droplet formation. (b) The pulses of three different lasers as a particle passes by each beam separated in space. Thus, a particle will pulse from each laser a few microseconds apart. This way, signals from each laser can be individually analyzed. (c) An alternative sorting system using a mechanical fluid-switching technique. In this system the waste stream is blocked momentarily to allow a desired cell to pass into the sorting pathway.

device to generate droplets. As described by van den Engh, it is necessary to match the nozzle diameter and sheath pressure to the sample particle size. Droplet generation frequency as well as high-speed electronics must also be carefully defined for each sample to obtain the stable droplet generation mandatory for high-speed cell sorting. The principle that governs the generation of droplets has been characterized by Kachel et al. (1977), whereby the wavelength of the undulations is:

$$\lambda = v/f,$$

where λ is the undulation wavelength (m), v the stream velocity (m s^{-1}), and f the modulation frequency (1 s^{-1}).

The system is optimized for maximum droplet generation when $\lambda = 4.5d$ (d = exit orifice = jet diameter). Thus, from the equation above, the optimal generation frequency is given by $f = v/4.5d$. If a system is designed to accommodate this optimal droplet formation, the jet velocity is proportional to the square root of the jet pressure, as demonstrated by Pinkel and Stovel (1985). Thus, to sort at 20 000 particles per second (or 20 000 Hz) such that each drop is separated from the next by 4.5 stream diameters and flowing at 10 m s^{-1}, an optimal system would make the distance between drops 200 μm. As the number of drops sorted increases, the diameter must decrease, with the obvious conclusion that the speed of high-speed sorters will eventually be partially regulated by the size of the particle to be sorted and by the velocity to which the stream can be taken without destroying the sample. This is particularly important for biological particles such as cells. High-speed sorters are essentially sorters that are designed to operate at sort speeds in the range of 20 000 to 100 000 particles per second. A higher pressure must be placed on the sample stream.

When system rates exceed 40 000 cells per second, the analysis time becomes a key issue and is obviously the limiting factor, since complex analysis must precede the sorting decision. Therefore the maximum speed of droplet formation is not the limiting factor in the design of a high-speed flow cytometer. As discussed in van den Engh (2000), the primary issue is the high pressures that must be used to create ultra high-speed droplet formation. For instance, at a droplet frequency of 250 000 per second the jet pressure must approach 500 psi (pounds per square inch; see Chapter 1 for conversion to SI units), a significantly higher value than can be designed safely in most systems. If the pressure is limited to around 100 psi, a droplet rate of around 100 000 is closer to the realistic maximum. This then is the real limitation to current high-speed sorting systems.

Since it is impossible to predict exactly when any particle is going to pass the interrogation point, Poisson statistics enter the equation. This adds uncertainty into the analysis, and as discussed previously, it is crucial to ensure that no measurements take place as two or more cells try to pass the interrogation point simultaneously. This implies that there is a narrow relationship between particle abundance and coincidence likelihood.

Cell sorting has become a very important component of FCM and is currently used in many fields. The isolation of CD34 human hematopoietic stem cells, whereby flow-sorted cells are specifically purified for transplantation purposes, in particular has revolutionized capabilities in transplantation (Andrews et al. 1989). It goes without saying that to perform such a sort all components of the instrument that come into contact with the samples must be sterile. A similar issue occurs when sorting plant protoplasts (see Chapter 10).

Another important issue is the potential danger involved in sorting hazardous samples because of the generation of aerosols in the operation of a flow cytometer. This is particularly the case with human samples that may be infected with HIV or (more commonly) hepatitis virus, and can lead to considerable tension between operators and researchers wanting to sort materials from infected patients. If the sampling must be performed, it is necessary to employ complex biosafety systems to reduce the potential of infection. There is also a significant literature dealing with the dangers posed by microbes and carcinogenic molecules such as the fluorescent dyes commonly used to label cells (Nicholson 1994).

2.7
Calibration Issues

Because FCM is defined as a quantitative technology, calibration standards are important. Some standards were primarily developed by Schwartz and Fernández-Repollet (1993) and others to allow reproducibility of clinical assays. Schwartz developed the concept of "molecule equivalents of soluble fluorescein" (MESF units). Using a mixture of beads with known numbers of fluorescent molecules, it is possible to create a standard curve based on a least-squares regression based on the median fluorescence intensity of each bead population. This value is then converted into MESFs (Fig. 2.7) from which comparisons can be made between different instruments or for the same instrument on different days. Future instruments will most likely provide data in units such as MESFs rather than in the "arbitrary unit" frequently observed in publications. It would seem highly desirable to provide more quantitative data for comparison purposes.

2.8
Conclusions

Flow cytometry has made a significant impact on many fields because it is one of the few technologies, if not the only one, that can evaluate so many parameters on such small samples in such short time periods and at the single-cell level. The principle of evaluating each and every cell or particle that passes through one or several light sources and then producing a highly correlated data set is specific to FCM. The combination of the fluorochromes available now and the ability to physically separate cells by the process of cell sorting provides some unique

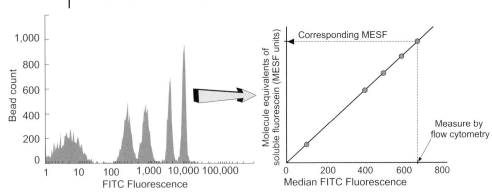

Fig. 2.7 Calibration beads with fluorescent molecules attached to their surface can be used to create quantitative measures by flow cytometry. In the histogram on the left there are five peaks, the lowest peak being the negative control (no fluorescent molecule bound to the beads) and the other four peaks representing four different levels of fluorescence intensity. From this histogram and the data given by the manufacturer, a standard curve (on the right) can be obtained for quantitation of the particles being labeled with this probe. FITC = fluorescein isothiocyanate.

characteristics. It is almost 40 years since FCM first demonstrated its importance in medical research. Since that time, well over 60 000 publications have highlighted its usefulness. It emerged as one of the most important technologies in the early 1980s upon the recognition of AIDS. The ability of FCM to identify and quantify the T-cell population subsets CD4 and CD8 lymphocytes appointed it as a most important technology in the diagnosis of HIV and the monitoring of AIDS patients. Similarly, the ability of FCM to make complex multivariate analyses of bone marrow to identify CD34$^+$ cells and subsequently sort and purify them has been a vital resource in transplantation immunology.

Currently, FCM is used routinely in a number of fields ranging from basic cell biology to genetics, immunology, molecular biology, and environmental science. It makes possible the detection of cells at the single-cell level based on their optical properties (light scatter and fluorescence emission). It is thus possible to detect them, quantify them, obtain information on the individual characteristics of each single cell analyzed, and to gain information about the heterogeneity of the physiological states within a population. FCM can be applied to either prokaryotes or eukaryotes, autotrophic or heterotrophic cells whose fluorescence is natural (from intrinsic fluorescent molecules such as photosynthetic pigments for instance) or induced by artificial physiological or taxonomic fluorescent probes added to the samples. The size of the particles can vary from that of viruses (<0.2 μm) to larger cells hundreds of microns in diameter. The time of analysis is very short and thus a very high number of cells can be analyzed, which provides strong statistics and gives results representative of the structure of the whole population.

In many cases it is also possible to physically separate (sort) the cells of interest from a heterogeneous mixture according to their optical properties. In the majority of cases, cells are kept viable after the sorting process, making it possible to harvest them and supplement their analysis by further techniques (molecular biology). With current technology it is possible to sort and purify a very large number of cells (10^6–10^8) from complex heterogeneous mixtures in a reasonable time (hours).

During the last 15 years or so, FCM techniques have been applied to the analysis of plant cells. The goals of these studies covered several aspects of plant cells ranging from nucleic acid content (both DNA and RNA), karyotyping, and transgene expression, to cell counting, chloroplasts, cell membranes, and cell wall regeneration, as well as mitochondrial activity, secondary metabolite accumulation, or sorting of cells or subcellular organelles of interest (for review see Yanpaisan et al. 1999 and other chapters in this volume).

Progress in fluorescent probe development combined with flow cytometry creates new opportunities to study cells. Technology developments such as the application of multispectral detection will change the basis for detection and analysis of cellular systems (Robinson 2004). Many of these are barely achievable, if not impossible, by more conventional methods. For example, FCM has now been closely linked with new opportunities in proteomics – in fact, flow cytometry clearly possesses the ability to apply complex phenotypic analysis and subsequent cell sorting for application of purified populations to proteomic analysis in semi-automated systems for protein profiling (Bernas et al. 2006). This will help to bring answers to a great diversity of scientific applications ranging from simple cell detection and counting to more complex molecular biology, biochemistry, physiology, and ecology.

References

Anonymous **1990**, *Cytometry* 11, 323–332.
Andrews, R. G., Singer, R. G., Bernstein, I. D. **1989**, *J. Exp. Med.* 169, 1721–1731.
Bagwell, C. B., Adams, E. G. **1993**, *Ann. NY Acad. Sci.* 677, 167–184.
Bernas, T., Gregori, G., Asem, E. K., Robinson, J. P. **2006**, *Mol. Cell. Proteomics* 5, 2–13.
Bonner, W. A., Hulett, H. R., Sweet, R. G., Herzenberg, L. A. **1972**, *Rev. Scient. Instr.* 43, 404–409.
Coulter, W. H. **1956**, *Proc. Natl Elect. Conf.* 12, 1034–1042.
Crosland-Taylor, P. J. **1953**, *Nature* 171, 37–38.
De Rosa, S. C., Roederer, M. **2001**, *Clinics Lab. Med.* 21, 697–712.
Dittrich, W., Göhde, W. **1969**, *Zeit. Naturforsch.* 24b, 360–361.
Fulwyler, M. J. **1965**, *Science* 150, 910–911.
Herweijer, H., Stokdijk, W., Visser, J. W. M. **1988**, *Cytometry* 9, 143–149.
Kachel, V., Kordwig, E., Glossner, E. **1977**, *J. Histochem. Cytochem.* 25, 774–780.
Kamentsky, L. A., Melamed, M. R. **1967**, *Science* 156, 1364–1365.
Kamentsky, L. A., Melamed, M. R., Derman, H. **1965**, *Science* 150, 630–631.
Moldavan, A. **1934**, *Science* 80, 188–189.
Murphy, R. F., Chused, T. M. **1984**, *Cytometry* 5, 553–555.
Nicholson, J. K. A. **1994**, *Cytometry* 18, 55–59.
Papanicolaou, G. N., Traut, R. **1941**, *Am. J. Obstet. Gynecol.* 42, 193–206.

Petersen, S. E. **1983**, *Cytometry* 3, 305–307.

Pinkel, D., Stovel, R., **1985**, Flow chambers and sample handling in *Flow Cytometry, Instrumentation and Data Analysis*, ed. M. A. Van Dilla, P. N. Dean, O. D. Laerum, M. R. Melamed, Academic Press, New York, USA, pp. 77–126.

Robinson, J. P. **2004**, *Biophoton. Internat.* 2004, 36–40.

Robinson, J. P., Durack, G., Kelley, S. **1991**, *Cytometry* 12, 82–90.

Schwartz, A., Fernández-Repollet, E. **1993**, *Ann. NY Acad. Sci.* 677, 28–39.

Seamer, L. C., Bagwell, C. B., Barden, L., Redelman, D., Salzman, G. C., Wood, J. C., Murphy, R. F. **1997**, *Cytometry* 28, 118–122.

Shapiro, H. M. **1983**, *Cytometry* 3, 227–243.

Shapiro, H. M. **2003**, *Practical Flow Cytometry*, 4th edn, Wiley-Liss, New York, NY, USA.

Steen, H. B. **1983**, *Histochem. J.* 15, 147–160.

Steen, H. B., Lindmo, T. **1979**, *Science* 204, 403–404.

Sweet, R. G. **1965**, *Rev. Scient. Instr.* 36, 131–136.

van den Engh, G. **2000**, High speed cell sorting in *Emerging Tools for Single-Cell Analysis*, ed. G. Durack, J. P. Robinson, Wiley-Liss, New York, USA, pp. 21–48.

Yanpaisan, W., King, N. J. C., Doran, P. M. **1999**, *Biotechnol. Adv.* 17, 3–27.

3
Flow Cytometry with Plants: an Overview

Jaroslav Doležel, Johann Greilhuber, and Jan Suda

Overview

The rise of plant flow cytometry since 1973 is testimony to the impact of a single elementary methodological innovation – the use of a razor blade instead of enzymes for isolation of nuclei. From 1983 onwards, this innovation led to an avalanche of applications of sophisticated instrumentation originally designed for biomedical research, and rarely before employed in the plant sciences. Owing to the physical size of the cell nucleus, its stainability with nucleic acid dyes and its genetic hegemony, nuclear DNA flow cytometry still dominates the field. Modern, small and affordable multiparameter instruments that measure fluorescence and light scatter allow convenient and rapid ploidy screening of living and dried plant samples, high precision genome size and endopolyploidy measurements, as well as cell cycle studies. Simultaneous measurement of side scatter allows improved analysis by discriminating clean nuclear fractions from samples corrupted with fluorescent debris particles which stick to the nuclei. The unambiguous detection of aneuploidy requires a high resolution technology, and is possible only under favorable cytogenetic circumstances. The estimation of the AT/GC ratio in nuclear DNA with a pair of fluorochromes remains a special field of research yielding sometimes controversial results. Since the nature of (multicellular) plant architecture is such that it is difficult to isolate individual cells, cell studies are often carried out with protoplasts (e.g. analysis of gene expression using fluorescent proteins). The study of subcellular processes (photosynthesis with plastids, respiration with mitochondria, membrane studies, and apoptosis-associated events) is mostly carried out using isolated organelles. When particles become too small to be analyzed as individuals, or when multiple antibodies are to be analyzed at one time, the innovative bead-based technologies can be applied. Thus, in phytosanitary screening with the microsphere immunoassay, dual-fluorochrome-tagged beads loaded with various covalently bound antibodies for microbial pests are employed. As up to 100 different bead classes can be made discernible by tagging them with various proportions of fluorochromes, a large number of different microbes can theoretically be quantified. The method can

Flow Cytometry with Plant Cells. Edited by Jaroslav Doležel, Johann Greilhuber, and Jan Suda
Copyright © 2007 WILEY-VCH Verlag GmbH & Co. KGaA, Weinheim
ISBN: 978-3-527-31487-4

also be used for the analysis of other molecules and viruses. The autofluorescence of various photopigments together with light scatter of the cells makes phytoplankton (together with abiotic particles and zooplankton) amenable to flow cytometry *in situ*. In this area of research, the main approach is the discrimination of particles by their bi- or multivariate fluorescence/scatter pattern or Coulter-sizing, and their association with classified organisms. Time-resolved pulse analysis even makes cell counting of linear algal colonies (e.g. diatoms) possible. Buoyant or submersed vehicles with purpose-designed flow cytometers are put to service for the control of environmental changes in the sea. A hitherto generally deplored disadvantage of flow cytometry is a lack of visualization of the objects being measured, but this has been overcome by integrated video imaging in recent instrumental developments. It is only a matter of time before this innovation becomes standard in bench-top flow cytometers which stand in botany laboratories.

3.1
Introduction

Flow cytometry (FCM) is a unique tool combining a powerful analytical capability to measure simultaneously many optical parameters of cells and subcellular particles, with the potential to isolate purified populations of selected particles. Both analysis and sorting can be carried out at high speed so that statistically relevant samples can be measured and minor subpopulations detected. The ability to isolate large numbers of particles of the same type then provides enough material for subsequent manipulations, including further growth and culture, as well as analysis with biochemical and genomics techniques.

The development of FCM in the late 1960s and the 1970s was largely stimulated by a need for automatic and high throughput analysis of human cells. As the available hardware and software was unable to achieve this by analyzing the images, evaluation of the optical parameters of particles passing through a narrow stream of liquid was an elegant alternative (see Chapter 1). Measured particles could be run at high speeds and, unlike the images of individual particles, pulses of scattered light and fluorescence generated by each particle were easy to collect and analyze by the extant optical and electronic systems. In fact, the pulses were analyzed by the so-called multichannel pulse height analyzers, which were originally developed to characterize radioisotopes. This explains why the classes in FCM histograms and cytograms have been called channels.

The absence of visual information on particles measured by FCM may be seen as a disadvantage, especially when compared with currently available laser-scanning cytometers and image analyzers. However, the growing battery of fluorescent dyes, antibodies and other reagents facilitate a plethora of FCM assays invaluable in pathology, oncology, immunology, and hematology, among others, making FCM very popular both in research and clinical laboratories. Some flow cytometers are even capable of capturing images of particles measured in flow (Kachel and Wietzorrek 2000).

Although originally not intended, the option of evaluating cells during their passage through a liquid stream offered the possibility of modifying flow cytometers for flow sorting. The so-called flow sorters then facilitated isolation of subpopulations of large numbers of particles based on their optical parameters. Even today this is hard to achieve with any other system, including the slide-based systems, which only permit manipulation with a very limited number of particles using either mechanical microdissection or laser capture. Another benefit of the flow-based design which had not been foreseen, is that it facilitates automatic sampling from liquid environments like the sea, lakes and sewage waters (see Chapter 13), something that would be hard to achieve with slide-based systems.

The dominance of biomedical research shaped the design of the first generations of commercially produced flow cytometers and until now, most instruments available on the market did not consider other applications. This, together with relatively high costs did not promote the use of FCM in plant sciences. After the pioneering paper of Heller (1973), the first uses of FCM in plant sciences were only reported in 1980s and their numbers did not increase markedly until the early 1990s. Despite a rather slow start, the past two decades of development resulted in a well-established and flourishing area with many applications in basic and applied research as well as industry. Plant FCM is now an actively developing research discipline progressively introducing innovative and unprecedented applications.

3.2
Fluorescence is a Fundamental Parameter

Fluorescence is the most common optical parameter in plant FCM. It is a physical phenomenon in which a substance, either organic or inorganic, absorbs light of a certain color (i.e. of certain wavelength and energy) and emits light of another color, with a lower energy and thus at longer wavelength (a part of the excitation energy is always lost). Dye molecules capable of such light (energy) conversion are termed fluorochromes. It is important that the difference in absorbed and emitted light wavelengths (called Stokes' shift to honor George Stokes who in 1852 first documented fluorescence in a solution of quinine) is sufficient to allow their separation using optical filters.

Detection of fluorescent signals from stained particles in liquid suspension is possible due to two mechanisms. The first involves accumulation of a fluorochrome in the stained region, making a contrast in dye concentration between the target and the background, while the second involves an increase in quantum efficiency (i.e. the ratio between the number of emitted and absorbed photons) when a fluorochrome molecule finds itself in a particular environment. For example, the quantum efficiency of the DNA-selective stains propidium iodide and

DAPI increases about 30- and 100-fold, respectively, after binding to nucleic acid as compared to an aqueous solution.

The number of fluorochromes employed in plant FCM continues to increase, and they can be broadly classified into three categories: (i) nucleic acid dyes (e.g. DAPI, propidium iodide, and Hoechst dyes); (ii) protein dyes, including fluorescent labels for antibodies (e.g. fluorescein, allophycocyanin, Texas red); and (iii) functional probes for *in-situ* monitoring of cellular activity (e.g. cyanine and oxonol dyes).

3.3
Pushing Plants through the Flow Cytometer

The high costs and the focus on analyzing human and animal cells certainly did not encourage the rapid spread of FCM in the plant sciences. But at least equally important obstacles were, and in many ways continue to be, the difficulties associated with sample preparation, as a consequence of the plant body design and cell structure. Plants are built of complex three-dimensional tissues of various types of irregularly shaped cells with rigid cell walls, which are held together by extracellular matrix. However, a sample for FCM must be in a form of liquid suspension of single particles. Additional problems are posed by the chemical composition of cells. Plant cells produce a vast array of secondary metabolites which may interfere with staining of particular cell constituents and/or exhibit autofluorescence, thus hampering quantification of signals from fluorescent probes. Provided that analysis at the cellular level is not required, a frequent alternative is to isolate the organelles of interest and treat them separately.

3.3.1
Difficulties with Plants and their Cells

Land plants (Embryophyta) are multicellular organisms in which free viable cells are rarely observed with the exception of microspores in land plants, pollen grains in seed plants, and sperms in bryophytes and pteridophytes. Conversion of solid plant tissues to single cells is possible using appropriate hydrolytic enzymes, such as pectinases which digest the pectin lamella between cells. However, due to a variety of secondary anatomical modifications accompanying differentiation, cells can be isolated from some tissues only, young tissues often being the best candidates. As a matter of fact, a single-cell suspension cannot be prepared from every plant tissue and in every plant species. The same holds true for higher plant cells cultured *in vitro*, which usually grow in clumps that must be treated like solid plant tissues to release single cells.

There are many reasons why even success in isolating single cells is not sufficient to facilitate FCM analysis. Plant cells often reach a size approaching or even exceeding the diameter of orifices in the flow chambers of many flow cytometers which usually range from 50 to 100 µm. The second difficulty arises from the fact that plant cells are not circular. Irregularly-shaped objects disturb laminar flow, the particles then do not follow the same trajectory and interact with the excitation light beam differently, causing variation in output signals.

But this is not the end of the difficulties with plant cells. Other problems are due to low permeability of the cell wall for various reagents. Moreover, many cellular structures may bind fluorescent probes non-specifically, and fluorescence emanating from autofluorescent pigments may be difficult to distinguish from measured signals. Secondary metabolites often present in plant cells may interfere with staining and/or the fluorescence of the fluorochromes. Although some of these problems such as the cell wall permeability and autofluorescence may be overcome by applying cytological fixatives, many remain. Consequently, with certain exceptions, intact cells from land plants generally cannot be analyzed reliably using FCM. The situation with plankton FCM, which analyzes mostly single-celled organisms, is different insofar as the primary aim is to recognize particles and associate these with organisms through optical parameters such as autofluorescence and light scatter properties, and not to quantitatively evaluate the cellular characteristics (see Chapter 13).

The first published FCM analysis of plant cells (Heller 1973) focused on the nuclear DNA content of *Vicia faba* (field bean; Fabaceae). In order to obtain measurable samples, acetic ethanol-fixed root tips were digested with pectinase, mechanically disintegrated, sieved to eliminate tissue debris and cellular clumps, pepsin treated, and stained with ethidium bromide. The histograms exhibited two broad peaks representing unreplicated and replicated nuclei, and a significant amount of debris. A disadvantage of this method is the need to optimize the enzymatic treatment and the difficulty of isolating single cells from differentiated tissues. In fact, this approach has not been used frequently, and its most successful application is probably the analysis of DNA content in cultured plant cells (Pfosser 1989).

3.3.2
Protoplasts are somewhat "Easier" than Intact Cells

As many problems with plant cells are due to their rigid cell wall, an obvious solution is to remove it. Cells devoid of their walls are called protoplasts. They are characterized by circular shape and therefore do not disturb the laminar flow. Protoplasts can be prepared from plant tissues and cells by hydrolyzing the components of cell wall under isotonic conditions. It should be noted, however, that the preparation of protoplasts is even more difficult and laborious than the preparation of cells. Moreover, only a few types of tissues (typically leaf mesophyll) are suitable, and the isolated protoplasts are fragile and require a careful handling.

Despite being more suitable for FCM than intact cells, protoplasts share many of their disadvantages. These include the large size, low permeability of the plasma membrane for some compounds, autofluorescence, presence of secondary metabolites, and non-specific binding of fluorescent probes. If viable protoplasts are not needed, some of these difficulties may be overcome by appropriate fixation. As it happens, some disadvantages may be turned into advantages. For example, autofluorescence of chlorophyll may be useful for identifying protoplasts from autotrophic tissues containing chloroplasts (Buiteveld et al. 1998) or to measure the content of a secondary metabolite (Brown et al. 1984). To conclude, the use of protoplasts provides a solution if FCM analysis and/or sorting of whole plant cells is required. Different options should be pursued in other cases.

3.3.3
Going for Organelles

The suitability of protoplasts for analysis of cellular organelles such as the nucleus, is compromised by their acentric position in the cell. Because of this, even the regular movement of protoplasts in the flow will not guarantee the same position of all the organelles to be measured with respect to the excitation light beam and optical detector system. Moreover, light may scatter on internal components of protoplasts and disturb the analysis. It is then logical to isolate the organelles and analyze them separately out of the interfering cellular environment.

Lysis of intact protoplasts in a hypotonic buffer is a gentle procedure for isolating cell organelles. The integrity of isolated organelles may be secured by the appropriate chemical composition of the isolation buffer. This approach has been used for example, to isolate cell nuclei (Bergounioux et al. 1988a). The advantage of this procedure is that the samples are "clean" (i.e. free of large cellular debris), and the analysis of organelles can be performed with a high resolution. The main disadvantage is that protocols for the isolation of intact protoplasts are only available for a limited number of plant tissues and species, and the situation is unlikely to change to any great extent in future. In addition, the methodology is laborious and time consuming.

It was the ingenious idea of David Galbraith to isolate cell nuclei by mechanical homogenization of plant tissues that revolutionized plant FCM and, after a lag phase, catalyzed its rapid spread. Suddenly, a method was available that could be used with many (though not all) types of plant samples. Moreover, the method was rapid and convenient. It was enough to chop a small amount of plant tissue using a razor blade in a suitable buffer solution and filter the crude homogenate through a nylon mesh to obtain a measurable suspension of cell nuclei (Galbraith et al. 1983; see Chapter 4 for details on sample preparation).

In addition to stimulating the spread of plant FCM, the main effect of Galbraith's protocol was that it focused attention on the analysis of cell nuclei. All other targets were more laborious and difficult to prepare and analyze, and hence did not become so popular. Fortunately, the analysis of nuclei, which harbor most of the cell's hereditary material, provides a wealth of information on the genome. Recently, the Galbraith's protocol has also been shown to be suitable for dry herbarium vouchers (Suda and Trávníček 2006), significantly expanding the use of FCM in plant taxonomy, population biology, and ecology.

Attractive applications of DNA flow cytometry stimulated further research on the composition of nuclei isolation buffers, fluorescent staining of DNA and, of course, standardization (see Chapter 4). As the original chopping method has not always been considered appropriate, alternative methods were pursued. For example, Sgorbati et al. (1986) developed a protocol for isolation of nuclei from formaldehyde-fixed tissues. The fixation process helped in preserving nuclear integrity and made them suitable for immunofluorescent staining of incorporated 5-bromo-2′-deoxyuridine (Levi et al. 1987). A need to analyze nuclei isolated from small seeds not amenable to chopping led to the development of a protocol in which the nuclei are isolated by crushing the seeds between two sheets of sandpaper (see Chapter 6).

3.4
Application of Flow Cytometry in Plants

FCM has been used extensively as an analytical and preparative tool in a number of applications, and different experimental approaches have been adopted. It is not the purpose of this chapter to provide an extensive review but rather we aim to provide a general survey. Chapters 4–17 of this book describe important applications and methods in detail and Chapter 18 provides quantitative data on plant DNA flow cytometry.

3.4.1
Microspores and Pollen

Microspore cultures *in vitro* have been used both to produce haploid plants and to study embryogenesis. Using FCM, some authors revealed a correlation between light-scatter and fluorescence properties of cultured microspores and their embryogenic potential. In induced microspore cultures of *Brassica napus* (Brassicaceae), Schulze and Pauls (1998) identified, using fluorescein-diacetate staining, a subpopulation of enlarged viable cells which were absent in non-induced cultures. Embryos developed from induced cells only after their sorting into a culture medium. Subsequently, Schulze and Pauls (2002) used Calcofluor white to stain cellulose in cell walls of cultured microspores of *B. napus*. The results showed that FCM analysis of cellulose could be used to track the development of embryogenic cells. Tyrer (1981) observed slight differences in light-scatter properties of maize pollen expressing a wild-type phenotype (starchy) and a mutant (waxy) phenotype. Similar differences were also observed in autofluorescence after excitation at 488 nm.

3.4.2
Protoplasts

Protoplasts have been used in a number of experiments as a substitute for intact plant cells that cannot be easily run through a flow cytometer, and also to generate somatic hybrids (see Chapter 10). The latter application was facilitated by the fact that viable protoplasts can be sorted under sterile conditions and cultured further (Galbraith et al. 1984). As the diameter of a plant protoplast may reach 100 μm, the use of nozzles with large orifices (100–200 μm) was required. According to Harkins and Galbraith (1987), the use of such larger orifices dictates lower pressure of sheath fluid and the droplet generation frequency, resulting in inferior protoplast analysis and sort rates.

3.4.2.1 Physiological Processes

The studies on isolated protoplasts focused on targets that included the estimation of chlorophyll content (Galbraith et al. 1988), cell wall synthesis in cultured protoplasts using Calcofluor white which stains cellulose, and cell cycle kinetics after measuring DNA content in fixed protoplasts (Galbraith and Shields 1982; Meadows 1983). Gantet et al. (1990) used a lipophilic probe to study membrane fluidity and order in leaf epidermal and mesophyll cells of *Lupinus albus* (Fabaceae) to evaluate the effect of water stress in susceptible and resistant genotypes. Isolated protoplasts also provide a suitable system to study programmed cell death or apoptosis. Among others, O'Brien et al. (1997, 1998) studied Annexin V binding, nucleosomal fragmentation and chromatin degradation in protoplasts of *Nicotiana plumbaginifolia* (Solanaceae) induced to apoptosis. More recent studies on the analysis of apoptosis in plant protoplasts were published by Lei et al. (2003), Weir et al. (2003), and Cronje et al. (2004).

3.4.2.2 Secondary Metabolites

Plants can produce and accumulate (usually in vacuoles) a wide range of secondary metabolites with protective or signaling functions. They may be a focal point of biotechnology processes and their presence, distribution, and concentration may be analyzed with FCM (Yanpaisan et al. 1999). The most comprehensive data apparently concern *Catharanthus roseus* (Apocynaceae) where FCM has been used to quantify both alkaloid (Brown et al. 1983) and anthocyanin (Hall and Yeoman 1987) content in isolated protoplasts. Other examples include determination of anthocyanin and berberine levels in *Aralia cordata* (Araliaceae; Sakamoto et al. 1994) and *Coptis japonica* (Ranunculaceae; Hara et al. 1989), respectively. Detection of some secondary products (e.g. berberine) is possible because of their autofluorescence whereas others require staining with an appropriate fluorochrome (e.g. FITC). In the latter case, fluorescence quenching is often used as a measure of metabolite concentration (Sakamoto et al. 1994). As significant variation among individual cell lines has regularly been detected, with only some subpopulations capable of producing high levels of secondary metabolites, FCM and sorting may be used for screening, identification, and selection of highly productive cells.

3.4.2.3 Gene Expression

Regulation of gene expression determines the growth and development of any plant organisms. In order to understand these processes and the intricate network of gene expression and interaction, tools for characterizing gene expression in particular cell types and tissues are needed. An elegant approach is to use transgenic plants in which the expression of a fluorescent protein (e.g. green fluorescent protein, GFP) is under control of a tissue-specific promoter (see Chapter 17). Preparation of protoplasts from a desired part of a plant body and subsequent flow-sorting of protoplasts expressing the fluorescent protein provides a purified population of cells for RNA analysis. Birnbaum et al. (2003) used this strategy to analyze the expression of more than 22 000 genes in different tissues of *Arabidopsis thaliana* (Brassicaceae) root at three different zones from the tip (root tip, elongation zone and maturation zone). This work was extended by Nawy et al. (2005) who analyzed gene expression in the quiescent center of *Arabidopsis* root (see also Birnbaum et al. 2005).

3.4.2.4 Somatic Hybrids

One of the problems in constructing somatic hybrids after protoplast fusion has been the identification of heterokaryons. An attractive solution is to label the two parental protoplast populations with two different viable fluorescent dyes (e.g. fluorescein isothiocyanate and rhodamine isothiocyanate) prior to fusion and to sort the heterokaryons containing both dyes (Afonso et al. 1985). The method has been used to obtain interspecific somatic hybrids within the family Brassicaceae (Fahleson et al. 1988), and in the Solanaceae genera *Nicotiana* (Afonso et al. 1985) and *Solanum* (Waara et al. 1998).

3.4.2.5 DNA Transfection

Provided the molecules and particles of interest can be fluorescently labeled, flow cytometry can be used to follow their binding to and uptake into protoplasts. This strategy was used for example by Millman and Lurquin (1985) to study the interaction of spheroplasts of *Escherichia coli* and *Agrobacterium tumefaciens* with plant protoplasts. Tagu et al. (1987) used FCM to optimize the conditions for uptake of plasmids stained with ethidium bromide into protoplasts of *Petunia hybrida* (Solanaceae). In a similar study, Blackhall et al. (1995) assessed electroporation-induced uptake of FITC-labeled dextran into protoplasts of rice. Maddumage et al. (2004) described a protocol for delivery of plasmid DNA into protoplasts of apple. They compared two methods to detect transformed protoplasts and discussed the effectiveness of flow cytometry techniques in this throughput method.

3.4.3
Cell Nuclei

Nuclei and their DNA are the most frequent targets for FCM studies (Chapter 4). Nuclei suspensions are prepared by lysing protoplasts or, more frequently, by mechanical homogenization of a tissue (e.g. by chopping using a razor blade) in an appropriate buffer (Galbraith et al. 1983). Fresh plant material is usually employed. There is limited experience regarding the stoichiometry of binding fluorochromes to DNA and DNA peak quality when using fixed tissues, but under controlled conditions gentle formaldehyde fixation has proven practical (Doležel et al. 1992). For work with some types of dry material (e.g. seeds), a certain method of grinding can be useful (see Chapter 6). Prior to analysis, large tissue fragments and cellular clumps should be filtered out, e.g. using a nylon mesh. Biparametric (fluorescence plus light scatter) cytograms enable *in silico* selection (gating) of fairly pure nuclear fractions, so that cumbersome physical purification procedures prior to analysis can be circumvented.

3.4.3.1 Ploidy Levels

Chromosome counting in dividing cells is an unambiguous way to determine the ploidy, or the number of basic chromosome sets in cell nuclei. However, this is time consuming and tissues containing dividing cells may not be readily available. Hence, alternatives which do not require mitotically-active tissues and chromosome preparations were always sought. But, none of the early options, such as the estimation of leaf stomata density and size and pollen grain size, were found to be sufficiently reliable (cf. Schwanitz 1953; Stebbins 1971). As the nuclear DNA content correlates with ploidy, a high-throughput solution was provided by the FCM estimation of DNA content. Nevertheless, because ploidy level is inferred only indirectly using FCM, karyological and cytometrical results should be distinguished from each other, with the latter designated by the prefix DNA, i.e. DNA ploidy (Suda et al. 2006).

The first ploidy applications focused on crop plants (de Laat et al. 1987) and plant breeding still dominates this field. Three main areas of use include: (i) characterization of ploidy in available plant material (e.g. in gene banks); (ii) control of ploidy stability at various steps in breeding programmes, including *in vitro* cultures; (iii) screening for desired cytotypes after ploidy manipulation and hybridization, including the detection of mixoploids (cytochimeras).

FCM ploidy screening plays an important role especially in the breeding of polyploid crops such as banana (*Musa* spp., Musaceae). Cultivated forms of banana are seed-sterile diploids, triploids and tetraploids of different genomic composition, combining gene pools of *M. acuminata* and *M. balbisiana*. Using FCM, rapid and reliable ploidy estimation is possible, which facilitates the correct classification of accessions held in germplasm collections (Bartoš et al. 2005; Pillay et al. 2006). Similar germplasm characterization has been carried out in *Medicago* (Fabaceae; Brummer et al. 1999), *Dioscorea* (Dioscoreaceae; Egesi et al. 2002), and various grasses (Johnson et al. 1998; Tuna et al. 2001).

FCM is particularly well suited to the determination of ploidy variation in samples with low mitotic indices such as cultures of plant cells *in vitro*, where chromosome counting cannot provide a representative picture of a heterogeneous population of cells. Actually, assessment of ploidy stability *in vitro* was among the first FCM applications in plant breeding (e.g. Ramulu and Dijkhuis 1986). Recently, *in vitro* cultures have been routinely utilized in clonal propagation via somatic embryogenesis, and to produce haploids from anthers, isolated microspores and egg cells, which usually represent an intermediate step in the development of homozygous dihaploid plants for further breeding (Eeckhaut et al. 2006). Verification of ploidy is required both in regenerants and after polyploidization of haploids. Another important industrial application is ploidy manipulation and subsequent screening for desired cytotypes. As this task requires examination of large populations, it only became practical with the advent of FCM (Koutoulis et al. 2005; Thao et al. 2004; Van Duren et al. 1996). In some plants, stable ploidy chimeras exhibit favorable features and such individuals may also be rapidly detected (Geoffriau et al. 1997; Zonneveld and Van Iren 2000). It is important that FCM ploidy screening can be performed at early ontogenetic stages, thus saving both space and time (Vainola 2000).

Although agriculture and horticulture lead the way in ploidy studies, the last decade has also seen significant progress in wild plant research (see Chapters 5 and 6). The pioneering work dealing with ploidy heterogeneity in natural populations of *Andropogon gerardii* (Poaceae) had already appeared in 1987 (Keeler et al. 1987). In plant biosystematics, ecology, and population biology, FCM is primarily used to address questions of phenotypic manifestation, spatial distribution, and evolutionary significance of genome duplication (see Chapter 5). Because representative sampling is possible, novel insights into the extent of intra- and interpopulation ploidy variation, niche differentiation, and ecological preferences of individual cytotypes can be gained. Recently, increased attention has attracted surveys of interactions between particular cytotypes and other trophic levels (e.g. pollinators, parasites, symbionts), triggered by the discovery that polyploidy may have important influences on the structure and diversification of terrestrial communities (Thompson et al. 1997). Such conclusions clearly highlight the indispensability of cytotype determination (most effectively realized by FCM) in any experimental study that may involve heteroploid plant samples.

3.4.3.2 Aneuploidy

Aneuploidy is the most common type of chromosomal aberration in the plant kingdom. At the same time, aneuploids represent valuable cytogenetic material for the study of chromosomal evolution (breakage, reunion, and rearrangement), phenotypic manifestation of chromosome gain or loss, and their effects on fitness and evolutionary success of a particular chromosome race as well as for genome mapping. Terminologically, aneuploidy (such as trisomy or monosomy) should be distinguished from dysploidy (Tischler 1937), which is a change in chromosome number due to karyotypic rearrangements (see also Rieger et al. 1991).

It is hardly surprising that most of our knowledge about aneuploids comes from investigation of economically important crops. The sensitivity of FCM was proven sufficient to detect individual rye chromosomes and/or, chromosome arms in wheat–rye addition lines (Pfosser et al. 1995), and monosomic individuals of *Triticum aestivum* (Poaceae; Lee et al. 1997). Similarly, deviations in nuclear DNA content signaled the presence of aneuploidy in *Asparagus officinalis* (Asparagaceae; Ozaki et al. 2004), *Humulus* (Cannabaceae; Šesek et al. 2000), *Lolium* (Poaceae; Barker et al. 2001), and *Musa* (Musaceae; Roux et al. 2003).

In contrast to a rather trivial task of estimating DNA ploidy levels, FCM detection of aneuploidy (or more precisely DNA aneuploidy) is a demanding task. An essential prerequisite is a high-resolution analysis because such differences in DNA content can only be discriminated in simultaneous FCM runs, which are at least twice the coefficient of variation of G_0/G_1 peaks (cf. Doležel and Göhde 1995). It must be emphasized that conventional chromosome counting should follow any suspicion of chromosomal heterogeneity inferred from FCM data in order to elucidate its nature.

3.4.3.3 B Chromosomes

Another type of chromosomal variation, reported in about 1300 plant species, concerns the presence of B (= accessory, supernumerary) chromosomes (Trivers et al. 2004). They are dispensable, often heterochromatic and sometimes demonstrably selfish genetic elements, which do not pair with A complement members. They are found in some (but not all) individuals of a species and may even vary within the individual. When present in higher numbers, their effect on the phenotype is negative. In FCM research, their presence may result in apparent intraspecific or even intra-plant genome size variation that is usually difficult to interpret. The most thoroughly studied plant group with respect to B chromosomes is the *Boechera holboellii* complex (Brassicaceae). FCM analyses of leaf tissues revealed a bimodal distribution of nuclear fluorescence, corresponding to the absence/presence of accessory chromosomes (Sharbel et al. 2004). Interestingly, supernumerary chromosomes at the diploid level seemed to trigger apomictic reproduction (Sharbel et al. 2005).

3.4.3.4 Sex Chromosomes

In a small number of dioecious plants, sex chromosomes have been recognized as a causative agent of gender separation. Ainsworth (2000) compiled a list containing only six plant families, namely Arecaceae (*Phoenix dactylifera*), Cannabaceae (*Humulus lupulus*, *H. japonicus*, and *Cannabis sativa*), Caryophyllaceae (*Silene dioica* and *S. latifolia*), Cucurbitaceae (*Coccinia dioica*), Loranthaceae (*Viscum* spp.), and Polygonaceae (*Rumex* sect. *Acetosa*). The existence of a sex-specific chromosomal heteromorphism provides a unique opportunity to discriminate between male and female individuals even at very early stages of their ontogenesis. Costich et al. (1991) and Vagera et al. (1994) demonstrated in *Silene* that such a task can be accomplished by estimating nuclear DNA amounts. Plants of both sexes were analyzed separately and their C-values compared. A significant improvement appeared in the study of Doležel and Göhde (1995) who analyzed nuclei isolated from both sexes of *S. latifolia* simultaneously and observed two non-overlapping DNA peaks. Their FCM assay actually represents one of the most dramatic examples of high-precision DNA analysis, with coefficients of variation as low as 0.53%.

3.4.3.5 Cell Cycle and Endopolyploidy

The relationship between nuclear DNA content and the position of a cell within the cell cycle provides an opportunity to study cell cycle kinetics. As the analysis of DNA content is relatively straightforward, DNA flow cytometry has been used in numerous studies that included cell cycle analysis in various tissues and organs, including seeds (Śliwińska et al. 1999). Flow cytometry has been used to evaluate the extent of cell cycle synchronization and the position of a synchronized population within the cell cycle (Doležel et al. 1999; Peres et al. 1999). Synchronized populations facilitate the analysis of cell cycle regulation (Roudier et al. 2000) and the effect of various compounds on cell cycle progression (Binarová et al. 1998).

The popularity of DNA flow cytometry in the analysis of the cell cycle should not mask the fact that the method suffers from rather low resolution. In addition to the fact that the monoparametric method does not detect cells in mitotic (M-) phase and non-cycling (quiescent) cells, deconvolution of DNA-content histograms relies on the assumption that all cells have the same cell cycle duration. The method is particularly prone to errors when analyzing perturbed populations. Better precision can be obtained after multiparametric analyses. The most frequently-used methods are based on the incorporation of the thymidine analog 5-bromo-2'-deoxyuridine (BrdU) into newly synthesized DNA. The incorporated BrdU is detected either by using specific antibodies (Lucretti et al. 1999) or based on quenching the fluorescence of a DNA stain Hoechst 33258 (Glab et al. 1994). While the latter approach requires continuous incubation with BrdU, short BrdU pulses can be followed using immunofluorescence detection. Less frequently used multiparametric methods of cell cycle analysis involve simultaneous measurement of DNA and RNA (Bergounioux et al. 1988b), DNA and total nuclear protein (Sgorbati et al. 1989) or DNA and proliferating cell nuclear antigen (Citterio et al. 1992).

Endopolyploidization (i.e. occurrence of elevated ploidy levels of cells within an organism) occurs frequently in differentiating tissues of plants through rounds of DNA replication without mitosis, most often even without a trace of a mitotic prophase. To analyze systemic endopolyploidy, FCM is undoubtedly an ideal analytical tool (Galbraith et al. 1991) and has since been used in a number of studies (see Joubes and Chevalier 2000). Recently, Barow and Meister (2003) carried out an extensive study of the occurrence and degree of endopolyploidy in angiosperms, addressing organ-specific differences and association with taxonomic affiliation and ecological requirements (see Chapter 15).

3.4.3.6 Reproductive Pathways

Knowledge of reproductive processes involved in embryo and seed formation is essential in many fields of plant sciences, including plant breeding, the seed industry, and evolutionary and biosystematics studies. However, such investigations are hampered by a tiny size and difficult accessibility of the gametophytic generation, and, until recently, only time-consuming and/or inconvenient experimental approaches have been available.

The occurrence of double fertilization can be inferred from the ploidy of embryo and endosperm. This provides an opportunity to apply FCM to determine the reproductive pathway that gave rise to a particular seed, simply by measuring ploidy separately in both tissues. Unfortunately, this approach suffered from laborious sample preparation and the need for accurate experiment timing (Naumova et al. 1993). It was Fritzk Matzk with his co-workers who developed a rapid and efficient protocol for screening the reproductive mode using FCM in mature seeds and opened a new era in plant reproduction biology (Matzk et al. 2000). The authors demonstrated in ripe seeds of both dicots and monocots, that only the embryo and the aleuron layer of the endosperm contain intact nuclei at maturity while other maternal tissues (e.g. seed coat) are free of nuclei at this stage and will thus not distort the results. Considering the type of male and female gametes (reduced versus unreduced), embryo origin (zygotic versus parthenogenetic), and endosperm origin (pseudogamous versus autonomous route), 12 different modes of seed formation can theoretically be reconstructed (see Chapter 6).

The so-called flow cytometric seed screen has been adopted in many plant laboratories and used to characterize mode of reproduction in *Hypericum* (Hypericaceae; Matzk et al. 2001, 2003), *Coprosma* (Rubiaceae; Heenan et al. 2003), *Boechera* (Brassicaceae; Naumova et al. 2001), and *Paspalum* (Poaceae; Cáceres et al. 2001). Discrimination of apomictic mutants in sexual plants, quantification of the apomictic/sexual progeny ratio in species with facultative apomixis, and the assessment of the contribution of unreduced gametes, rank among the most common contemporary applications. The methodology has an immense potential in breeding programmes focused on the introduction of apomixis into crop plants, in the seed industry to test the origin of progeny as well as in plant systematics to understand the sources of phenotypic variation and gain insights into underlying microevolutionary processes.

3.4.3.7 Nuclear Genome Size

Vascular plants exhibit remarkable variation in genome size (Bennett and Leitch 2005; Chapter 7). Differing nearly 2000-fold, their 1C-values range from only about 0.065 pg in *Genlisea aurea* and *G. margaretae* (Lentibulariaceae; Greilhuber et al. 2006) to 127.4 pg in *Fritillaria assyriaca* (Liliaceae). A valuable source of information on plant genome size is the Plant DNA C-values database accessible on-line at http://www.kew.org/genomesize/homepage.html, which holds data for approximately 1.8% of land species. The interest in genome size has primarily been fueled by the fact that nuclear DNA content can itself affect various phenotypic characters, phenology, and ecological behavior, collectively explained by the nucleotype theory (Bennett 1972; see Chapter 7 for a review of nucleotypic correlations).

The presence and extent of intraspecific variation in genome size remained controversial for over 20 years. Interestingly, many of the early reports were dismissed by subsequent investigations using best practice methodology (Greilhuber 2005). Perhaps the most blatant and almost surely erroneous case of intraspecific DNA content variation refers to *Collinsia verna* (Scrophulariaceae) with up to 288% reported divergence among individual accessions (Greenlee et al. 1984). Over time, several sources of artifactual variation in genome size have been identified: (i) instrumental or methodological errors; (ii) interference of secondary metabolites in DNA staining with potential seasonal fluctuation (e.g. Walker et al. 2006); (iii) differences in measurements between different laboratories (Doležel et al. 1998); and (iv) taxonomic heterogeneity of the material under investigation (Murray 2005). As a result, the concept of stable genome size within species has gained broader support. Examples have nevertheless been accumulating in recent years of a certain genome size variation in FCM assays despite meticulous methodology (Obermayer and Greilhuber 2005; Šmarda and Bureš 2006). Species adapted to various climates or occurring in diverse habitats over vast geographic areas (possibly cosmopolitan), spatially isolated populations, crops under long-standing selection by humans, and allopolyploids with multiple origins represent other candidates where variation in genome size could potentially be detected. Nevertheless, current knowledge indicates that this phenomenon is certainly much less common and prominent than previously thought.

Genome size also attracts the ever-increasing attention of plant taxonomists (see Chapter 5). In this field, FCM assays open up undreamt-of possibilities for untangling complex homoploid alliances owing to the ability to discriminate among taxa with the same number of chromosomes but a different genome size. The knowledge that the amount of nuclear DNA may vary considerably even among closely related taxa whilst showing remarkable constancy within most species (Bennett et al. 2000; Greilhuber 1998, 2005; Murray 2005) constitutes a rationale for employing genome size as an important taxonomic marker. Such promise has recently materialized in taxa delimitation at specific (e.g. Mishiba et al. 2000), subspecific (Dimitrova et al. 1999), as well as sectional (Zonneveld 2001) levels.

Another fascinating, though still sporadic taxonomic application of genome size is the assessment of genomic constitution in allopolyploid taxa (i.e. polyploid plants combining genomes from at least two different parental species). Works on wheat (Lee et al. 1997) and bananas (Lysák et al. 1999) are illustrative examples of this. Finally, many valuable assays utilizing differences in nuclear DNA content deal with the identification of homoploid hybrids. Although quite challenging, increasing numbers of papers are being published that document the eligibility of FCM to achieve this objective (Jeschke et al. 2003; Mahelka et al. 2005; Morgan-Richards et al. 2004).

3.4.3.8 DNA Base Content

DNA base composition (i.e. the proportion of AT/GC base pairs) may provide a closer insight into genome organization as compared to the amount of nuclear DNA itself (see Chapter 8). FCM assay is based on comparing fluorescence intensities of nuclei stained with intercalating (e.g. propidium iodide) and base-specific (e.g. DAPI with AT preference or mithramycin with GC preference) fluorochromes (Dagher-Kharrat et al. 2001; Ellul et al. 2002; Schwencke et al. 1998; Siljak-Yakovlev et al. 2002). While the pioneering experiments assumed a linear correlation between fluorescence intensity of base-specific dyes and the proportion of particular bases (Doležel et al. 1992), a more sophisticated formula was later derived (Godelle et al. 1993). However, recent studies question even this model (Barow and Meister 2002; Meister 2005). Over 250 estimates of AT/GC ratio covering more than 200 different plant species are now available (see Chapter 8). While showing close similarities in lower taxonomic categories (e.g. species and subspecies), differences in base composition often exist among plant families. Doubtful values notwithstanding, the actual proportion of AT bases in vascular plants ranges from 52.8% in *Zea mays* (Poaceae) to 65.3% in *Allium cepa* (Alliaceae). Siljak-Yakovlev et al. (1996) reported on an interesting application of AT/GC ratio estimation to distinguish between male and female plants of date palm (*Phoenix dactylifera*). If reliable, the assay would enable identification of female plants at a very early stage of growth.

3.4.3.9 Chromatin Composition

Some plant species like maize, wheat and many bulbous and rhizomatous taxa such as Hyacinthaceae (*Scilla, Othocallis*), Liliaceae (*Tulipa, Fritillaria*) and Trilliaceae (*Trillium*) exhibit intraspecific variation in the amount of condensed chromatin (heterochromatin). As different DNA fluorochromes exhibit different sensitivities to chromatin compactness (Darzynkiewicz et al. 1984), a rational choice of fluorochrome pairs should allow estimation of the amount of condensed chromatin. Rayburn et al. (1992) were the first to use FCM in this type of assay. DAPI, which binds to a minor groove of DNA, is largely insensitive to chromatin compactness while propidium iodide (PI), which intercalates into double-stranded DNA, shows sensitivity to chromatin structure. By comparing DAPI/PI fluorescence intensities, the authors were able to estimate the proportion of heterochromatin in maize. The same approach was used by Rayburn et al. (1997) to assess the amount of heterochromatin in maize inbreds and their F1 hybrids.

3.4.3.10 Sorting of Nuclei

The isolation of nuclei by flow sorting provides exciting opportunities for the analysis of genome structure and gene expression. Plant cells are characterized by the presence of cytoplasmic organelles, mitochondria and chloroplasts, carrying their own DNA. During evolution, some of their DNA sequences were transferred to the nucleus. However, in some studies it is important to ascertain whether a particular sequence is localized in the nucleus or in the cytoplasm. FCM sorting supplies clean fractions of nuclei that can be used for PCR with specific primers to answer this question. Šafář et al. (2004a) used flow-sorted nuclei of banana (*Musa balbisiana*) to isolate high molecular weight DNA and construct a genomic BAC (bacterial artificial chromosome) library. The use of purified nuclei not only solved the problems of isolating DNA due to the excess of phenolic compounds but also avoided library contamination with cytoplasmic DNA. The same approach was used to construct a BAC library from *Lupinus angustifolius* (Kasprzak et al. 2006).

The analysis of gene expression using flow-sorted protoplasts (see Section 3.4.2.3) requires that protoplasts are prepared from a tissue of interest. This may not always be possible and Galbraith (2003) proposed analyzing RNA transcripts within isolated nuclei. The advantage is that cell nuclei can be isolated by chopping from almost any plant tissue. Labeling nuclei from cells expressing a particular gene is achieved by accumulating a reporter fluorescent protein in the nucleus by including nuclear localization signal (NLS) in a translational fusion. Labeled nuclei are then sorted and their RNA analyzed (see Chapter 17 for details).

3.4.4
Mitotic Chromosomes

The first report in plants dates back to 1984, when De Laat and Blaas (1984) described chromosome analysis and sorting in *Haplopappus gracilis* (Asteraceae). Since that time, chromosome analysis (flow karyotyping) has been reported in 18 plant species (see Chapter 16). Although the early experiments involved the preparation of samples (suspensions of intact chromosomes) from protoplasts, current methodology is based on mechanical homogenization of synchronized meristem root tips (Doležel et al. 1992). Flow karyotyping involves staining chromosomes in suspension with a DNA fluorochrome (typically PI or DAPI). The resulting distribution of fluorescence intensities is called flow karyotype. Due to the similarity in size, only a few chromosomes within a karyotype can usually be discriminated as single peaks on a flow karyotype. This limits the application of flow karyotyping for analysis of structural and numerical chromosome changes. Despite this, a number of reports confirms the utility of flow karyotyping to detect chromosome polymorphism (Kubaláková et al. 2002; Lee et al. 2002), translocations (Doležel and Lucretti 1995; Vrána et al. 2000), deletions (Gill et al. 1999; Kubaláková et al. 2002), alien additions (Kubaláková et al. 2003), as well as numerical changes (Kubaláková et al. 2003; Lee et al. 2000).

By dissecting nuclear genomes to small and defined fractions, chromosome sorting has a great potential for gene cloning and genome sequencing. Flow-sorted chromosomes have been used for PCR with specific primers to integrate physical and genetic maps (Kovářová et al. 2007; Neumann et al. 2002), for physical mapping of DNA sequences using FISH (Janda et al. 2006; Kubaláková et al. 2002) and PRINS (Kubaláková et al. 2001), analysis of chromosomal proteins (Binarová et al. 1998; Ten Hoopen et al. 2000), and construction of subgenomic DNA libraries. While short insert DNA libraries are useful sources of molecular markers for saturation of genetic maps at specific regions (Požárková et al. 2002; Román et al. 2004), large insert libraries (e.g. BAC libraries) facilitate development of ready-to-sequence physical clone-based maps and positional gene cloning, especially in large genomes rich in repetitive DNA and in polyploid genomes (e.g. in wheat). Such libraries have already been produced from wheat chromosomes (Janda et al. 2004; Šafář et al. 2004b) and wheat chromosome arms (Janda et al. 2006).

3.4.5
Chloroplasts

The ability to evaluate single particles and purify subpopulations makes flow cytometry attractive for analysis of chloroplast structure and function. Ashcroft et al. (1986) and Schroeder and Petit (1992) employed FCM to analyze forward and side scatter, and used emitted fluorescence to characterize intact chloroplasts and thylakoids. The examination of intact maize chloroplasts revealed two populations exhibiting different chlorophyll fluorescence intensities, probably representing bundle sheath and mesophyll chloroplasts (Kausch and Bruce 1994). Pfündel and Meister (1996) characterized chloroplast thylakoids obtained from mesophyll and bundle sheath cells of maize. Due to the differences in chlorophyll fluorescence spectra, it was possible to purify both types of thylakoids for further analysis. Cho et al. (2004) found that reduced expression of gyrase in chloroplasts of *Nicotiana benthamiana* resulted in a lower number of chloroplasts per cell, which exhibited higher DNA contents than control chloroplasts.

3.4.6
Mitochondria

In addition to analyzing mitochondria in living plant cells (protoplasts), only a few studies employed FCM to study isolated plant mitochondria. Petit et al. (1986) measured binding of concavalin A and wheat germ agglutinin to mitochondria of potato. While the fluorescein-labeled wheat agglutinin bound only weakly, about half of the mitochondria showed specific binding of fluorescein-labeled concavalin A. Subsequently, Petit (1992) demonstrated that a fluorescent dye Rhodamine 123 was suitable to monitor changes in membrane potential. Variation in light-scatter properties of isolated mitochondria was also analyzed, and it was concluded that modifications of the inner structure rather than changes in mitochondria size were responsible for the pattern observed.

3.4.7
Plant Pathogens

Although not explored extensively, FCM can be used as a convenient and rapid tool to detect plant pathogens (e.g. viruses, bacteria, and fungi), and to assess their viability (see Chapter 9). As in other applications, FCM excels in its ability to analyze large population samples in a short time and to discriminate subpopulations.

Identification of specific viruses, bacteria and fungal spores can be achieved after immunofluorescent staining of a pathogen in a crude plant extract (Chitarra and van den Bulk 2003). The number of pathogens that can be detected is, however, limited by the number of suitable antibody–fluorochrome conjugates. This problem has been solved by developing multiplex assays that involve microspheres conjugated to antibodies. Originally, the microspheres conjugated to different antibodies were discriminated based on their size (Iannelli et al. 1997). A recent approach used beads that are stained with two fluorochromes at different ratios. Up to 100 different bead types are now available, theoretically allowing the detection of the same number of different pathogens (Joos et al. 2000; Vignali 2000). In fact, the limiting factor for simultaneous detection of a large number of pathogens is the lack of suitable antibodies and not the number of different beads. A possible solution involves DNA-based microsphere assays (see Chapter 9).

3.4.8
Aquatic Flow Cytometry

Successful use of FCM is not limited to terrestrial ecosystems and vascular plants but the technique has also found a myriad of applications in aquatic sciences, investigating both prokaryotic cyanobacteria and eukaryotic algae (i.e. non-vascular plants). It is their occurrence in the form of isolated cells which makes algae suitable for FCM analyses and this triggered pilot analyses in the early days of plant FCM. Since the first attempts to characterize algal cells cultivated in laboratory conditions (e.g. Paau et al. 1978), FCM has become one of the key instruments in contemporary (phyto)plankton research and has been used to address a wide array of questions, including plankton composition, its dynamics, and physiological state. In addition, the last decade has seen a clear departure from laboratory-based measurements to *in-situ* studies, employing a variety of (semi)automated moored or submerged flow cytometers and wireless data transfer (see Chapter 13).

Ecological research on phytoplankton is aimed primarily at elucidating algal community composition, dynamics, and physiology. FCM applications involve the estimation of phytoplankton biomass, either total or distinguishing a few basic groups (Campbell and Vaulot 1993; Li et al. 1993), cell cycle studies in order to assess specific growth rates and to trace the wax and wane of key species (Veldhuis et al. 1997), and characterization of physiological conditions and metabolic activity of phytoplankton cells (Berges and Falkowski 1998). Another flourishing field is biodiversity assessment in aquatic habitats. Generally, much less is known about life diversity in water than in terrestrial ecosystems. FCM has proved successful in detection and characterization of planktonic microorganisms that are difficult or almost impossible to observe with conventional microscopy. Two notable achievements were the discovery of cyanobacterium *Prochlorococcus marinus*, considered to be the most abundant living organism (Chisholm et al. 1988), and *Ostreococcus tauri* (Chlorophyta, Prasinophyceae), which is amongst the smallest free eukaryotes (Courties et al. 1994).

3.5
A Flow Cytometer in Every Laboratory?

We have every reason to believe that FCM will be used more and more frequently in plant research and industry. Progress in electronics and the development of solid state light sources, among other advancements, allows production of compact instruments at affordable prices. The new generation of analyzers can simultaneously measure several optical parameters, occupy little space, and are easy to operate. This however does not mean that they can produce the results automatically and from every sample. The quality of samples will play as important role as before. The same holds true for the ability of an operator to set up the machine for a particular assay, check its performance, use proper controls, and correctly interpret the results of the analysis.

The ease with which any flow cytometer can generate "some" results, i.e. frequency distributions of selected optical parameters, may be deceiving and lead to the production of artifacts. It is a reality that the literature on plant FCM includes questionable results. Reliable data cannot be obtained without highly trained operators and/or their qualified supervision. Unfortunately, the operators in biomedical laboratories, who represent a majority of the experts in the field, work with completely different types of samples and can hardly advise their fellow workers about plant samples. Samples prepared from plants are typically less concentrated and the target population of particles often represents a minority within a population of various, often fluorescing debris particles (Chapter 4). Consequently, different optical set-ups and signal gating strategies need to be used. In some extreme cases, highly modified or even specialized instruments may be required, as in phytoplankton analysis.

Commercially produced cytometers range from low-cost compact one-parameter instruments suitable for DNA content analysis, some of them portable, to expensive multiparameter instruments equipped with multiple excitation light sources and with the ability to sort up to four populations simultaneously and at high speed. The purchase of a flow cytometer should be considered carefully in the light of its planned use and the availability of adequately trained operator(s). Our experience tells us that a flow cytometer should be used regularly in order to guarantee its proper functioning, namely the fluidics. This is especially critical for large instruments and flow sorters, and it might be a mistake to invest in a flow cytometer that will be used only occasionally. For botanical laboratories working mainly with DNA FCM, an instrument with a green laser as the excitation light source (suitable for propidium iodide), fluorescence and side scatter as parameters, and a video control of the sample stream seems to be the minimal but rational choice.

3.6
Conclusions and Future Trends

We are witnessing a growing interest in flow cytometry and anticipate a further burst of activity using this technique in plant sciences, including both laboratory- and field-based disciplines, as well as in industrial applications. The methodology promises quantitative and qualitative advances in our understanding of plant growth, development and function at subcellular, cellular and organismal levels.

In particular, FCM facilitates a wide range of parameters to be simultaneously recorded at ever increasing rates. In contrast to other techniques that generate average values for the whole population of measured particles, values unique to every single object are recorded, paving the way for the dissection of complex populations of particles. Multiple cellular properties can be simultaneously quantified and their relationships established more reliably than in separate experiments, thus providing better understanding of factors involved in cell functioning.

Generally, any cellular and organelle characteristic can be measured using FCM provided a specific light-scattering property is identified and/or a fluorescence signal correlated with the desired attribute is available. Due to high numbers of scored particles and their random selection, both statistical precision and unbiased distribution of measured parameters are guaranteed, and detection of rare events is possible. The ability to establish complex analytical and sorting gates contributes to the popularity of FCM. In contrast to other separating methods based on a single parameter decision, any combination of parameters can be set here as a sorting criterion. Finally, the technique is suitable to analyze an overwhelming spectrum of particles, ranging from multicellular organisms (e.g. colony-forming algae) to isolated cells, nuclei, chromosomes, bacteria and viruses.

The above-mentioned advantages outweigh some FCM limitations. FCM measurements are often referred to as zero resolution measurements, that is, no morphological information about the particles analyzed is provided. Although it is possible to obtain certain information related to particle volume, diameter, or external texture, it is only an oversimplification of the complexity of forms, shapes, and structures observed in biological material. Also, the need for the analysis of a large number of particles may pose problems in some cases. A combination of microscopy and flow cytometry is technically feasible through integrated video imaging (e.g. Kachel and Wietzorrek 2000; see also Chapter 13). Although this is not within reach for the majority of plant scientists, it is only a matter of time before this equipment becomes standard in botany laboratories.

Although it is not a safe business to predict the future, we expect that the highest number of applications of plant flow cytometry will continue to be the estimation of nuclear DNA content for ploidy and genome size (compare Chapter 18). The main areas of use will include plant taxonomy and population biology, biotechnology research, breeding and the seed industry. In contrast to DNA flow cytometry, other applications are currently less frequent and we consider them to be under-explored. The ability of FCM to measure many parameters simultaneously in large numbers of cells and subcellular organelles and to purify subpopulations of particles makes the technique an ideal tool in many research areas as indicated in this chapter and as elaborated in more detail in Chapters 4–18 of this book. It remains to be seen whether plant scientists will discover the full potential of FCM and use it for their own benefit.

References

Afonso, C. L., Harkins, K. R., Thomas-Compton, M., Krejci, A., Galbraith, D. W. **1985**, *Biotechnology*, 3, 811–816.

Ainsworth, C. **2000**, *Ann. Bot.* 86, 211–221.

Ashcroft, R. G., Preston, C., Cleland, R. E., Critchley, C. **1986**, *Photobiochem. Photobiophys.* 13, 1–14.

Barker, R. E., Kilgore, J. A., Cook, R. L., Garay, A. E., Warnke, S. E. **2001**, *Seed Sci. Technol.* 29, 493–502.

Barow, M., Meister, A. **2002**, *Cytometry* 47, 1–7.

Barow, M., Meister, A. **2003**, *Plant, Cell Environ.* 26, 571–584.

Bartoš, J., Alkhimova, O., Doleželová, M., De Langhe, E., Doležel, J. **2005**, *Cytogenet. Genome Res.* 109, 50–57.

Bennett, M. D. **1972**, *Proc. Roy. Soc. Lond. Series B – Biol. Sci.* 181, 109–135.

Bennett, M. D., Leitch, I. J. **2005**, *Ann. Bot.* 95, 45–90.

Bennett, M. D., Bhandol, P., Leitch, I. J. **2000**, *Ann. Bot.* 86, 859–909.

Berges, J. A., Falkowski, P. G. **1998**, *Limnol. Oceanogr.* 43, 129–135.

Bergounioux, C., Perennes, C., Brown, S. C., Sarda, C., Gadal, P. **1988a**, *Protoplasma* 142, 127–136.

Bergounioux, C., Perennes, C., Brown, S. C., Gadal, P. **1988b**, *Cytometry* 9, 84–87.

Binarová, P., Hause, B., Doležel, J., Dráber, P. **1998**, *Plant J.* 14, 751–757.

Birnbaum, K., Shasha, D. E., Wang, J. Y., Jung, J. W., Lambert, G. M., Galbraith, D. W., Benfey, P. N. **2003**, *Science* 302, 1956–1960.

Birnbaum, K., Jung, J. W., Wang, J. Y., Lambert, G. M., Hirst, J. A., Galbraith, D. W., Benfey, P. N. **2005**, *Nature Methods* 2, 1–5.

Blackhall, N. W., Finch, R. P., Power, J. B., Cocking, E. C., Davey, M. R. **1995**, *Protoplasma* 186, 50–56.

Brown, S. C., Renaudin, J. P., Prevot, C. **1983**, *Biol. Cell* 49, A3–A3.

Brown, S. C., Renaudin, J. P., Prevot, C., Guern, J. **1984**, *Physiol. Vég.* 22, 541–554.

Brummer, E. C., Cazcarro, P. M., Luth, D. **1999**, *Crop Sci.* 39, 1202–1207.

Buiteveld, J., Suo, Y., Van Lookeren Campagne, M. M., Creemers-Molenaar, J. **1998**, *Theor. Appl. Genet.* 96, 765–775.

Cáceres, M. E., Matzk, F., Busti, A., Pupilli, F., Arcioni, S. **2001**, *Sex. Plant Reprod.* 14, 201–206.

Campbell, L., Vaulot, D. **1993**, *Deep-Sea Res. Part I – Oceanogr. Res. Papers* 40, 2043–2060.

Chisholm, S. W., Olson, R. J., Zettler, E. R., Goericke, R., Waterbury, J. B., Welschmeyer, N. A. **1988**, *Nature* 334, 340–343.

Chitarra, L. G., van den Bulk, R. W. **2003**, *Eur. J. Plant Pathol.* 109, 407–417.

Cho, H. S., Lee, S. S., Kim, K. D., Hwang, I., Lim, J. S., Park, Y. I., Pai, H. S. **2004**, *Plant Cell* 16, 2665–2682.

Citterio, S., Sgorbati, S., Levi, M., Colombo, B. M., Sparvoli, E. **1992**, *J. Cell Sci.* 102, 71–78.

Costich, D. E., Meagher, T. R., Yurkow, E. J. **1991**, *Plant Mol. Biol. Rep.* 9, 359–370.

Courties, C., Vaquer, A., Troussellier, M., Lautier, J., Chretiennotdinet, M. J., Neveux, J., Machado, C., Claustre, H. **1994**, *Nature* 370, 255–255.

Cronje, M. J., Weir, I. E., Bornman, L. **2004**, *Cytometry* 61A, 76–87.

Dagher-Kharrat, M. B., Grenier, G., Bariteau, M., Brown, S., Siljak-Yakovlev, S., Savouré, A. **2001**, *Theor. Appl. Genet.* 103, 846–854.

Darzynkiewicz, Z., Traganos, F., Kapuscinski, J., Staiano-Coico, L., Melamed, M. R. **1984**, *Cytometry* 5, 355–363.

De Laat, A. M. M., Blaas, J. **1984**, *Theor. Appl. Genet.* 67, 463–467.

De Laat, A. M. M., Göhde, W., Vogelzang, M. J. D. C. **1987**, *Plant Breeding* 99, 303–307.

Dimitrova, D., Ebert, I., Greilhuber, J., Kozhuharov, S. **1999**, *Plant Systemat. Evol.* 217, 245–257.

Doležel, J., Göhde, W. **1995**, *Cytometry*, 19, 103–106.

Doležel, J., Lucretti, S. **1995**, *Theor. Appl. Genet.* 90, 797–802.

Doležel, J., Sgorbati, S., Lucretti, S. **1992**, *Physiol. Plant.* 85, 625–631.

Doležel, J., Greilhuber, J., Lucretti, S., Meister, A., Lysák, M. A., Nardi, L., Obermayer, R. **1998**, *Ann. Bot.* 82(Suppl. A), 17–26.

Doležel, J., Číhalíková, J., Weiserová, J., Lucretti, S. **1999**, *Methods Cell Sci.* 21, 95–107.

Eeckhaut, T., Leus, L., Van Huylenbroeck, J. **2006**, *Acta Physiol. Plant.* 27, 743–750.

Egesi, C. N., Pillay, M., Asiedu, R., Egunjobi, J. K. **2002**, *Euphytica* 128, 225–230.

Ellul, P., Boscaiu, M., Vicente, O., Moreno, V., Roselló, J. A. **2002**, *Ann. Bot.* 90, 345–351.

Fahleson, J., Rahlén, L., Glimelius, K. **1988**, *Theor. Appl. Genet.* 76, 507–512.

Galbraith, D. W. **2003**, *Comp. Funct. Genomics* 4, 208–215.

Galbraith, D. W., Shields, B. S. **1982**, *Physiol. Plant.* 55, 25–30.

Galbraith, D. W., Harkins, K. R., Maddox, J. R., Ayres, N. M., Sharma, D. P., Firoozabady, E. **1983**, *Science* 220, 1049–1051.

Galbraith, D. W., Afonso, C. L., Harkins, K. R. **1984**, *Plant Cell Rep.* 3, 151–155.

Galbraith, D. W., Harkins, K. R., Jefferson, R. A. **1988**, *Cytometry* 9, 75–83.
Galbraith, D. W., Harkins, K. R., Knapp, S. **1991**, *Plant Physiol.* 96, 985–989.
Gantet, P., Hubac, C., Brown, S. C. **1990**, *Plant Physiol.* 94, 729–737.
Geoffriau, E., Kahane, R., Bellamy, C., Rancillac, M. **1997**, *Plant Sci.* 122, 201–208.
Gill, K. S., Arumuganathan, K., Lee, J. H. **1999**, *Theor. Appl. Genet.* 98, 1248–1252.
Glab, N., Labidi, B., Qin, L. X., Trehin, C., Bergounioux, C., Meijer, L. **1994**, *FEBS Lett.* 353, 207–211.
Godelle, B., Cartier, D., Marie, D., Brown, S. C., Siljak-Yakovlev, S. **1993**, *Cytometry* 14, 618–626.
Greenlee, J. K., Rai, K. S., Floyd, A. D. **1984**, *Heredity* 52, 235–242.
Greilhuber, J. **1998**, *Ann. Bot.* 82 (Suppl. A), 27–35.
Greilhuber, J. **2005**, *Ann. Bot.* 95, 91–98.
Greilhuber, J., Borsch, T., Müller, K., Worberg, A., Porembski, S., Barthlott, W. **2006**, *Plant Biol.* 8, 770–777.
Hall, R. D., Yeoman, M. M. **1987**, *J. Exp. Bot.* 38, 1391–1398.
Hara, Y., Yamagata, H., Morimoto, T., Hiratsuka, J., Yoshioka, T., Fujita, Y., Yamada, Y. **1989**, *Planta Medica* 2, 151–154.
Harkins, K. R., Galbraith, D. W. **1987**, *Cytometry* 8, 60–71.
Heenan, P. B., Molloy, B. P. J., Bicknell, R. A., Luo, C. **2003**, *NZ J. Bot.* 41, 287–291.
Heller, F. O. **1973**, *Ber. Deutsch. Bot. Gesellsch.* 86, 437–441.
Iannelli, D., D'Apice, L., Cottone, C., Viscardi, M., Scala, F., Zoina, A., Del Sorbo, G., Spigno, P., Capparelli, R. **1997**, *J. Virol. Methods* 69, 137–145.
Janda, J., Bartoš, J., Šafář, J., Kubaláková, M., Valárik, M., Číhalíková, J., Šimková, H., Caboche, M., Sourdille, P., Bernard, M., Chalhoub, B., Doležel, J. **2004**, *Theor. Appl. Genet.* 109, 1337–1345.
Janda, J., Šafář, J., Kubaláková, M., Bartoš, J., Kovářová, P., Suchánková, P., Pateyron, S., Číhalíková, J., Sourdille, P., Šimková, H., Faivre-Rampant, P., Hřibová, E., Bernard, M., Lukaszewski, A., Doležel, J., Chalhoub, B. **2006**, *Plant J.* 47, 977–986.
Jeschke, M. R., Tranel, P. J., Rayburn, A. L. **2003**, *Weed Sci.* 51, 1–3.
Johnson, P. G., Riordan, T. P.,

Arumuganathan, K. **1998**, *Crop Sci.* 38, 478–482.
Joos, T. O., Schrenk, M., Hopfl, P., Kroger, K., Chowdhury, U., Stoll, D., Schorner, D., Durr, M., Herick, K., Rupp, S., Sohn, K., Hammerle, H. **2000**, *Electrophoresis* 21, 2641–2650.
Joubes, J., Chevalier, C. **2000**, *Plant Mol. Biol.* 43, 735–745.
Kachel, V., Wietzorrek, J. **2000**, *Sci. Mar.* 64, 247–254.
Kausch, A. P., Bruce, B. D. **1994**, *Plant J.* 6, 6767–779.
Kasprzak, A., Šafář, J., Janda, J., Doležel, J., Wolko, B., Naganowska, B. **2006**, *Cell. Mol. Biol. Lett.* 11, 396–407.
Keeler, K. H., Kwankin, B., Barnes, P. W., Galbraith, D. W. **1987**, *Genome*, 29, 374–379.
Koutoulis, A., Roy, A. T., Price, A., Sherriff, L., Leggett, G. **2005**, *Sci. Horticult.* 105, 263–268.
Kovářová, P., Navrátilová, A., Macas, J., Doležel, J. **2007**, *Biol. Plant.* 51, 43–48.
Kubaláková, M., Vrána, J., Číhalíková, J., Lysák, M. A., Doležel, J. **2001**, *Methods Cell Sci.* 23, 71–82.
Kubaláková, M., Vrána, J., Číhalíková, J., Šimková, H., Doležel, J. **2002**, *Theor. Appl. Genet.* 104, 1362–1372.
Kubaláková, M., Valárik, M., Bartoš, J., Vrána, J., Číhalíková, J., Molnár-Láng, M., Doležel, J. **2003**, *Genome* 46, 893–905.
Lee, J. H., Yen, Y., Arumuganathan, K., Baenziger, P. S. **1997**, *Theor. Appl. Genet.* 95, 1300–1304.
Lee, J. H., Arumuganathan, K., Chung, Y. S., Kim, K. Y., Chung, W. B., Bae, K. S., Kim, D. H., Chung, D. S., Kwon, O. C. **2000**, *Molecules Cells* 10, 619–625.
Lee, J. H., Arumuganathan, K., Kaeppler, S. M., Park, S. W., Kim, K. Y., Chung, Y. S., Kim, D. H., Fukui, K. **2002**, *Planta* 215, 666–671.
Lei, X. Y., Liao, X. D., Zhang, G. Y., Dai, Y. R. **2003**, *Acta Bot. Sin.* 45, 944–948.
Levi, M., Sparvoli, E., Sgorbati, S., Chiatante, D. **1987**, *Physiol. Plant.* 71, 68–72.
Li, W. K. W., Zohary, T., Yacobi, Y. Z., Wood, A. M. **1993**, *Marine Ecol. – Prog. Series* 102, 79–87.
Lucretti, S., Nardi, L., Trionfetti Nisini, P., Moretti, F., Gualberti, G., Doležel, J. **1999**, *Methods Cell Sci.* 21, 155–166.

Lysák, M. A., Doleželová, M., Horry, J. P., Swennen, R., Doležel, J. **1999**, *Theor. Appl. Genet.* 98, 1344–1350.

Maddumage, R., Fung, R. M. W., Weir, I., Ding, H., Simons, J. L., Allan, A. C. **2004**, *Plant Cell, Tissue Org. Cult.* 70, 77–82.

Mahelka, V., Suda, J., Jarolímová, V., Trávníček, P., Krahulec, F. **2005**, *Folia Geobot.* 40, 367–384.

Matzk, F., Meister, A., Schubert, I. **2000**, *Plant J.* 21, 97–108.

Matzk, F., Meister, A., Brutovská, R., Schubert, I. **2001**, *Plant J.* 26, 275–282.

Matzk, F., Hammer, K., Schubert, I. **2003**, *Sex. Plant Reprod.* 16, 51–58.

Meadows, M. G. **1983**, *Plant Sci. Lett.* 28, 337–348.

Meister, A. **2005**, *J. Theor. Biol.* 232, 93–97.

Millman, R. A., Lurquin, P. F. **1985**, *J. Plant Physiol.* 117, 431–440.

Mishiba, K. I., Ando, T., Mii, M., Watanabe, H., Kokubun, H., Hashimoto, G., Marchesi, E. **2000**, *Ann. Bot.* 85, 665–673.

Morgan-Richards, M., Trewick, S. A., Chapman, H. M., Krahulcová, A. **2004**, *Heredity* 93, 34–42.

Murray, B. G. **2005**, *Ann. Bot.* 95, 119–125.

Naumova, T. N., den Nijs, A. P. M., Willemse, M. T. M. **1993**, *Acta Bot. Neerland.* 42, 299–312.

Naumova, T. N., der Laak, J. van, Osadtchiy, J., Matzk, F., Kravtchenko, A., Bergervoet, J., Ramulu, K. S., Boutilier, K. **2001**, *Sex. Plant Reprod.* 14, 195–200.

Nawy, T., Lee, J. Y., Colinas, J., Wang, J. Y., Thongrod, S. C., Malamy, J. E., Birnbaum, K., Benfey, P. N. **2005**, *Plant Cell* 17, 1908–1925.

Neumann, P., Požárková, D., Vrána, J., Doležel, J., Macas, J. **2002**, *Chrom. Res.* 10, 63–71.

Obermayer, R., Greilhuber, J. **2005**, *Heredity* 95, 91–95.

O'Brien, I. E. W., Reutelingsperger, C. P. M., Holdaway, K. M. **1997**, *Cytometry* 29, 28–33.

O'Brien, I. E. W., Murray, B. G., Baguley, B. C., Morris, B. A. M., Ferguson, I. B. **1998**, *Exp. Cell Res.* 241, 46–54.

Ozaki, Y., Narikiyo, K., Fujita, C., Okubo, H. **2004**, *Plant Sex. Reprod.* 17, 157–164.

Paau, A. S., Oro, J., Cowles, J. R. **1978**, *J. Exp. Bot.* 29, 1011–1020.

Peres, A., Ayaydin, F., Nikovics, K., Gutierrez, C., Horvath, G. V., Dudits, D. N., Feher, A. **1999**, *J. Exp. Bot.* 50, 1373–1379.

Petit, P. X. **1992**, *Plant Physiol.* 98, 279–286.

Petit, P. X., Diolez, P., Muller, P., Brown, S. C. **1986**, *FEBS Lett.* 196, 65–70.

Pfosser, M. **1989**, *J. Plant Physiol.* 134, 741–745.

Pfosser, M., Amon, A., Lelley, T., Heberle-Bors, E. **1995**, *Cytometry* 21, 387–393.

Pfündel, E. E., Meister, A. **1996**, *Cytometry* 23, 97–105.

Pillay, M., Ogundiwin, E., Tenkouano, A., Doležel, J. **2006**, *Afr. J. Biotechnol.* 5, 1224–1232.

Požárková, D., Koblížková, A., Román, B., Torres, A. M., Lucretti, S., Lysák, M. A., Doležel, J., Macas, J. **2002**, *Biol. Plant.* 45, 337–345.

Ramulu, K. S., Dijkhuis, P. **1986**, *Plant Cell Rep.* 5, 234–237.

Rayburn, A. L., Auger, J. A., McMurphy, L. M. **1992**, *Exp. Cell Res.* 198, 175–178.

Rayburn, A. L., Bashir, A., Biradar, D. P. **1997**, *Maydica* 42, 393–399.

Rieger, R., Michaelis, A., Green, M. M. **1991**, *Glossary of Genetics – Classical Molecular*, 5th edn, Springer-Verlag, Berlin, Heidelberg, New York.

Román, B., Satovic, Z., Požárková, D., Macas, J., Doležel, J., Cubero, J. I., Torres, A. M. **2004**, *Theor. Appl. Genet.* 108, 1079–1088.

Roudier, F., Fedorova, E., Gyorgyey, J., Feher, A., Brown, S., Kondorosi, A., Kondorosi, E. **2000**, *Plant J.* 23, 73–83.

Roux, N., Toloza, A., Radecki, Z., Zapata-Arias, F. J., Doležel, J. **2003**, *Plant Cell Rep.* 21, 483–490.

Šafář, J., Noa-Carrazana, J. C., Vrána, J., Bartoš, J., Alkhimova, O., Lheureux, F., Šimková, H., Caruana, M. L., Doležel, J., Piffanelli, P. **2004a**, *Genome* 47, 1182–1191.

Šafář, J., Bartoš, J., Janda, J., Bellec, A., Kubaláková, M., Valárik, M., Pateyron, S., Weiserová, J., Tušková, R., Číhalíková, J., Vrána, J., Šimková, H., Faivre-Rampant, P., Sourdille, P., Caboche, M., Bernard, M., Doležel, J., Chalhoub, B. **2004b**, *Plant J.* 39, 960–968.

Sakamoto, K., Iida, K., Koyano, T., Asada, Y., Furuya, T. **1994**, *Planta Medica* 60, 253–259.

Schroeder, W. P., Petit, P. X. **1992**, *Plant Physiol.* 100, 1092–1102.

Schulze, D., Pauls, P. K. **1998**, *Plant Cell Physiol.* 39, 226–234.

Schulze, D., Pauls, P. K. **2002**, *New Phytologist* 154, 249–254.

Schwanitz, F. **1953**, *Der Züchter* 23, 17–44.

Schwencke, J., Bureau, J. M., Crosnier, M. T., Brown, S. C. **1998**, *Plant Cell Rep.* 18, 346–349.

Šesek, P., Sustar-Vozlic, J., Bohanec, B. **2000**, *Pflügers Archiv – Eur. J. Physiol.* 439 (Suppl), R16–R18.

Sgorbati, S., Levi, M., Sparvoli, E., Trezzi, F., Lucchini, G. **1986**, *Physiol. Plant.* 68, 471–476.

Sgorbati, S., Sparvoli, E., Levi, M., Chiatante, D. **1989**, *Physiol. Plant.* 75, 479–484.

Sharbel, T. F., Voigt, M. L., Mitchell-Olds, T., Kantama, L., de Jong, H. **2004**, *Cytogenet. Genome Res.* 106, 173–183.

Sharbel, T. F., Mitchell-Olds, T. M., Dobeš, C., Kantama, L., de Jong, H. **2005**, *Cytogenet. Genome Res.* 109, 283–292.

Siljak-Yakovlev, S., Benmalek, S., Cerbah, M., De la Pena, T. C., Bounaga, N., Brown, S. C., Sarr, A. **1996**, *Sex. Plant Reprod.* 9, 127–132.

Siljak-Yakovlev, S., Cerbah, M., Coulaud, J., Stoian, V., Brown, S. C., Zoldoš, V., Jelenič, S., Papeš, D. **2002**, *Theor. Appl. Genet.* 104, 505–512.

Śliwińska, E., Jing, H. C., Job, C., Job, D., Bergervoet, J. H. W., Bino, R. J., Goot, S. P. C. **1999**, *Seed Sci. Res.* 9, 91–99.

Šmarda, P., Bureš, P. **2006**, *Ann. Bot.* 98, 665–678.

Stebbins, G. L. **1971**, *Chromosomal Evolution in Higher Plants*, Edward Arnold Ltd., London.

Suda, J., Trávníček, P. **2006**, *Cytometry* 69A, 273–280.

Suda, J., Krahulcová, A., Trávníček, P., Krahulec, F. **2006**, *Taxon* 55, 447–450.

Tagu, D., Bergounioux, C., Perennes, C., Brown, S., Muller, P., Gadal, P. **1987**, *Plant Sci.* 51, 215–223.

Ten Hoopen, R., Manteuffel, R., Doležel, J., Malysheva, L., Schubert, I. **2000**, *Chromosoma* 109, 482–489.

Thao, N. T. P., Ozaki, Y., Okubo, H. **2004**, *J. Jpn. Soc. Horticult. Sci.* 73, 63–65.

Thompson, J. N., Cunningham, B. M., Segraves, K. A., Althoff, D. M., Wagner, D. **1997**, *Am. Natural.* 150, 730–743.

Tischler, G. **1937**, *J. Ind. Bot. Soc.* 16, 165–169.

Trivers, R., Burt, A., Palestis, B. G. **2004**, *Genome* 47, 1–8.

Tuna, M., Vogel, K. P., Arumuganathan, K., Gill, K. S. **2001**, *Crop Sci.* 41, 1629–1634.

Tyrer, H. W. **1981**, *Environ. Health Perspect.* 37, 137–142.

Vagera, J., Paulíková, D., Doležel, J. **1994**, *Ann. Bot.* 73, 455–459.

Vainola, A. **2000**, *Euphytica* 112, 239–244.

Van Duren, M., Morpurgo, R., Doležel, J., Afza, R. **1996**, *Euphytica* 88, 25–34.

Veldhuis, M. J. W., Kraay, G. W., Van Bleijswijk, J. D. L., Baars, M. A. **1997**, *Deep-Sea Res. Part I – Oceanogr. Res. Papers* 44, 425–449.

Vignali, D. A. A. **2000**, *J. Immunol. Methods* 243, 243–255.

Vrána, J., Kubaláková, M., Šimková, H., Číhalíková, J., Lysák, M. A., Doležel, J. **2000**, *Genetics* 156, 2033–2041.

Waara, S., Nyman, M., Johannisson, A. **1998**, *Euphytica* 101, 293–299.

Walker, D. J., Monino, I., Correal, E. **2006**, *Environ. Exp. Bot.* 55, 258–265.

Weir, I. E., Pham, N.-A., Hedley, D. W. **2003**, *Cytometry* 54A, 109–117.

Yanpaisan, W., King, N. J. C., Doran, P. M. **1999**, *Biotechnol. Adv.* 17, 3–27.

Zonneveld, B. J. M. **2001**, *Plant Systemat. Evol.* 229, 125–130.

Zonneveld, B. J. M., Van Iren, F. **2000**, 3, 176–185.

4
Nuclear DNA Content Measurement

Johann Greilhuber, Eva M. Temsch, and João C. M. Loureiro

Overview

This chapter reviews essential aspects of the flow cytometric studies of plant DNA contents, starting with a discussion of the recently updated revised terminology for presenting nuclear DNA amounts. Plants have a relatively complicated life cycle with alternation of generations and nuclear phases, they exhibit somatic polyploidization during ontogenetic differentiation and generative polyploidization during evolution. The terms "holoploid genome size" and "monoploid genome size", and their acronyms C-value and Cx-value, respectively, are promoted as elements of a precise terminology for unambiguous data presentation. DNA amounts can be presented relative to a reference species (standard) or in absolute units of picograms or base pairs, for which the correct conversion factor is specified. The methodological aspects of preparing samples for DNA content measurements are discussed with special consideration of standardization and the interfering role of secondary metabolites. Internal standardization with a plant standard is regarded as the most important approach to minimizing the effect of fluorescence inhibitors and balancing out all technical variations which occur during an experiment. It is accepted that a consensus on a set of standard species covering the whole range of C-values has still not been achieved. Some rules are outlined for assuring data quality and sufficiently detailed data presentation. As far as the methodological side of measuring DNA amounts is concerned, it is expected that important future research developments will occur in the field of preparative improvements to overcome stoichiometric errors, the utilization of dormant diaspores and conserved tissues for flow cytometry, and that a reliable plant standard species will be established in addition to guidelines for internal calibration.

4.1
Introduction

Estimation of DNA content in cell nuclei is one of the important applications of flow cytometry (FCM) in plant sciences. Although first results on plant material

Flow Cytometry with Plant Cells. Edited by Jaroslav Doležel, Johann Greilhuber, and Jan Suda
Copyright © 2007 WILEY-VCH Verlag GmbH & Co. KGaA, Weinheim
ISBN: 978-3-527-31487-4

(root tips of *Vicia faba*, Fabaceae) with the then novel methodology were reported in the early 1970s (Heller 1973), it was not before the introduction of the ingenious chopping method for isolation of plant nuclei by Galbraith et al. (1983), that FCM became widely accepted as a convenient approach for measuring DNA contents and genome size. Galbraith et al. (1983) circumvented cumbersome protoplasting or enzymatic isolation of nuclei by simply chopping up with a razor blade fresh leaves of tobacco (*Nicotiana tabacum*, Solanaceae) and a number of other plant species in an appropriate buffer plus detergent and then sieving out large particles, whereby enough nuclei were released to yield clear histograms upon FCM. This paper was also notable in applying internal standardization with chicken red blood cells (CRBCs) for genome size determination (although the GC-specific fluorochrome mithramycin was used for staining DNA, which overestimates DNA amount in GC-rich genomes (Doležel et al. 1992)). The DNA content of the standard was determined chemically. In those early days, the cost of instruments, which were not easy to operate, was the main reason why they were not used outside the field of biomedical sciences (cf. Chapter 1). Today, there are affordable instruments are on the market, so that even small botany laboratories are increasingly using FCM.

The advantages of flow cytometry over static cytometry are clear: speed of preparation and data gathering, and higher precision due to high numbers of nuclei measured and possibly also due to a more homogeneous staining of isolated nuclei in suspension. A mysterious disadvantage of static cytometry (i.e. mainly Feulgen densitometry), which is explained neither by notoriously small sample sizes nor by technical difficulties, is the plain fact, that many published results are unreliable for unknown and untraceable reasons (Greilhuber 2005). This is apparently not the case to a comparable extent with FCM data. An advantage of static cytometry is the absence of debris, because only nuclei are measured. Problems common to both technologies are bias caused by variation in chromatin compactness and the interference of secondary metabolites with the staining process. Presently it seems that the latter source of error is specific to plants, but, as phenolic compounds are involved and these also occur in animals (e.g. phenoloxidases play a role in melanin production), the problem may exist with zoological material as well, but remained unrecognized. There is also another particularity of FCM: the nuclei are measured without visual selection, what may be judged as being more objective than selecting nuclei in the microscope by eye. However, in critical cases light-microscopic evidence must be obtained for unequivocal interpretation of FCM results, for instance when the histogram peak of unreplicated nuclei is small and could be overlooked, or when genome size is very small and debris is abundant.

It is the purpose of the present chapter to discuss basic problems associated with FCM work on nuclear DNA content in plants. The biological significance of genome size and variation in DNA content is discussed in Chapters 5, 7, 9 and 15, and genetic aspects are covered by Chapters 6, 9, 14, 16 and 17. The first plant DNA flow cytometry database (FLOWER) is presented in Chapter 18. A particularly useful review on plant DNA flow cytometry is the publication by Doležel and Bartoš (2005).

4.2
Nuclear DNA Content: Words, Concepts and Symbols

Swift (1950) introduced the symbol "C", meaning the "constant" of DNA content, which is represented in multiples in nuclei of various tissues of an organism (see Bennett and Smith 1976; Greilhuber et al. 2005). Bennett and Smith (1976) defined the C-value (i.e. the 1C-value!) as the "DNA content of the unreplicated haploid chromosome complement". To avoid the ambiguity of terms such as "genome size" and "nuclear DNA content" or "basic nuclear DNA content" or "amount", Bennett et al. (1998) restricted "genome size" to the monoploid genome, while "C-value" continued to refer to the DNA content of the complete chromosome complement. But it was soon felt that this restricted use would entirely eliminate the established and phonetically pleasing term "genome size" from the discourse, because often the degree of polyploidy is unknown, genomic reconstructions in polyploids reshuffled ancestral genomes, and possibly all plants have experienced one or more polyploidizations in their ancestry (Wendel 2000).

Greilhuber et al. (2005) thus presented a slightly modified and complete terminology, which was guided (i) by accepting an explicit link between genomic DNA content designations and the chromosome numbers n (the haplophasic or meiotically reduced number) and x (the basic chromosome number of a polyploid series), and (ii) by striving at linguistic consistency in using full terms and their acronyms. At the same time the well-established symbol C had to remain unchanged. The term genome size thus retains its everyday meaning as a covering term usable in titles, introductory and concluding phrases. The adjectives "monoploid" and "holoploid" distinguish between genome size of the monoploid genome (= the single genome with x chromosomes, of which there are two per unreplicated nucleus in a diploid individual and several in a polyploid individual) and the complete, that is, holoploid genome. The respective abbreviations are C-value for the holoploid genome and Cx-value for the monoploid genome (the letter x refers to the basic chromosome number x). Quantitative data are given with numerical prefix, as 1C-, 2C-, 1Cx-, 2Cx-values and so on. A summary of the terminology is presented in Table 4.1.

Plants in particular are more complicated than most animals owing to their complex life cycle with alternation of generations and alternation (or not) of nuclear phases, and the frequently occurring generative and somatic polyploidy. Thus, the application of an unambiguous terminology is essential but not always adhered to in publications. This can lead to confusion.

There are basically four different kinds of DNA copy number status.

4.2.1
Replication–Division Phases

Replication–division phases of the mitotic nuclear cycle are related to its G_1, S and G_2 phases (cf. Chapter 14, Fig. 14.1). Replication and division lead to changes in DNA content expressed in terms of C. For instance, mitotically active nuclei

Table 4.1 Genome size terminology (from Greilhuber et al. 2005).

Genome status	Monoploid	Holoploid
Chromosome number designation	x	n
Covering term for genomic DNA content	Genome size	Genome size
Kinds of genome size	Monoploid genome size	Holoploid genome size
Short terms	Cx-value	C-value
Short terms quantified	1Cx, 2Cx, etc.	1C, 2C, etc.

in a haplophasic moss gametophyte cycle between 1C and 2C, in a diplophasic angiosperm root tip between 2C and 4C, and in a triploid endosperm between 3C and 6C.

4.2.2
Alternation of Nuclear Phases

Alternation of nuclear phases (not to be confused with alternation of generations!) is associated with meiotic reduction and fertilization (in angiosperms including endosperm fertilization). The nuclear phase status is denoted using the letter n. n indicates the meiotically reduced, haplophasic chromosome number, $2n$ the unreduced, diplophasic number, and $3n$, $5n$, and so forth the endospermic chromosome numbers. The DNA content levels are indicated using the letter C, 1C usually being the lowest level recognized, such as in an unreplicated nucleus in a haploid moss gametophyte, or a sperm nucleus of an animal. 1C levels can also be calculated from higher C-levels by dividing the DNA amount by the corresponding ploidy level. Thus, it is not necessary to measure haplophasic unreplicated nuclei to determine a 1C-value of a seed plant.

4.2.3
Generative Polyploidy Levels

Generative polyploidy levels refer to the presence of one, two, or more monoploid genomes (each with chromosome number x) in the complete, holoploid genome with chromosome number n (Greilhuber et al. 2005), which characterize single individuals, populations or taxa. The level of generative polyploidy is indicated by the letter x. A diploid angiosperm species has $2n = 2x$, a tetraploid $2n = 4x$, and so on. But note, that a plant of a haploid moss species has $n = x$ while a plant of a diploid species has $n = 2x$ (the haplophase dominates; see Chapter 12). A symbol was needed for presenting not only C-values, but also the amounts of DNA in the monoploid genomes involved and their multiples. Consequently, Cx was introduced, 1Cx being the amount of DNA of an unreplicated monoploid genome (see above and Table 4.1; Greilhuber et al. 2005). Cx-values will usually be average values unless the monoploid genomes constituting a holoploid genome can be measured separately.

4.2.4
Somatic Polyploidy

Somatic polyploidy is caused by endocycles of replication or by mitotic restitution (breakdown of mitosis in various stages) in somatic tissues (compare Chapter 15). The degree of polyploidy and the amount of DNA in such nuclei can be given as C-levels. It would be misleading here to present DNA amounts on the basis of n, because this denotes a chromosome number, and chromosomes can be unreplicated or replicated. For example, an endopolyploid root cell interphase nucleus in *Arabidopsis thaliana* (Brassicaceae) with 1C = 0.16 pg or 157 Mbp ($n = 5$, $2n = 10$) with a DNA content (not genome size!) of 2.56 pg is in 16C. From this value it is not evident, whether the nucleus is octoploid or 16-ploid. However, microscopically a spontaneous mitotic telophase nucleus with 80 chromatids and in 16C can be termed 16-ploid, while the preceding prophase nucleus in 32C would have shown 80 prophase chromosomes, thus being also 16-ploid. For comparative purposes it is possible to indicate the number of (endo)reduplication rounds to reach a certain C-level, as Barow and Meister (2003) used it for comparing different tissues in a number of angiosperm species, that is, 2C nuclei receive cycle value 0, 4C receive value 1, 8C receive value 2, and so forth. For tissues and plant organs averaged cycle values can so be given.

These rules have not only theoretical but also practical significance, for example, in labeling histograms of DNA content. A diagrammatic example of how flow histograms of different cytotypes would be labeled is presented in Fig. 4.1. In Chapter 6, Fig. 6.3, the Cx symbol is used to label histogram peaks in the flow cytometric seed screen of mixed samples of tetraploid *Hypericum perforatum* (Hypericaceae). Previously, Śliwińska and Lukaszewska (2005) analyzed polysomaty in di-, tri- and tetraploid sugarbeet, and labeled the G_1 peaks 2C, 3C and 4C, respectively, the G_2 peaks 4C, 6C, and 8C, respectively, and so on. Now that the Cx symbol is available, it is not advisable to label the G_1 peaks of di-, tri- and tetraploid individuals of a higher plant species as 2C, 3C and 4C, because all are in 2C. But it is correct to label these peaks with 2Cx, 3Cx and 4Cx (compare Fig. 4.1). The G_2 peaks of these plants would be correctly labelled 4Cx, 6Cx and 8Cx, and so on. Any individual of zygotic origin starts at 2C, be it diploid, triploid or whatever, because it starts at 2n. This avoids an infinite progression in C-levels with the advent of higher levels of generative polyploidy. For indicating these, x and Cx exist. Likewise, haplophasic individuals such as haplophasic sporophytes and gametophytes start at 1C, notwithstanding that in some cytogenetic traditions (not followed here) haplophasic sporophytes and animals such as male hymenoptera are given the chromosome number $2n$ (cf. John 1990).

Schween et al. (2003) used the C and G symbols in combination to indicate DNA amounts in the moss *Physcomitrella* (Funariaceae), so that the 1C peak was

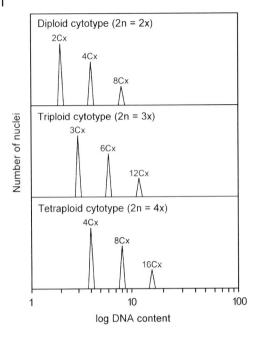

Fig. 4.1 Diagrammatic sketch of labeling peaks on DNA content histograms of cytotypes of different ploidy using the C/Cx-terminology to describe nuclear DNA contents (Greilhuber et al. 2005). Note that in each cytotype the first peak is to be regarded the 2C-peak of that cytotype. For further explanation see text.

identified as "1CG$_1$", the 2C peak "1CG$_2$", and the 4C peak "2CG$_2$ or 4CG$_1$". Here, C was obviously used in the sense of n, which should be avoided (see above).

4.3
Units for Presenting DNA Amounts and their Conversion Factors

Nuclear DNA amounts can be presented relative to the DNA content of biological standard nuclei (%, ratio), as mass units (usually picograms, pg), or as number of base pairs (bp, Mbp, Gbp). Although pg have long been used as the preferred units, with photometric methods mass is measured indirectly at best. Rather it is the relative number of base pairs, which is estimated, provided the DNA stain binds stoichiometrically and without base-dependent bias. Therefore, more recently the prevailing convention for presenting the amount of DNA is by specifying the number of base pairs. It should be noted that molecular biologists often use base number (kb, Mb, Gb) instead of base pair number, meaning DNA length instead of mass. As DNA is a double-stranded molecule, a misunderstanding can cause a two-fold error in calculating DNA content. Thomas et al. (2001) made this mistake when calculating the size of the human genome; however the error was corrected by Doležel et al. (2003). Presenting DNA amounts as the number of base pairs (bp) rather than bases is unequivocal and is therefore recommended.

Surprisingly enough, partially incorrect or poorly-supported conversion factors for pg into bp number and vice versa have been used for a long time and are even being used today. A factor of 0.965×10^9 to convert pg into base pair number has been in use until recently (Bennett and Smith 1976) with reference to Straus (1971), who reported "5.8 pg or 5.6×10^9 nucleotide pairs" for the frog, *Rana pipiens*, but did not give a conversion factor. Cavalier-Smith (1985, Preface, p. x) presented (without a derivation) a correct factor of 0.98×10^9, which was rounded up to the second decimal place. A derivation of the factor has been published recently (Doležel et al. 2003), which is as follows:

$$\text{DNA content (bp)} = (0.978 \times 10^9) \times \text{DNA content (pg)}$$

$$\text{DNA content (pg)} = \text{DNA content (bp)}/(0.978 \times 10^9)$$

Table 4.2 gives the relative weights of nucleotide pairs, $AT = 615.3830$ and $GC = 616.3711$, whereby the loss of one H_2O molecule during the formation of one phosphodiester linkage is taken into account. Note, that GC differs from AT only 1.0016-fold in weight, so that negligible bias is introduced in using mass units instead of base pair number. At physiological pH the proton is dissociated from the phosphate of any nucleotide. Assuming a 1:1 ratio of AT to GC and disregarding modified nucleotides, the mean molecular weight of one nucleotide pair is 615.8771. Multiplying the relative molecular weight by the atomic mass unit 1u, which equals 1/12 of a mass of ^{12}C, that is, 1.660539×10^{-27} kg, the mean weight of one nucleotide pair can be calculated to be 1.023×10^{-9} pg. 1 pg of DNA thus represents 0.978×10^9 base pairs.

Table 4.2 Relative molecular weights of nucleotides.

Nucleotide	Chemical formula	Relative molecular weight
2'-deoxyadenosine 5'-monophosphate	$C_{10}H_{14}N_5O_6P$	331.2213
2'-deoxythymidine 5'-monophosphate	$C_{10}H_{15}N_2O_8P$	322.2079
2'-deoxyguanosine 5'-monophosphate	$C_{10}H_{14}N_5O_7P$	347.2207
2'-deoxycytidine 5'-monophosphate	$C_9H_{14}N_3O_7P$	307.1966

Calculated with the following standard atomic weights:
$A_r(H) = 1.0079$, $A_r(C) = 12.0107$, $A_r(N) = 14.0067$, $A_r(O) = 15.9994$, $A_r(P) = 30.9738$. Standard atomic weights are scaled to nuclide ^{12}C with $A_r(^{12}C) = 12$ and rounded to four decimals. (From Doležel et al. 2003).

4.4
Sample Preparation for Flow Cytometric DNA Measurement

4.4.1
Selection of the Tissue

In principle, every tissue containing vital nuclei should be suitable for measurement of nuclear DNA content with FCM, but the presence or absence of endogenous fluorescence inhibitor substances and coatings of debris (see below) primarily influences the quality of the results. Generally, fresh almost fully expanded leaves are preferable. Very young leaves may be less suitable because of their higher content of inhibitors. It is preferable to use colorless plant organs rather than those colored by anthocyan (a fluorescence inhibitor, see below). If results are unsatisfactory, other tissues are worth considering. The light regime during plant cultivation will influence the synthesis of flavonoids, anthocyans and other phenolics, and should be selected so as to minimize the production of these substances (see Section 4.6). This effect has unintentionally been shown by Price and Johnston (1996). Nevertheless, little is known about the effect of light during cultivation with regard to FCM, and targeted studies are required. Optimal light for plant growth may not necessarily be optimal for nuclear DNA flow cytometry.

There are several investigations indicating the suitability of dry seed material for determination of nuclear DNA content by FCM. Normal seed contains a diplophasic embryo and depending on the taxon may also contain endosperm (basically haplophasic endosperm in gymnosperms and most frequently triplophasic, but occasionally diplophasic and pentaplophasic endosperm in sexual angiosperms, and other levels in hybrid situations and in apomicts; see Chapter 6). Bino et al. (1992, 1993) followed the replication levels in germinating seeds of a number of plant species and observed triploid endosperm in dry seed of *Cichorium endivia* and *Lactuca sativa* (both Asteraceae), *Solanum melongena* and *Lycopersicon esculentum* (both Solanaceae) and *Spinacia oleracea* (Chenopodiaceae/APG: Amaranthaceae), the latter two species exhibiting only the 6C-level (Bino et al. 1993). Matzk et al. (2000, 2001, 2003, 2005) analyzed the relative nuclear DNA content in dry seeds of *Arabidopsis thaliana*, *Hypericum perforatum*, *Poa annua* (Poaceae) and other angiosperms for reproduction mode screening with considerable success (the Flow Cytometric Seed Screen, FCSS; see Chapter 6). The technical side of the approach used by Matzk is remarkable, that is, dry seeds or parts thereof are crushed between two sheets of sand-paper, rinsed off with DAPI buffer, and measured. Baranyi and Greilhuber (1996) and Baranyi et al. (1996) measured the genome size of some poorly-germinating pea accessions using ethidium bromide and hypocotyl and root samples from briefly hydrated seed. Śliwińska et al. (2005) found that hypocotyls from non-hydrated seeds of *Brassica napus* (Brassicaceae) and several other crop species gave more reliable results than leaf tissue. Thus, this approach should be widely tested for studies of genome size which require intercalating dyes. On the one hand it is surprising that chromatin from dormant tissue can be easily stained with fluorochromes.

However, on the other hand it is possible that dry cells release less nucleases into the nuclear isolation solution than turgid cells from soft tissue (Chapter 6) and that certain organs such as hypocotyl contain fewer inhibitors or that dry tissues release less of them into solution (see Section 4.6). For optimal results it seems to be essential to first crush the dry tissue and then to immediately stain in buffer and measure the fluorescence (see Chapter 6). This is reminiscent of the behavior of herbarium material subjected to FCM; in this case the best results were obtained by chopping up the sample in DAPI staining buffer without pre-soaking (Suda and Trávníček 2006; Chapter 5). Targeted investigations into the time scale on which such measurements can be performed with different categories of seed, are desirable. Measurements can even be done with non-germinable seeds (Chapter 6), but if so, how old should such seeds be? And what are the reasons for quality decay with respect to DNA structure? The "seminal approach" has the potential to open a new era for biodiversity-oriented genome size studies (cf. Chapter 7), but the particularities of the material (e.g. replication levels and endopolyploidy in the embryo, spontaneous hybridization, fertilization with unreduced gametes and apomixis; cf. Chapter 6) will need to be carefully considered.

4.4.2
Reagents and Solutions

Researchers involved in the early work with plant FCM isolated protoplasts with hydrolytic enzyme mixtures, lysed the protoplasts and stained them with a fluorochrome, mainly DAPI. Doležel et al. (1989) give examples of *Zea mays* (Poaceae) and *Medicago sativa* (Fabaceae) callus and leaf material.

Today, the method of preparing a suspension of nuclei for measurement follows the ingeniously simple procedure of Galbraith et al. (1983). It consists basically of (i) chopping up the plant material with a sharp razor blade to release nuclei into isolation buffer or buffer component, (ii) sieving the homogenate to remove large particles, and (iii) staining the nuclei in (buffered) suspension with the fluorochrome of choice. RNase should be added, if intercalating dyes such as ethidium bromide (EB) or propidium iodide (PI) rather than the base-specific minor grove-binding Hoechst dyes and DAPI (AT specific), or mithramycin, olivomycin and chromomycin (GC specific) are used. It is important to use PI or EB to quantify the DNA content without biasing the results with the base content (Doležel et al. 1992). A saturation curve of PI is shown in Fig. 4.2, indicating that PI concentrations between 50 and 150 mg l^{-1} are appropriate. A similar result was obtained by Loureiro et al. (2006a) for *Pisum sativum* isolated with four different buffers. The steps can be carried out in sequence or can be combined so that chopping, staining and RNase digestion are completed in one or two steps (i.e. the chopping buffer also contains the RNase, or in addition the dye). RNase addition may often show no effect due to the low RNA content, in leaves for instance, and thus may seem dispensable, but is essential with tissues rich in RNA such as meristems and seeds, and is also for principal reasons an established step in the procedure. It should be noted that chopping up the tissue in the stain solution, as

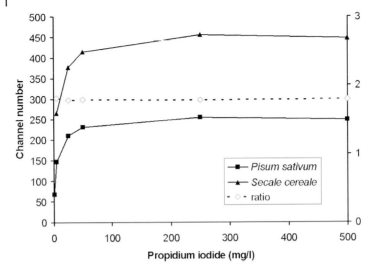

Fig. 4.2 Propidium iodide saturation curve. Nuclei were isolated from co-chopped leaves of *Pisum sativum* "Kleine Rheinländerin" and *Secale cereale* "Elect" in Otto buffer component I. The isolate was divided into 0.4-ml aliquots, which were treated with RNase at 37 °C for 30 min and immediately stored in the refrigerator. The aliquots were then stained with Otto buffer component II supplemented with 0.5, 5, 25, 50, 250 and 500 mg l^{-1} propidium iodide and measured with a flow cytometer (Partec PA II) after a 1-h incubation at 7 °C. (Original by E. M. Temsch).

is sometimes practised, increases the likelihood of skin and laboratory contamination of the sample and also increases the number of disposables that would need to be treated as toxic waste. Also RNase spills can be problematic in some laboratories. It should therefore be carefully considered whether a small gain in time outweighs laboratory safety (but note the recommendations on work with dry material, see above).

4.4.2.1 Isolation Buffers and DNA Staining

Various isolation buffers are used in plant FCM (Table 4.3). Staining is carried out at neutral or slightly basic pH and there is some detailed information available on the effect of pH on DNA specificity for the stain Hoechst 33258. Hilwig and Gropp (1975) showed that in cytological preparations at pH 2, nucleoli and cytoplasm, probably the RNA, are stained as well as chromatin DNA, while at pH 7 only chromatin is stained. Slides stained at pH 2 lost the non-specific DNA staining if mounted with pH 7 buffer, and did not regain it at pH 2 unless re-stained. Other proton concentrations were not tested. For DAPI even less information is available, despite its wide use in cytogenetics and its high level of biochemical evaluation (Kapuscinski 1995). In chromosome cytology, DAPI staining of DNA is generally carried out at pH 7, and this is also the case in plant FCM. However, Wen et al. (2001) in a study on dye concentration and pH in biomedical DNA measurements, found in tumor and mouse cell lines the best CVs (coeffi-

Table 4.3 Ten most popular non-commercial nuclear isolation buffers in plant DNA flow cytometry. Buffers are arranged in decreasing order of preference according to the FLOWer database (see Chapter 18).

Buffer	Composition[a]	References
Galbraith's	45 mM $MgCl_2$; 30 mM sodium citrate; 20 mM MOPS; 0.1% (v/v) Triton X-100; pH 7.0	Galbraith et al. (1983)
$MgSO_4$	9.53 mM $MgSO_4.7H_2O$; 47.67 mM KCl; 4.77 mM HEPES; 6.48 mM DTT; 0.25% (v/v) Triton X-100; pH 8.0	Arumuganathan and Earle (1991)
LB01	15 mM Tris; 2 mM Na_2EDTA; 0.5 mM spermine.4HCl; 80 mM KCl; 20 mM NaCl; 15 mM β-mercaptoethanol; 0.1% (v/v) Triton X-100; pH 7.5	Doležel et al. (1989)
Otto's[b]	Otto I: 100 mM citric acid monohydrate; 0.5% (v/v) Tween 20 (pH approx. 2–3) Otto II: 400 mM $Na_2PO_4.12H_2O$ (pH approx. 8–9)	Otto (1990), Doležel and Göhde (1995)
Tris.$MgCl_2$[c]	200 mM Tris; 4 mM $MgCl_2.6H_2O$; 0.5% (v/v) Triton X-100; pH 7.5	Pfosser et al. (1995)
Baranyi's[b]	Baranyi solution I: 100 mM citric acid monohydrate; 0.5% (v/v) Triton X-100 Baranyi solution II: 400 mM $Na_2PO_4.12H_2O$; 10 mM sodium citrate; 25 mM sodium sulfate	Baranyi and Greihuber (1995)
Bergounioux's	"Tissue culture salts" supplemented with 700 mM sorbitol; 1.0% (v/v) Triton X-100; pH 6.6	Bergounioux et al. (1986)
Rayburn's	1 mM hexylene glycol; 10 mM Tris; 10 mM $MgCl_2$; 0.5% (v/v) Triton X-100; pH 8.0	Rayburn et al. (1989)
Bino's	200 mM mannitol; 10 mM MOPS; 0.05% (v/v) Triton X-100; 10 mM KCl; 10 mM NaCl; 2.5 mM DTT; 10 mM spermine.4HCl; 2.5 mM $Na_2EDTA.2H_2O$; 0.05% (w/v) sodium azide; pH 5.8	Bino et al. (1993)
De Laat's	15 mM HEPES; 1 mM EDTA $Na_2.2H_2O$; 0.2% (v/v) Triton X-100; 80 mM KCl; 20 mM NaCl; 15 mM DTT; 0.5 mM spermine.4HCl; 300 mM sucrose; pH 7.0	de Laat and Blaas (1984)

[a] Final concentrations are given. MOPS, 4-morpholinepropane sulfonate; DTT, dithiothreitol; Tris, tris-(hydroxymethyl)-aminomethane; EDTA, ethylenediaminetetraacetic acid; HEPES, 4-(hydroxymethyl)piperazine-1-ethanesulfonic acid. For details on the buffer preparation see the original reference(s).
[b] pH of the buffers is not adjusted.
[c] The original recipe and reference for Tris.$MgCl_2$ is presented. Several minor modifications have been made so far, nonetheless, the basic composition remains stable.

cients of variation) and least debris at pH 6, while at pH 8 the histograms had already collapsed. At pH 7, in the mouse cell line MAT-B1 the histogram was still highly resolved, while in the line P388/R84 a significant decay in quality was observed. This is difficult to explain and stands in contradiction to the results of Otto et al. (1981). Studies on the effects of pH on staining intensity, histogram quality and DNA specificity in plant FCM are thus urgently required.

PI and EB stain DNA above pH 4, with some increase at higher pH as shown for EB by Le Pecq and Paoletti (1967). The buffer should also provide ionic strength for PI and EB to stain the nucleic acid quantitatively (Le Pecq and Paoletti 1967). If nuclei are isolated at acidic pH in citric acid plus detergent (Otto procedure; Otto et al. 1981), the dye must be added in basic solution (Na_2HPO_4) so that a final neutral pH is achieved (first used with unfixed plant nuclei by Doležel and Göhde (1995), then slightly modified by Baranyi and Greilhuber (1995), and later called the "two-step procedure" by Doležel et al. (1998)).

Isolation buffers, in addition to releasing nuclei from the cytoplasm in sufficient quantities, must also maintain nuclear integrity throughout the experiment, protect DNA from degradation by endonucleases and permit stoichiometric DNA staining. From about 26 different isolation formulas described, six are commonly used in plant DNA flow cytometry (Loureiro et al. 2006a; Table 4.3). Their usual components include: (i) organic buffer substances (e.g. Tris, MOPS and HEPES) to stabilize the pH of the solution (usually set between 7.0 and 8.0, which is compatible with common DNA fluorochromes); (ii) non-ionic detergents (e.g. Triton X-100, Tween 20) to release and clean nuclei, and decrease the aggregation affinity of nuclei and debris (note that ionic detergents such as sodium dodecyl sulfate would change the fluorescence properties of the dye molecule; Kapuscinski 1995); (iii) chromatin stabilizers (e.g. $MgCl_2$, $MgSO_4$, spermine); (iv) chelating agents (e.g. EDTA, sodium citrate) to bind divalent cations, which serve as nuclease cofactors; and (v) inorganic salts (e.g. KCl, NaCl) to achieve proper ionic strength (Doležel and Bartoš 2005).

"Otto's buffer", which is in fact the well-known McIlvaine's buffer system (e.g. Rauen 1964, pp. 92, 95) plus detergent, was first introduced to FCM in combination with DAPI by Otto et al. (1981) for ethanol-fixed mouse cells, which were resuspended in 0.2 M citric acid plus 0.5% Tween 20, adjusted to pH 7.4 and stained. With regard to this technique Otto et al. (1981) refer to Pinaev et al. (1979), who isolated non-fixed HeLa chromosomes in 0.1 M citric acid plus 0.1 M sucrose plus 0.5% Tween 20. Ulrich and Ulrich (1991) used Otto's buffer for nuclei isolation from living plant tissue, but fixed the nuclei in acetic ethanol; staining and analysis was again carried out in Otto's buffer with very narrow CVs obtained. Otto's buffer system plus DAPI was first used for unfixed plant nuclei by Doležel and Göhde (1995) for sex identification in *Melandrium* (Caryophyllaceae) and basically (with minor modification) also by Baranyi and Greilhuber (1995) to demonstrate the lack of variance of genome size in *Pisum sativum* (Fabaceae). This buffer system was obviously the essence of a commercial Partec buffer (solutions A and B) with proprietary composition in the early 1990s. It consists of two components, citric acid plus detergent ("Otto I") for nuclei isolation, and the basic Na_2HPO_4 plus fluorochrome ("Otto II"), which is added to the isolate for staining at neutral pH. Baranyi and Greilhuber (1996) first modified and applied this system for EB and PI staining (with some non-essential additions; J. Greilhuber and E. M. Temsch, unpublished data). Otto's buffer differs essentially from other buffers, because the first step combines isolation of nuclei with mild fixation and possibly some histone removal.

The other buffers (Table 4.3) work *a priori* at near-neutral pH and are based on popular organic buffer substances such as MOPS (Galbraith et al. 1983), Tris (Doležel et al. 1989; Pfosser et al. 1995) and HEPES (Arumuganathan and Earle 1991). With these buffers it is intended to keep the nuclei in an intact or even sub-vital state. Chromatin stabilizers such as Mg^{2+} (Galbraith et al. 1983) or spermine (Bino's buffer, Doležel's LB01 buffer) are added. Mannitol and sucrose are used to provide isotony. Chelators such as EDTA bind metal ions and thus block DNase activity (DNases need Mg^{2+} and Mn^{2+}). Citrate acts as a chelator as well. Thus, Mg salts as stabilizers combined with chelators as DNase inhibitors seems to make little sense. Some buffers contain mercaptoethanol, sulphite, ascorbic acid and dithiothreitol as reductants, and PVP to bind tannins (see below).

The different buffer characteristics and the cytosolic compounds released upon chopping up the tissue can affect sample and measurement quality. Comparative analyses of buffers are therefore required, but such studies have seldom been undertaken.

Recently, Loureiro et al. (2006a) compared four common and chemically different lysis buffers, namely Galbraith's buffer (Galbraith et al. 1983), LB01 (Doležel et al. 1989), Otto's buffer (Doležel and Göhde 1995) and Tris.$MgCl_2$ (Pfosser et al. 1995), taking into consideration the following parameters: fluorescence yield of nuclei in suspension, CVs of G_1 peaks, forward and side scatter, amount of debris, and the number of particles released from the sample tissue. Samples were prepared from fresh leaf tissue of seven plant species covering a wide range of genome sizes (1.30–26.90 pg/2C), differing in tissue structure and being either easy to prepare (*Pisum sativum*, *Vicia faba* and *Lycopersicon esculentum*) or more challenging (*Oxalis pes-caprae*, Oxalidaceae, complicated by acidic cell sap; *Celtis australis*, Ulmaceae, complicated by mucilage, *Festuca rothmaleri*, Poaceae, complicated by xeromorphic, and *Sedum burrito*, Crassulaceae, complicated by succulent leaves).

The buffers performed differently, although with acceptable results in most cases. Excellent results (high fluorescence yield, high nuclei yield, low CV, little debris) were obtained only with some buffers for some species. *Oxalis pes-caprae* with very acidic cell sap worked only with Otto's and Galbraith's buffer. Spermine (in LB01) seems to be a better chromatin stabilizer than $MgSO_4$, and MOPS (in Galbraith's buffer) seems to be a better buffer substance than Tris (evident in the acidic *O. pes-caprae*). A higher concentration of detergent (0.5% Triton X-100) was essential for the improved performance of Tris.$MgCl_2$ buffer in *Celtis australis* which contains a high level of mucilage. Generally, the results obtained with Otto's buffer were excellent (nuclei had high relative fluorescence intensity and the lowest CV values) in many species. An exception was the grass *Festuca roth-*

maleri, a technically difficult taxon to work with, which produces less satisfactory results with Otto's buffer and Tris.MgCl$_2$. Loureiro et al. (2006a) even found that for a given species the analysis of scatter properties (FS and SS) of nuclei provides a "fingerprint" of each buffer.

The finding that LB01 buffer, which contains Tris as the buffer substance, performed very well while Tris.MgCl$_2$ buffer yielded the least satisfactory results (with exceptions), shows that it is probably not the buffer substance itself which makes a good isolation buffer, but its concentration and the additives such as chromatin stabilizers and antioxidants, ionic strength, and detergent concentration.

Which buffer is preferable? Loureiro et al. (2006a) showed that of the four lysis buffers used, none gave consistently good results with all seven species tested. Although LB01 and Otto's buffer are recommended as the first choice, it is worthwhile testing various buffers to identify the best one for a given material. Notably, Loureiro et al. (2006a) also documented some slight differences in relative fluorescence yield depending on which buffer was used. This would mean that it may be the buffer which causes some divergence between laboratories in the estimation of genome size of the same material. The reasons for this divergence are therefore unclear and deserve investigation.

4.5
Standardization

It is self-evident and long known in DNA cytometry, that data can seldom be used straight from the machine (Bennett and Smith 1976). To make data widely comparable and thus useful, there must be some reference, that is, a biological sample having known parameters of interest, with which the unknown sample is compared. This reference material is known as the *standard*. The standard may already be present endogenously, such as in cases where in the same test material a certain type of nuclei functions as the reference for other nuclei (*endogenous standard*, for example in endopolyploidy studies). Otherwise, the standard must be added. Standardization can be performed at different levels of stringency and with different aims.

4.5.1
Types of Standardization

There are different meanings attached to the word "standard". Often a set of rules for executing a method or preparing a reagent is called the standard, but in our context standard mostly means biological material included in the procedure to compensate for the technical variables and imponderables as far as possible, so that the true relationship between the unknown and the standard is revealed and universal comparability is (hopefully) achieved. Fluorescent beads are an example of an abiotic or physical reference for instrument setting and are included in tests to calibrate the instrument gain or to serve as a staining-insensitive landmark in histograms.

The *biological standard* is a biological material with similar characteristics to those of the unknown sample, which can be measured in the same way, so that comparison and conversion of data is possible and a reference material is appointed for forthcoming experiments. Application can be as external or internal.

The *external biological standard* is not included in the sample to be measured, but the conditions of sample preparation are kept as similar as possible for both the unknown sample and the standard. In *Glycine max* (Fabaceae) DNA content studies have been undertaken in which the instrument was calibrated in the morning for a certain peak position of the external standard (a soybean cultivar), and for the rest of the day a number of cultivars were measured at constant machine settings. It was assumed that variation in peak position up to 1.12-fold indicated differences in DNA content, as opposed to technical fluctuations (Graham et al. 1994; Rayburn et al. 1997). It is clear that such an assumption would have been more justified had the standard always been co-processed with the sample (cf. Table 4.4). Other authors using the latter approach could not confirm this variation (Greilhuber and Obermayer 1997; Obermayer and Greilhuber 1999). External standardization is acceptable when the demands of precision are not high, as in DNA ploidy screening.

The *internal biological standard* is included in the same experiment to guarantee as far as possible identical conditions for the unknown sample and the standard during the whole procedure of preparation, staining and evaluation. Here, of particular relevance are the secondary metabolites of plants (often phenolics) which bind to chromatin. Acting as a steric barrier for fluorochrome binding they modify peak shape and position (Price et al. 2000; the Report on the IBC Workshop on Genome Size in Bennett and Leitch 2005; Loureiro et al. 2006b). If the secondary metabolites do act in these ways, then they ought to influence both the unknown and the standard, in as similar a manner as possible. Consequently, if the standard and sample are chopped up together then the standard should be inhibited by the secondary metabolites to a similar degree as the sample, so that the calibrated value of the unknown is more or less rectified (with emphasis on *more or less*). This is also the basis of a test for inhibitors (see below).

Some authors have used a type of standardization that is intermediate between external and internal standardization, i.e. isolating standard and sample independently and mixing the isolates, or adding the standard to the stained sample isolate after having cleaned the stained sample by centrifugation and replacing the old dye with a fresh one (Johnston et al. 1999). The standard is then stained in an environment free from the inhibitors present in solution, but which have already influenced the sample nuclei during staining. Not surprisingly, its peak quality may be better, but the relationship to the unknown sample peak is not any more authentic. Such a procedure may be termed *pseudo-internal* standardization (Noirot et al. 2005) and is approaching external standardization.

Table 4.4 Covariation of DNA content values upon internal standardization in *Secale cereale* "Elect" (the unknown) and *Pisum sativum* "Kleine Rheinländerin" (the standard) (Otto procedure, propidium iodide staining at 50 mg l^{-1} overnight). One co-isolate was divided into two aliquots (tubes a and b) and measured in steps as indicated. AU, 2C peak position at gain 551 on the Partec PAII. Conversion to pg was based on 1C = 4.38 pg for *P. sativum*. For details see text.

Time (min)	P. sativum 2C, AU	S. cereale 2C, AU	Ratio	S. cereale 1C, pg
0	54.65[a]	98.96[a]	1.811	7.931
7	55.06[a]	99.75[a]	1.812	7.935
15	55.69[a]	100.83[a]	1.811	7.930
24	56.52[a]	102.00[a]	1.805	7.904
37	56.86[a]	103.85[a]	1.826	8.000
42	52.17[b]	94.84[b]	1.818	7.962
46	52.27[b]	94.86[b]	1.815	7.949
52	52.83[b]	95.93[b]	1.816	7.953
57	49.95[b]	90.48[b]	1.811	7.934
63	50.63[b]	91.65[b]	1.810	7.929
Mean	53.66	97.32	1.813	7.943
SD			0.006[c]	0.026[c]
CV (%)			0.321[c]	0.321[c]
SD	2.43	4.45	0.117[d]	0.512[d]
CV (%)	4.53	4.57	6.435[d]	6.435[d]

[a] test tube a.
[b] test tube b.
[c] SD and CV based on co-chopped ratios.
[d] SD and CV based on ratio of species sums.

4.5.2
Requirement of Internal Standardization – a Practical Test

The importance of internal standardization is highlighted by the test shown in Table 4.4, in which *Secale cereale* "Elect" (Poaceae), the "unknown sample", is compared with *Pisum sativum* "Kleine Rheinländerin", the standard. One co-chopped isolate was divided in two parts (tubes) and processed. Each tube was measured five times in sequence. While the absolute variation in arbitrary units was up to 1.138-fold in pea and 1.148-fold in rye, variation of the rye/pea ratio reached a maximum of 1.012-fold, at a coefficient of variation of 0.3% between runs. The resulting rye/pea ratio of 1.813 differs only slightly from the average 1.779-fold found in *S. cereale* "Dankovske" by four laboratories in a ring-study on plant standards (Doležel et al. 1998) and coincides with the 1.813-fold found by laboratory 3 in the quoted study (with a different operator and using a different type of lamp-based instrument). Had single absolute values been used in an arbitrary manner, up to 1.307-fold variation could have been stated for the unknown.

It should be noted that the non-standardized variation within pea and rye reported here is in the range of the "intraspecific variation" between cultivars described in studies where the authors did not use internal standardization (Graham et al. 1994; Rayburn et al. 1997). Therefore, internal standardization is a necessity even when no fluorescence inhibitors are present. There are variables in the procedure of isolation, staining and measurement, which without internal standard could be controlled only with difficulty. Such variables include temperature and time of staining, dye concentration, pH shifts due to cell sap, and quantity of material.

4.5.3
Choice of the Appropriate Standard Species

Standard species should fulfil several criteria.

4.5.3.1 Biological Similarity

The researcher should be able to prepare the standard material synchronously together with the unknown sample, and the materials should be biologically similar. Fixed chicken red blood cells (CRBCs), human leucocytes or salmon sperm can thus hardly be regarded as an ideal internal standard for determination of genome size in plants. CRBCs are commercially available or are self-prepared, fixed and stored, often for years at low temperatures. Such material then often has a history different from the plant samples to be tested. There are no targeted studies known which could have proven the full reliability of this type of material, but there are indications that caution is appropriate. Johnston et al. (1999) report $2C = 2.49$ pg for CRBCs kept at Texas A&M University, and 3.02 pg for chicken cells kept at Arizona University (a beetle *Tetraodes* sp., Caraboidea, with $2C = 1.0$ pg was the standard). This is a 1.21-fold variation which seems to have been reproducible in their study. The genome of a male chicken (with ZZ constitution) is 2.7% larger than that of a female (with ZW constitution) (Tiersch et al. 1989). Galbraith et al. (1983) provided a more recent chemical determination of the DNA content of CRBCs and arrived at $2C = 2.33 \pm 0.22$ pg (mean $\pm$ SD, $N = 7$), meaning a 95% confidence interval between 2.167–2.493 pg. Bennett et al. (2003) co-ran chicken and *Arabidopsis thaliana* and estimated about 15% less DNA in the 2C peak of the bird than in the 16C peak of the plant (2.569 pg), indicating $2C = 2.233$ pg for chicken. This value is lower than commonly accepted values between 2.33 and 2.5 pg (cf. Bennett et al. 2003), but is within the 95% confidence interval of the chemically-determined value given by Galbraith et al. (1983). Based on the data by Tiersch et al. (1989), male human leucocytes should have $2C = 6.278$ pg, because the chicken/human ratio is 0.3557. The sex of the two chicken samples ($2C = 2.45$ and 2.53 pg) was not given by Tiersch et al. (1989), but their mean values are used here.

The data of Bennett et al. (2003) indicate a CRBC/*Arabidopsis* ratio of 6.960, whereas the data of Ozkan et al. (2006) indicate a value of 5.224. Whilst Bennett et al. (2003) compared the genome size of these organisms and co-prepared their material, Ozkan et al. (2006) primarily compared the genome size of di- and tetraploid *A. thaliana* lines and used CRBCs as a reference for staining intensity without explicitly mentioning co-preparation. This 1.33-fold discrepancy may be at least partly caused by a staining artifact of the CRBCs.

4.5.3.2 Genome Size

The standard species should be different in genome size from the unknown sample, but not too different to avoid instrumental problems with linearity. The peaks of the standard should not overlap with the peaks of the unknown sample. NB at high N, say 1500, and normal distribution, the range of a sample can be estimated by SD × 6, where 99.7% of the values are included (Sachs 1978, p. 79). The difference between the standard and the sample should thus be equal to or exceed the threefold sum of both standard deviations. The minimum difference should be about 20% when the CV is about 3%. Linearity problems with FCM are the main reason why a single DNA standard species in plants cannot be sufficient for the nearly 2000-fold range in C-values.

4.5.3.3 Nature of the Standard

Ideally the standard species should be free of fluorescence inhibitors, and its preparation should be unproblematic so that its analysis should result in narrow peaks. Thus, colored or mucous-containing plants or plant organs appear *a priori* to be inappropriate. Infected plants should be rejected, because they may be stimulated to produce inhibitors. A procedure for checking for inhibitors is given below.

4.5.3.4 Availability

Permanent availability of seed or plant material should be guaranteed for continuous experimental work. Seeds should germinate easily. Opinions differ with regard to the strictness which should be applied to selecting standards. Some authors favor a few elite standards (i.e. selected breeds of a few species; see below). For instance, M. D. Bennett et al. (personal communication) recommend for the future a mutant of *Arabidopsis thaliana* which has no flavonoids (inhibitors), and whose endopolyploid nuclei can be used as reference points in addition to the 2C and 4C peaks (cf. Chapter 7). Other authors assume a more pragmatic standpoint. We believe that laboratories which have no resources for breeding standard species themselves can obtain suitable material from reliable distributors. This material can then be calibrated with elite standards. For example, a variety of vegetable pea common in a country (e.g. *Pisum sativum* "Kleine Rheinländerin" in Austria) can be calibrated with *P. sativum* "Minerva Maple", a standard used and recommended by Bennett and Smith (1976), or with *P. sativum* "Ctirad", as suggested by Doležel et al. (1998). But note that *P. sativum* "Minerva Maple" is a field pea with colored flowers and possibly higher phenolic content than vegetable peas. Greilhuber and Ebert (1994), Baranyi and Greilhuber (1995, 1996) and Baranyi et al. (1996) have shown that the genome size of P. sativum is stable worldwide. These authors concluded this from the fact that land races and even wild accessions from extremely different climates did not differ in C-value from highbred cultivars. Why should authors be restricted to a certain pea line of limited availability, when probably any vegetable pea (i.e. the white-flowering variety) will fulfil the same criteria? Likewise, the genome size of *Glycine max* is apparently universally stable (Greilhuber and Obermayer 1997, 1998a; Obermayer and Greilhuber 1999). Recent reports of some marginal variation between lines (Chung et al. 1998; Rayburn et al. 2004) should be reconsidered in the light of the effect of fluorescence inhibitors. In the case of Rayburn et al. (2004), the low variation found (ca. 3%) may rather depend on the anthocyans present in the hypocotyls used for the measurements and in addition the results were not confirmed using rigorous statistical testing (only the LSD test was applied). Chromosomally engineered and hybrid strains of modern cereal varieties, and also onions, should be used cautiously. It is more meaningful to use old-established lines.

4.5.3.5 Cytological Homogeneity
The standard and sample should be cytologically fairly homogeneous. Seedlings from aged seeds can be problematic because of mitotic aberrations.

4.5.3.6 Accessibility
Standards used should be accessible to other researchers, that is, should be distributed upon request in sufficient quantity.

4.5.3.7 Reliability of C-Values
A reliable C-value should be established, optimally based on measurements by different laboratories. This is a sensitive point, because in fact only one C-value for a plant standard evaluated using a method yielding absolute amounts of DNA is generally accepted. This is *Allium cepa* (Alliaceae), whose nuclear DNA content per root tip meristem cell (expectedly corresponding to roughly 3C) has been chemically determined as 54.3 pg by Sparrow and Miksche (1961) and was re-calculated as 2C = 33.55 pg by Van't Hof (1965) who took into account the relative lengths of the mitotic cycle phases. This value agrees well with chemically determined values obtained from animals and humans using the Feulgen cytophotometric comparison (Greilhuber et al. 1983). Almost all other trustworthy C-values for plants are based on cytometric comparisons with plants and lastly with onion, or with human and animals, for which chemical estimates exist. The old chemical estimates in the human vary between 1C = 3.0 and 3.5 pg (Métais et al. 1951; Vendrely and Vendrely 1949). Many authors arbitrarily used the higher value for their calibrations, although a value of 3.1–3.2 pg may be closer to the truth (Doležel et al. 2003; Greilhuber et al. 1983). In one important recent investigation (Bennett et al. 2003), the size of a completely sequenced genome size was already known, that is, of the nematode *Caenorhabditis elegans*, which was used for FCM

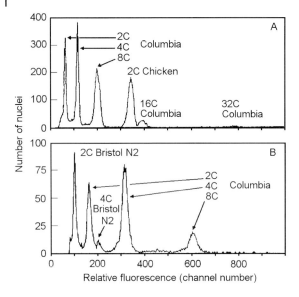

Fig. 4.3 Simultaneously prepared and measured propidium-iodide stained nuclear suspensions of *Arabidopsis thaliana* "Columbia" and chicken (a) and *Caenorhabditis elegans* "Bristol N2" (b), respectively. The positions of chicken 2C relative to *A. thaliana* 16C and of *C. elegans* 4C versus *A. thaliana* 2C give an indication of the genome size of *A. thaliana* and chicken on the basis of a *C. elegans* 1C-value of 100 Mbp. For details see text. (From Bennett et al. 2003 with permission).

comparison with *Arabidopsis thaliana* "Columbia". For this important plant species, a value of 1C = 157 Mbp was estimated using FCM, based on 1C = 100 Mbp for this worm (Fig. 4.3). This example clearly showed the fragility of the value of 125 Mbp published by the Arabidopsis Genome Initiative (2000), which significantly underestimated the non-sequenced DNA harbored in the heterochromatin (Bennett et al. 2003). But note that *A. thaliana* collected in the wild was meanwhile reported to vary by about 10% in genome size between accessions (Schmuths et al. 2004). There is clearly a need for in-depth analyses of genome sizes of plant standard species to arrive at agreed absolute values.

4.5.4
Studies on Plant Standards

Doležel et al. (1998) were the first to compare a set of nine different standard species of defined cultivars or lines in four laboratories with PI and also with DAPI, and laser and lamp-based flow cytometers, and with Feulgen scanning densitometry. The species were compared in a cascade-like manner starting from *Allium cepa* (assumed to be 2C = 33.55 pg) down to *Arabidopsis thaliana*, with a mean result of 2C = 0.37 pg, while 0.321 pg is the expected value reported by Bennett

Table 4.5 Ratios of C-values and relative standard deviations ($N = 10$) estimated for pairs of species by four laboratories (L1–L4). Nuclei were isolated simultaneously and stained with propidium iodide. A.c. *Allium cepa*, V.f. *Vicia faba*, S.c. *Secale cereale*, H.v. *Hordeum vulgare*, P.s. *Pisum sativum*, Z.m. *Zea mays*, G.m. *Glycine max*, R.s. *Raphanus sativus*, A.t. *Arabidopsis thaliana*. (Adapted from Doležel et al. 1998).

	Ratio of C-values (CV%)							
	V.f./ A.c.	S.c./ V.f.	H.v./ S.c.	P.s./ H.v.	Z.m./ P.s.	G.m./ Z.m.	R.s./ G.m.	A.t./ R.s.
L1[a]	0.778 (0.9)	0.613 (1.0)	0.647 (0.6)	0.874 (1.0)	0.639 (3.3)	0.469 (6.6)	0.506 (1.2)	0.310 (1.0)
L4[a]	0.792 (3.5)	0.606 (2.8)	0.661 (0.8)	0.869 (0.9)	0.658 (2.9)	0.519 (0.8)	0.464 (0.6)	0.302 (0.3)
L2[b]	0.776 (1.3)	0.595 (0.8)	0.638 (0.8)	0.863 (0.8)	0.609 (1.3)	0.441 (1.6)	0.462 (1.7)	0.300 (0.7)
L3[b]	0.752 (2.1)	0.586 (1.4)	0.632 (0.8)	0.879 (0.5)	0.586 (0.5)	0.438 (0.8)	0.465 (1.9)	0.313 (3.5)
Mean ratio	0.774	0.600	0.645	0.870	0.623	0.467	0.474	0.306
Largest difference between laser cytometers (%)	1.8	1.1	2.1	0.6	2.9	9.6	8.3	2.6
Largest difference between lamp cytometers (%)	3.1	1.5	0.9	1.8	3.8	0.7	0.6	4.2
Largest difference (all instruments) (%)	5.1	4.4	4.4	1.8	10.9	15.6	8.7	4.2

[a] Laser-based instruments.
[b] Lamp-based instruments.

et al. (2003). Feulgen DNA measurements with $2C = 0.326$ pg closely approached this value. The four laboratories produced strongly correlated data although the types of cytometer used differed in that laser instruments seemed to slightly underestimate the larger genomes. Nevertheless, some critical differences between laboratories were noticed (Table 4.5). Ratios of single species pairs differed by up to 15.9% (mean 6.9%), which was higher than anticipated. Laser instruments produced results which differed by up to 9.6% (mean 3.6%), and with lamp-based instruments the results differed by up to 3.8% (mean 2.1%; Table 4.5). These differences are difficult to explain but may be related to instrument-specific linearity bias, differences in the growth conditions of the plants, the use of different plant parts and perhaps also to the use of different buffers, which according to Loureiro et al. (2006a) can influence the various species investigated somewhat differently (see above).

Johnston et al. (1999) conducted a study on plant standards for FCM involving two laboratories, using among other crop species *Pisum sativum* "Minerva Maple", *Hordeum vulgare* "Sultan" (Poaceae), *Vicia faba* "GS011", and *Allium cepa* "Ailsa Craig". This study revealed problems with CRBC variability, and compared with Doležel et al. (1998) generally yielded somewhat higher 2C-values for *P. sativum* (9.56 vs. 8.75 pg), *H. vulgare* (11.12 vs. 10.04 pg), and *V. faba* (26.66 vs. 25.95 pg). The value for *Allium cepa* was accepted to be 33.55 pg. As already mentioned, in this study, the beetle *Tetraodes* sp. (2C = 1.0 pg) was the primary standard; it served for two chicken accessions whose 2C-values were quite different i.e. 2.49 and 3.01 pg. Of these, the higher (and probably too high) value of 3.01 pg was used for calibrating *H. vulgare*, which was then used to calibrate the remaining species (Johnston et al. 1999). It seems that assuming a too high value for the chicken is the main reason for the higher plant DNA values given by Johnston et al. (1999) compared to Doležel et al. (1998).

4.5.5
Suggested Standards

A widely used standard is *Pisum sativum*, but the absolute values which have been assigned to it are divergent; this is in sharp contrast to the findings of Baranyi and Greilhuber (1995, 1996) that the genome size of *P. sativum* is stable worldwide. *Pisum sativum* has the advantage of being intermediate in genome size among angiosperms, poor in or devoid of inhibitors, well established for genome size stability, and neither rich in nor completely devoid of heterochromatin. It is easily available and germinates fast, and responds equally well to different isolation buffers (Loureiro et al. 2006a). Therefore, it has all the qualifications of a primary standard, against which secondary standard species can be calibrated. Its 1C-value is presently best taken as 4.38 pg or 4.284 Gbp, which is the mean value obtained by four laboratories using laser and lamp-based flow cytometers (Doležel et al. 1998). A very similar 1C-value of 4.42 pg has been measured with Feulgen densitometry by comparison with *Allium cepa* (Greilhuber and Ebert 1994). Marie and Brown (1993) report an almost 5% lower value (i.e. 4.185 pg/1C) for *P. sativum* "Express Long", when calibrated with *Petunia hybrida* "PxPc6" (1.425 pg/1C, Solanaceae), which had been calibrated with female CRBCs (1.165 pg/1C). Doležel et al. (1998) assumed 1C = 4.545 pg for *Pisum sativum* after calibration against human leucocytes with 1C = 3.5 pg, but the latter value seems to be the upper limit for the human (see above).

Thus, there is great interest in a unique standard which fulfils all demands – the "plant gold standard", against which all other plant standards can be calibrated. An *Arabidopsis thaliana* mutant with knocked-out flavonoid production is being reviewed as a potential standard (M. D. Bennett et al., personal communication), in which the 2C, 4C, 8C, and 16C peaks could be used, the first peak representing 0.321 pg DNA (314 Mbp), the final peak, 2.569 pg (2.512 Mbp).

However, reduced peak height at the higher C-levels may limit the use of this species as a standard. While such a ladder meets the most frequent 2C-values in angiosperms, higher C-values need other standards. It seems, that a set of standard species covering the whole range of DNA content in angiosperms cannot be circumvented. Unfortunately, consensus on a unified set of standard species with agreed C-values has not been achieved.

An overview of species used in the literature for standardization is presented in Chapter 18 (Table 18.2), and occasionally large variations of assumed C-values are recognized. A list of nine species and the values obtained by four laboratories are presented by Doležel et al. (1998). These data also give the impression of some variation between teams, notwithstanding the application of best practice rules.

4.6
Fluorescence Inhibitors and Coatings of Debris

Although the interference of secondary metabolites with staining procedures had been recognized for some time in cytophotometry (Greilhuber 1986), it was not until Noirot et al. (2000) and Price et al. (2000) published their findings that this effect was taken seriously in plant FCM. Until recently, this interference was thought to be fluorescence inhibition, but research carried out in the meantime appears to suggest that there are additional effects such as the aggregation of minor particles with nuclei that also play a role in this interference and can even lead to an apparent increase in nuclear fluorescence (Loureiro et al. 2006a). The role of autofluorescing metabolites is still hypothetical and needs investigation. Therefore, we distinguish here between *inhibitors* and *coatings of debris*, the latter being particles of endogenous substances sticking to the nuclei, resulting in a deterioration of the quality of the FCM histogram peaks without necessarily decreasing the overall nuclear fluorescence.

4.6.1
What are Fluorescence Inhibitors and Coatings of Debris?

The chemical identities of fluorescence inhibitors are poorly explored, but in many instances phenolic substances possessing active hydroxyl groups (providing free electrons capable of forming hydrogen bonds) are most probably involved. Such compounds can consist of glycosylated or non-glycosylated monomers (e.g. anthocyans, flavonoids), oligomers, and polymers. Condensable tannins and the hydrolyzable tannins (mainly gallotannins and ellagitannins) are the more widely known types of the polymers. In the reduced state, these phenolics often show little or no color, and they form strong hydrogen bonds with carboxyl groups of proteins and probably also with DNA (Walle et al. 2003). Polyhydroxyphenols (phenolics with two or more active hydroxyl groups) can crosslink proteins. Tannins (for tanning leather) are large polymeric molecules which are able to cross-link the collagen fibres of skin (Endres 1961); these bonds can be disrupted with 8 M urea. Also heat, high pH and the compound Dioxan (ethylendioxid, $C_4H_8O_2$) can act as tannin strippers. When hydroxyphenols are oxidized, a quinone structure is formed which often results in browning or coloring of the compound. Such quinones are highly reactive species themselves and form covalent bonds with carboxyl groups (Endres 1961). Such bonds are irreversible, while hydrogen bonds are reversible. When nuclear suspensions turn brown or show precipitation, the presence of phenolics is evident. Workers have added antioxidants such as β-mercaptoethanol (a component of Doležel's LB01 buffer, see Table 4.3; cf. also Chapter 18), ascorbic acid or sodium metabisulphite to the isolation buffer to keep any phenolics (which are reductants themselves, that is, are easily oxidized) in their reduced state (e.g. Bharathan et al. 1994). Any hydrogen bonds could then hopefully be maintained in their reversible state and disrupted by the addition of a competitor. An example of such competitors is the low-molecular weight polyvinylpyrrolidones (PVPs); for reasons of viscosity the lower molecular weight classes (e.g. PVP-10, PVP-40) are used in FCM. Note that the monomer, vinylpyrrolidone, is highly hazardous, while the polymer is harmless. PVPs are not reductants but their amide groups are available for binding with inhibitors, in competition with those of the proteins and DNA (Gustavson 1963). PVPs are used in biochemistry, whenever problems caused by secondary plant metabolites occur, especially in protein electrophoresis and in DNA extraction procedures (e.g. Friar 2005). PVPs can reactivate enzymes which have been inactivated by phenolics (Schneider and Hallier 1970) and are widely used in beverage production as an absorbent for tannins. It seems reasonable to combine a PVP with antioxidants in nuclear isolation buffers to allow the phenolics to be stripped from proteins and DNA *before* they become oxidized. Once oxidized, phenolics, as quinones, bind covalently and practically irreversibly to the carboxyl groups, a situation which should be prevented. Bharathan et al. (1994) observed positive effects of PVP on histogram quality. Yokoya et al. (2000) found that a minimum of 10 g l^{-1} of PVP-40 greatly improved the quality and fluorescence intensity of DAPI-stained co-processed preparations of parsley, as the standard, and roses, while parsley alone was unaffected. This was attributed to the phenolics in the leaves of roses, which also influenced the standard to the same degree but in that case were absorbed by the PVP.

The effect of cytosol on PI fluorescence in *Coffea* (Rubiaceae) was demonstrated by Noirot et al. (2000, 2002, 2003, 2005). Cytosol from *Coffea* leaves and defined components such as the phenolic chlorogenic acid, reduced the fluorescence yield of *Petunia hybrida* nuclei which was used as the non-phenolic standard. Elevating the temperature of nuclear isolates before staining changed the relative fluorescence values of *Coffea* and *Petunia* by decompaction of chromatin which enhances fluorochrome binding. Addition of caffeine was able to partly restore the fluorescence yield of quenched *Petunia* nuclei (Noirot et al. 2003), which may be explained by the known gallotannin-binding property of caffeine.

That phenolics bind to DNA is clearly evident from results with purified DNA. A binding mechanism for phenolic monomers has been proposed by Sarma and Sharma (1999), who observed the direct complexation of cyanidin with calf thymus DNA, suggesting that it was the positively-charged cyanidin molecule which associates with the negatively-charged phosphate groups of the DNA backbone. Walle et al. (2003) investigated the binding of quercetin to protein and DNA using human intestinal and hepatic cells as the targets, and demonstrated the covalent binding of quercetin to the DNA following peroxidase-induced oxidation. The covalent binding of quercetin to protein (75–125 pmol mg^{-1}) was stronger than that to DNA (5–15 pmol mg^{-1}).

Ellagic acid is a highly efficient DNA-binding polyhydroxyphenol and belongs to a class of hydrolyzable tannins known as ellagitannins. It is abundant in certain fruits, for example, in strawberries and raspberries, and has anticancer activity, which can be explained by its anti-methylation properties resulting from a double-helical DNA affinity binding mechanism, rather than by an oxidant-scavenging mechanism (Dixit and Gold 1986).

Whitley et al. (2003) administered ^{14}C-labeled ellagic acid to cultured intestinal human cells and found a rapid, intense and irreversible binding to macromolecules. Proteins were crosslinked (which was not found to the same extent with quercetin; Walle et al. 2003), whereby irreversible binding required oxidation of ellagic acid. However, five times more ellagic acid was bound by DNA (5020 pmol mg^{-1} DNA) than by proteins (982 pmol mg^{-1} protein). This binding to DNA was irreversible but did not require oxidation of ellagic acid. Ellagic acid seems to be firmly bound to DNA by an intercalation mechanism (Whitley et al. 2003). From the foregoing it appears that ellagic acid could be a major factor in nuclear fluorescence quenching as observed with FCM.

Another class of phenolic compounds of concern are the coumarins, which intercalate into DNA and cause ApT adducts and crosslinks after UV irradiation (Sastry et al. 1992). Walker et al. (2006) associated variable DNA values in *Bituminaria bituminosa* (Fabaceae) with temperature-dependent variation of furanocoumarins in this species.

There are reports that phenolics, such as flavonoids and flavanols, are present *in vivo* within plant nuclei (Feucht et al. 2004). It appears probable that the finding of conspicuous flavanol content (evidenced by dark-blue coloring) of plant nuclei (of trees such as conifers, *Coffea* and *Prunus*) after *in vivo* application of the DMACA reagent (i.e. 1 g 4-dimethylaminocinnamaldehyde dissolved in 100 ml 1.5 N sulfuric acid) is an artifact, although the authors put forward arguments for *in vivo* binding (Feucht et al. 2004; and the preceding literature). While cells die and cell membranes break down, especially under acidic conditions, vacuole-located condensable tannins penetrate all surrounding tissue and are attached conspicuously to nuclei and chromosomes. Note that at the same time tannins act as a strong fixative, that is, the nuclei retain their shape. This is what also occurs in such plants during fixation with acidic-alcoholic fixatives or during hydrolysis of unfixed cells in hydrochloric acid (Greilhuber 1986). Clearly, *in vivo* binding of phenolics to nuclear chromatin can only be proven by analysis of living cells.

However, there is evidence in *Arabidopsis thaliana* that flavonoids are located in purportedly living cells not only in the cytoplasm, but also within nuclei. Flavonoids were stained with the fluorescent reagent diphenylboric acid 2-aminoethylester (DPBA) and appeared in nuclei in plasmolyzed cells in the root elongation zone (Peer et al. 2001). Plasmolysis was obviously elicited to test the vital status of the cells but it is not clear if the cells were alive at the time the photographs were taken. Saslowsky et al. (2005) showed flavonoid localization with DPBA in all protoplasm including nuclei of root cells, but did not mention viability. The reality of phenolic *in vivo* binding to nuclei is of importance for plant FCM and needs to be corroborated on a broad scale.

4.6.2
Experiments with Tannic Acid

Tannic acid is the glycoside of gallic acid and a common water-soluble hydrolyzable tannin or gallotannin, which is useful in heuristic experiments to investigate staining interference in FCM. Loureiro et al. (2006a) applied tannic acid in 13 concentrations (0.25–3.5 mg ml^{-1}) to nuclear suspensions of *Pisum sativum* and *Zea mays* prepared with four buffers, and checked the preparations with epi-fluorescence microscopy. Side and forward scatter properties were cytometrically monitored in addition to PI fluorescence. With increasing tannic acid concentration, nuclei to which debris of low fluorescence was attached could be visualized. This caused an increase in fluorescence and side scatter. A population of clumps of debris then appeared in the absence of any nuclei; the clumps of debris fluoresced more weakly than the nuclei and were of higher optical complexity. Finally, the highest tannin concentrations provoked a general precipitation of the sample. The buffers exhibited some differences in performance with tannic acid, and it is likely that this was due to higher concentration or greater efficiency of the detergent. Figure 4.4 shows examples of the so-called tannic acid effect in *P. sativum*. Figure 4.5 presents FCM diagrams from pigmented young leaves of *Rumex pulcher* (Polygonaceae) plus *P. sativum* showing a comparable effect. The side scatter discloses the fraction of nuclei with attached debris as a tail. The "poor quality" of such peaks is largely a consequence of the characteristics of the material. For genome size measurement in such cases, modal values should be taken instead of means, or rigorous gating should be applied (Fig. 4.5c), if more suitable parts of the plant are not available. When the clean nuclei can be sorted out on the scattergram, physical purification of nuclei is unnecessary.

The studies of mechanisms of fluorescence distortion are in their infancy, but from the information available it is likely that the bound inhibitors and debris attached to nuclei can have two main effects. First, they may provide steric barriers to fluorochrome binding and thereby cause fluorescence reduction. This presumably results in a left-hand shoulder or tail, or a shift of the whole peak to the left, if all nuclei are affected. Such a tail may be confluent with non-nuclear particle

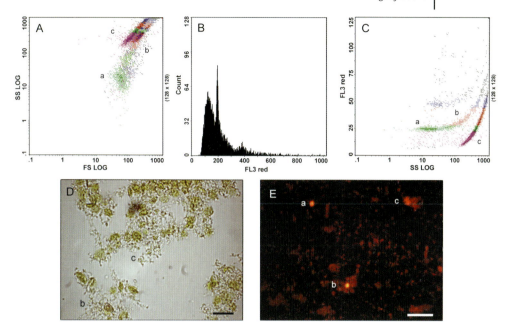

Fig. 4.4 The effect of tannic acid applied to *Pisum sativum* nuclei in suspension. Nuclei were isolated in Tris.MgCl$_2$ buffer, incubated for 15 min with 1.75 mg ml^{-1} tannic acid (TA), and stained for 5 min with propidium iodide (PI). (a) Forward scatter (logarithmic scale, FS·log) versus side scatter (logarithmic scale, SS·log) scattergram; (b) PI fluorescence intensity (FL3 red) histogram; (c) SS·log versus FL3 red scattergram; (d) bright field image after addition of TA (bar = 10 μm); (e) fluorescence image after addition of TA (bar = 20 μm, image overexposed to highlight particles with low fluorescence); a, not inhibited G$_1$/G$_0$ nuclei, b, nuclei coated with debris exhibiting enhanced fluorescence, c, fluorescent particles without nuclei. (a, c) Magenta: particles without nuclei; green, clean G$_1$/G$_0$ nuclei; brown, coated nuclei with enhanced fluorescence; blue, G$_2$ nuclei; gray, larger particles. (From Loureiro et al. 2006a with permission).

aggregates, as shown in Figs 4.4 and 4.5. Second, secondary metabolites may bind to nuclei and attract fluorescing debris, whereby a halo of low-fluorescing particles is created. This *coating of debris* leads to a right-hand tail or shoulder of the nuclear peaks and affects sample and standard nuclei in the same way (Figs 4.4 and 4.5). It is possible, that very large polymeric polyphenols do not penetrate the nuclei but attach externally, thus leading to more of an increase than a decrease in fluorescence. Nothing is known about other possible fluorescence quenching mechanisms, such as energy transfer.

Simple tests for the presence of phenolics are required. Such tests exist, but need to be adapted to the requirements of FCM, that is, nuclear isolates need to be tested for the presence of gallotannins, condensable tannins, ellagitannins, stilbenes, flavonoids, flavanols, coumarins, and so on. The dark-blue coloring of

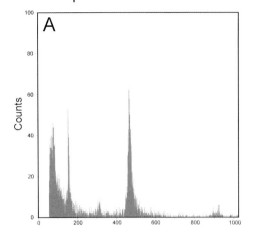
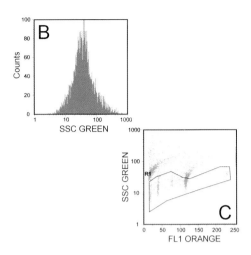
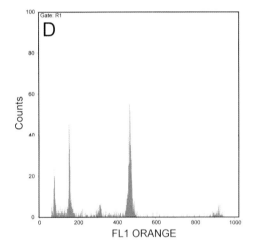
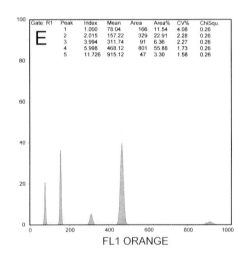

Fig. 4.5 Preparation of a very young *Rumex pulcher* leaf and *Pisum sativum* as standard, exhibiting unsatisfactory quality of the histogram and tannin-like scattergram effects (cf. Fig. 4.3). Otto's buffer, propidium iodide (PI) staining. (a) PI fluorescence histogram; (b) side scatter histogram; (c) PI fluorescence/side scatter scattergram with gating; (d) gated PI fluorescence histogram; (e) histogram with software-generated Gaussian peaks and peak parameters (peaks 1–3 belong to *R. pulcher*, peaks 4 and 5 to *P. sativum*). (Original by E. M. Temsch).

gallotannins and the green coloring of non-hydrolyzable catechin tannins obtained with ferrichloride are well known (e.g. Endres 1962). DPBA (diphenylboric acid-2-aminoethyl ester) for flavonoids (Markham 1982) and the Folin-Ciocalteu reagent for total polyphenol content (Singleton et al. 1999; Snell and Snell 1953) could also be promising reagents.

4.6.3
A Flow-cytometric Test for Inhibitors

There are examples in the literature which indicate that fluorescence inhibitors were probably involved but were at first not identified as the reason for unexpected results. Wakamiya et al. (1993) measured 19 *Pinus* species (Pinaceae) using megagametophyte and embryo tissue of *P. eldarica*. Instead of finding a 1:2-ratio between gametophyte 1C (haploid) and embryo 2C (diploid), the ratio in *P. eldarica* was 1:1.74. With Feulgen scanning densitometry the ratio was 1:1.72. However, *Pinus* embryos have tannin cells, which cause reduced staining both with Feulgen and fluorochromes, while gametophytes may have less or none. Michaelson et al. (1991), Price and Johnston (1996), and Price et al. (1998) were confronted with unprecedented DNA content variation (unorthodox genome size variation *sensu* Greilhuber 1998) in *Helianthus annuus* (sunflower; Asteraceae). At first they interpreted this variation as developmentally controlled genome downsizing and proposed the role of light quality (Price and Johnston 1996; Price et al. 1998). Later, Price et al. (2000) identified this variation as being caused by fluorescence inhibitors and described a simple test to disclose their effect. The test is based on the observation that inhibitors are released into the isolation buffer when the tissue is chopped up, and also interact with the standard nuclei. Therefore, it is necessary to compare the fluorescence intensity of the standard nuclei isolated alone with that of standard nuclei isolated together with the unknown sample. In cases where the fluorescence of the co-chopped standard appears reduced compared to the lone-chopped standard, this difference is likely to be an effect of the released inhibitor. In this way J. S. Johnston et al. (personal communication; see Bennett and Leitch 2005) elegantly demonstrated that the anthocyan, cyanidin-3-rutinoside acted as a fluorescence inhibitor in *Poinsettia* (Euphorbiaceae), in which this compound is present in red bracts but absent in green leaves.

Clearly, upon co-chopping the unknown sample is at least as strongly inhibited as the standard, if not more so. The latter could occur through the co-localization of nucleus and inhibitor in the same cell at the moment of chopping, while the standard nuclei can only be influenced by diluted inhibitor. It is thus recommended that both materials should be chopped up in a sandwich-like fashion rather than sequentially (J. Loureiro et al., unpublished results).

4.7
Quality Control and Data Presentation

The unsatisfactory situation with much of the data that had been gathered with static cytophotometry (see Greilhuber 1998, 2005) should be a warning that similar problems with FCM data should be avoided following best practice rules (cf. Chapters 5 and 7). From the foregoing it is clear that proper standardization and observation of inhibitors and coatings of debris are paramount. The highest ac-

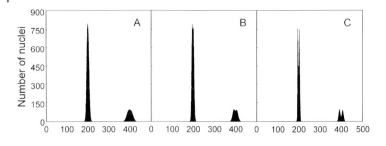
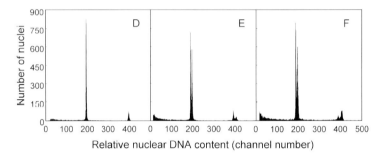

Fig. 4.6 High resolution histograms from male and female *Melandrium album* and *M. rubrum*. (a–c) Theoretical model distributions assuming a 3.7% difference between sexes at peak CVs of 3% (a), 2% (b), and 1% (c); only at CVs of 1% or lower is a clear separation obtained. (d–f) Typical histograms obtained from female *M. album* with CV = 0.53% (d), from female and male *M. album* with CV = 0.56% and 0.61%, respectively (e), and from female and male *M. rubrum* with CV = 0.70% and 0.64%, respectively (f). (From Doležel and Göhde 1995 with permission).

ceptable CV in a study, say 3 or 5%, should be set in advance. The CVs obtained should be given in some detail in the publication. Even small modern instruments measure more than one parameter. Side scatter is of much help in recognizing and eventually eliminating suspicious populations of particles. High resolution studies need stringent criteria (Suda 2004; Chapter 5). The full power of FCM has been exploited by Doležel and Göhde (1995), when the sex-difference in male and female *Melandrium album* and *M. rubrum* with XX/XY sex determination mechanism was visualized in joint preparations (Fig. 4.6). The most convincing test for true differences in DNA content is the appearance of two separate peaks in co-processed joint runs. However, this separation requires a difference of peak means of more than twice the standard deviation (Doležel and Göhde 1995).

There are rules of thumb for the required number of nuclei and the acceptable peak quality (cf. Chapter 5). Instruments are usually set to stop at 5000 to 10 000 counts, but these include G_2 and polyploid peak nuclei, perhaps S-phase nuclei, and debris depending on the sample quality and on the lower and upper level setting. Relevant peaks in general should represent ≥ 1300 nuclei. The high number of nuclei is desirable because of some fluctuation in values during a run, which

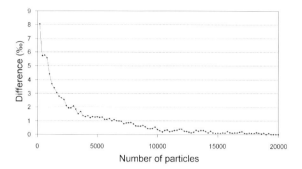

Fig. 4.7 Determination of the required number of counts to obtain stable peak position. Ten species were co-chopped and stained with DAPI (4 μg ml^{-1}, 10 min at room temperature) using the nearest standard species after Doležel et al. (1998). Conditions were: the species with lower genome size positioned at channel 200, 30 particles per second, about equal peak heights of standard and unknown sample, three replications per species on different days. As the measurements progressed, peak ratios were recorded at intervals of 200 counts, and after 20 000 particles the deviation from the end-value was measured. After 3000 and 7000 particles, this deviation is less than 0.2% and 0.1%, respectively. The species were: *Anthoxanthum alpinum* 2x, *Campanula patula* 4x, *Galeobdolon luteum* 2x, *Hieracium pilosella* 4x, *Oxycoccus palustris* 4x, *Pimpinella saxifraga* 4x, *Sorbus eximia* 4x, *Tragopogon pratensis* 2x, *Vicia cracca* 4x. (Courtesy of J. Suda).

should be averaged (Fig. 4.7). At a CV of 3% and 1300 nuclei the relative SE is ~0.1%. If two such peaks differ by 0.4% in position they are already statistically different ($P \leq 0.05$). Researchers should be aware that the number of nuclei per run and its CV are often not sufficiently decisive to insure the precision of a result. Independent repeats should be carried out, and the variance of these will give the measure of precision of a DNA-content determination at the level above the single preparation. The number of dependent and independent repeats should be stated in the publication. Note that statisticians regard an $N = 4$ as the lowest number of samples for meaningful statistics to be applied. Furthermore, if slightly but significantly different samples are found during an experiment, the difference should be confirmed by re-comparing these samples or accessions in independent tests. This approach has been extensively used to demonstrate the invariance of genome size in *Pisum sativum* (Baranyi and Greilhuber 1995, 1996; Baranyi et al. 1996), *Glycine max* (Greilhuber and Obermayer 1997, 1998a; Obermayer and Greilhuber 1999), and *Arachis hypogaea* (Fabaceae; Temsch and Greilhuber 2000), and to support much lower variation in *Cajanus cajan* (Fabaceae) than had been claimed previously (Greilhuber and Obermayer 1998b). Šmarda and Bureš (2006) considerably substantiated their finding of intraspecific genome size variation in *Festuca pallens* (Poaceae) by comparing results on the same accessions obtained in spring and autumn of different years and obtained with DAPI and PI, which were all highly significantly correlated.

When genome sizes are correlated with other parameters, for instance altitude above sea level of the locality of collection or mean annual precipitation, independence of the correlated data has to be ascertained. For example, the genome sizes of populations within a species are not independent at the level of genera. A correlation of C-values with another parameter over an area of a genus with several species should be carried out using the mean values of the species, and not of the populations. The principle that forms the basis of all quality control and best practice rules is that the reader of the publication should be able to understand what has been done by the investigator.

4.8
Future Directions

There are two broad and anastomosing avenues of plant nuclear DNA content research. On the one hand we see the application of techniques available to biological questions of genome size variation such as inter- and intraspecific variation, its functional meaning, possible selective factors, directed changes in evolution, and application to systematics. These topics are covered in Chapters 5, 6, 7, 12 and 15. On the other hand we see the research in optimizing techniques, from which these biology-oriented research avenues are profiting.

Methodologically, the identification of fluorescence inhibitors and compounds causing debris-coatings, the mechanisms of their action, the degree of their effect, and finding remedies is a big challenge. Loureiro et al. (2006a) give an example of the importance of parallel light-microscopic analyses. We can conclude that plant phenolics are certainly among the secondary metabolites which constitute the main part of the problem. A set of phytochemical tests for the presence of phenolics in nuclei isolates needs to be worked out. The inhibitor test for peak shifting (Price et al. 2000) should become standard in plant FCM work. The problem of minor intraspecific DNA content variation could be much better evaluated if such tests were routinely carried out.

Another technical challenge is the utilization of conserved (dried, fixed) plant material for FCM. There are thousands and thousands of fixed cytological samples in deep freezers in botany laboratories all over the world. DNA in fixed cell nuclei stored in ethanol at low temperatures (≤ -20 °C) remains stable over many years (Greilhuber and Temsch 2001). Routine techniques for applying meaningful FCM to this kind of material seem to be realistic but are still not available.

Recent work (Suda and Trávníček 2006) has shown the feasibility of carrying out FCM on herbarium material in certain plant groups for up to 2 years at least (cf. Chapter 5). This opens up new perspectives for analyzing field-collected material. It is a common experience that for the botanist the shortage of time in the field is the major obstacle to preparing suitably fixed cytological samples for densitometric DNA content studies. FCM of herbarium or silica gel-dried material (now used routinely for DNA studies) could become a popular alternative to densitometry. Research into the reasons for the decay in the quality of dried material, and how to slow it down or to overcome it, is urgently needed.

The recent publications reporting the use of dormant seed material for FCM (Matzk et al. 2000, 2003; Śliwińska et al. 2005) are encouraging and should be widely expanded to assess their general applicability. It is possible that in certain cases dry but living or subvital nuclei, if appropriately prepared, suffer to a lesser extent from cytosolic inhibitors than living leaf tissue.

Finally, a stringent and generally agreed list of standard plant species for the whole range of C-values in plant FCM has still not been achieved and should be worked out. This requires the concerted work of several laboratories, similar to the work carried out by Doležel et al. (1998). There is presently only one suitable organism whose genome has been sufficiently sequenced to serve as a general gold standard: the worm *Caenorhabditis elegans* "Bristol N2" with certified 96.893 Mbp (0.9907 pg) per 1C, which amounts to an estimated total of 100 Mbp (Bennett et al. 2003). In the chicken and human there are still uncertainties as to the precise absolute genome size (cf. Table 18.2). For *Arabidopsis thaliana*, $1C = 157$ Mbp or 0.1605 pg presently seems to be the best estimate (Bennett et al. 2003).

Acknowledgments

The authors thank Karin Vetschera and Harald Greger (Vienna) for help with phytochemical literature and useful discussions, and J. Doležel and J. Suda for valuable comments. J. Suda is especially thanked for his permission to include Fig. 4.7. The Austrian Science Fund is acknowledged for grants to J. G. (projects P9593-BIO to P14607-B03). J. L. is supported by fellowship FCT/BD/9003/2002.

References

Arabidopsis Genome Initiative **2000**, *Nature* 408, 796–815.

Arumuganathan, K., Earle, E. D. **1991**, *Plant Mol. Biol. Reporter* 9, 208–218.

Baranyi, M., Greilhuber, J. **1995**, *Plant Systemat. Evol.* 194, 231–239.

Baranyi, M., Greilhuber, J. **1996**, *Theor. Appl. Genet.* 92, 297–307.

Baranyi, M., Greilhuber, J., Święcicki, W. K. **1996**, *Theor. Appl. Genet.* 93, 717–721.

Barow, M., Meister, A. **2003**, *Plant, Cell Environ.* 26, 571–584.

Bennett, M. D., Leitch, I. J. **2005**, *Plant DNA C-values database*, http://www.rbgkew.org.uk/genomesize/homepage.html.

Bennett, M. D., Smith, J. B. **1976**, *Phil. Trans. Roy. Soc. Lond. Series B – Biol. Sc.* 274, 227–274.

Bennett, M. D., Leitch, I. J., Hanson, L. **1998**, *Ann. Bot.* 82, 121–134.

Bennett, M. D., Leitch, I. J., Price, H. J., Johnston, J. S. **2003**, *Ann. Bot.* 91, 547–557.

Bergounioux, C., Perennes, C., Miege, C., Gadal, P. **1986**, *Protoplasma* 130, 138–144.

Bharathan, G., Lambert, G., Galbraith, D. W. **1994**, *Am. J. Bot.* 81, 381–386.

Bino, R. J., De Vries, J. N., Kraak, H. L., van Pijlen, J. G. **1992**, *Ann. Bot.* 69, 231–236.

Bino, R. J., Lanter, S., Verhoeven, H. A., Kraak, H. L. **1993**, *Ann. Bot.* 72, 181–187.

Cavalier-Smith, T. (ed.) **1985**, *The Evolution of Genome Size*, John Wiley & Sons, Chichester, New York, Brisbane, Toronto, Singapore.

Chung, J., Lee, J.-H., Arumuganathan, K., Graef, G. L., Specht, J. E. **1998**, *Theor. Appl. Genet.* 96, 1064–1068.

de Laat, A. M. M., Blaas, J. **1984**, *Theor. Appl. Genet.* 67, 463–467.

Dixit, R., Gold, B. **1986**, *Proc. Natl Acad. Sci. USA* 83, 8039–8043.

Doležel, J., Bartoš, J. **2005**, *Ann. Bot.* 95, 99–110.

Doležel, J., Göhde, W. **1995**, *Cytometry* 19, 103–106.

Doležel, J., Binarová, P., Lucretti, S. **1989**, *Biol. Plant.* 31, 113–120.

Doležel, J., Sgorbati, S., Lucretti, S. **1992**, *Physiol. Plant.* 85, 625–631.

Doležel, J., Greilhuber, J., Lucretti, S., Meister, A., Lysák, M. A., Nardi, L., Obermayer, R. **1998**, *Ann. Bot.* 82, 17–26.

Doležel, J., Bartoš, J., Voglmayr, H., Greilhuber, J. **2003**, *Cytometry* 51A, 127–128.

Endres, H. **1961**, *Leder* 12, 294–297.

Endres, H. **1962**. Gerbstoffe: chemisch technologischer Teil in *Die Rohstoffe des Pflanzenreichs*, 5th edn, ed. C. von Regel, J., Cramer, Weinheim, pp. 1–162.

Feucht, W., Treutter, D., Polter, J. **2004**, *Plant Cell Rep.* 22, 430–436.

Friar, E. A. **2005**, *Methods Enzymol.* 395, 3–14.

Galbraith, D. W., Harkins, K. R., Maddox, J. M., Ayres, N. M., Sharma, D. P., Firoozabady, E. **1983**, *Science* 220, 1049–1051.

Graham, M. J., Nickell, C. D., Rayburn, A. L. **1994**, *Theor. Appl. Genet.* 88, 429–432.

Greilhuber, J. **1986**, *Can. J. Genet. Cytol.* 28, 409–415.

Greilhuber, J. **1998**, *Ann. Bot.* 82, 27–35.

Greilhuber, J. **2005**, *Ann. Bot.* 95, 91–98.

Greilhuber, J., Ebert, I. **1994**, *Genome* 37, 646–655.

Greilhuber, J., Obermayer, R. **1997**, *Heredity* 78, 547–551.

Greilhuber, J., Obermayer, R. **1998a**, Genome size variation and maturity group in the soybean, *Glycine max* in *Current Topics in Plant Cytogenetics Related to Plant Improvement*, ed. T. Lelley, WUV Universitätsverlag, Vienna, pp. 290–296.

Greilhuber, J., Obermayer, R. **1998b**, *Plant Systemat. Evol.* 212, 135–141.

Greilhuber, J. Temsch, E. M. **2001**, *Acta Bot. Croat.* 60, 285–298.

Greilhuber, J., Volleth, M., Loidl, J. **1983**, *Can. J. Genet. Cytol.* 25, 554–560.

Greilhuber, J., Doležel, J., Lysák, M. A., Bennett, M. D. **2005**, *Ann. Bot.* 95, 255–260.

Gustavson, K. H. **1963**, *Leder* 14, 27–34.

Heller, F. O. **1973**, *Ber. Deutsch. Bot. Gesell.* 86, 437–441.

Hilwig, I., Gropp, A. **1975**, *Exp. Cell Res.* 91, 457–460.

John, B. **1990**, *Meiosis*. Cambridge University Press, Cambridge, New York.

Johnston, J. S., Bennett, M. D., Rayburn, A. L., Galbraith, D. W., Price, H. J. **1999**, *Am. J. Bot.* 86, 609–613.

Kapuscinski, J. **1995**, *Biotechn. Histochem.* 70, 220–233.

Le Pecq, J. B., Paoletti, G. **1967**, *J. Mol. Biol.* 27, 87–106.

Loureiro, J., Rodriguez, E., Doležel, J., Santos, C. **2006a**, *Ann. Bot.* 98, 515–527.

Loureiro, J., Rodriguez, E., Doležel, J., Santos, C. **2006b**, *Ann. Bot.* 98, 679–689.

Marie, D., Brown, S. C. **1993**, *Biol. Cell* 78, 41–51.

Markham, K. R. **1982**, *Techniques of Flavonoid Identification*, Academic Press, London.

Matzk, F., Meister, A., Schubert, I. **2000**, *Plant J.* 21, 97–108.

Matzk, F., Meister, A., Brutovská, R., Schubert, I. **2001**, *Plant J.* 26, 275–282.

Matzk, F., Hammer, K., Schubert, I. **2003**, *Sex. Plant Reprod.* 16, 51–58.

Matzk, F., Prodanovic, S., Bäumlein, H., Schubert, I. **2005**, *Plant Cell* 17, 13–24.

Métais, P., Cuny, S., Mandel, P. **1951**, *Comptes-rendus Soc. Biol.* 145, 1235–1238.

Michaelson, M. J., Price, H. J., Johnston, J. S., Ellison, J. R. **1991**, *Am. J. Bot.* 78, 1238–1243.

Noirot, M., Barre, P., Louarn, J., Duperray, C., Hamon, S. **2000**, *Ann. Bot.* 86, 309–316.

Noirot, M., Barre, P., Louarn, J., Duperray, C., Hamon, S. **2002**, *Ann. Bot.* 89, 385–389.

Noirot, M., Barre, P., Duperray, C., Louarn, J., Hamon, S. **2003**, *Ann. Bot.* 92, 259–264.

Noirot, M., Barre, P., Duperray, C., Hamon, S., De Kochko, A. **2005**, *Ann. Bot.* 95, 111–118.

Obermayer, R., Greilhuber, J. **1999**, *Ann. Bot.* 84, 259–262.

Otto, F. J. **1990**, DAPI staining of fixed cells for high-resolution flow cytometry of nuclear DNA in *Methods in Cell Biology*, (vol. 33), ed. H. A. Crissman, Z. Darzynkiewicz, Academic Press, New York, pp. 105–110.

Otto, F. J., Oldiges, H., Göhde, W., Jain, V. K. **1981**, *Cytometry* 2, 189–191.

Ozkan, H., Tuna, M., Galbraith, D. W. **2006**, *Plant Breeding* 125, 288–291.

Peer, W. A., Brown, D. E., Tague, B. W., Muday, G. K., Taiz, L., Murphy, A. S. **2001**, *Plant Physiol.* 126, 536–548.

Pfosser, M., Amon, A., Lelley, T., Heberle-Bors, E. **1995**, *Cytometry* 21, 387–393.

Pinaev, G., Bandyopadhyay, D., Glebov, O., Shanbhag, V., Johansson, G., Albertsson, P. A. *Experimental Cell Research* **1979**, *124*, 191–203.

Price, H. J., Johnston, J. S. **1996**, *Proc. Natl Acad. Sci. USA* 93, 11264–11267.

Price, H. J., Morgan, P. W., Johnston, J. S. **1998**, *Ann. Bot.* 82, 95–98.

Price, H. J., Hodnett, G., Johnston, J. S. **2000**, *Ann. Bot.* 86, 929–934.

Rauen, H. M. (ed.) **1964**, *Biochemisches Taschenbuch, Zweiter Teil*, 2nd edn, Springer Verlag, Berlin, Göttingen, Heidelberg, New York.

Rayburn, A. I., Auger, J. A., Benzinger, E. A., Hepburn, A. G. **1989**, *J. Exp. Bot.* 40, 1179–1183.

Rayburn, A. L., Biradar, D. P., Bullock, D. G., Nelson, R. L., Gourmet, C., Wetzel, J. B. **1997**, *Ann. Bot.* 80, 321–325.

Rayburn, A. L., Biradar, D. P., Nelson, R. L., McCloskey, R., Yeater, K. M. **2004**, *Crop Science* 44, 261–264.

Sachs, L. **1978**, *Angewandte Statistik*, 5th edn, Springer-Verlag, Berlin, Heidelberg, New York.

Sarma, A. D., Sharma, R. **1999**, *Phytochemistry* 52, 1313–1318.

Saslowsky, D. E., Warek, U., Winkel, B. S. J. **2005**, *J. Biol. Chem.* 280, 23735–23740.

Sastry, S. S., Spielmann, H. P., Hearst, J. E. **1992**, *Adv. Enzymol.* 66, 85–148.

Schmuths, H., Meister, A., Horres, R., Bachmann, K. **2004**, *Ann. Bot.* 93, 317–321.

Schneider, V., Hallier, U. W. **1970**, *Planta* 94, 134–139.

Schween, G., Gorr, G., Hohe, A., Reski, R. **2003**, *Plant Biol.* 5, 50–58.

Singleton, V. L., Orthofer, R., Lamuela-Raventos, R. M. **1999**, *Methods Enzymol.* 299, 152–178.

Śliwińska, E., Lukaszewska, E. **2005**, *Plant Science* 168, 1067–1074.

Śliwińska, E., Zielińska, E., Jedrzejczyk, I. **2005**, *Cytometry* 64A, 72–79.

Šmarda, P., Bureš, P. **2006**, *Ann. Bot.* doi:10.1093/aob/mcl150.

Snell, F. D., Snell, C. T. **1953**, *Colorimetric Methods of Analysis*, (vol. 3), 3rd edn, Van Nostrand Co., Inc., Toronto, New York, London.

Sparrow, A. H., Miksche, J. P. **1961**, *Science* 134, 282–283.

Straus, N. A. **1971**, *Proc. Natl Acad. Sci. USA* 68, 799–802.

Suda, J. **2004**, *An Employment of Flow Cytometry into Plant Biosystematics*, PhD Thesis, Charles University in Prague. http://www.ibot.cas.cz/fcm/suda/presentation/disertation.pdf.

Suda, J., Trávníček, P. **2006**, *Cytometry* 69A, 273–280.

Swift, H. H. **1950**, *Physiol. Zool.* 23, 169–198.

Temsch, E. M., Greilhuber, J. **2000**, *Genome* 43, 826–830.

Thomas, R. A., Krishan, A., Robinson, D. M., Sams, C., Costa, F. **2001**, *Cytometry* 43, 2–11.

Tiersch, T. R., Chandler, R. W., Wachtel, S. S., Elias, S. **1989**, *Cytometry* 10, 706–710.

Ulrich, I., Ulrich, W. **1991**, *Protoplasma* 165, 212–215.

Van't Hof, J. **1965**, *Exp. Cell Res.* 39, 48–58.

Vendrely, R., Vendrely, C. **1949**, *Experientia* 5, 327–329.

Wakamiya, I., Newton, R. J., Johnston, J. S., Price, H. J. **1993**, *Am. J. Bot.* 80, 1235–1241.

Walker, D. J., Monino, I., Correal, E. **2006**, *Environ. Exp. Bot.* 55, 258–265.

Walle, T., Vincent, T. S., Walle, U. K. **2003**, *Biochem. Pharmacol.* 65, 1603–1610.

Wen, J., Krishan, A., Thomas, R. A. **2001**, *Cytometry* 43, 12–15.

Wendel, J. F. **2000**, *Plant Mol. Biol.* 42, 225–249.

Whitley, A. C., Stoner, G. D., Darby, M. V., Walle, T. **2003**, *Biochem. Pharmacol.* 66, 907–915.

Yokoya, K., Roberts, A. V., Mottley, J., Lewis, R., Brandham, P. E. **2000**, *Ann. Bot.* 85, 557–561.

5
Flow Cytometry and Ploidy: Applications in Plant Systematics, Ecology and Evolutionary Biology

Jan Suda, Paul Kron, Brian C. Husband, and Pavel Trávníček

Overview

Since the late 1980s, publications that include flow cytometry (FCM) have increased dramatically in plant biology. An important use of this technology has been to estimate variation in genome copy number. Genome copy number variation is inherent in the cell cycle, occurs as alternation of nuclear phases associated with the sporophytic–gametophytic alternation of generations, and is widespread within and among taxonomic groups of vascular plants. It can be both an important genome marker as well as a focus of study for many plant biologists. By estimating relative nuclear DNA content, FCM offers a faster and more convenient method of detecting ploidy than other methods (chromosome counts, Feulgen microdensitometry), and enables large-scale surveys on the landscape, population, individual, and tissue scales. In systematics and evolutionary biology, FCM is an essential tool for quantifying spatial and temporal patterns of ploidy variation and identifying cryptic taxonomic structure. With high resolution and success on gametes as well as somatic tissue, FCM also offers insights into questions of process as well as pattern. Research has focused on mating systems, unreduced gamete production, interactions among intraspecific cytotypes, and dynamics of hybrid zones. FCM has become an essential tool in the population biologist's kit. In combination with other molecular and phenotypic approaches, it promises qualitative advances in our understanding of genome multiplication and the population biology of vascular plants.

5.1
Introduction

In the last two decades, the use of flow cytometry (FCM) in plant science has rapidly expanded, particularly in evolutionary biology, ecology, and systematics. By far, the dominant use of FCM in these fields is in the estimation of nuclear DNA content in absolute terms or in relative units, as an indicator of ploidy level.

Use of FCM to characterize plant samples in ways other than DNA content (e.g. side and forward scatters, particle volume) is relatively scarce. The emphasis on DNA content has been driven primarily by increasing interest in the magnitude and causes of genome size variation and demand for effective and rapid cytotype discrimination. To date, ecologists and evolutionary biologists have exploited only a fraction of the potential of FCM, but have already made major advances with far-reaching implications.

Our objective here is to evaluate the use of FCM for estimating ploidy in vascular plants and highlight research programs in systematics and evolutionary biology that are being advanced by this technology. Use of FCM for estimating absolute genome size is described in Chapters 4 and 7 and thus will not be discussed here. We begin with some practical considerations. FCM has many advantages over other methods; however, these benefits are realized only with awareness of potential pitfalls and technical guidelines specific to measuring ploidy. For the remainder of the chapter, we examine a selection of taxonomic and evolutionary applications for ploidy analysis using FCM and identify new avenues for future growth. For each research area, we discuss the benefits as well as the specific challenges of using flow cytometry.

5.2
Practical Considerations

5.2.1
Relative DNA Content, Ploidy and Flow Cytometry

In FCM-based studies of DNA content, the measured parameter is the fluorescence of isolated particles (mostly nuclei) stained with a DNA-specific fluorochrome. The fluorescence intensity of these nuclei is compared to some appropriate reference (e.g. an internal standard), and the relative fluorescence interpreted as DNA content. When the reference point is a standard of known genome size (e.g. pg/2C), this yields an estimate of absolute DNA content, in picograms, for the test nuclei (see Chapters 4 and 7). However, in many applications, the information of interest is not absolute genome size, but rather the amount of DNA relative to the reference material. One such example is the determination of the ploidy of individuals, tissues, or cells, in which DNA content of a sample is compared to a reference of known ploidy and expressed as multiples of a single chromosome complement.

A fundamental assumption in using FCM to assign ploidy is that increments of DNA content correspond in a predictable way to increments in chromosome number. Errors in interpretation can occur when this assumption fails, as in species with holokinetic chromosomes (see Section 5.2.4.1), or in cases where gross changes in DNA content per chromosome have occurred following polyploidization (Gregory and Hebert 1999; Leitch and Bennett 2004; Soltis and Soltis 1999;

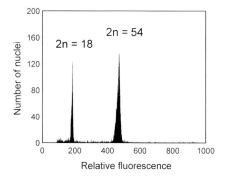

Fig. 5.1 Simultaneous FCM analysis of DAPI-stained nuclei isolated from karyologically verified diploid (2n = 2x = 18) and hexaploid (2n = 6x = 54) cytotypes of *Aster amellus* (Asteraceae). The plants differ threefold in the number of chromosomes but only 2.54-fold in the nuclear DNA content (peaks were located on channels 190 and 483, respectively). Pentaploidy may thus be erroneously inferred for the latter cytotype using FCM measures alone. (Partec PA II cytometer equipped with a mercury arc lamp. Reproduced from Suda et al. 2006, with permission).

Song et al. 1995; Wendel 2000). For this reason, comparisons of evolutionarily distant species (e.g. belonging to different sections within a genus) may lead to inaccurate estimates of ploidy. Moreover, errors (ploidy underestimation or overestimation) may sometimes be introduced even when comparing different cytotypes of the same species, unless FCM measures are combined with chromosome counts (Fig. 5.1). Combining DNA content measurements with chromosome counts not only reduces the probability of error, but may also provide other insights into genome size evolution (Dart et al. 2004; Obermayer and Greilhuber 2006; Ramírez-Morillo and Brown 2001). When the relationship between chromosome complement and DNA content is not directly verified, the prefix "DNA" (i.e. DNA ploidy, DNA aneuploidy) should designate FCM results (Hiddemann et al. 1984; Suda et al. 2006).

5.2.2
General Guidelines for Ploidy-level Studies

FCM has several advantages over other methods of measuring ploidy, including: (i) rapid sample preparation so dozens of samples can be processed per day, (ii) the ability to measure mitotically inactive cells from a broad variety of tissues, (iii) non-destructive sampling, enabling investigation of rare and endangered species, (iv) rapid detection of mixed samples or endopolyploidy (occurrence of nuclei of enhanced ploidy levels within an individual caused by endoreduplication or endomitosis), and (v) relatively low operating costs. To ensure that these advantages of FCM are realized in a consistent way, efforts have been made in recent years to promote universal guidelines for the measurement of nuclear DNA content in plants (Bennett et al. 2000a; Doležel and Bartoš 2005). These include: (i) the use of histograms obtained after the analysis of ≥ 5000 nuclei and with DNA peaks with coefficients of variation (CVs) ≤ 3%, (ii) tissue preparation methods (i.e. chopping) and tissue types (i.e. fresh leaves) that will yield these values, (iii) replication of individuals across three different days, (iv) stains that are not base-pair specific (e.g. propidium iodide, PI), and (v) internal plant standards that are as close as possible to the study species in DNA content, without overlap (Doležel and Bartoš 2005; Johnston et al. 1999; Marie and Brown 1993; Chapter 4). These recommendations have been driven by the requirements of genome size studies and, at their most basic level, are intended to ensure that such estimates are accurate (i.e. reflect the true DNA content) and precise (i.e. have a variance low enough to allow for the resolution of small differences in absolute DNA content).

For many applications that use relative DNA content for ploidy determination, such accurate measures of genome size are unnecessary or secondary at best, and the level of resolution required may be lower than that demanded in genome size estimates. The high standards of genome size studies may guide protocols, but it would be a mistake to abandon approaches useful in evolutionary biology and systematics because these standards cannot be achieved in every case. Large-scale population surveys, for example, may be greatly facilitated if preserved tissue can be used, but preservation techniques almost always result in a decrease in peak quality, reflected in higher CVs and certain shifts in absolute peak position (see Section 5.2.3.1). Non-destructive tissue sampling or sampling of small structures (e.g. seeds) is a strength of FCM, but nuclei numbers below 5000 may be a consequence, and replication within an individual may not be possible. These violations of recommended practice may not be an issue in all cases: Figures 5.2 and 5.3 illustrate two cases in which violations of standard recommendations nevertheless yielded useful results. In both cases, relaxation of sampling standards may be sufficient as ploidy is the main concern; moreover, classification of near-euploids with euploids can be tolerated, and small deviations from chromosome number/DNA content linearity can be ignored. Conversely, these kinds of surveys of DNA content may be useful for identifying individuals that are potentially aneuploid; however, higher nuclei counts and replication within individuals would be required to confirm their identity and estimate the deviation in chromosome number (Roux et al. 2003).

Guidelines for peak CVs, replication within individuals, and similarity in DNA content of the target species and the standard, all depend on the importance of distinguishing small differences in DNA content. For differences on the scale of whole sets of chromosomes, CV values between 5 and 10% can be tolerated (especially when using preserved tissues), and replication within individuals, while always advisable, may be reduced when fine scale distinctions are not required. Acquiring nuclei numbers (per peak) in excess of 1000 does not significantly affect peak position measures in many cases (P. Kron and B. C. Husband, unpublished data; Chapter 4), and numbers below 1000 may even yield useful information (Figs. 5.2 and 5.3). Higher total nuclei numbers only become critical when counts in small, secondary peaks are of interest, as in unreduced gamete estima-

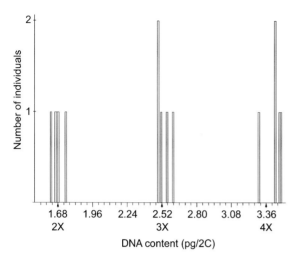

Fig. 5.2 Distribution of absolute DNA contents among individuals in a population of *Chamerion latifolium* (Onagraceae; $N = 15$). DNA contents were calculated relative to an internal standard of fresh *Epilobium hirsutum* leaf (Onagraceae; 0.88 pg/2C, P. Kron and B. C. Husband, unpublished data). Minor divisions on the x-axis correspond to increments of the DNA content of one average *C. latifolium* chromosome. Average chromosome DNA content was estimated based on the assumption that the three discrete clusters correspond to diploid ($2n = 2x = 36$), triploid ($2n = 3x = 54$) and tetraploid ($2n = 4x = 72$) plants, and arrows on the x-axis indicate the expected position of euploid individuals, based on this same estimate. Nuclei were stained with propidium iodide after extraction from field-desiccated leaves; extraction was carried out with a tissue homogenizer (FastPrep™) rather than chopping. The DNA content shown for each plant is the mean of two replicates, run on separate days. In the FCM histograms, CVs of nuclei peaks ranged from 2.6 to 5.4%, and numbers of nuclei per peak ranged from 77 to 2 387 (P. Kron and B. C. Husband, unpublished data).

tion, cell cycle analysis, or studies of endopolyploidy. DNA content standards may be selected based on their utility in screening for multiple ploidy levels on the same instrument scale, rather than based on how close they are in DNA content to any particular ploidy. Provided that an internal standard is used, co-chopped with the test species, effects of inhibition (Price et al. 2000), and base-pair specific staining (Doležel and Bartoš 2005) are unlikely to be important, assuming that inhibition effects or base-pair ratios do not vary significantly with ploidy level. Base-specific fluorochromes (e.g. DAPI) often yield histograms with high resolution and may therefore be useful not only for ploidy screening but also in assays aimed at detecting small differences in base composition/nuclear DNA content, such as discrimination between euploids and aneuploids or between different homoploid species (see Section 5.3.2). As base-pair composition generally varies only slightly among closely related taxa (Barow and Meister 2002; Chapter 8), differences in DAPI fluorescence intensity most likely reflect variation in nuclear DNA content.

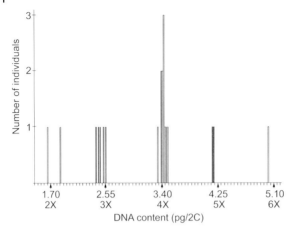

Fig. 5.3 Distribution of absolute DNA contents among seeds collected from a wild population of *Malus coronaria* (Rosaceae, 2n = 4x = 68; N = 24). DNA contents were calculated relative to an internal standard of *Epilobium hirsutum* leaf. Minor divisions on the x-axis correspond to the DNA content of one average *M. coronaria* chromosome. Average chromosome DNA content was estimated based on the assumption that three discrete clusters (gray bars) correspond to three pure *M. coronaria* cytotypes: diploids (apomictic seeds; 2n = 2x = 34), tetraploids (sexual – reduced gametes; 2n = 4x = 68), and hexaploids (sexual – reduced + unreduced gamete; 2n = 6x = 102). Arrows on the x-axis indicate the expected position of euploid individuals, based on this same estimate. Dark bars near the 3x and 5x positions are hybrids between *M. coronaria* and *M. ×domestica* (2n = 2x = 34). The positions of the hybrids are consistent with a lower genome size in *M. ×domestica*. Nuclei were stained with propidium iodide after extraction from a single cotyledon of an ungerminated seed, and each individual was run only once. In the FCM histograms, CVs of nuclei peaks ranged from 2.5 to 4.6%, and nuclei numbers per peak ranged from 145 to 3311 (P. Kron and B. C. Husband, unpublished data).

5.2.3
Use of Alternative Tissues

5.2.3.1 Preserved or Dormant Tissue

The often-cited need to use fresh tissue for FCM (Doležel and Bartoš 2005) may substantially limit the power of the technique in field research. For example, for many species and applications, analyzing tissue shortly after collection makes the investigation of samples from distant localities difficult. In some cases, rapid transport of living material packed in humid paper tissues and kept at a low temperature is a solution. However, under certain circumstances (e.g. long-term expeditions to remote areas), alternative approaches are needed. While recent advances in portable flow cytometers suggest future options (Doležel and Bartoš 2005), the use of dormant or preserved tissues has greater practical potential in the short term.

One solution is to collect and transport seeds and then grow plantlets close to the FCM facility (Suda et al. 2005). Dormant seeds may sometimes be used for DNA content estimation (Śliwińska et al. 2005; Fig. 5.3). Nevertheless, care must be taken when interpreting histograms as: (i) nuclei of two different ploidy levels (embryo + endosperm) may co-occur in the seed, (ii) a large proportion of embryonic nuclei may be arrested in the G_2 phase of the cell cycle, leading to incorrect estimates and (iii) the ploidy level of seeds may differ from that of the maternal plants due to heteroploid hybridization, involvement of non-reduced gametes, or haploid parthenogenesis (Śliwińska et al. 2005; Fig. 5.3). This last consideration is of particular importance when the goal is to infer the DNA content of the parents, and in such cases, sampling seed families is advisable.

Few studies have used fixed plant material for estimating DNA content (in contrast to routine investigation of fixed or frozen cells in animal FCM research; Gregory 2005). Successful analysis of fixed nuclei was reported in some graminoid species after staining with ethidium bromide + olivomycin (Hülgenhof et al. 1988) or DAPI (Jarret et al. 1995). In *Actinidia deliciosa* (kiwi fruit, Actinidiaceae), purified meristematic nuclei were stored in 30% glycerol in a freezer for 9 months without any appreciable loss of integrity, although propidium iodide fluorescence declined by 5–7% (Hopping 1993). Similarly, purified pea nuclei remained intact for several weeks of storage at $-20\,°C$ (Chiatante et al. 1990). Nuclei fixation generally did not dramatically increase CVs, although aggregation was often a side effect. In addition to demonstrating the use of preserved extracted nuclei, Sgorbati et al. (1986) showed that chopping of preserved whole tissue (formaldehyde-fixed leaves and roots) could yield a large number of nuclei suitable for DAPI flow cytometry. Despite these successes, such potentially helpful approaches have gained little attention from field-oriented plant researchers, primarily due to limited storage time after which successful FCM investigation was feasible (mostly a few weeks), and the complexity of some protocols that hampered their completion in the field.

The use of desiccated plant material in FCM has only recently received attention, despite the fact that the preparation of herbarium vouchers is a well-established form of sample preservation in field botany. Suda and Trávníček (2006) initiated a long-term study aimed at the potential use of desiccated plant material for DAPI flow cytometry, and found that the majority of tested species produced distinct peaks with reasonable CVs after 9 months of storage at room temperature. After 20 months, DNA ploidy estimation was still feasible in 43 out of 60 samples. Nuclei isolated from desiccated tissues often experienced a decrease in fluorescence intensity after long-term storage; nevertheless, this shift did not compromise result reliability. Moreover, preserving desiccated tissue in a deep freezer substantially extended their FCM lifetime (to 4 years at least), reduced fluorescence shift, and maintained high histogram resolution (Fig. 5.4). Successful use of FCM for DNA ploidy estimation was also reported in 6-months to 2-year-old herbarium specimens of Central European fescues (Šmarda et al. 2005), and in 4–5-year-old fescue vouchers from northern South America (Šmarda and Stančík 2006). It is possible that some aspects of standard herbarium voucher preparation may act against the maintenance of nuclear quality that is essential in FCM work (e.g. the rate of drying in plant presses). An approach that may avoid some of these problems is to rapidly dry leaves in the field using desiccant. Leaves of *Chamerion angustifolium* (fireweed) and *C. latifolium* (Onagra-

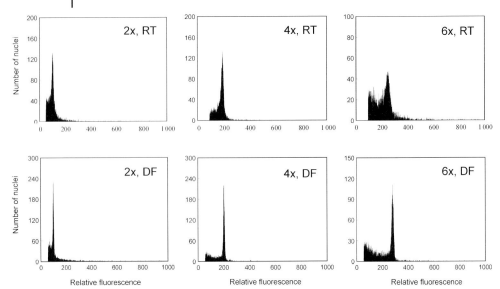

Fig. 5.4 Fluorescence histograms of DAPI-stained nuclei isolated from herbarium vouchers of diploid (2n = 2x = 24), tetraploid (2n = 4x = 48), and hexaploid (2n = 6x = 72) cytotypes of *Vaccinium* sect. Oxycoccus (Ericaceae) stored for 3 years at room temperature (RT) and in a deep freezer (DF). Mean channel positions of individual peaks were: 104 (2x, RT), 101 (2x, DF), 199 (4x, RT), 207 (4x, DF), 257 (6x, RT), and 285 (6x, DF). The instrument was calibrated prior to the analyses so that a peak of the room temperature-stored tetraploid plant is located on channel 200. Note marked differences in peak quality between RT- and DF-preserved material. (Partec PA II cytometer equipped with a mercury arc lamp; J. Suda and P. Trávníček, unpublished data).

ceae) dried in this way yield peaks comparable in quality to fresh leaves after a few weeks (Fig. 5.2) and even months of storage, and in some cases, tissue has yielded peaks useful for DNA ploidy distinction after at least 3 years storage at room temperature (P. Kron and B. C. Husband, unpublished data). Similarly, up to 2-year-old silica-dry samples of *Juncus biglumis* (Juncaceae) still allowed reliable detection of nuclear DNA content variation (Schönswetter et al. 2007). The FCM analysis of herbarium vouchers and other desiccated tissue is thus a very beneficial approach that simplifies sample transportation from remote areas, facilitates retrospective DNA ploidy determination in already desiccated vouchers, and allows the postponement of analyses if the capacity of a laboratory is saturated.

The use of frozen tissue has received even less attention than desiccation. Nsabimana and van Staden (2006) successfully determined the ploidy of banana clones (*Musa* spp.) using leaf tissue frozen at −70 °C, although with fewer nuclei, more debris, and higher CV's than fresh tissue. Fresh leaves of *C. angustifolium* that were rolled in 1.5-ml microcentrifuge tubes and frozen at −80 °C for up to 3 years also yielded peaks of sufficient quality to assign DNA ploidy, although CVs were often (but not always) between 5 and 10%, and some downward shifting of fluorescence occurred (P. Kron and B. C. Husband, unpublished data).

5.2.3.2 Pollen

While FCM has been used extensively to study the DNA content and ploidy of somatic tissues, its use to directly examine gametes has been less common, possibly in part because of the technical challenges involved (Pan et al. 2004). Reports of FCM work on female gametes appear to be lacking, but there are several ploidy-based studies of pollen, focusing on pollen development and unreduced (2n) pollen production (Bino et al. 1990; Boluarte 1999; Jacob et al. 2001; Owen et al. 1988; Pan et al. 2004; Pichot and Maâtaoui 2000; Sugiura et al. 1998, 2000; van Tuyl et al. 1989). Most of this work has involved extracted nuclei or sperm cells, rather than intact pollen, for two reasons: (i) autofluorescence and/or non-specific staining of the pollen exine, or arrangement of nuclei within the pollen grain, result in poorly resolved, overlapping fluorescence peaks that do not correspond in a predictable way to nuclear DNA content (Boluarte 1999; Owen et al. 1988; P. Kron and B. C. Husband, unpublished data), and (ii) particle size limitations for some machines and species.

Extracting sufficient numbers of intact nuclei from pollen is often difficult, and the effectiveness of different methods varies from species to species. The most common method used for somatic tissue, chopping, does not always work with pollen, although it has been used successfully on some species (Bino et al. 1990; Pichot and Maâtaoui 2000; Sugiura et al. 1998, 2000; van Tuyl et al. 1989, for germinated pollen). Other approaches include: crushing or squashing (Jacob et al. 2001; Pichot and Maâtaoui 2000), bursting in hypotonic solutions (Zhang et al. 1992), freezing pollen in buffer (P. Kron and B. C. Husband, unpublished), and sonication (Pan et al. 2004). More studies using these and other methods for non-FCM applications are cited in de Paepe et al. (1990), but it is not clear whether the quality of such preparations would be suitable for FCM.

The interpretation of DNA content results may be complicated in pollen by the fact that vegetative, generative, and sperm nuclei can be structurally and morphologically quite different, and as a result, may take up nuclear stains differently (de Paepe et al. 1990). Such differential staining can yield deviations from expected fluorescence ratios, as in *Chamerion angustifolium*, where 2n generative nuclei have approximately 1.7-times the fluorescence of 1n vegetative nuclei when stained with PI (P. Kron et al., unpublished data; Fig. 5.5). This effect will likely be more pronounced with nuclear stains that are sensitive to chromatin structure, such as ethidium bromide (EB) and PI, and less of a problem with other stains, such as DAPI (Doležel and Bartoš 2005). In other binucleate species in which DAPI was used, only slight deviations from 2:1 appear to be present in most cases (Bino et al. 1990; Pichot and Maâtaoui 2000; van Tuyl et al. 1989), and in trinucleate pollen, 1n sperm and 1n vegetative nuclei appear to have essentially the same fluorescence (Bino et al. 1990; Pan et al. 2004; Sugiura et al. 1998, 2000). However, in all published cases, the level of statistical detail presented is insufficient to fully examine this phenomenon.

When the focus of a study is unreduced pollen (see Section 5.4.4), it is not always necessary to quantify the proportion of unreduced nuclei with great accuracy, as for example when the objective is to simply identify individuals that are

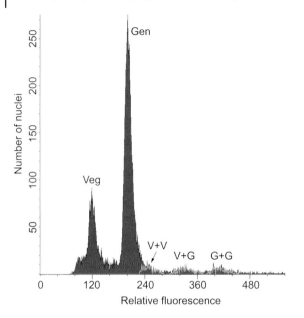

Fig. 5.5 Fluorescence histogram of propidium iodide-stained nuclei isolated from Chamerion angustifolium (Onagraceae) pollen. Veg, vegetative nuclei (1C); Gen, generative nuclei (2C); V + V, doublets and/or unreduced (2C) vegetative nuclei; V + G, doublets of 1 vegetative (1C) and 1 generative (2C) nucleus; G + G, doublets and/or unreduced (4C) generative nuclei. The ratio of the generative to vegetative peak fluorescence is 1.7:1, rather than the expected 2:1 (P. Kron et al., unpublished data).

producers of large numbers of 2n pollen (van Tuyl et al. 1989). When precise measures of the frequency of unreduced nuclei are required, complications may arise due to the overlap of same-ploidy peaks, as in binucleate pollen with 2n generative nuclei from reduced pollen, and 2n vegetative nuclei from unreduced pollen (van Tuyl et al. 1989). Differential staining of generative and vegetative nuclei by PI may work in the researcher's favor in such cases, as the deviation from expected 2:1 ratios reduces this kind of overlap (Fig. 5.5). In all species, distinguishing true unreduced pollen nuclei from contaminating somatic tissue nuclei and doublets is critical, especially when low numbers are being measured. Careful technique and microscopic examination can be used to rule out somatic nuclei, but eliminating doublets can be difficult.

Distinguishing doublets from large particles with doubled DNA, using the ratio of fluorescence signal width to area, is possible in some cases, but it is dependent on the relationship of particle (nucleus, cell) size to the beam width (Sharpless et al. 1975), and therefore does not work well on some machines. It may be possible to rule out the presence of doublets based on an absence of triplets (Bino et al. 1990; Pan et al. 2004), but this may not be sufficient if doublets are present at low frequencies (and triplets, therefore, at even lower frequencies). More generally, microscopic examination of samples is often recommended as a way of detecting doublets (e.g. Bino et al. 1990; Pan et al. 2004), although doublets at low frequency may be missed.

In some cases, the biology of the system under study may provide means to identify doublets. Along with other evidence, Pan et al. (2004) concluded that small 2C peaks detected in mature pollen were unreduced pollen nuclei, not doublets, because pollen at an earlier developmental stage, prepared in the same way, lacked this peak. In general, however, the quality of unreduced gamete estimates are dependent upon careful sample preparation to minimize doublets, peak analysis methods that take aggregates into consideration, and the presence of sufficient numbers of nuclei to allow for precise measurements of their number.

5.2.4
Other Considerations/Pitfalls

5.2.4.1 Holokinetic Chromosomes (Agmatoploidy)

Changes in chromosome number may not always be reflected in changes in DNA content, as in plant species with holokinetic chromosomes (i.e. possessing a diffuse kinetochore). The most important implication for FCM assays is the ability of such chromosomes to undergo fusion or fragmentation without any abnormalities in mitotic division. Consequently, variation in chromosome number is associated with changes in chromosome size but genome size remains more or less constant. This phenomenon (so-called agmatoploidy) is most thoroughly documented in two monocotyledonous families, Cyperaceae and Juncaceae, but is known to exist also in other plant groups, such as genera *Chionographis* (Melanthiaceae), *Cuscuta* (Convolvulaceae), *Drosera* (Droseraceae), and *Myristica* (Myristicaceae). Kuta et al. (2004) performed an extensive karyological and FCM study on the three representatives of the genus *Luzula* (Juncaceae), and concluded that despite major fluctuation in the number of somatic chromosomes (e.g. 2n = 12–84 in *L. multiflora*), intraspecific DNA content variation was only negligible (no more than 4.7%). Using FCM alone would obscure much of this chromosomal variability, and only the inclusion of conventional karyological treatment (i.e. chromosome counting) would provide a clear picture of chromosome number.

5.2.4.2 DNA Content Variation within Individuals

One of the strengths of FCM is the relative ease with which endopolyploidy can be detected (see Chapter 15), but endopolyploidy may also bias DNA ploidy estimation in some species. Provided that nuclei with 2C DNA content constitute only a minor fraction, they may easily remain unnoticed on the FCM histogram. Consequently, the DNA ploidy level may be erroneously inferred from the nuclei that have already undergone reduplication and possess the elevated DNA amount.

In addition, bizarre FCM histograms are obtained when nuclei isolated from somatic tissues (i.e. roots, stems, leaves) of some temperate orchids, such as *Dactylorhiza*, *Gymnadenia*, and *Orchis*, are run (J. Suda et al., unpublished data). They are composed of several peaks arranged in an endopolyploidy-like fashion. However, the increase in nuclear DNA content deviates from double and actual peak ratios vary from about 1.35 (*Orchis tridentata*) to 1.95 (various *Ophrys*). This coef-

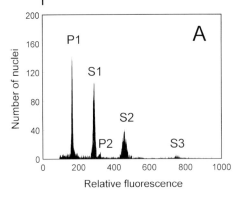
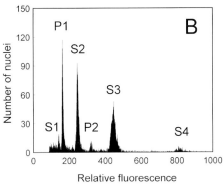

Fig. 5.6 Fluorescence histograms of nuclei isolated from leaf tissue of karyologically verified tetraploid, 2n = 4x = 40 (panel A), and octoploid, 2n = 8x = 80 (panel B), plants of Gymnadenia conopsea (Orchidaceae), with peaks arranged in endopolyploidy-like fashion. The species contains several classes of nuclei showing a non-proportional increase in nuclear DNA content (peak ratios are 1:1.74:3.15 and 1:1.58:2.60 for tetraploid and octoploid cytotypes, respectively). Nuclei were stained with propidium iodide and Pisum sativum cv. Ctirad was used as the internal standard. P1 and P2, nuclei of *Pisum* with 2C and 4C DNA contents, respectively; S1, nuclei of *Gymnadenia* with 2C DNA content; S2–S4, *Gymnadenia* nuclei with elevated DNA contents. S1 peak of the tetraploid cytotype is often inconspicuous and may easily be overlooked. (Partec CyFlow cytometer equipped with a Cobolt Samba™ 532-nm laser; J. Suda et al., unpublished data).

ficient seems to be species-specific and perfectly stable among different tissues. Although true endopolyploidy in orchids has been repeatedly documented (e.g. Fukai et al. 2002; Lee et al. 2004), the reason for the observed anomaly remains unknown. Presumably, it may be related to differential replication (under- or overreplication) of some, possibly heterochromatin, regions (e.g. Nagl 1987). An alternative and non-exclusive hypothesis concerns nuclear hypertrophy triggered by mycorrhizal infection (Barroso and Pais 1990; Peterson et al. 1998). If such underreplication/nuclei hypertrophy is accompanied by a large variation in ploidy levels and a hardly distinguishable peak of 2C nuclei (such as in *Gymnadenia*, Orchidaceae), precise DNA ploidy estimation using FCM becomes very difficult (Fig. 5.6).

5.3
Applications in Plant Systematics

5.3.1
Systematics of Heteroploid Taxa

Perhaps the most obvious application of FCM is in resolving taxonomic complexities in groups with variation in ploidy. Chromosome doubling is widespread in vascular plants (Leitch and Bennett 1997) and has detectable effects on phenotypic and reproductive traits (Levin 2002; Otto and Whitton 2000). Consequently, ploidy is often an important criterion guiding taxonomic delineation of plants. Allopolyploids, which are derived through doubling of hybrid genomes, are often morphologically distinct and reproductively isolated from their diploid progenitors. As a result, allopolyploidy is a well-recognized mode of speciation and allopolyploids are frequently circumscribed at the species level. A well-known example involves two tetraploid goatsbeard species (*Tragopogon mirus*, *T. miscellus*; Asteraceae) colonizing western N. America, which are distinguished taxonomically from their three diploid ancestors (*T. dubius*, *T. pratensis*, and *T. porrifolius*) (Soltis and Soltis 1993). Autopolyploids, formed through the multiplication of genomes from a single species, are less common than allopolyploids (but see Soltis and Soltis 1993, 1999). Because they often resemble their diploid ancestors morphologically, autopolyploids traditionally have not been recognized taxonomically or else are identified at a subspecific level (e.g. Mosquin 1967). However, closer scrutiny of selected autopolyploids and their diploid progenitors has revealed morphological discontinuities and reproductive incompatibilities that often warrant recognition as distinct species (Soltis et al., 2007). In an early example, Hagerup (1927) distinguished tetraploid *Empetrum hermaphroditum* (crowberry, Ericaceae) from its diploid ancestor, *E. nigrum*, to recognize their morphological differences.

FCM offers a rapid and precise method for identifying taxa of different ploidy. It is of particular value in polyploid complexes that show considerable phenotypic plasticity or lack distinct morphological characters (e.g. parasites, many graminoids, and geophytes). Because of the rate at which samples can be processed and screened, FCM is also valuable as an exploratory tool in groups that are in need of taxonomic revision. Table 5.1 summarizes some polyploid alliances native to Central Europe where taxonomic problems may be resolved by flow cytometry.

FCM has recently been used to examine the systematics and taxonomy of a number of plant groups, such as *Vaccinium* subg. *Oxycoccus* (Ericaceae; Suda and Lysák 2001), *Centaurea jacea* aggregate (Asteraceae; Vanderhoeven et al. 2002), *Lamium* subg. *Galeobdolon* (Lamiaceae; Rosenbaumová et al. 2004), and *Saxifraga rivularis* agg. (Saxifragaceae; Guldahl et al. 2005). In each case, extensive morphological variation has led to a dispute concerning species concepts and boundaries. However, new, more robust classifications have been proposed based on cytotype diversity, phenotype, and frequency of inter-ploidy crosses. To resolve heteroploid taxonomies, joint use of FCM and multivariate morphometrics have proven particularly beneficial. Together, these analyses have provided reliable insights into the range and organization of phenotypic variation and have often allowed researchers to distinguish morphologically indiscernible taxa.

In addition to the power of FCM for resolving complex taxonomies at the species level, the technique is at least as powerful below the species rank. At this taxonomic level, morphological discontinuities are often inconsistent and reproductive isolation ambiguous or incomplete, making identifications difficult. Here, ploidy variation may be the most reliable method of distinguishing between taxa.

Table 5.1 Selected closely related polyploid alliances from Central European flora with particular ploidy levels corresponding to different taxonomic entities (compiled from Kubát et al. 2002). Flow cytometry provides a reliable method for distinguishing between taxa as well as their hybrids based on estimating relative nuclear DNA content (DNA ploidy level).

Alliance	Diploid	Tetraploid	Hexaploid	Octoploid
Achillea millefolium agg. (Asteraceae), $n = 9$	*A. setacea* *A. asplenifolia*	*A. collina* *A. tanacetifolia* *A. pratensis*	*A. millefolium*	*A. pannonica*
Alisma plantago-aquatica agg. (Alismataceae), $n = 7$	*A. plantago-aquatica*	*A. lanceolatum*	–	–
Arenaria serpyllifolia agg. (Caryophyllaceae), $n = 10$	*A. leptoclados*	*A. serpyllifolia* *A. patula*	–	–
Ficaria verna agg. (Ranunculaceae), $n = 8$	*F. calthifolia*	*F. verna* ssp. *bulbifera*	–	–
Galium mollugo agg. (Rubiaceae), $n = 11$	*G. mollugo*	*G. album*	–	–
Galium palustre agg. (Rubiaceae), $n = 12$	*G. palustre*	–	–	*G. elongatum*
Glyceria (Poaceae), $n = 10$	*G. declinata* *G. nemoralis*	*G. fluitans* *G. notata*	–	–
Leucanthemum vulgare agg. (Asteraceae), $n = 9$	*L. vulgare*	*L. ircutianum*	*L. margaritae*	–
Myosotis palustris agg. (Boraginaceae), $n = 11$	*M. nemorosa*	–	*M. palustris* *M. brevisetacea*	*M. caespitosa*
Nasturtium officinale agg. (Brassicaceae), $n = 8$	–	*N. officinale*	*N.* ×*sterile*	*N. microphyllum*
Papaver dubium agg. (Papaveraceae), $n = 7$	–	*P. confine* *P. lecoqii*	*P. dubium*	–
Polygonum aviculare agg. (Polygonaceae), $n = 10$	–	*P. arenastrum*	*P. aviculare* *P. rurivagum*	–
Spergularia rubra agg. (Caryophyllaceae), $n = 9$	*S. echinosperma*	*S. rubra*	–	–
Valeriana officinalis agg. (Valerianaceae), $n = 7$	*V. officinalis*	*V. stolonifera*	–	*V. excelsa*
Veronica hederifolia agg. (Plantaginaceae), $n = 9$	*V. triloba*	*V. sublobata*	*V. hederifolia*	–
Viola reichenbachiana agg. (Violaceae), $n = 10$	*V. reichenbachiana*	*V. riviniana*	–	–

Indeed, Walker et al. (2005) found that in *Atriplex halimus* (Chenopodiaceae/APG: Amaranthaceae), the diploid and tetraploid cytotypes corresponded to the subspecies *halimus* and *schweinfurthii*, respectively. A parallel situation exists in *Ulex europaeus* (Fabaceae), with tetraploid subspecies *lactebracteatus* and hexaploid *europaeus* (Misset and Gourret 1996), and *Chamerion angustifolium*, in which diploid and polyploid (4x, 6x) cytotypes correspond to subsp. *angustifolium* and *circumvagum*, respectively (Mosquin 1967). In *Eleocharis palustris* (Cyperaceae), subsp. *vulgaris* may be easily distinguished from the nominate subspecies by its polyploid nature (Bureš et al. 2004a).

5.3.1.1 Detecting Rare Cytotypes

Knowledge of the full range of ploidy variation, including rare cytotypes, is necessary for constructing robust taxonomic treatments and evolutionary interpretations of plant groups. Incidence of rare cytotypes is also important for the purpose of crop and ornamental plant breeding, where rare ploidies may have economic value or may be used as a source of germplasm. However, acquiring such data requires large sample sizes and therefore is particularly amenable to flow cytometry.

In general, the use of FCM has changed our perception of the magnitude of chromosomal and ploidy variation in wild species. For example, Bennert et al. (2005) reported the first case of triploidy in the ancient spore-producing genus *Equisetum* (horsetail, Equisetaceae), which has long been considered cytologically uniform and diploid. Using FCM, triploids have also been reported in *Chamerion angustifolium* (Husband and Schemske 1998), *Cirsium rivulare* (thistle, Asteraceae; Bureš et al. 2004b), *Lamium* subg. *Galeobdolon* (Rosenbaumová et al. 2004), *Draba lonchocarpa* (Brassicaceae; Grundt et al. 2005), *Pimpinella saxifraga* (Apiaceae; K. Mozolová et al., unpublished data), and *Vaccinium uliginosum* agg. (I. G. Alsos et al., unpublished data). Increasing evidence for triploids has led researchers to re-examine the importance of such odd-ploidies in the evolutionary dynamics of these complexes (Husband 2004; Yamauchi et al. 2004).

Similarly, FCM is leading to increased detection of novel high-ploidy cytotypes. Heptaploids in *Rubus ursinus* (Rosaceae; Meng and Finn 2002), octoploids in *Sesleria heufleriana* (Poaceae; Lysák and Doležel 1998), nonaploids in the *Elytrigia repens – E. intermedia* alliance (Poaceae; Mahelka et al. 2005), and hexaploids in *Lythrum salicaria* (purple loosestrife, Lythraceae; Kubátová et al. 2007) are all examples of this. Collectively, the above-mentioned examples indicate that ploidy diversification in the natural environment is much more prolific than previously thought. Indeed, novel cytotype(s) have been detected in virtually all angiosperm alliances subjected to detailed investigation in our laboratories.

5.3.1.2 Phylogenetic Inference

In addition to being used for identifying taxa and detecting rare cytotypes, FCM may be useful for reconstructing relationships and developing phylogenetic hypotheses in taxonomic groups with polyploids. The direction of polyploid evolution is, with few exceptions (e.g. haploid parthenogenesis), unidirectional, leading from diploidy to higher ploidies. Based on this assumption (aneuploid changes notwithstanding), it could be predicted that polyploid genomes have arisen more recently in mixed-ploidy clades. Therefore, ploidy estimates may be used to establish relationships within taxonomic groups. It should, however, be noted that the value of ploidy alone as a phylogenetic marker should be used cautiously and ideally in concert with independent sources of phylogenetic data, as there are numerous reports of polyploidy having multiple origins, even within species (Soltis and Soltis 1993, 1999).

Although little explored as yet, FCM can also be useful for tracing progenitors of polyploid taxa through genome size analysis (Leitch and Bennett 1997). For example, in allopolyploid, triploid banana cultivars, Lysák et al. (1999) demonstrated about 12% difference in DNA amount between their component genomes A (donated from *Musa acuminata*; Musaceae) and B (donated from *M. balbisiana*), and proposed that comparative analysis of genome size may be helpful in identifying putative diploid progenitors of cultivated triploid *Musa* clones. Similarly, hexaploid wheat comprises three different diploid genomes: the D genome (2C = 5.05 pg) seems to contain less DNA than both A (2C = 6.15 pg) and B (2C = 6.09 pg) genomes (Lee et al. 1997). In addition, differences in genome size within *Hieracium* subg. *Pilosella* (Asteraceae; Bräutigam and Bräutigam 1996) can be used as a clue for clarification of species relationships, identification of putative parents and genomic constitution in hybridogenous taxa (J. Suda et al., unpublished data). However, the general efficacy of this application will depend on the extent of homoeologous crossing over and genome restructuring, which may cause the component genomes in the allopolyploid to diverge quantitatively from the ancestral diploids.

5.3.2
Systematics of Homoploid Taxa

An added benefit of FCM over traditional chromosome squashes is that it can be used to differentiate between taxa with the same chromosome number (i.e. homoploid taxa) but different DNA amount (genome size). Nuclear DNA content can vary markedly even among closely related homoploid species (e.g. Cerbah et al. 2001; Zonneveld 2001) while showing considerable uniformity within species (Bennett et al. 2000b; Greilhuber 1998, 2005; Murray 2005). This variation may be associated with variation in chromosome size and amount of non-coding and repetitive DNA, often modified uniformly across all chromosomes (Levin 2002). Regardless of the cause, it can be a marker for phylogenetic relatedness and morphology in some genera (Levin 2002). It should be noted, however, that the methods of FCM necessary for distinguishing related homoploid taxa must be more stringent than when assessing ploidy, as the differences in DNA content can often be small.

Some examples where genome size measures have helped to delineate taxonomic categories include species of *Agapanthus* (Agapanthaceae; Zonneveld and Duncan 2003), *Galanthus* (Amaryllidaceae; Zonneveld et al. 2003), *Gasteria*

(Asphodelaceae; Zonneveld and van Jaarsveld 2005), *Lactuca* (lettuce, Asteraceae; Doležalová et al. 2002; Koopman 2000), and *Petunia* (Solanaceae; Mishiba et al. 2000), subspecies of *Crepis foetida* (Asteraceae; Dimitrova et al. 1999), sections of *Helleborus* (Ranunculaceae; Zonneveld 2001), and subgenera of *Equisetum* (Obermayer et al. 2002).

5.4
Applications in Plant Ecology and Evolutionary Biology

5.4.1
Spatial Patterns of Ploidy Variation

Since its initial introduction to evolutionary biology, FCM has mostly been used to describe patterns of variation in ploidy within and among natural populations. Traditionally, these studies have been conducted by plant systematists interested in the taxonomic implications of chromosome number and have relied on conventional karyological techniques, which are technically challenging (especially with large chromosome numbers) and time-consuming (e.g. Stuessy et al. 2004). With FCM, ploidy variation can now be surveyed over large spatial scales and involve large sample sizes. For example, recent geographic studies routinely gathered DNA ploidy data from >1000 individuals (e.g. Baack 2004; Burton and Husband 1999; Husband and Sabara 2003). To this degree, FCM has revolutionized the field of cytogeography and is changing our perception of the magnitude of ploidy variation and its dynamic nature in the wild.

The extensive surveys of ploidy facilitated by FCM have fueled a number of research problems in population biology. Researchers have been able to more fully explore the distribution patterns and extent of ecological overlap between diploids and their polyploid derivatives. Specifically, FCM made it possible to better characterize regions of allopatry (Baack 2004; Ohi et al. 2003) as well as contact zones between multiple ploidies (Hardy et al. 2000; Husband and Schemske 1998; Liebenberg et al. 1993; Suda et al. 2004). These results have raised questions about the underlying historical and selective mechanisms maintaining these patterns (Felber-Girard et al. 1996; Johnson et al. 2003; Petit et al. 1999; Renno et al. 1995; van Dijk and Bakx-Schotman 1997). In addition to describing variation across entire geographic ranges, FCM has enabled researchers to map fine-scale distributions of ploidies within individual populations (Husband and Schemske 1998; Keeler et al. 1987 – the first article using FCM in field botany; Suda 2003; Weiss et al. 2002). This work is generally revealing greater cytotype variability and more hybrid cytotypes in natural populations than previously recognized.

5.4.1.1 Invasion Biology
Researchers are using FCM in geographic surveys to understand the determinants of biological invasions. There is some evidence that invasive behavior and

spread of alien species may be positively correlated with ploidy level (Bleeker and Matthies 2005). This hypothesis is being tested by comparing diversity of ploidy variation in native and adventive parts of a species' range. One of the earliest papers to explore this issue using FCM was by Amsellem et al. (2001), although the authors were not able to confirm their working hypothesis concerning ploidy divergence in *Rubus alceifolius*. In contrast, *Senecio inaequidens* (Asteraceae; Lafuma et al. 2003) and *Lythrum salicaria* (Kubátová et al. 2007) both show more diversity in their native ranges. The geographic pattern of ploidy variation is even more dramatic in the eastern Asian genus *Reynoutria* (Polygonaceae). The number of chromosomal races found in primary versus secondary areas was as follows: four to one in *R. japonica* ssp. *japonica*, three to three (but with only one corresponding) in *R. sachalinensis*, and one to three in *R.* ×*bohemica* (Mandák et al. 2003). The latter two species thus vividly exemplify rapid genesis of novel cytotypes in the territory of their secondary distribution. Without FCM, however, much of this variation would remain concealed.

5.4.2
Evolutionary Dynamics of Populations with Ploidy Variation

The increased number of reports concerning mixed-ploidy populations as a result of FCM has generated much interest in the evolutionary dynamics of polyploidy. These populations are of scientific value because they provide conditions similar to the early stages of polyploid evolution. When polyploids first arise, they will by necessity occur as rare cytotypes in diploid populations. Although the polyploids in existing populations may have diverged somewhat from their original form, mixed populations still offer opportunities to study interactions between cytotypes through competition for abiotic resources, pollinator behavior, and mating (Petit et al. 1999). In addition, mixed populations are of interest because, in general, the co-existence of two or more ploidies (with strong postzygotic incompatibilities) runs counter to most theoretical predictions (Felber 1991; Levin 1975) and therefore offers insights into the mechanisms of sympatric speciation.

Research on polyploid evolution using mixed populations has focused on two processes: (i) formation of new polyploids from diploids, and (ii) establishment of polyploids, both of which have benefited from FCM. Studies of polyploid formation have addressed primarily the mechanisms by and rates at which new polyploids are produced. Few estimates for natural populations exist but generally this work involves screening the ploidy of many seed offspring from natural or controlled pollinations to test indirectly for unreduced gamete production and assess the frequency of new polyploids generated from diploids (Husband 2004). Crosses between diploid plants will sometimes yield triploids and occasionally tetraploids, presumably formed through the union of unreduced gametes. Crosses between triploids and diploids can in turn yield tetraploids, suggesting that triploid hybrids can produce gametes with one, two or three chromosome sets and thereby facilitate the formation of tetraploids (triploid bridge; Ramsey and Schemske 1998). Rates of polyploid formation may also be predicted indirectly from the frequency of unreduced gamete production, which can be estimated with FCM (see Section 5.4.4 and Chapter 6).

FCM has also been used to explore the evolutionary forces governing the establishment of polyploids once they have been formed. The single largest evolutionary force acting against polyploid persistence is minority cytotype exclusion, which is a reproductive disadvantage operating against rare cytotypes (Husband 2000; Levin 1975). Several different mechanisms may counter this negative effect on polyploids, such as assortative mating through flowering time divergence (Bretagnolle and Thompson 1995; Petit et al. 1997), ecological differentiation (Johnson et al. 2003; Suda et al. 2004; J. Ramsey, unpublished data), pollinator fidelity (Segraves and Thompson 1999), self-fertilization or clonal growth (Quarin et al. 2001), and competitive superiority of polyploids (Baack 2005). One of the most comprehensive research programs on the evolutionary dynamics of mixed-ploidy populations has involved the plant, *Chamerion angustifolium* (as summarized by Husband and Sabara 2003). In a suite of studies, assortative mating between diploids and tetraploids in this species was assessed by measuring five pre-zygotic and two post-zygotic reproductive isolating barriers in the zone of sympatry. Of these, FCM was used to address the role of spatial isolation (Husband and Sabara 2003; H. Sabara and B. C. Husband, unpublished data), flowering asynchrony (Husband and Schemske 2000), pollinator fidelity (Husband and Sabara 2003; Husband and Schemske 2000), gametic competition (Husband et al. 2002), and selection against triploid hybrids (Burton and Husband 2000). These detailed studies of polyploid dynamics would not be possible without FCM.

5.4.3
Ploidy Level Frequencies at Different Life Stages (Temporal Variation)

In general, screening different life stages for ploidy has the potential to identify specific stages where ploidy-level differences may impact fitness. A number of studies have shown discrepancies between 2n gamete production (based on pollen size and morphology) and occurrence of polyploids in progeny (e.g. El Mokadem et al. 2002). When polyploid frequencies in progeny are different from those predicted by unreduced gamete production, one possibility is a favoring or a filtering out of polyploids, either pre- or post-zygotically. While such results may provide insights about polyploidy origin and establishment, they depend on the reliable estimation of unreduced gamete production, which can be improved using FCM (see Section 5.4.4).

Studies can potentially be carried out in natural populations, screening seed crop, seedlings and adults to make inferences about life stage-specific ploidy fitnesses. It may be difficult to make comparisons between developmental stages in which nuclei numbers are low or technically difficult to extract (e.g. comparing unreduced gamete production rates to expected ploidy ratios in any stage prior to seedling), but not necessarily impossible. In many species, screening seeds for DNA ploidy level is feasible, although depending on size, bulking of samples may be required (Krahulcová and Suda 2006; Śliwińska et al. 2005). In *Malus cor-*

onaria (Rosaceae), DNA ploidy can be assigned using half of a seed (Fig. 5.3), allowing the other half to be used for genetic tests (P. Kron and B. C. Husband, unpublished data). Work is currently underway to test for differences in ploidy frequencies in seeds and seedlings, and strong differences have been detected between frequencies of polyploids, hybrids and parthenogenic haploids in seeds and mature trees.

5.4.4
Reproductive Pathways

Changes in ploidy can arise through processes occurring at various stages in the reproductive pathway, including the production of unreduced gametes, asexual seed production, and hybridization. FCM has provided insights into these processes by allowing for the screening of nuclei of differing ploidies at various life stages (gametes, seeds, seedlings, and mature plants), and for doing so across multiple individuals in populations. It is important to keep in mind, however, that with multiple and interacting processes generating ploidy changes, offspring ploidy alone may be insufficient to identify the actual reproductive pathway. For example, in species that produce both unreduced gametes and asexual seeds, the latter may develop from reduced or unreduced egg cells and have the same or half of the DNA content of the mother (e.g. *Rubus* spp.; Einset 1951). Similarly, hybrids between heteroploids, though easily detectable based on DNA content if they arise from the union of two reduced gametes, may remain unrecognized when one gamete is reduced and one is unreduced. For these reasons, FCM screening of progeny to infer reproductive pathways is ideally supplemented with the use of genetic markers, especially when two or more modes of asexual seed production, hybridization and unreduced gamete production are present.

5.4.4.1 Unreduced Gametes and Polyploidy
Polyploids frequently arise sexually via the production of unreduced gametes (Bretagnolle and Thompson 1995; Ramsey and Schemske 1998). The ability to identify plants that produce unreduced gametes, and to estimate the frequency of their production, is therefore useful both in a horticultural context (e.g. in generating new varieties; Owen et al. 1988) and in the study of polyploid ecology and evolution in natural populations (Bretagnolle and Thompson 1995; Thompson and Lumaret 1992).

There are relatively few studies in which the nuclear DNA content of pollen was directly measured using FCM, and none involving ovules. With one exception (Jacob et al. 2001), all pollen studies reported the presence of some pollen that was at least potentially 2n (Bino et al. 1990; Boluarte 1999; Owen et al. 1988; Pan et al. 2004; Pichot and Maâtaoui 2000; Sugiura et al. 1998, 2000; van Tuyl et al. 1989). A traditional approach to estimating unreduced pollen production is based on pollen morphology (e.g. size and pore number; reviewed by Bretagnolle and Thompson 1995), but variation in these traits is not necessarily ploidy-based (e.g. Dajoz et al. 1995). When ploidy–morphology associations are demonstrated, they tend to be imperfect, with discrepancies between frequency estimates based on FCM and morphology (Pichot and Maâtaoui 2000; Sugiura et al. 2000; van Tuyl et al. 1989). It is not clear which approach would yield more accurate results for any given system. Sugiura et al. (2000), by sorting *Diospyros* (Ebenaceae) pollen by size and then measuring each group's DNA content with FCM, showed that there was size overlap between two ploidy groups, suggesting that nuclear fluorescence discriminates ploidy better than size. Similarly, Owen et al. (1988) found that the larger size class of pollen (separated on the basis of forward scatter) contained some pollen with the lower (1C) fluorescence. Conversely, van Tuyl et al. (1989) suggested that pollen preparation for FCM may favor certain pollen sizes, and thereby bias the FCM estimate, although in this particular study, overlap in size classes could also explain the discrepancy found.

In addition to providing estimates of unreduced gamete production, FCM has provided insights into the development of both reduced and unreduced pollen. In binucleate pollen, the normal generative nucleus was found to be resting in the 2C stage, and approximately equal numbers of 1n (vegetative) and 2n (generative) nuclei were detected in several species, including *Lilium* (Liliaceae; Bino et al. 1990; van Tuyl et al. 1989), *Rosa* (Rosaceae; Jacob et al. 2001), and *Chamerion* (P. Kron and B. C. Husband, unpublished data). Trinucleate species have been found consistently to have a single main fluorescence peak corresponding to vegetative and sperm nuclei in the 1C state (Bino et al. 1990 (*Zea*; Poaceae and *Dendranthema*; Asteraceae), Sugiura et al. 1998 (*Diospyros*), Pan et al. 2004 (*Brassica*; Brassicaceae), P. Kron et al., unpublished data (*Rumex*; Polygonaceae)). Pichot and Maâtaoui (2000) found that in the gymnosperm genus *Cupressus*, the single nucleus was in 2C at pollen maturation and in 1C at the binucleate stage following germination.

Pan et al. (2004) extended these insights beyond normal (reduced) pollen development by examining mature trinucleate pollen of *Brassica*, as well as microspores at the uninucleate and binucleate stages. They found that in mature pollen 12% of nuclei had a 2C DNA content, compared to none in the uninucleate microspores, indicating that diploidization of some nuclei had taken place, and that this process was post-meiotic and likely due to failure of the second mitosis.

In plants in which neopolyploids are known to arise through unreduced gamete production, FCM has been used to indirectly estimate the frequency of unreduced gametes by screening somatic tissue of progeny, from both homoploid and heteroploid crosses, for DNA ploidy (Bretagnolle and Thompson 1995; Crespel and Gudin 2003; El Mokadem et al. 2001, 2002). FCM is well suited to the large sample sizes needed to detect low rates of unreduced gamete production, but this general approach can underestimate the production of these gametes in cases where they have lower rates of fertilization, or when polyploid progeny have lower early survival. Conversely, unreduced gamete production may be overestimated when such pollen is favored (El Mokadem et al. 2002). For these reasons, direct measures of the nuclear DNA content of gametes should provide better estimates of unreduced gamete frequencies, while screening of progeny arrays at different life stages has the potential to provide information about how unreduced gamete production translates into the establishment of polyploids in populations.

5.4.4.2 Asexual Seed Production

While it is not always the case, asexually produced offspring can differ in ploidy from their mother, for example, when seeds develop from unfertilized ovules (polyhaploids; Bicknell and Koltunow 2004). FCM can provide a relatively easy way to detect such asexual seeds (Krahulcová et al. 2004; Fig. 5.3). Other forms of apomixis can also be discriminated. In seeds with sufficient endosperm, the DNA content of both endosperm and embryo can be compared, and inferences drawn about the developmental pathways leading to particular ratios (Matzk et al. 2000). This is discussed in greater detail in Chapter 6.

5.4.4.3 Hybridization

Hybridization between heteroploid taxa usually lead to the production of progeny with ploidy levels differing from those of the parents. The use of FCM to screen progeny for ploidy may therefore be a valuable tool in identifying hybrids, particularly when: (i) recently diverged taxa are not readily distinguishable using morphological markers, (ii) karyotyping is impractical (e.g. in species with high chromosome numbers), or (iii) viable hybrids are rare, and thus large numbers of offspring must be screened. FCM has been used to conduct large-scale surveys of natural populations to test for the presence of heteroploid hybrids in *Chamerion* (Husband and Schemske 1998), *Galax* (Diapensiaceae; Burton and Husband 1999), *Empetrum* (Suda 2002), *Polypodium* (Polypodiaceae; Bureš et al. 2003), and *Ranunculus* (Ranunculaceae; Baack 2004). Husband and Sabara (2003) also estimated rates of hybridization directly from FCM surveys of seed progeny collected from natural populations. They reported the frequency of hybrids in progeny arrays and found that it was more than three times higher in seeds than in the adult generation, suggesting strong selection against hybrids in this system (H. Sabara and B. C. Husband, unpublished data).

Hybridization between homoploid taxa can sometimes generate descendants of new ploidy levels (allopolyploids; Kihara and Ono 1926). It is worth noting, however, that even when there is no change in ploidy level, identification of homoploid hybrids is possible if there is adequate divergence (mostly 6–8% at least) in genome size between the parental taxa. Morgan-Richards et al. (2004) used FCM to identify interspecific hybrids between two introduced *Hieracium* subg. *Pilosella* species in New Zealand whose karyotypes could not be distinguished by standard microscopic studies. Using FCM, a high proportion of hybrid individuals, previously overlooked or misidentified due to their weak morphological differentiation, was also found in the hexaploid *Elytrigia repens – E. intermedia* alliance (Mahelka et al. 2005). Similarly, fluorescence values near the mid parent values supported the existence of homoploid crosses in *Alstroemeria* (Alstroemeriaceae; Buitendijk et al. 1997), weedy *Amaranthus* (Amaranthaceae; Jeschke et al. 2003), and *Oxalis* (Oxalidaceae; Emshwiller 2002). Yet another example is the *Dryopteris dilatata* (Dryopteridaceae) alliance comprising two tetraploid taxa in Central Europe. De-

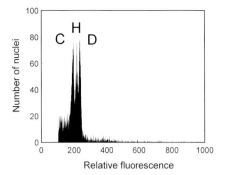

Fig. 5.7 Flow cytometric evidence of interspecific hybridization in homoploid alliance. DAPI-stained nuclei of parental species (C = *D. carthusiana*, D = *D. dilatata*) were analyzed together with those of the putative hybrid (H = *D.* ×*deweveri*). Peak ratios are 1:1.12:1.22. The number of chromosomes (2n = 164) was identical for all the taxa under investigation. (Partec PA II cyto-meter equipped with a mercury arc lamp; R. Holubová et al., unpublished data).

spite their identical (and high) number of chromosomes, the tetraploids differ by ca. 20% in genome size, allowing not only species separation but also reliable detection of hybrid individuals (Fig. 5.7).

It is important to note that hybrid identification may not always be straightforward. An illustrative example is the genus *Cirsium*, which is well known for its disposition to form natural interspecific crosses. Based on DNA amount data for 12 species and the same number of hybrids, Bureš et al. (2004b) concluded that the genome size of crosses may not be located exactly halfway between the values of their putative parents, but usually closer to the species with the smaller genome and, occasionally, even lower than either of the parents. Selective elimination of certain parts of the genome after hybridization has been documented in some species (*Helianthus*, Asteraceae; Baack et al. 2005) and was proposed as a feasible explanation for the observed discrepancy.

5.4.5
Trophic Level Interactions and Polyploidy

Large-scale cytotype surveys, made possible by FCM, have motivated studies of interactions between plants and other trophic levels (symbionts, pollinators, and herbivores). Since polyploidization is often accompanied by changes in morphology, phenology, physiology or secondary compound content, it is likely that plant–animal or plant–fungal interactions within natural communities will also be affected. In fact, differences in cytotype resistance to pathogens or herbivores has long been known and fruitfully exploited in agriculture (see Levin 2002).

Until recently, only one wild species, *Heuchera grossulariifolia* (Saxifragaceae), has been studied in sufficient detail with respect to herbivory (Thompson et al. 1997, 2004). Based on FCM, this species consisted of a mosaic of diploid and tetraploid populations occurring sympatrically or parapatrically over short geographical distances. Several *Heuchera* populations were attacked by the seed parasite *Greya politella* (Lepidoptera), which pollinates host plants while laying eggs in their flowers. Interestingly, tetraploids were found to experience significantly higher levels of herbivory than neighboring diploids. Two other moth species feeding on *Heuchera* also showed cytotype-specific preferences: *Greya piperella* more commonly attacked diploids, while *Eupithecia misturata* preferred tetraploid individuals. In addition, sympatric 2x and 4x *Heuchera* plants differed in their attractiveness for pollinating insects (Segraves and Thompson 1999). These results indicate that interacting organisms may be able to differentiate among plant ploidy levels (even when any morphological difference easily perceptible by human eyes is apparently lacking), and polyploidy may thus have important influences on the structure and diversification of terrestrial communities.

5.5
Future Directions

Flow cytometry has become an essential tool in the suite of contemporary analytical techniques used in evolutionary and systematic research on vascular plants. Thus far, FCM has allowed researchers to estimate ploidy more rapidly and on a larger scale than previously possible. Consequently, it has facilitated research on the taxonomy, phylogenetics, evolutionary ecology and reproductive biology of heteroploid taxa, and has provided novel insights into the magnitude and geographic organization of ploidy variation, and the evolutionary forces acting on this variation. Through these contributions, FCM has and will continue to advance our understanding of more general concepts such as mating system function and evolution, regulation of genetic variation, microevolution, and speciation.

Although FCM has made a large impact on plant population biology, in many ways its full value has yet to be realized. To date, most uses of FCM are those of pattern identification, through descriptive surveys of ploidy variation. This approach will likely continue for some time and can be further developed in new directions. A clear understanding of the fine and broad scale patterns of ploidy variation is available for only a relatively few species. Taxonomic assays exploiting differences in ploidy level and/or genome size at various geographic scales are also much needed. Although this kind of research is still in its infancy, it certainly constitutes an appropriate platform for combining FCM with other modern methods such as molecular markers, multivariate morphometrics and GIS (geographic information systems). When used in this way, FCM may provide a tool for identification when conventional taxonomy cannot easily be applied. For example, ecologists have had difficulty identifying roots collected from a soil sample or pollen collected from bees or the air column. Because of its success on a variety of tissues, FCM may provide one of a number of methods for linking these plant parts to specific species.

In addition, we expect FCM to become increasingly valued for interpreting patterns of variation through analysis of evolutionary processes. For example, FCM offers hope for a greater understanding of the timing and rate of polyploid formation in natural populations. Several mechanisms (sexual and somatic) may account for the formation of polyploids from diploids. Although sexual polyploidization via unreduced gametes is the current consensus by most biologists, the fact remains that the role of unreduced gametes in natural populations has only been studied indirectly. FCM will most likely soon become the leading technique to fill this gap, by providing estimates of ploidy (and unreduced gamete frequency) in pollen and frequencies of polyploid offspring in the earliest stages of seed maturation on a large scale. Another role for FCM involves the mechanisms of adaptive evolution through polyploidy. In an effort to understand the association between ploidy, phenotype and fitness, plant biologists are now focusing on the molecular (Adams et al. 2003; Song et al. 1995) and population biology (J. Ramsey, unpublished data; H. Sabara and B. C. Husband, unpublished data) of newly synthesized (neo)polyploids. Generalities are few at this early stage but some results indicate that polyploids undergo dramatic changes in gene composition, arrangement and expression in the few generations following genome multiplication. FCM analyses of genome size and ploidy in neo- and extant polyploids will provide additional insights into the genomic consequences of polyploidy and may help to explain phenomena such as genome downsizing that appears to occur after genome multiplication (Leitch and Bennett 2004).

Finally, we anticipate further developments in the implementation of FCM. The last decade has seen significant advances in protocols for the analysis of somatic as well as gametic tissues, and for fresh as well as preserved tissue. Development of simple, reliable, and universal protocols as well as methods of long-term sample storage are clearly desirable for field research. A complementary avenue, more relevant to manufacturers, will concern the instrumentation. It is likely that a new generation of low-cost, more compact and more stable field flow cytometers will be developed and launched into the market within a few years. Such a step would undoubtedly accelerate taxonomic and ecological research in geographic regions where current settings may still be prohibitive. Considering the ubiquity of polyploid variation within and among individuals and the breadth of applications of FCM being developed in plant systematics, population biology, and ecology, we believe that a flow cytometry facility will soon become an integral part of almost every plant research center.

Acknowledgments

We would like to acknowledge the Czech Science Foundation, Grant Agency of the Academy of Sciences of the Czech Republic and Grant Agency of the Charles University for funding to JS and PT, and the Natural Science and Engineering Research Council of Canada, Canadian Food Inspection Agency, Canadian Foundation for Innovation and the Canada Research Chair programs for funding to BCH.

References

Adams, K. L., Cronn, R., Percifield, R., Wendel, J. F. **2003**, *Proc. Natl Acad. Sci. USA* 100, 4649–4654.

Amsellem, L., Chevallier, M.-H., Hossaert-McKey, M. **2001**, *Plant Systemat. Evol.* 228, 171–179.

Baack, E. J. **2004**, *Am J. Bot.* 91, 1783–1788.

Baack, E. J. **2005**, *Am. J. Bot.* 92, 1827–1835.

Baack, E. J., Whitney, K. D., Rieseberg L. H. **2005**, *New Phytologist* 167, 623–630.

Barow, M., Meister, A. **2002**, *Cytometry* 47, 1–7.

Barroso, J., Pais, M. S. **1990**, *New Phytologist* 115, 93–98.

Bennert, W., Lubienski, M., Körner, S., Steinberg, M. **2005**, *Ann. Bot.* 95, 807–815

Bennett, M. D., Bhandol, P., Leitch, I. J. **2000a**, *Ann. Bot.* 86, 859–909.

Bennett, M. D., Johnston, S., Hodnett, G. L., Price, H. J. **2000b**, *Ann. Bot.* 85, 351–357.

Bicknell, R. A., Koltunow, A. M. **2004**, *Plant Cell* 16, 228–245.

Bino, R. J., van Tuyl, J. M., de Vries, J. N. **1990**, *Ann. Bot.* 65, 3–8.

Bleeker, W., Matthies, A. **2005**, *Heredity* 94, 664–670.

Boluarte, T. **1999**, *Bulk Segregant Analysis for Anther Culture Response and Leptine Content in Backcross Families of Diploid Potato*, PhD Thesis, Department of Horticulture, Virginia Polytechnic Institute and State University.

Bräutigam, S., Bräutigam, E. **1996**, *Folia Geobotan. Phytotaxon.* 31, 315–321.

Bretagnolle, F., Thompson, J. D. **1995**, *New Phytologist* 129, 1–22.

Buitendijk, J. H., Boon, E. J., Ramanna, M. S. **1997**, *Ann. Bot.* 79, 343–353.

Bureš, P., Tichý, L., Wang, Y.-F., Bartoš, J. **2003**, *Preslia* 75, 293–310.

Bureš, P., Rotreklová, O., Stoneberg Holt, S. D., Pikner, R. **2004a**, *Folia Geobotan.* 39, 235–257.

Bureš, P., Wang, Y.-F., W., Horová, L., Suda, J. **2004b**, *Ann. Bot.* 94, 353–363.

Burton, T. L., Husband, B. C. **1999**, *Heredity* 82, 381–390.

Burton, T. L., Husband, B. C. **2000**, *Evolution* 54, 1182–1191.

Cerbah, M., Mortreau, E., Brown, S., Siljak-Yakovlev, S., Bertrand, H., Lambert, C. **2001**, *Theor. Appl. Genet.* 103, 45–51.

Chiatante, D., Brusa, P., Levi, M., Sgorbati, S., Sparvoli, E. **1990**, *Physiol. Plant.* 78, 501–506.

Crespel, L., Gudin, S. **2003**, *Euphytica* 133, 65–69.

Dajoz, I., Mignot, A., Hoss, C., Till-Bottraud, I. **1995**, *Am. J. Bot.* 82, 104–111.

Dart, S., Kron, P., Mable, B. K. **2004**, *Can. J. Bot.* 82, 185–197.

De Paepe, R., Koulou, A., Pham, J. L., Brown, S. C. **1990**, *Plant Sci.* 70, 255–265.

Dimitrova, D., Ebert, I., Greilhuber, J., Kozhuharov, S. **1999**, *Plant Systemat. Evol.* 217, 245–257.

Doležalová, I., Lebeda, A., Janeček, J., Číhalíková, J., Křístková, E., Vránová, O. **2002**, *Genet. Resources Crop Evol.* 49, 383–395.

Doležel, J. Bartoš, J. **2005**, *Ann. Bot.* 95, 99–110.

Einset, J. **1951**, *Am. J. Bot.* 38, 768–772.

El Mokadem, H., Crespel, L., Meynet, J., Gudin, S., Jacob, Y. **2001**, *Acta Horticult.* 547, 289–296.

El Mokadem, H., Crespel, L., Meynet, J., Gudin, S. **2002**, *Euphytica* 125, 169–177.

Emshwiller, E. **2002**, *Ann. Bot.* 89, 741–753.

Felber, F. **1991**, *J. Evol. Biol.* 4, 195–207.

Felber-Girard, M., Felber, F., Buttler, A. **1996**, *New Phytologist* 133, 531–540.

Fukai, S., Hasegawa, A., Goi, M. **2002**, *HortScience* 37, 1088–1091.

Gregory, T. R. **2005**, *Ann. Bot.* 95, 133–146.

Gregory, T. R., Hebert, P. D. N. **1999**, *Genome Res.* 9, 317–324.

Greilhuber, J. **1998**, *Ann. Bot.* 82 (Suppl. A), 27–35.

Greilhuber, J. **2005**, *Ann. Bot.* 95, 91–98.

Grundt, H. H., Obermayer, R., Borgen, L. **2005**, *Bot. J. Linn. Soc.* 147, 333–347.

Guldahl, A. S., Gabrielsen, T. M., Scheen, A. C., Borgen, L., Steen, S. W., Spjelkavik, S., Brochmann, C. **2005**, *Flora* 200, 207–221.

Hagerup, O. **1927**, *Dansk Botanisk Arkiv* 5, 1–17.

Hardy, O. J., Vanderhoeven, S., de Loose, M., Meerts, P. **2000**, *New Phytologist* 146, 281–290.

Hiddeman, W., Schumann, J., Andreef, M., Barlogie, B., Herman, C. J., Leif, R. C., Mayall, B. H., Murphy, R. F., Sandberg, A. A. **1984**, *Cytometry* 5, 445–446.

Hopping, M. E. **1993**, *NZ J. Bot.* 31, 391–401.
Husband, B. C. **2000**, *Proc. Roy. Soc. Lond. Series B – Biol. Sci.* 267, 217–223.
Husband, B. C. **2004**, *Biol. J. Linn. Soc.* 82, 537–546.
Husband, B. C., Sabara, H. A. **2004**, *New Phytologist* 161, 703–713.
Husband, B. C., Schemske, D. W. **1998**, *Am. J. Bot.* 85, 1688–1694.
Husband, B. C., Schemske, D. W. **2000**, *J. Ecol.* 88, 689–701.
Husband, B. C., Schemske, D. W., Burton, T. L., Goodwillie, C. **2002**, *Proc. Roy. Soc. Lond. Series B – Biol. Sci.* 269, 2565–2571.
Hülgenhof, E., Weidhase, R. A., Schlegel, R., Tewes, A. **1988**, *Genome* 30, 565–569.
Jacob, Y., Priol, V., Ferrero, F. **2001**, *Acta Horticult.* 547, 383–385.
Jarret, R. L., Oziasakins, P., Phatak, S., Nadimpalli, R., Duncan, R., Hiliard, S. **1995**, *Genet. Resources Crop Evol.* 42, 237–242.
Jeschke, M. R., Tranel, P. J., Rayburn, A. L. **2003**, *Weed Sci.* 51, 1–3.
Johnson, M. T. J., Husband, B. C., Burton, T. L. **2003**, *Int. J. Plant Sci.* 164, 703–710.
Johnston, J. S., Bennett, M. D., Rayburn, A. L., Galbraith, D. W., Price, H. J. **1999**, *Am. J. Bot.* 86, 609–613.
Keeler, K. H., Kwankin, B., Barnes, P. W., Galbraith, D. W. **1987**, *Genome* 29, 374–379.
Kihara, H., Ono, T. **1926**, *Zeit. Zellforsch. Mikroskop. Anat.* 4, 475–481.
Koopman, W. J. M. **2000**, *Euphytica* 116, 151–159.
Krahulcová, A., Suda, J. **2006**, *Biol. Plant.* 50, 457–460.
Krahulcová, A., Papoušková, S., Krahulec, F. **2004**, *Hereditas* 141, 19–30.
Kubát, K., Hrouda, L., Chrtek, J. Jun., Kaplan, Z., Kirschner, J., Štěpánek, J. (eds.) **2002**, *Key to the Flora of the Czech Republic*, Academia, Praha.
Kubátová, B., Trávníček, P., Bastlová, D., Čurn, V., Suda J. **2007**, *J. Biogeogr.* (in press).
Kuta, E., Bohanec, B., Dubas, E., Vizintin, L., Przywara, L. **2004**, *Genome* 47, 246–256.
Lafuma, L., Balkwill, K., Imbert, E., Verlaque, R., Maurice, S. **2003**, *Plant Systemat. Evol.* 243, 59–72.
Lee, H.-C., Chiou, D.-W., Chen, W.-H., Markhart, A. H., Chen, Y.-H., Lin, T.-Y. **2004**, *Plant Sci.* 166, 659–667.
Lee, J.-H., Yen, Y., Arumuganathan, K., Baenziger, P. S. **1997**, *Theor. Appl. Genet.* 95, 1300–1304.
Leitch, I. J., Bennett, M. D. **1997**, *Trends Plant Sci.* 2, 470–476.
Leitch, I. J., Bennett, M. D. **2004**, *Biol. J. Linn. Soc.* 82, 651–663.
Levin, D. A. **1975**, *Taxon* 24, 35–43.
Levin, D. A. **2002**, *The Role of Chromosomal Change in Plant Evolution*, Oxford University Press, Oxford.
Liebenberg, H., Lubbinge, J., Fossey, A. **1993**, *S. African J. Bot.* 59, 305–310.
Lysák, M. A., Doležel, J. **1998**, *Caryologia* 51, 123–132.
Lysák, M. A., Doleželová, M., Horry, J. P., Swennen, R., Doležel, J. **1999**, *Theor. Appl. Genet.* 98, 1344–1350.
Mahelka, V., Suda, J., Jarolímová, V., Trávníček, P., Krahulec, F. **2005**, *Folia Geobotan.* 40, 367–384.
Mandák, B., Pyšek, P., Lysák, M. A., Suda, J., Krahulcová, A., Bímová, K. **2003**, *Ann. Bot.* 92, 265–272.
Marie, D., Brown, S. C. **1993**, *Biol. Cell* 78, 41–51.
Matzk, F., Meister, A., Schubert, I. **2000**, *Plant J.* 21, 97–108.
Meng, R. G., Finn, C. **2002**, *J. Am. Soc. Hort. Sci.* 127, 767–775.
Mishiba, K.-I., Ando, T., Mii, M., Watanabe, H., Kokubun, H., Hashimoto, G., Marchesi, E. **2000**, *Ann. Bot.* 85, 665–673.
Misset, M. T., Gourret, J. P. **1996**, *Botan. Acta* 109, 72–79.
Morgan-Richards, M., Trewick, S. A., Chapman, H. M., Krahulcová, A. **2004**, *Heredity* 93, 34–42.
Mosquin, T. **1967**, *Evolution* 21, 713–719.
Murray, B. G. **2005**, *Ann. Bot.* 95, 119–125.
Nagl, W. **1987**, *Prog. Bot.* 49, 181–191.
Nsabimana, A., van Staden, J. **2006**, *South African Journal of Botany* 72, 302–305.
Obermayer, R., Greilhuber, J. **2006**, *Plant Systemat. Evol.* 256, 201–208.
Obermayer, R., Leitch, I. J., Hanson, L., Bennett, M. D. **2002**, *Ann. Bot.* 90, 209–217.
Ohi, T., Kajita, T., Murata, J. **2003**, *Am. J. Bot.* 90, 1645–1652.
Otto, S. P., Whitton, J. **2000**, *Annu. Rev. Genet.* 34, 401–437.
Owen, H. R., Veilleux, R. E., Haynes, F. L., Haynes, K. G. **1988**, *Am. Potato J.* 65, 131–139.

Pan, G., Zhou, Y., Fowke, L. C., Wang, H. **2004**, *Plant Cell Rep.* 23, 196–202.
Peterson, R. L., Uetake, Y., Zelmer, C. **1998**, *Symbiosis*, 25, 29–55.
Petit, C., Lesbros, P., Ge, X., Thompson, J. D. **1997**, *Heredity* 79, 31–40.
Petit, C., Bretagnolle, F., Felber, F. **1999**, *Trends Ecol. Evol.* 14, 306–311.
Pichot, C., El Maâtaoui, M. **2000**, *Theor. Appl. Genet.* 101, 574–579.
Price, H. J., Hodnett, G., Johnston, J. S. **2000**, *Ann. Bot.* 86, 929–934.
Quarin, C. L., Espinoza, F., Martinez, E. J., Pessino, S. C., Bovo, O. A. **2001**, *Sex. Plant Reprod.* 13, 243–249.
Ramírez-Morillo, I. M., Brown, G. K. **2001**, *Systemat. Bot.* 26, 722–726.
Ramsey, J., Schemske, D. W. **1998**, *Annu. Rev. Ecol. Systemat.* 29, 467–501.
Renno, J. F., Schmelzer, G. H., De Jong, J. H. **1995**, *Plant Systemat. Evol.* 198, 89–100.
Rosenbaumová, R., Plačková, I., Suda, J. **2004**, *Plant Systemat. Evol.* 244, 219–244.
Roux, N., Toloza, A., Radecki, Z., Zapata-Arias, F. J., Doležel, J. **2003**, *Plant Cell Rep.* 21, 483–490.
Schönswetter, P., Suda, J., Popp, M., Weiss-Schneeweiss, H., Brochmann, C. **2007**, *Mol. Phylogenet. Evol.* 42, 92–103.
Segraves, K. A., Thompson, J. N. **1999**, *Evolution* 53, 1114–1127.
Sgorbati, S., Levi, M., Sparvoli, E., Trezzi, F., Lucchini, G. **1986**, *Physiol. Plant.* 68, 471–476.
Sharpless, T., Traganos, F., Darzynkiewicz, Z., Melamed, M. R. **1975**, *Acta Cytolog.* 19, 577–581.
Śliwińska, E., Zielinska, E., Jedrzejczyk, I. **2005**, *Cytometry* 64A, 72–79.
Šmarda, P., Stančík, D. **2006**, *Plant Biol.* 8, 73–80.
Šmarda, P., Müller, J., Vrána, J., Kočí, K. **2005**, *Biologia* 60, 25–36.
Soltis, D. E., Soltis P. S. **1993**, *Crit. Rev. Plant Sci.* 12, 243–273.
Soltis, D. E, Soltis, P. S. **1999**, *Trends Ecol. Evol.* 14, 348–352.
Soltis, D. E., Soltis, P. S., Schemske, D. W., Hancock, J. F., Thompson, J. N., Husband, B. C., Judd, W. S. **2007**, *Taxon* in press.
Song, K., Lu, P., Tang, K., Osborn, T. C. **1995**, *Proc. Natl Acad. Sci. USA* 92, 7719–7723.
Stuessy, T. F., Weiss-Schneeweis, H., Keil, D. J. **2004**, *Am. J. Bot.* 91, 889–898.
Suda, J. **2002**, *Ann. Bot. Fennici* 39, 133–141.

Suda, J. **2003**, *Nordic J. Bot.* 22, 593–601.
Suda, J., Lysák, M. A. **2001**, *Folia Geobotan.* 36, 303–320.
Suda, J., Trávníček, P. **2006**, *Cytometry* 69A, 273–280.
Suda, J., Malcová, R., Abazid, D., Banaš, M., Procházka, F., Šída, O., Štech, M. **2004**, *Folia Geobotan.* 39, 161–171.
Suda, J., Kyncl, T., Jarolímová, V. **2005**, *Plant Systemat. Evol.* 252, 215–238.
Suda, J., Krahulcová, A., Trávníček, P., Krahulec, F. **2006**, *Taxon* 55, 447–450.
Sugiura, A., Tao, R., Ohkuma, T. **1998**, *HortScience* 33, 149–150.
Sugiura, A., Ohkuma, T., Choi, Y. A., Tao, R. **2000**, *J. Am. Soc. Hort. Sci.* 125, 609–614.
Thompson, J. D., Lumaret, R. **1992**, *Trends Ecol. Evol.* 7, 302–307.
Thompson, J. N., Cunningham, B. M., Segraves, K. A., Althoff, D. M., Wagner, D. **1997**, *Am. Naturalist* 150, 730–743.
Thompson, J. N., Nuismer, S. L., Merg, K. **2004**, *Biol. J. Linn. Soc.* 82, 511–519.
Van Dijk, P., Bakx-Schotman, T. **1997**, *Mol. Ecol.* 6, 345–352.
Van Tuyl, J. M., de Vries, J. N., Bino, R. J., Kwakkenbos, T. A. M. **1989**, *Cytologia* 54, 737–745.
Vanderhoeven, S., Hardy, O., Vekemans, X., Lefèbvre, C., de Loose, M., Lambinon, J., Meerts, P. **2002**, *Plant Biol.* 4, 403–412.
Walker, D. J., Moñino, I., González, E., Frayssinet, N., Correal, E. **2005**, *Bot. J. Linn. Soc.* 147, 441–448.
Weiss, H., Dobeš, C., Schneeweiss, G. M., Greimler, J. **2002**, *New Phytologist* 156, 85–94.
Wendel, J. F. **2000**, *Plant Mol. Biol.* 42, 225–249.
Yamauchi, A., Hosokawa, A., Nagata., H., Shimoda, M. **2004**, *American Naturalist* 164, 101–112.
Zhang, G., Campenot, M. K., McGann, L. E., Cass, D. D. **1992**, *Plant Physiol.* 99, 54–59.
Zonneveld, B. J. M. **2001**, *Plant Systemat. Evol.* 229, 125–130.
Zonneveld, B. J. M., Duncan, G. D. **2003**, *Plant Systemat. Evol.* 241, 115–123.
Zonneveld, B. J. M., van Jaarsveld, E. J. **2005**, *Plant Systemat. Evol.* 251, 217–227.
Zonneveld, B. J. M., Grimshaw, J. M., Davis, A. P. **2003**, *Plant Systemat. Evol.* 241, 89–102.

6
Reproduction Mode Screening

Fritz Matzk

Overview

Investigations into the mode of reproduction are difficult to conduct experimentally in angiosperms because the gametophytic generation comprises just a few cells, and these are deeply embedded within the ovary during a short part of the life cycle. The traditional embryological methods, clearing techniques, chromosome counting, and progeny tests are time consuming and not suitable for large sample sizes. The substitution of light microscopic chromosome counting by flow cytometric ploidy analyses in order to screen for deviating ploidy in progenies which originated from selfing or interploidy crossing has increased the efficiency of mode of reproduction analyses. More detailed conclusions about the reproductive mode can be reached by comparing the ploidy of embryo and endosperm cells. Seeds with a 2C embryo (2C representing nuclear DNA amount; for terminology see Introduction) and 3C endosperm can be shown to have developed sexually; 2C or 3C embryos associated with 4C (autonomous) or 5C (pseudogamous) endosperm indicate unreduced embryo sacs; and 1C or 2C embryos combined with 3C, 4C or 5C endosperm show evidence of parthenogenesis. Until recently, the different tissues from immature seeds had to be separated, but the novel flow cytometric seed screen (FCSS) uses whole dormant seeds. Different events during sporogenesis and embryo and endosperm development can be reconstructed once the DNA contents of embryo and endosperm nuclei of ripe seeds are known. The FCSS is a simple and powerful tool for reproduction mode screening applicable in both monocots and dicots. The advantages and the first significant results are described in this chapter. Flow cytometric analyses of ripe seeds also have the potential to become a valuable tool for other defined purposes.

6.1
Introduction

The alternation between the sporophyte and the gametophyte is an essential part of the plant life cycle, associated in most cases with an alternation of the nuclear

Flow Cytometry with Plant Cells. Edited by Jaroslav Doležel, Johann Greilhuber, and Jan Suda
Copyright © 2007 WILEY-VCH Verlag GmbH & Co. KGaA, Weinheim
ISBN: 978-3-527-31487-4

phases (ploidy). Most angiosperm species reproduce sexually. The sexual reproduction involves the formation of gametes with a reduced chromosome number, and double fertilization for separate embryo and endosperm formation. All reproductive processes, including mega- and micro-sporogenesis and gametogenesis, and double fertilization for embryo and endosperm formation, are regulated by independent genetic controls.

Mutations that change or interrupt the process of sexual reproduction in some way have been identified in several plant species. These include: the formation of unreduced male and/or female gametes in alfalfa (Barcaccia et al. 1995), barley (Finch and Bennett 1979), maize (Barrell and Grossniklaus 2005; Golubovskaya et al. 1992) and potato (Ramanna 1979); fertilization-independent embryo or endosperm formation in *Arabidopsis* (Brassicaceae; Chaudhury et al. 1997; Grossniklaus et al. 1998; Ohad et al. 1996), barley (Hagberg and Hagberg 1980), maize (Tyrnov and Enaleeva 1983) and wheat (Matzk et al. 1995); and defects or the abortion of spores, gametophytes or gametes in *Arabidopsis* (reviewed in Pagnussat et al. 2005). In nature such mutations could be starting material for evolutionary changes.

Although sexual reproduction is dominant in nature, asexual seed formation is also widespread. Apomixis (gametophytic apomixis and adventitious embryony) has been found in at least 220 genera, and of these, 126 are known to use gametophytic apomixis (diplospory and apospory; Carman 1997). Apomictic reproduction results in seeds containing embryos with a maternal genotype (also from highly heterozygous hybrids). Apomeiosis and parthenogenesis are the basic components of gametophytic apomixis. The endosperm develops in most cases after fertilization of the central cell, and only in a few cases autonomously. Most of the apomictic species are facultative apomicts which means that sexual and apomictic processes occur simultaneously. Apomixis has recently become an important topic for both science and seed industry (Jefferson and Bicknell 1996; Vielle-Calzada et al. 1996). New hypotheses about the genetic control of apomixis have been postulated and confirmed (Matzk et al. 2005; van Dijk et al. 1999). Extensive international research programs for harnessing apomixis in crop plants are in progress (reviewed in Matzk et al. 1997). The major benefit of an "apomixis technology" (Spillane et al. 2004) to agriculture would be the widespread use of fixation of hybrid vigor in crop plants.

Fundamental knowledge of the molecular regulation of reproduction in plants has expanded rapidly during the last decade (for comprehensive reviews see Boavida et al. 2005; Drews et al. 1998; Grossniklaus and Schneitz 1998). However, further information is necessary if we are to manipulate the mode of reproduction experimentally. Efficient screening methods of the individual processes of reproduction are an essential prerequisite for successful discrimination of reproductive mutations, and for studies of the evolution, inheritance, and engineering of apomixis. Flow cytometry (FCM) can help us to solve some of these problems.

In this chapter, the nuclear DNA contents of embryo and endosperm cells are represented by C-values (where C is the "Constant" which occurs in multiples in defined tissues of the organisms; see also Chapters 4 and 7). Following the terminology suggested by Greilhuber et al. (2005), the terms "C" and "Cx" are distinguished. In unspecified cases (plants or species in general), "2C" is used to designate DNA content of the unreplicated, non-reduced (diplophasic) chromosome complement ($2C \rightarrow 2n = ?x$). Although embryos, if viewed as the starting point of the life cycle, ought to receive a 2n chromosome number and a DNA 2C-value for presynthetic G_1 nuclei, here the ratio of the DNA content of embryo and endosperm nuclei is used to characterize the mode of reproduction of the mother plant, and thus the seeds are largely viewed as the endpoint of the life cycle (Fig. 6.1). Therefore, compared to the DNA content of the mother, embryos receive a 1C (autonomous development of reduced egg cells), 2C (sexual or apomictic de-

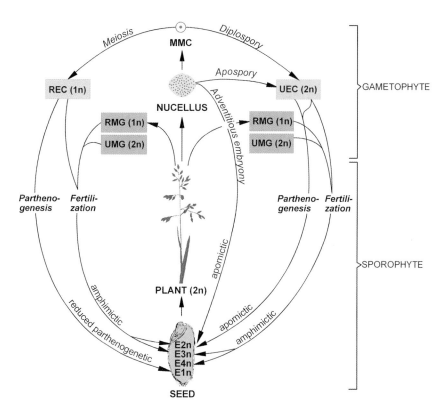

Fig. 6.1 Different pathways of embryo formation within the life cycle of angiosperms (reduced or unreduced gamete formation, and fertilization-dependent or -independent embryo formation) yield different ploidy levels of embryos. Note that the x-level of n is not specified. E1n, haploid embryo arose from reduced egg cell parthenogenetically; E2n, diploid embryo from sexual or apomictic pathway; E3n, triploid embryo from unreduced egg cell fertilized by reduced pollen or reduced egg cell fertilized by unreduced pollen; E4n, tetraploid embryo from unreduced egg cell fertilized by unreduced pollen; RMG, reduced male gametes; UMG, unreduced male gametes; MMC, megaspore mother cell; REC, reduced egg cell; UEC, unreduced egg cell.

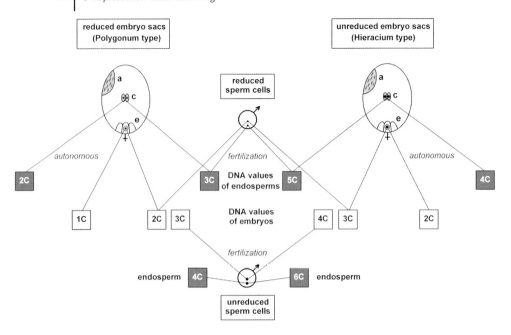

Fig. 6.2 C-values of unreplicated embryo and endosperm nuclei depending on whether the female and/or male gametes were reduced or unreduced and, whether the embryo and/or endosperm developed autonomously or after fertilization. a, antipodals; c, central cell with two polar nuclei; e, egg apparatus with egg cell and two synergids. Figure reprinted from Matzk et al. (2000) with permission.

velopment) or 3C-value (development of unreduced egg cells after fertilization or fertilization of reduced egg cell by unreduced pollen; Fig. 6.2 and Table 6.1). In specified cases, for example the tetraploid *Hypericum perforatum* (Hypericaceae $2n = 4x = 32$), corresponding diplophasic embryo nuclei would receive a "2Cx", "4Cx" and "6Cx" (Tables 6.2 and 6.3). The Cx-values correspond to the chromosomal ploidy of the cells.

6.2
Analyses of the Mode of Reproduction

6.2.1
Traditional Techniques

The serial microdissection of ovaries was the means traditionally employed to assess the mode of reproduction. It is a very time-consuming method, starting with dissection and prefixation of the ovaries at a defined developmental stage, followed by clearing, post-fixation, hydration in an ethanol series, embedding in paraffin or some other media, sectioning with a microtome, and finally staining

Table 6.1 Reproductive pathways theoretically differentiated by differences in the C-values of embryo and endosperm nuclei.

Pathway	C-values[a] embryo + (endosperm)	Embryo sac	Pollen	Egg cell	Central cell	Progenies[b]
1	1 + (2)	Reduced	–	Autonomous	Autonomous	M_I, segregating
2	1 + (3)	Reduced	Reduced	Autonomous	Fertilized	M_I, segregating
3	1 + (4)	Reduced	Unreduced	Autonomous	Fertilized	M_I, segregating
4	2 + (2)	Reduced	Reduced	Fertilized	Autonomous	B_{II}, segregating
5	2 + (3)	Reduced	Reduced	Fertilized	Fertilized	B_{II}, segregating
6	2 + (4)	Unreduced	–	Autonomous	Autonomous	M_{II}, maternal
7	2 + (5)	Unreduced	Reduced	Autonomous	Fertilized	M_{II}, maternal
8	2 + (6)	Unreduced	Unreduced	Autonomous	Fertilized	M_{II}, maternal
9	3 + (4)	Reduced	Unreduced	Fertilized	Fertilized	B_{III}, segregating
10	3 + (5)	Unreduced	Reduced	Fertilized	Fertilized	B_{III}, segregating
11	4 + (4)	Unreduced	Unreduced	Fertilized	Autonomous	B_{IV}, segregating
12	4 + (6)	Unreduced	Unreduced	Fertilized	Fertilized	B_{IV}, segregating

[a] 2C (relating to 2n) of embryo cells corresponds with 2C of somatic cells of the mother plant, and is independent of the real generative ploidy level; endosperm values in parentheses.
[b] For explanation of B_{II}, B_{III}, B_{IV}, M_I and M_{II} see Section 6.2.2.

and viewing all sections under a light microscope. This method is too laborious for routine analyses or screenings of large numbers of individuals.

In Panicoideae, the clearing technique is often sufficient on its own to detect apospory (Herr 1971; Young et al. 1979). Aposporous embryo sacs of several Panicoideae species do not have antipodals, whereas the antipodal apparatus is easily recognizable in meiotic embryo sacs after clearing the ovules (Burson 1997; Chen and Kozono 1994; Sherwood et al. 1994). The procedure involves dissection and fixation of the ovaries or ovules, clearing, and then viewing the ovules with differential interference optics.

The so-called auxin test (Matzk 1991a, 1991b) enables the identification of autonomous embryo formation, and discriminates among individual plants as to their potential for parthenogenetic versus fertilization-dependent embryo formation. This rapid and accurate method works well in pseudogamous Pooideae species. The procedure involves spraying or dipping the inflorescences into an aqueous solution of synthetic auxins at the developmental stage immediately before anthesis. Fifteen days or more after anthesis, the ovaries/caryopses can be examined with a dissecting microscope to determine whether embryos are present (parthenogenesis) or absent (sexual).

The degree of apomictic versus non-apomictic seed formation may be determined by the frequency of maternal or aberrant individuals in the progeny of a heterozygous mother plant. In such progeny tests, the mother is compared with

6 Reproduction Mode Screening

Table 6.2 Divergent reproductive pathways found in 113 populations of *Hypericum perforatum* by FCSS analyses with 50 bulked seeds each (modified from Matzk et al. 2001).

Reproductive type	Cx-values[a] embryo + (endosperm)	Embryo type(s)[b]	Mode(s) of reproduction	No. of populations	Pathways from Table 6.1[c]
1	2 + (3)	B_{II}	Obligate sexual (reduced double fertilized; 2x parents)	2	5
2	4 + (10)	M_{II}	Obligate apomictic (unreduced pseudogamous; 4x parents)	2	7
3	4 + (8 + 10)	M_{II}	Obligate apomictic (endosperm unreduced pseudogamous + autonorrous; 4x parents)	1	6 + 7
4	4 + (6 + 10)	$M_{II} + B_{II}$	Facultative apomictic/sexual (4x parents)	26	5 + 7
5	4 + 6 + (6? + 10)	$M_{II} + B_{II}? + B_{III}$	Facultative apomictic?/sexual?/unreduced double fertilized (4x parents)	27	5? + 7? + 10
6	4 + 6 + (6? + 10 + 15)	$M_{II} + B_{II}? + B_{III}$	Facultative apomictic?/sexual?/unreduced double fertilized (4x + 6x mother and 4x father plants)	7	5? + 7? + 10
7	2 + 4 + (6 + 10)	$M_{II} + B_{II} + M_{I}$	Facultative apomictic?/sexual/reduced parthenogenetic (4x parents)	20	2 + 5 + 7?
8	2 + 4 + 6 + (6 + 10)	$M_{II} + B_{II} + M_{I} + B_{III}$	Facultative apomictic?/sexual/reduced parthenogenetic/unreduced double fertilized (4x parents)	17	2 + 5 + 7? + 10

9	$2+4+6+(6+10+15)$	$M_{II} + B_{II} + M_I + B_{III}$	Facultative apomictic?/sexual/reduced parthenogenetic/unreduced double fertilized ($4x + 6x$ mother and $4x$ father plants)	8	$2 + 5 + 7? + 10$
10	$4+5+6+(6?+9+10)$	$M_{II} + B_{II}? + B_{III}$	Facultative apomictic?/sexual?/unreduced double fertilized ($4x$ mother and $2x + 4x$ father plants)	1	$5? + 7? + 10$
11	$2+4+5+6+(6+9+10)$	$M_{II} + B_{II} + M_I + B_{III}$	Facultative apomictic?/sexual/reduced parthenogenetic/unreduced double fertilized ($4x$ mother and $2x + 4x$ father plants)	2	$2 + 5 + 7? + 10$

[a] Endosperm values in parentheses; in some cases the small 6Cx endosperm peak may be superimposed with a high 6Cx embryo peak, and therefore, the sexual path remains unproven (?) unless an additional 2Cx embryo peak reveals reduced embryo sacs. The Cx-values correspond with ploidy.
[b] For explanation see Section 6.2.2.
[c] ?, confirmation of this pathway is only possible by single seed analyses (see Table 6.3). Pathways with autonomous endosperm formation in reduced embryo sacs (4Cx), theoretically differentiated in Table 6.1, are not considered here because they were never found in large series of single seed analyses.

Table 6.3 FCSS analyses with single seeds for a further specification of the facultative reproduction mode determined previously with bulked seeds in *Hypericum perforatum*.

Reproductive type Table 6.2	Seed samples[a]		Cx-values of embryo + (endosperm)	Facultative mode of reproduction	Pathways from Table 6.1
5	50 bs:		4 + 6 + (6? + 10)	Facultative apomictic?/sexual?/unreduced double fertilized	5? + 7? + 10
	50 ss:	Either	4 + (6) and 6 + (10)	Sexual and aposporous	5 + 10
		or	4 + (10) and 6 + (10)	Apomictic and aposporous	7 + 10
		or	4 + (6) and 4 + (10) and 6 + (10)	Sexual and apomictic and aposporous	5 + 7 + 10
7	50 bs:		2 + 4 + (6 + 10)	Facultative apomictic?/sexual/reduced parthenogenetic	2 + 5 + 7?
	50 ss:	Either	2 + (6) and 4 + (10)	Reduced parthenogenetic and apomictic	2 + 7
		or	2 + (6) and 4 + (6) and 4 + (10)	Reduced parthenogenetic, sexual and apomictic	2 + 5 + 7
8	50 bs:		2 + 4 + 6 + (6 + 10)	Facultative apomictic?/sexual/reduced parthenogenetic/unreduced double fertilized	2 + 5 + 7? + 10
	50 ss:	Either	2 + (6) and 4 + (6) and 6 + (10)	Reduced parthenogenetic, sexual and aposporous	2 + 5 + 10
		or	2 + (6) and 4 + (10) and 6 + (10)	Reduced parthenogenetic, apomictic and aposporous	2 + 7 + 10
		or	2 + (6) and 4 + (6) and 4 + (10) and 6 + (10)	Reduced parthenogenetic, sexual, apomictic and aposporous	2 + 5 + 7 + 10

[a] bs, bulked seeds; ss, single seeds.

the progeny plants with respect to morphological characters, chromosome numbers, and/or molecular markers.

Both the sexual and apomictic pathways result in progenies having a chromosome number identical to the mother. Different ploidy levels of the embryos or progeny plants may be a consequence of deviations from the regular sexual or apomictic seed formation (Fig. 6.1). Plants originating via the sexual and apomictic pathways can be discriminated from those derived from reduced parthenogenetic and fertilized unreduced gametes using traditional microscopic chromosome counting. Light microscopic analyses are very laborious, but FCM is an alternative and efficient replacement.

6.2.2
Ploidy Analyses of Progenies Originating from Selfing or Crossing

In most FCM investigations, nuclear suspensions are produced by chopping leaves from young plantlets with a sharp razor blade in an isolation buffer, which is designed to stabilize the nuclei. After filtration, the nuclei are transferred into staining buffer and the nuclear DNA content is estimated using a flow cytometer (Bennett and Leitch 1995; Bharathan et al. 1994; Doležel et al. 1998). To compare the ploidy of young progeny plants with that of adult mother plants, the youngest leaves or any other fresh somatic tissue from the mother may be used.

For classification of the progeny plants, the following nomenclature system (adapted and expanded from Rutishauser (1967) and Matzk et al. (2005)) will be used: a B_{II} plant results from fertilization of a reduced egg cell by a reduced pollen (sexual, n + n, where B stands for "bastard" = hybrid); a B_{III} plant results from fertilization of an unreduced egg cell by a reduced male gamete (maternal B_{III}, 2n + n), or from a reduced egg cell fertilized by an unreduced male gamete (paternal B_{III}, n + 2n); a B_{IV} plant results from fertilization of an unreduced egg cell by an unreduced pollen (2n + 2n); an M_I plant results from the parthenogenetic development of a reduced egg cell (n + 0, where M stands for "maternal"); and an M_{II} plant results from the parthenogenetic development of an unreduced egg cell (2n + 0, apomictic).

6.2.2.1 Identification of B_{III}, B_{IV} and M_I Individuals after Selfing or Intraploidy Pollinations

B_{III} plants and M_I plants are characterized by increased (mother + $\frac{1}{2}$ father or $\frac{1}{2}$ mother + father) or decreased ($\frac{1}{2}$ mother) ploidy, respectively. In self progenies it is not possible to determine whether the B_{III} plants originate from unreduced male or from unreduced female gametes. Also, B_{II} and M_{II} individuals cannot be discriminated by simple ploidy analyses, as both have the same chromosome number (nuclear DNA content) as the mother.

The frequency of B_{III} and M_I plants was determined by FCM ploidy analyses of progenies originating from *in vitro* regenerated mother plants of *Hypericum perforatum* (Brutovská et al. 1998), and in self progenies of *Poa pratensis* (Poaceae; Huff and Bara 1993). B_{IV} plants (unreduced egg cell fertilized by unreduced pollen) do not often occur; a few individuals were identified by FCM in *P. pratensis* (Huff and Bara 1993; Matzk et al. 2005) and *H. perforatum* (F. Matzk, unpublished data). Similarly, parthenogenetically developed individuals (M_I: n + 0) were discriminated by ploidy analyses from the actual F_1 hybrids (B_{II}: n + n) of interspecific crosses in *Actinidia* (Actinidiaceae; Chat et al. 1996). In this case, the phenotypic identification of the F_1 was not possible due to the small morphological differences between the parents.

6.2.2.2 Crossing of Parents with Different Ploidy or with Dominant Markers

Discrimination between B_{II} and M_{II} plants, as well as between B_{III} hybrids originating from unreduced male or unreduced female gametes, would be possible by crossing parents with different ploidy, or using pollinators with homozygous dominant markers. However, the prerequisites for this approach, including the availability and the crossability of suitable parents, are not always realized. Seed mortality frequently occurs in interploidy crosses, and for detailed analyses, a combination of different techniques would be necessary.

After interploidy crosses, B_{II} plants are characterized by an intermediate 2C-value of the parents or by a dominant marker of the father, and M_{II} plants can be recognized by a 2C-value identical to that of the mother or by the lack of a marker belonging to the father. Discrimination between these two types of progeny is an essential prerequisite for apomixis research.

Since obligate sexual plants are frequently missing in apomictic species, crosses between related sexual and apomictic species with different chromosome numbers have been used for quantification of the different reproductive pathways in facultative apomictic plants, and for analyses of the inheritance of apomixis. Such approaches may be problematic, because free recombination can be depressed by the selection of functional gametes or seeds after interspecific pollination, and by meiotic barriers in the F_1 plants (Bicknell and Koltunow 2004; Matzk et al. 2005).

Five pathways of embryo formation were found by FCM ploidy analyses in *Hieracium rubrum* (hexaploid) after crossing with the related tetraploid *H. pilosella* (Asteraceae; Krahulcová et al. 2004). A comparison of these data with results obtained by a novel screening method (cf. Section 6.3) for the same apomictic plants is in progress (see Krahulcová et al. 2004). Even discrimination between maternal and paternal B_{III} hybrids was possible in *Hypericum* using interploidy crosses and chromosome counting by light microscopy (Lihová et al. 2000), and with FCM (Brutovská et al. 1998).

Interploidy crosses with subsequent FCM ploidy analyses were also applied in order to screen for unreduced pollen producers in *Dactylis glomerata* (Poaceae; Maceira et al. 1992), *Vaccinium* (Ericaceae; Ortiz et al. 1992), and *Lotus* (Fabaceae; Negri et al. 1995) species. A high percentage of tetraploid, non-maternal progenies originated from a cross between a $4x$ mother and a $2x$ father plant, indicating that the male parent has a high capacity for unreduced pollen formation.

Plants with the homozygous dominant alleles for blue aleurone were used as markers to indicate normal sexual reproduction in wheat (Morrison et al. 2004). By means of markers, which are at present frequently replaced by molecular markers, parthenogenetic development can be excluded or maternal identity confirmed.

DNA fingerprinting (RAPD and AFLP markers) was used in alfalfa to verify the rare occurrence of the complete apomictic pathway in a mutant with a high capacity for diplospory (Barcaccia et al. 1997). In *Hypericum perforatum* the maternal plants (M_{II}) were discriminated from segregants (B_{II}, B_{III} and M_I) by AFLP, RAPD and RFLP fingerprints (Arnholdt-Schmitt 2000; Halušková and Čellárová 1997). By combination of DNA fingerprinting with light microscopic chromosome counting it has also been possible to differentiate between B_{II} and B_{III} hybrids or M_I and M_{II} plants in *H. perforatum* (Mayo and Langridge 2003). FCM ploidy analyses, in combination with RAPD fingerprinting, were applied for a detailed determination of the genetic origin of aberrant progenies derived from facultatively apomictic mother plants of *Poa pratensis* (Huff and Bara 1993). Such detailed studies were laborious, since molecular markers had to be developed, and a relatively complicated procedure for FCM (involving young, fully expanded leaves which were subjected to washing, chopping in buffer, centrifugation and re-suspension in staining buffer) was used for analyzing both mother and progeny plants.

6.2.3
Flow Cytometric Analyses of the Relative DNA Content of Microspores or Male Gametes

Direct screening for unreduced pollen producers should be applicable using FCM determination of the relative DNA content of microspores or sperm cells (Bino et al. 1990). This approach would not be as time-consuming as the interploidy crosses mentioned above. Various procedures for the preparation of anthers or pollen for FCM studies have been described, and include: direct chopping, mechanical crushing, osmotic shocking, and ultrasonic treatment (Pan et al. 2004). However, analyses of the DNA content of male spores or gametes remain problematic since the isolation of identical cell types and cell cycle phases is not easy (see also Chapter 5). The DNA content of haplophasic, unreplicated sperm nuclei should be 1C, although a small number are sometimes 2C (G_2 phase). In *Lilium* (Liliaceae), for instance, the reduced but replicated generative nucleus of mature pollen has a 2C and only the vegetative nucleus a 1C-value (van Tuyl et al. 1989). Using whole anthers at the immature developmental stage, the occurrence of additional 2C and 4C (G_1 and G_2) peaks originating from the somatic tapetum cells might be expected. As a consequence of endopolyploidization and endomitosis, even higher C-values may occur.

When the actual cell type and cell cycle phase are not carefully checked, the experiments may produce misleading results, which is what happened in Feulgen-stained sections of anthers in buffelgrass (*Pennisetum ciliare*, Poaceae; Sherwood 1995). The DNA contents of the pollen mother cell, dyad, tetrad, 1-nucleate pollen, generative cell, and persistent tapetum cell nuclei varied between 4C-, 2C- and 1C-values in sexual as well as in apomictic genotypes.

The identification of unreduced male gamete formation by means of FCM analyses of the DNA content of pollen has been successful in interspecific *Lilium* hybrids (van Tuyl et al. 1989) and in *Cupressus dupreziana* (Cupressaceae; Pichot and El Maâtaoui 2000).

6.2.4
The Ploidy Variation of Embryo and Endosperm Depending on the Reproductive Mode

The mode of reproduction in plants is characterized by a specific ploidy ratio of embryo and endosperm cells. Depending on whether the embryo sacs are reduced or unreduced and whether or not the egg and/or central cells are fertilized (by reduced or unreduced male gametes), different ploidy levels occur in the nuclei of seed cells, as shown in Fig. 6.2. Diploid sexual plants form a diploid embryo and triploid endosperm. Triploid embryos originating from unreduced male or unreduced female gametes are combined with tetraploid or pentaploid endosperm, respectively. A pentaploid endosperm is evidence for unreduced embryo sacs. Other deviations from the normal sexual pathway may also alter the ploidy of embryo and/or endosperm cells. Therefore, reproductive events can be reconstructed from the interrelationship between the DNA content of the nuclei in embryo and endosperm cells.

For detailed analysis of the mode of reproduction in *Hypericum perforatum*, *Boechera* (formerly *Arabis*) *holboellii* (Brassicaceae) and *Ranunculus auricomus* (Ranunculaceae), light microscopic chromosome counting of embryo and endosperm cells has already been carried out (Böcher 1951; Noack 1939; Rutishauser 1967). The ploidy of embryo and endosperm cells can be determined more quickly, however, using FCM both in angiosperms (Grimanelli et al. 1997; Kowles et al. 1994; Naumova et al. 1993) and gymnosperms (Pichot et al. 1998; Wyman et al. 1997). In these studies, fresh ovaries or immature seeds (mitotically active tissues) were used, the embryo and/or endosperm tissues were dissected, and their ploidy was determined separately. The procedure of dissecting the tissues was very laborious and could only be performed during a short period of the plant's development, so data were obtained only for separate processes of reproduction. In cases where complete fresh ovules are prepared for FCM analyses (Naumova et al. 1993), nuclei from three different tissues are involved (embryo, endosperm, and maternal somatic cells) and the B_{II} or M_{II} embryo nuclei cannot be discriminated from nuclei of the maternal tissue.

6.3
A Recent Innovative Method: the Flow Cytometric Seed Screen (FCSS)

The recently developed flow cytometric seed screen (FCSS) allows the reconstruction of reproductive pathways from mature (dry) seeds, that is, whether reduced or unreduced female and/or male gametes, a zygotic or parthenogenetic embryo development, and a pseudogamous or autonomous endosperm formation were involved in the seed formation (Matzk et al. 2000). The FCSS is suitable to screen for mutants with deviations from the normal process of sexual reproduction, to classify the mode of reproduction in natural populations or in species with hitherto unknown breeding systems, to identify pure sexual or apomictic genotypes from facultative apomictic species, and to analyze the inheritance of individual reproductive processes.

A prerequisite for the reconstruction of the mode of reproduction from mature seeds is the survival of endosperm nuclei in ripe seeds. It was shown (Matzk et al. 2000) that in both monocots and dicots only the aleuron has intact cell nuclei belonging to the endosperm. The seed-coat representing maternal tissues (testa, tegmen and pericarp of caryopses) does not contain any cell nuclei at this stage. Nor do the lemma and palea, which may be directly connected with the seeds of Pooideae species. Their resorption by apoptosis is complete at maturity (Matzk et al. 2000, 2001).

Mature and dormant seeds of monocots and dicots are suitable for FCM analyses without any specific pretreatment and without separation of the target tissue. Thus, a simple and powerful screening method is available, which can simultaneously identify different reproductive pathways of angiosperms based on the proportional DNA contents of embryo and endosperm nuclei, irrespective of the real ploidy level. A flow cytometer as normally used for routine DNA ploidy estimation in plants is sufficient to perform such reproduction mode analyses. No specific requirements are needed. The different reproductive pathways, which theoretically can be discriminated by the C-values of embryo and endosperm nuclei, are shown in Table 6.1 (cf. also Fig. 6.2).

Differentiation between embryo and endosperm DNA peaks within the histograms is possible in most cases. The number of nuclei is much higher in the embryo than in the aleuron layer, resulting in high embryo but only very small endosperm peaks (see Fig. 6.3). Often endopolyploidization occurs in endosperm as well as in embryo cells, and leads to additional peaks with multiple duplications of the basic peak values. The appearance of endoreduplication is variable between both different species and tissues within one species (see Chapter 15). Examples of a species-specific and tissue-specific regulation of endoreduplication include *Arabidopsis thaliana*, with multiple peaks from leaf nuclei and single peaks from embryo and endosperm nuclei (non-replicated), and *Zea mays* and *Tripsacum dactyloides* (both Poaceae), with the reverse appearance of a single peak from leaf nuclei (non-replicated) and multiple peaks from embryo and endosperm nuclei (endopolyploidization). The occurrence of endoreduplication varies even between the cell types within the embryo of some species (Bino et al. 1993). However, the additional peaks of embryo and endosperm cells are not relevant with respect to the characterization of the mode of reproduction.

6.3.1
Advantages and Limitations of the FCSS

The FCSS does not require prior genetic or molecular information or a specific constitution of the plants. The different events of sporogenesis, embryo and endosperm formation can be reconstructed simultaneously from mature seeds of virtually any plant, line, natural population or breeding stock. This mode of reproduction screen considers only the really functional gametes of the parents.

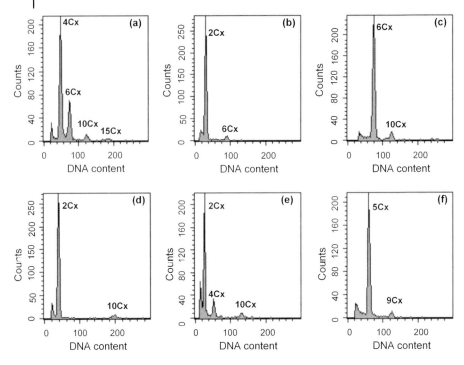

Fig. 6.3 Histograms from cell nuclei of bulked and single seed samples of *Hypericum perforatum* (Hypericaceae). (a) A bulked seed sample (50 seeds) from the facultative apomictic accession "Bremen 202", with tetraploid plants forming seeds from reduced double fertilized (4Cx embryo:6Cx endosperm), unreduced double fertilized (6Cx:10Cx), and unreduced pseudogamous embryo sacs (4Cx:10Cx); hexaploid pseudogamous plants were also present (6Cx:15Cx). (b) A single seed (accession "Berlin 1232") that arose by parthenogenetic development of the reduced egg cell (2Cx embryo) and fertilization of the reduced central cell (6Cx endosperm). (c) A single seed (accession "Münster 699") that arose from a fertilized, unreduced egg cell (6Cx embryo) and a fertilized, unreduced central cell (10Cx endosperm). (d) A single seed (accession "Chrest. 6/1") that arose from autonomous development of the reduced egg cell of the legitimate embryo sac (2Cx embryo) and the fertilized central cell of an aposporous embryo sac (10Cx endosperm). (e) A single seed (accession "Chrest. 6/1") with a twin embryo which occurs very rarely. One embryo (2Cx) arose from the reduced egg cell and the other from the unreduced egg cell (4Cx, embryo aborted?), both by autonomous development. Only the unreduced embryo sac formed an endosperm (10Cx). (f) A single seed of a diploid mother plant (accession "Ren. 54") that arose from chromosome doubling within the aposporous initial cell and subsequent development of a tetraploid, unreduced embryo sac; both the egg cell and the central cell were fertilized by haploid pollen (5Cx embryo:9Cx endosperm). Figure reprinted from Matzk et al. (2001) with permission.

To characterize the variation of the reproductive mode within small-seeded species, populations or plants, bulked seed samples (30 to 50 seeds) should be analyzed at first. In this way, 11 different types of reproduction were identified after analyzing 113 accessions from *Hypericum perforatum* (Table 6.2). The results of this study (Matzk et al. 2001) provide evidence for the high potential of the FCSS. Most of the accessions were tetraploid facultative apomicts, and rare obligate apomicts or diploid sexuals. Unreduced egg cells were frequently fertilized in several tetraploid populations, as indicated by a high 6Cx embryo and 10Cx endosperm peak. A 6Cx embryo peak associated with a 15Cx endosperm peak (Fig. 6.3a) indicated that hexaploid apomictic individuals (B_{III} plants) were already present in the population. Reduced parthenogenetic embryo formation also occurred frequently in association with other reproductive pathways, however (Table 6.2). Apospory and parthenogenesis were unlinked (Matzk et al. 2001).

For further precision of the facultative reproductive mode, the degree (percentage) of expression of apomeiosis, parthenogenesis and autonomous endosperm formation may be estimated by analyses of single or pairs of seeds (e.g. 100 seeds per genotype; Matzk et al. 2000, 2001, 2005). Several possibilities for subdivision of the reproductive mode in facultative apomicts by single seed analyses compared with bulked seed analyses are demonstrated for *H. perforatum* in Table 6.3. Genotypes with a new pathway of reproduction (Fig. 6.3f) or seeds containing an embryo descended from a reduced embryo sac and endosperm from an aposporous embryo sac (Fig. 6.3d) were identified by single seed analyses for the first time (Matzk et al. 2001).

The mode of reproduction can be reconstructed from single seeds even in the extremely small-seeded *Arabidopsis thaliana* (see Section 6.3.3). This may prove important for screening apomeiosis and parthenogenesis mutants in this model species of molecular genetics.

A very important difference between the FCSS and earlier FCM mode of reproduction analyses is the use of mature seeds instead of fresh tissues. The cells of ripe (dry) seeds are dormant and metabolically inactive. Reduced enzyme activities may result in a higher stability of the DNA during isolation, staining and storage of the nuclei. For this reason, reproducible and sharp histogram peaks have also been obtained (Fig. 6.3) after using a simplified sample preparation (see Section 6.3.3). Even from normal wheat flour the C-value of embryo nuclei, and from wheat bran the C-values of embryo and endosperm nuclei, can still be determined. Compared with fresh tissues from immature seeds, the analysis time window is not limited for the FCSS, and the cell cycle phase should be comparable for the specific cell types (embryo, endosperm) in all mature seeds.

An internal standard is not required for reproductive mode analyses with the FCSS, because an endogenous standard is present (cf. Chapter 4). Embryo and endosperm nuclei are isolated together and the ratio of the C-values of these two cell types reveals the pathway of reproduction. To simultaneously compare the genome sizes or DNA ploidy levels of the species, populations or individual plants investigated for mode of reproduction, an external standard may be sufficient (Matzk et al. 2003).

Moreover, seed analyses do not require time or space (glass house) for cultivation of plants. Even for older seeds which have lost their germination ability, such as those used in early experiments, the FCSS can still be applied to reconstruct the mode of reproduction.

The FCSS is not applicable if embryo and endosperm have the same ploidy (e.g. in some early angiosperm lineages; Williams and Friedman 2002), and different reproductive pathways cannot be discriminated if sexual and asexual seed formation yield identical ploidy levels for the embryo and the endosperm cells. This means, for example, that adventitious embryony or apospory of the *Panicum* type cannot be discriminated from the sexual pathway using the FCSS. It is furthermore not possible to differentiate between the two forms of apomeiosis (i.e. diplospory and apospory). An analysis to see whether unreduced egg cells have resulted from FDR (first division restitution) or SDR (second division restitution) requires the application of additional techniques, for example DNA fingerprinting or detection of deviating segregation ratios.

6.3.2
Applications of the FCSS

6.3.2.1 Botanical Studies

The breeding system of many species within the highly variable genus *Hypericum* has been elucidated using the FCSS over a short period of time. Apomictic reproduction was identified in 15 species after analyzing 71 *Hypericum* species, each represented by several accessions (Matzk et al. 2003). *Hypericum carpaticum* was classified as a separate hybridogenous species with facultative apomictic reproduction (Mártonfi 2001).

The variability of the reproductive mode within *H. perforatum* was studied by analyzing >125 accessions with the FCSS (Arzenton et al. 2006; Matzk et al. 2001). Nearly all populations were facultative apomicts with a varying degree of sexual versus apomictic reproduction. Somaclonal variation after *in vitro* regeneration of *H. perforatum* plants was discovered by RFLP fingerprinting (Halušková and Čellárová 1997). Within and between these somaclones, a large variation in mode of reproduction across four subsequent generations was identified using the FCSS (Koperdáková et al. 2004).

For several species of other genera (e.g. *Stipa pennata*, *S. pulcherrima* and *Potamophila parviflora*, all Poaceae) and mutants (*Arabidopsis thaliana*) hypothesized apomictic reproduction or candidates for individual asexual components have been rejected as a result of analyses with the FCSS (F. Matzk et al., unpublished data). Facultative apomictic reproduction with autonomous endosperm formation was, however, identified using the FCSS in *Coprosma robusta* (Rubiaceae) for the first time (Heenan et al. 2003). In *Paspalum simplex* (Poaceae) it was demonstrated that nearly all apomictic plants form B_{III} hybrids in low numbers, while autonomous endosperm formation was never observed (Cáceres et al. 2001).

6.3.2.2 Evolutionary Studies

Trends concerning co-evolution of mode of reproduction and genome size were elucidated by screening both traits in 71 species of the genus *Hypericum* with the FCSS (Matzk et al. 2003). It was found that the apomicts of the evolutionarily older section *Ascyreia* have significantly larger genomes than all other species, a result of both polyploidization and higher DNA content per chromosome. A similar situation may exist in *Hieracium* (Bicknell and Koltunow 2004).

Using the FCSS *Boechera holboellii* was shown to be a pseudogamous (Naumova et al. 2001) and not an autonomous apomict, as was previously reported (Böcher 1951). Moreover, in *B. holboellii* as well as in *Hypericum scabrum* it was found that in apomictic seed formation, exclusively unreduced male gametes fertilize the central cell. It could be speculated that these species originated through interspecific hybridization, and have escaped sterility by both apomixis and unreduced male gamete formation (Matzk et al. 2003; Naumova et al. 2001).

6.3.2.3 Genetical Analyses of Apomixis

Knowledge about the genetic basis of apomixis is limited (Bicknell and Koltunow 2004; Grossniklaus et al. 2001). Most previous studies on the inheritance of apomixis suffer from drawbacks, including small sample sizes, the consideration of apomixis versus sexual reproduction as qualitative traits, estimation of apomixis by correlated traits, insufficient discrimination between the individual components (apomeiosis and parthenogenesis), or the lack of consideration of possible multiple gene control for each component. The FCSS has helped to overcome these drawbacks and has led to the confirmation of a novel five-locus model (Fig. 6.4) of inheritance of apospsorous pseudogamous apomixis in *Poa pratensis* (Matzk et al. 2005). After quantifying the reproductive pathways of a large number of individuals in segregating progenies, it was demonstrated that four classes of expression (zero, low, intermediate and high) result from interactions between two genes, each controlling apospory and parthenogenesis. Thus it is hypothesized that the large variation in reproductive mode in other facultative apomicts may be caused by interactions between multiple major genes controlling apospory and parthenogenesis, rather than by environmental factors or modifier genes as has frequently been speculated (reviewed in Nogler 1984). The new results contradict earlier models of a monogenic control of apomixis (Savidan 2000).

Comprehensive studies of the inheritance and variability of apomixis in *Hypericum perforatum* using the FCSS are in progress (F. Matzk et al., unpublished data). A few examples of the FCSS histograms from *H. perforatum* seed analyses are shown in Fig. 6.3.

6.3.3
Methodological Implications

For general methodological information about the FCSS and its application in *Hypericum perforatum* and *Poa pratensis* see Matzk et al. (2000, 2001, 2005). Fur-

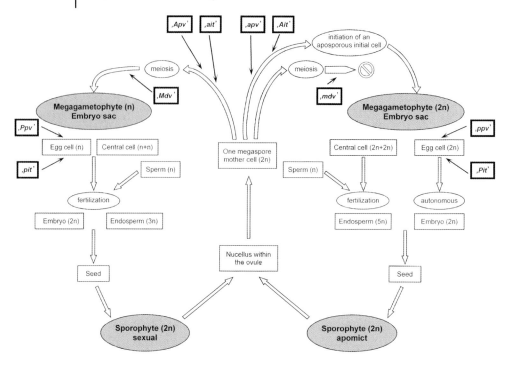

Fig. 6.4 Model of inheritance of apomixis in *Poa pratensis* (Poaceae). Five major genes (boxes with thick black borders) control the sexual (left side) and apomictic (right side) pathway: "*apospory prevention*" dominant alleles in sexuals ("*Ap*") and recessive alleles in apomicts ("*apv*"); "*apospory initiation*" dominant alleles in apomicts ("*Ait*") and null or recessive alleles in sexuals ("*ait*"); "*parthenogenesis prevention*" dominant alleles in sexuals ("*Ppv*") and recessive alleles in apomicts ("*ppv*"); "*parthenogenesis initiation*" dominant alleles in apomicts ("*Pit*") and null or recessive alleles in sexuals ("*pit*"); "*megaspore development*" dominant alleles in sexuals ("*Mdv*") and recessive alleles in apomicts ("*mdv*"). Figure reprinted from Matzk et al. (2005) with permission.

ther methodological progress was made after testing several procedures for preparing nuclear suspensions (with a razor blade, homogenizer, mortar, sand paper), different types of fixation and staining (separate and combined, with and without centrifugation), and six different DAPI-staining buffers (F. Matzk et al., unpublished data). The sharpest and most consistent peaks are obtained by crushing dry (metabolically inactive) seeds between two pieces of finely granulated sand paper (large seeds can be cut into several pieces with a razor blade or scalpel before crushing). Then the two pieces of sand paper are rinsed with 2–3 ml of DAPI-staining buffer. This simple and time-saving one-step procedure yields sharp embryo and endosperm peaks. The peak quality is reduced by hydration before crushing, or by direct chopping or homogenizing within the buffer. Dry crushing with sand paper followed by direct addition of staining buffer has already been applied in the author's laboratory for mode of reproduction analyses in about 150 plant species.

For most species, previously described staining buffers (Matzk et al. 2001, 2005) or DNA staining solutions from Partec (Münster, Germany) are suitable. In some cases, however, a modified staining buffer yields sharper peaks. In *Arabidopsis thaliana* or *Boechera holboellii*, for example, the following recipe for staining buffer may be recommended specifically for single seed analyses: 100 mM Tris-HCl, 5.3 mM $MgCl_2$, 86 mM NaCl, 30 mM sodium citrate, 1.5 mM Triton X-100, 0.003 mM 4′-6-diamidino-2-phenylindole, pH 7.0.

6.4
Flow Cytometry with Mature Seeds for other Purposes

Plant breeders working with artificial polyploids alongside natural diploids of the same species sometimes encounter problems with purity and stability of the ploidy level within the stocks (e.g. in *Lolium* and *Festuca* species from Poaceae, *Trifolium pratense* from Fabaceae, and *Beta vulgaris* from Chenopodiaceae/APG: Amaranthaceae). To recognize aneuploid or interploidy contamination in pollen or plants from the field, or in seeds during seed multiplication, the ploidy of a large number of progeny plants has hitherto been checked by FCM using young leaves. A determination of the relative DNA content of embryos within the relevant seed lots would be more time-saving than such progeny tests. In many cases, a large number of seeds can be pooled in one sample (bulked samples with 30 to 50 mature seeds are possible in small-seeded species).

Many cultivars of sugar beet are triploid F_1 hybrids. Through analyses of seed samples we can test whether all embryos are actually triploid ($3x$), and whether the seeds originated from a diploid mother pollinated with a tetraploid father (3Cx embryo + 4Cx endosperm) or from a tetraploid mother pollinated with a diploid father (3Cx embryo + 5Cx endosperm).

The identification and classification of fine fescues (*Festuca ovina*, *F. rubra* and others), with their close morphological resemblance to each other and the existence of numerous ecotypes, is a problem for turfgrass scientists, taxonomists and breeders. Confusion and difficulties in the identification of the species and subspecies result from controversial classifications, an ever-changing scientific nomenclature, and consequently the existence of many synonyms. However, a high variation in the ploidy or chromosome numbers ($2n = 14$ to 56) occurs among the species and subspecies. For this reason, the determination of the DNA ploidy level by FCM analyses of fresh tissues has become a powerful tool for identifying subspecies of fine fescue, assigning native accessions or primary breeding germplasm to their proper species categories (Huff and Palazzo 1998), and for the comparison of the genome sizes of other cool-season and warm-season turfgrass species (Arumuganathan et al. 1999). In all these cases the application of FCM seed analyses could be a better alternative to any other test. The use of ripe seeds was recommended recently for the estimation of the genome size of species on the basis of the relative DNA content of embryo nuclei (Sliwinska et al. 2005).

Before registration of a new cultivar, the germplasm has to be tested for homogeneity and stability in its progenies. The rules of the Bundessortenamt in Germany require that >90% maternal plants occur in progeny tests in order for the plant to be registered as an apomictic cultivar of the species *Poa pratensis*. This prerequisite for registration could be estimated by the FCSS more easily, more exactly, and more rapidly than by the currently-used progeny test in the field.

Sometimes, the certification of a seed lot is rejected because of inadmissible contamination with deviating ploidy levels or other plant species, which may have occurred during propagation in the field or during the harvesting and cleaning processes. This could be re-checked by direct FCM seed analyses without cultivation of the plants over a long period. Seed testing by FCM could become an additional requirement for seed certification of subspecies with different DNA content or species with germplasm of different ploidy. There are diverse possibilities for the utilization of FCM seed analyses in controlling registration, maintenance and seed propagation of cultivars in addition to seed certification and seed trade control.

6.5
Conclusions

Sexual reproduction in plants includes alteration of a sporophytic and a gametophytic generation associated with a reduction of the ploidy level during meiosis and an increase by subsequent fertilization. Apomixis bypasses meiosis and fertilization of egg cells. The traditional ways to study reproductive modes using light microscopy of squashed or microdissected tissues, progeny tests or similar are too laborious for routine analyses of large numbers of individual plants. In some cases they can be replaced by FCM ploidy analyses, which use fresh tissues such as leaves, ovaries or immature seeds from mother and progeny plants. The recently developed flow cytometric seed screen (FCSS) yields much more information about the reproductive behavior of individual plants compared with all other available tests. The mode of reproduction can be reconstructed from ripe seeds using the proportional DNA content of embryo and endosperm cells. It is an efficient tool for the simultaneous analysis of the different events of sporogenesis, embryo and endosperm formation in plant species, populations or individuals. Most of the shortcomings and difficulties connected with the traditional techniques are overcome with the FCSS, and the first novel results concerning the characterization of mode of reproduction in several species and the evolution and inheritance of apomixis have already been reported. Further progress in apomixis research can be expected. The FCSS will facilitate the introduction of apomixis into sexually reproduced crops as well as the breeding process with apomictic plants. Beyond that, FCM analyses using ripe seeds may become an important tool in controlling some steps of germplasm registration and seed propagation of cultivars. FCM seed tests are very useful in basic research, plant breeding and registration of cultivars or certification of seed lots.

Acknowledgment

The author thanks Ingo Schubert and Timothy F. Sharbel for critical reading of the manuscript and the anonymous reviewer as well as the book editors for helpful suggestions.

References

Arnholdt-Schmitt, B. **2000**, *Theor. Appl. Genet.* 100, 906–911.

Arumuganathan, K., Tallury, S. P., Fraser, M. L., Bruneau, A. H., Qu, R. **1999**, *Crop Sci.* 39, 1518–1521.

Arzenton, F., Barcaccia, G., Sharbel, T. F., Lucchin, M., Parrini, P. **2006**, *Heredity* 96, 322–334.

Barcaccia, G., Tosti, N., Falistocco, E., Veronesi, F. **1995**, *Theor. Appl. Genet.* 91, 1008–1015.

Barcaccia, G., Tavoletti, S., Falcinelli, M., Veronesi, F. **1997**, *Plant Breeding* 116, 475–479.

Barrell, P. J., Grossniklaus, U. **2005**, *Plant J.* 43, 309–320.

Bennett, M. D., Leitch, I. J. **1995**, *Ann. Bot.* 76, 113–176.

Bharathan, G., Lambert, G., Galbraith, D. W. **1994**, *Am. J. Bot.* 81, 381–386.

Bicknell, R. A., Koltunow, A. M. **2004**, *Plant Cell* 16, S228–S245.

Bino, R. J., van Tuyl, J. M., de Vries, J. N. **1990**, *Ann. Bot.* 65, 3–8.

Bino, R. J., Lanteri, S., Verhoeven, H. A., Kraak, H. L. **1993**, *Ann. Bot.* 72, 181–187.

Boavida, L. C., Vieira, A. M., Becker, J. D., Feijó, J. A. **2005**, *Int. J. Dev. Biol.* 49, 615–632.

Böcher, T. W. K. **1951**, *Danske Vidensk. Selskab Biol. Skrifter* 6, 3–59.

Brutovská, R., Čellárová, E., Doležel, J. **1998**, *Plant Sci.* 133, 221–229.

Burson, B. L. **1997**, *Crop Sci.* 37, 1347–1351.

Cáceres, M. E., Matzk, F., Busti, A., Pupilli, F., Arcioni, S. **2001**, *Sexual Plant Reprod.* 14, 201–206.

Carman, J. G. **1997**, *Biol. J. Linn. Soc.* 61, 51–94.

Chat, J., Dumoulin, P. Y., Bastard, Y., Monet, R. **1996**, *Plant Breeding* 115, 378–384.

Chaudhury, A. M., Ming, L., Miller, C., Craig, S., Dennis, E. S., Peacock, W. J. **1997**, *Proc. Natl Acad. Sci. USA* 94, 4223–4228.

Chen, L.-Z., Kozono, T. **1994**, *Cytologia* 59, 253–260.

Doležel, J., Greilhuber, J., Lucretti, S., Meister, A., Lysák, M. A., Nardi, L., Obermayer, R. **1998**, *Ann. Bot.* 82(Suppl. A), 17–26.

Drews, G. N., Lee, D., Christensen, C. A. **1998**, *Plant Cell* 10, 5–17.

Finch, R. A., Bennett, M. D. **1979**, *Heredity* 43, 87–93.

Golubovskaya, I., Avalkina, N. A., Sheridan, W. F. **1992**, *Dev. Genet.* 13, 411–424.

Grimanelli, D., Hernández, M., Perotti, E., Savidan, Y. **1997**, *Sexual Plant Reprod.* 10, 279–282.

Greilhuber, J., Doležel, J., Lysák, M. A., Bennett, M. D. **2005**, *Ann. Bot.* 95, 255–260.

Grossniklaus, U., Schneitz, K. **1998**, *Sem. Cell Dev. Biol.* 9, 227–238.

Grossniklaus, U., Vielle-Calzada, J.-P., Hoeppner, M. A., Gagliano, W. B. **1998**, *Science* 280, 446–450.

Grossniklaus, U., Nogler, G. A., van Dijk, P. J. **2001**, *Plant Cell* 13, 1491–1497.

Hagberg, A., Hagberg, G. **1980**, *Hereditas* 93, 341–343.

Halušková, J., Čellárová, E. **1997**, *Euphytica* 95, 229–235.

Heenan, P. B., Molloy, B. P. J., Bicknell, R. A., Luo, C. **2003**, *NZ J. Bot.* 41, 287–291.

Herr Jr., J. M. **1971**, *Am. J. Bot.* 58, 785–790.

Huff, D. R., Bara, J. M. **1993**, *Theor. Appl. Genet.* 87, 201–208.

Huff, D. R., Palazzo, A. J. **1998**, *Crop Sci.* 38, 445–450.

Jefferson, R. A., Bicknell, R. **1996**, The potential impacts of apomixis: a molecular genetics approach in *The Impact of Plant Molecular Genetics*, ed. B. W. S. Sobral, Birkhäuser, Boston, pp. 87–101.

Koperdáková, J., Brutovská, R., Čellárová, E. **2004**, *Hereditas* 140, 34–41.

Kowles, R. V., Yerk, G. L., Schweitzer, L., Srienc, F., Phillips, R. L. **1994**, Flow cytometry for endosperm nuclear DNA in *The Maize Handbook*, ed. M. Freeling, V. Walbot, Springer, New York, pp. 400–406.

Krahulcová, A., Papoušková, S., Krahulec, F. **2004**, *Hereditas* 141, 19–30.

Lihová, J., Mártonfi, P., Mártonfiová, L. **2000**, *Caryologia* 53, 127–132.

Maceira, N. O., de Haan, A. A., Lumaret, R., Billon, M., Delay, J. **1992**, *Ann. Bot.* 69, 335–343.

Mártonfi, P. **2001**, *Folia Geobot.* 36, 371–384.

Matzk, F. **1991a**, *Euphytica* 55, 65–72.

Matzk, F. **1991b**, *Sexual Plant Reprod.* 4, 88–94.

Matzk, F., Meyer, H.-M., Bäumlein, H., Balzer, H.-J., Schubert, I. **1995**, *Sexual Plant Reprod.* 8, 266–272.

Matzk, F., Oertel, C., Altenhofer, P., Schubert, I. **1997**, *Trends Agron.* 1, 19–34.

Matzk, F., Meister, A., Schubert, I. **2000**, *Plant J.* 21, 97–108.

Matzk, F., Meister, A., Brutovská, R., Schubert, I. **2001**, *Plant J.* 26, 275–282.

Matzk, F., Hammer, K., Schubert, I. **2003**, *Sexual Plant Reprod.* 16, 51–58.

Matzk, F., Prodanovic, S., Bäumlein, H., Schubert, I. **2005**, *Plant Cell* 17, 13–24.

Mayo, G. M., Langridge, P. **2003**, *Genome* 46, 573–579.

Morrison, L. A., Metzger, R. J., Lukaszewski, A. J. **2004**, *Crop Sci.* 44, 2063–2067.

Naumova, T. N., den Nijs, A. P. M., Willemse, M. T. M. **1993**, *Acta Bot. Neerland.* 42, 299–312.

Naumova, T. N., der Laak, J. van, Osadtchiy, J., Matzk, F., Kravtchenko, A., Bergervoet, J., Ramulu, K. S., Boutilier, K. **2001**, *Sexual Plant Reprod.* 14, 195–200.

Negri, V., Lorenzetti, S., Lemmi, G. **1995**, *Plant Breed.* 114, 86–88.

Noack, K. L. **1939**, *Zeit. Indukt. Abst.-Ver.* 76, 569–601.

Nogler, G. A. **1984**, Gametophytic apomixis in *Embryology of Angiosperms*, ed. B. M. Johri, Springer, Berlin, pp. 475–518.

Ohad, N., Margossian, L., Hsu, Y.-C., Williams, C., Repetti, P., Fischer, R. L. **1996**, *Proc. Natl Acad. Sci. USA* 93, 5319–5324.

Ortiz, R., Vorsa, N., Bruederle, L. P., Laverty, T. **1992**, *Theor. Appl. Genet.* 85, 55–60.

Pagnussat, G. C., Yu, H.-J., Ngo, Q. A., Rajani, S., Mayalagu, S., Johnson, C. S., Capron, A., Xie, L.-F., Ye, D., Sundaresan, V. **2005**, *Development* 132, 603–614.

Pan, G., Zhou, Y. M., Fowke, L. C., Wang, H. **2004**, *Plant Cell Rep.* 23, 196–202.

Pichot, C., El Maâtaoui, M. **2000**, *Theor. Appl. Genet.* 101, 574–579.

Pichot, C., Borrut, A., El Maâtaoui, M. **1998**, *Sexual Plant Reprod.* 11, 148–152.

Ramanna, M. S. **1979**, *Euphytica* 28, 537–561.

Rutishauser, A. **1967**, *Fortpflanzungsmodus und Meiose apomiktischer Blütenpflanzen (Protoplasmatologia Bd. 6: Kern- und Zellteilung F: Die Chromosomen in der Meiose; 3)*, Springer, Wien.

Savidan, Y. **2000**, *Plant Breed. Rev.* 18, 13–86.

Sherwood, R. T. **1995**, *Sexual Plant Reprod.* 8, 85–90.

Sherwood, R. T., Berg, C. C., Young, B. A. **1994**, *Crop Sci.* 34, 1490–1494.

Sliwinska, E., Zielinska, E., Jedrzejczyk, I. **2005**, *Cytometry* 64A, 72–79.

Spillane, C., Curtis, M. D., Grossniklaus, U. **2004**, *Nature Biotechnol.* 22, 687–691.

Tyrnov, V. S., Enaleeva, N. K. **1983**, *Dok. Akad. Nauk SSSR* 272, 722–725.

van Dijk, P. J., Tas, I. C. Q., Falque, M., Bakx-Schotman, T. **1999**, *Heredity* 83, 715–721.

van Tuyl, J. M., de Vries, J. N., Bino, R. J., Kwakkenbos, T. A. M. **1989**, *Cytologia* 54, 737–745.

Vielle Calzada, J.-P., Crane, C. F., Stelly, D. M. **1996**, *Science* 274, 1322–1323.

Williams, J. H., Friedman, W. E. **2002**, *Nature* 415, 522–526.

Wyman, J., Laliberté, S., Tremblay, M.-F. **1997**, *Am. J. Bot.* 84, 1351–1361.

Young, B. A., Sherwood, R. T., Bashaw, E. C. **1979**, *Can. J. Bot.* 57, 1668–1672.

7
Genome Size and its Uses: the Impact of Flow Cytometry

Ilia J. Leitch and Michael D. Bennett

Overview

The huge diversity in genome size encountered in plants is striking, with 1C-values ranging nearly 2000-fold from 0.06 to 127.4 pg. Understanding the biological and evolutionary significance of this variation has puzzled biologists for many decades. Over the years it has become increasingly clear that variation in DNA amount has significant consequences not only at the cellular, tissue and organism level but also at the evolutionary and ecological levels, influencing how, when and where a plant may grow and its chances of survival in a changing world. This chapter reviews (i) why the study of genome size is important (Sections 7.1–7.2); (ii) what is known about genome size in three algal groups (Chlorophyta, Rhodophyta and Phaeophyta) and across land plants (bryophytes, lycophytes, monilophytes, gymnosperms and angiosperms) (Sections 7.3–7.4); (iii) the evolutionary and ecological consequences of genome size variation (Section 7.5); and (iv) what impact flow cytometry has played in contributing to the knowledge and understanding of genome size (Section 7.6).

7.1
Introduction

In November 2005 the journal *Science* included a Netwatch item headed "Another Day, Another Genome" (vol. 310, p. 1255). It noted that "as of last week, scientists had polished off 319 genomes", but had "at least 1300 to go" for genome sequencing. This epitomizes the striking exponential growth in genome science which has occurred since the term genome was first used by Winkler (1920), and especially since the discovery of the double helix structure of DNA, and its biological significance, in the mid-1950s (Watson and Crick 1953).

There is considerable awareness by scientists, and even among the general population, of the story of the discovery of the genetic code leading to the epic projects to sequence the entire genomes of the first virus (Baer et al. 1984), bacterium

Flow Cytometry with Plant Cells. Edited by Jaroslav Doležel, Johann Greilhuber, and Jan Suda
Copyright © 2007 WILEY-VCH Verlag GmbH & Co. KGaA, Weinheim
ISBN: 978-3-527-31487-4

(Fleischmann et al. 1995), eukaryotic animal (Goffeau et al. 1996), insect (Adams et al. 2000), and flowering plant (Arabidopsis Genome Initiative 2000) to read their genotype in the book of life. There is, however, another parallel strand to this story which is less well known, and whose importance is much less appreciated. This second story concerns not the content and arrangement of information within the DNA sequence (= genotype), but rather the total amount of DNA the genome contains, as this can also have considerable influence (biological and otherwise) on the organism, independent of the encoded genetic information. This second effect of the DNA on an organism's phenotype is known as the "nucleotype" – a term coined by Bennett to define those conditions of the nuclear DNA which affect the phenotype independently of its encoded informational content (Bennett 1971, 1972). This chapter is concerned with this less well-known aspect of the genome and will discuss what is known about genome size in plants, some of the ecological and other consequences of genome size variation, and how flow cytometry (FCM) has contributed to the study of genome size in plants. Practical aspects concerning the measurement of genome size in plants, and a discussion on the terminology used is dealt with in detail in Chapter 4.

7.2
Why is Genome Size Important?

Based on available data, genome size in plants varies nearly 2000-fold from 0.065 pg to over 125 pg (see Section 7.4). This considerable variation has numerous important non-biological and biological implications.

From a practical perspective genome size has been a major factor in determining which organisms are selected for complete genome sequencing, as the amount of DNA influences both the time and cost of such projects. Figure 7.1 plots genome size against the date of publication of the first complete genome sequences for several different life forms, and shows the clear relationship between them during 1984–2001, imposed by the combined dual constraints of limited technology and research funds. Clearly for two decades genome size was a key limiting factor governing the selection of model organisms for complete genome sequencing projects. Despite technical advances which now mitigate this to some extent, such as the potential to increase the rate of sequencing 100-fold to ca. 25 Mb in 4 h (Margulies et al. 2005), genome size remains an important consideration, and a few organisms with very large genomes are now the subject of special projects designed specifically to compare sequences in taxa with extremely different sized genomes using specially developed sequencing strategies (e.g. methyl filtration; Bedell et al. 2005). The analysis of complex genomes is also facilitated by chromosome sorting, another application of FCM which is discussed in Chapter 16.

Knowledge of genome size has also been shown to be important in determining the success of various genetic fingerprinting approaches such as microsatellites and amplified fragment length polymorphisms (AFLPs), which are widely

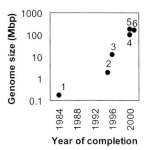

Fig. 7.1 Plot of 1C-value against the publication date of the first complete genome sequence for several different life forms between 1984 and 2001. 1, First virus sequenced 1984 (Epstein Barr, 1C = 0.17 Mbp); 2, first bacterium sequenced 1995 (*Haemophilus influenzae*, 1C = 1.8 Mbp); 3, first fungus sequenced 1996 (*Saccharomyces cerevisiae*, 1C = 12.1 Mbp); 4, first multicellular eukaryote sequenced 2000 (*Caenorhabditis elegans*, 1C = 100 Mbp); 5, first insect sequenced 2000 (*Drosophila melanogaster*, 1C = 180 Mbp); 6, first plant sequenced 2000 (*Arabidopsis thaliana*, 1C = 157 Mb).

used for analyzing population structure, gene flow, genetic diversity and so on (Costa et al. 2000; Fay et al. 2005; Garner 2002). For example, Fay et al. (2005) reported that genome size and ploidy level information were important for determining which protocol was most likely to yield informative data for population studies, and that above 1C > 15 pg the technique was usually uninformative. Genome size data are also valuable for designing various other molecular techniques such as determining the number of clones needed for complete genome coverage in a genomic library or for the development of efficient insertional mutagenesis strategies in large-scale genomic studies (Peer et al. 2003).

However, from the organism's perspective, the nucleotypic effects can be profound, with genome size having a predictive impact at many levels from the nucleus (e.g. chromosome size, nuclear volume) to the whole organism (e.g. determining where and when a plant may grow; reviewed in Bennett 1987; Bennett and Leitch 2005a). These biological consequences of genome size are discussed in Section 7.5.

7.3
What is Known about Genome Size in Plants?

Since the early 1950s, when the first genome size of a plant was estimated, more than 10 000 quantitative estimates of plant C-values have been made, covering over 5000 plant species (Table 7.1). While collected lists of genome size values have been published in hard copy for angiosperms since the early 1970s (e.g. Bennett 1972; Bennett and Leitch 1995, 1997, 2005b; Bennett and Smith 1976, 1991; Bennett et al. 1982, 2000), these were subsequently combined with genome size information for other plant groups and, from 2001, released electronically

Table 7.1 Minimum (Min.), maximum (Max.), mean, mode, and range (max./min.) of 1C DNA values in major groups of plants, together with the level of species representation of C-value data.

	No. of species with DNA C-values	Approx. no. of species recognized[a]	No. of species in Plant DNA C-values database[b]	% representation in the Plant DNA C-values database[b]	Min. (pg)	Max. (pg)	Mean (pg)	Mode (pg)	Range (Max./ Min.)
Algae									
Chlorophyta	91	6500	91	1.4	0.10	19.60	1.75	0.30	196
Rhodophyta	118	6000	118	1.9	0.10	1.40	0.43	0.20	14
Phaeophyta	44	1500	44	2.9	0.10	0.90	0.42	0.25	9
Bryophytes	176	18 000	176	~1.0	0.085	6.42	0.54	0.45	74
Pteridophytes									
Lycophytes	4	900	4	~0.4	0.16	11.97	3.81	n/a	75
Monilophytes	63	11 000	63	~0.6	0.77	72.68	13.58	7.80	95
Gymnosperms	207	730	207	~28.4	2.25	32.20	16.99	9.95	14
Angiosperms	~5000	250 000	4427	~1.8	0.065	127.40	6.30	0.60	~2000
All land plants	5672	280 000	5150	~1.8	0.065	127.40	6.46	0.60	~2000

[a] Numbers of species recognized taken from Kapraun (2005) for algae, Qiu and Palmer (1999) for bryophytes, lycophytes and monilophytes, Murray (1998) for gymnosperms, and Bennett and Leitch (1995) for angiosperms.
    C-value data for algae taken from Kapraun et al. (2004), for bryophytes, lycophytes, monilophytes, gymnosperms and for angiosperms from Bennett and Leitch (2005c) and Greilhuber et al. (2006).
[b] Plant DNA C-values database (release 4.0, Oct. 2005) (Bennett and Leitch 2005c).

as the Plant DNA C-values database (Bennett and Leitch 2005c, www.kew.org/ genomesize/homepage.html). The database aims to provide a one-stop, user-friendly electronic access to available plant DNA C-values. Release 4.0 (October 2005) currently contains C-values for 5150 species comprising 4427 angiosperms, 207 gymnosperms, 67 pteridophytes, 176 bryophytes, and 253 algae. As Table 7.1 shows, representation of C-value data in the database for the different plant groups is very varied, determined largely by the level of scientific and societal interest, and favoring model organisms and those of economic or other importance.

7.3.1
Angiosperms

Probably because of their fundamental value in agriculture, many angiosperms have been chosen for genome size studies since the 1950s. Consequently, angiosperms are probably the most studied higher order group with respect to genome size. Nevertheless, due to the large number of species recognized (at least 250 000) this corresponds to only ca. 1.8%. Whilst early studies tended to concentrate on temperate and crop plants, since 1995 there has been increased awareness of the need to make genome size data more representative of angiosperms as a whole. Improvements have been made in the systematic coverage following the active targeting of key systematic gaps identified at the first and second Plant Genome Size Meetings held at the Royal Botanic Gardens, Kew in 1997 and 2003, and by the formal constitution of an international group for genome size analysis named GESI (GEnome Size Initiative) to drive forward this vital process over the coming years (Bennett and Leitch 2005d). For example, at the 2003 meeting, with species representation standing at ca. 1.6% (i.e. ca. 4200 species), a specific target of estimating a further 1% (i.e. 2500) species in the next 5 years was set. In the most recent analysis, output of C-values was shown to be at a record high (ca. 290 first estimates for species per year; Bennett and Leitch 2005b). Species representation is thus expected to increase significantly over the next 5 years. At the family level, representation has also improved, increasing from 30 to over 50% between 1997 and 2003, largely due to targets set at the first Plant Genome Size Meeting (Hanson et al. 2001a, 2001b, 2003, 2005). At the genus level, over 8% (1126) of the 14 000 genera now have at least one C-value and this is targeted to rise to 10% by 2009 and may reach 15% by 2015. Moreover, generic representation may approach 100% in monocots as they are targeted for holistic genomic studies (including C-values) for the global Monocot Checklist Project (Govaerts 2004).

Representation of genome size data in some other key categories, however, remains poor. For example, despite highlighting the need to improve geographical representation over a decade ago (Bennett and Leitch 1995), gaping chasms still remain. With some exceptions, the sample is still dominated by crops and their wild relatives, model species grown for experimental use and other species growing near laboratories in temperate regions mainly in Western Europe and North America. There is still a dearth of data from taxa endemic to China, Japan, South America, and Africa. Similarly, although island floras are known to be rich in endemics, there has been no publication reporting C-values for any large island such as Borneo, New Guinea or Madagascar where 80% of the 12 000 described plant species are endemic (Robinson 2004). Other plant groups that were identified as poorly represented in terms of genome size include taxa from bog, fen, tundra, alpine, and desert environments and specific life forms such as parasitic, saprophytic and epiphytic species and their associated taxa.

7.3.2
Gymnosperms

With C-values for over 28% of the ca. 730 species, gymnosperms represent the plant group with best species representation. Further, they are also the first group to have complete familial representation (Leitch et al. 2001). Despite this, gaps at the generic level still remain and these were highlighted at the Genome Size

Workshop held in Vienna in July 2005. A full list of genera lacking C-values is available on the web at http://www.kew.org/genomesize/vienna05_report.pdf.

7.3.3
Pteridophytes

Phylogenetically, pteridophytes comprise two distinct groups: lycophytes (ca. 900 species), which are sister to all other vascular plants, and monilophytes (ca. 11 000 species), a monophyletic group comprising ferns and horsetails that is sister to seed plants (Pryer et al. 2001). Genome size data for these two groups are very poor at all taxonomic levels despite some carefully targeted studies which more than doubled familial representation from 10 to 25 (Hanson and Leitch 2002; Obermayer et al. 2002). To improve matters, a specific target was set at the 2003 Plant Genome Size Meeting (see "Full workshop report" at http://www.kew.org/genomesize/pgsm/index) to estimate genome size in a further 100 taxa with particular emphasis on leptosporangiate ferns – the most diverse group of land plants after angiosperms.

7.3.4
Bryophytes

Bryophytes (mosses, liverworts and hornworts) comprise a diverse group of ca. 18 000 species. However, apart from a few carefully targeted studies which surveyed genome size in 176 mosses by Temsch et al. (1998), Voglmayr (2000), and Greilhuber et al. (2003), reliable C-value data are scant and difficult to access. Further, published genome size estimates for liverworts and hornworts are still lacking although recent reports in abstracts from the 11th Meeting of Austrian Botanists (Greilhuber et al. 2004) and results presented at the Genome Size Workshop in Vienna (International Botanical Congress 2005) indicate that data for these important groups will soon be published and then made available in the Bryophyte DNA C-values database. To improve geographical representation of C-values, specific targets were set at the 2003 Plant Genome Size Meeting to estimate genome sizes for species from tropical and southern hemisphere floras and for rare taxa in the European flora.

7.3.5
Algae

While algae comprise a polyphyletic assemblage of organisms, collated data on genome size for any of these was not available until 2004 when release 3.0 of the Plant DNA C-values database included C-values for Chlorophyta, Rhodophyta and Phaeophyta for the first time. The main bulk of the data comes from work by Kapraun (2005) who assessed knowledge of C-values in these algal groups and concluded that species representation was poor (Table 7.1). Kapraun also highlighted that (i) an absence of data for Micromonadophyceae (an algal group considered to be ancestral to Chlorophyta, Rhodophyta, Phaeophyta, and all land plants), and (ii) very limited data for the charophycean lineage of Chlorophyta (the group considered sister to all land plants) should be addressed. Consequently, targets for increasing data in these groups have now been set.

7.4
The Extent of Genome Size Variation across Plant Taxa

Given the available data, it is clear that there is considerable variation in the genome size profiles between plant groups as shown in Fig. 7.2. Angiosperms are by far most variable with C-values ranging nearly 2000-fold from 0.065 pg in the carnivorous plant *Genlisea margaretae* (Lentibulariaceae; Greilhuber et al. 2005, 2006) to over 125 pg in tetraploid *Fritillaria assyriaca* (Liliaceae). Currently the least variable group is the Phaeophyta (brown algae) where reported C-values range just nine-fold from 0.1 to 0.9 pg. However, as discussed above, species representation for all but gymnosperms is poor, thus the range and distribution of C-values reported may not be entirely representative and may increase as new data are obtained. For example, the known range for bryophytes recently increased substantially from just 12-fold (based on 176 species' C-values) to 73-fold following the report of (i) 1C = 0.085 pg for the hornwort *Anthoceros agrestis*, Anthocerotaceae (this is less than half the size of the bryophyte with the previously smallest C-value of 0.2 pg), and (ii) 1C = 6.42 pg for the liverwort *Mylia taylorii*, Jungermanniaceae (three times larger than the previously largest bryophyte with

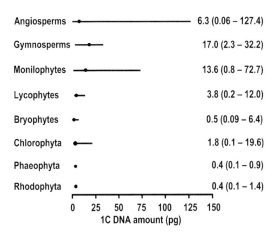

Fig. 7.2 Range of C-values in plant groups shown as a line with the mean 1C-value as a dot. The figures give the mean C-value in picograms followed by the minimum and maximum values in parentheses. Data for 4428 species taken from the Plant DNA C-values database (Bennett and Leitch 2005c; Greilhuber et al. 2006).

2.1 pg). Even angiosperms, which are most studied from a C-value perspective, can throw up surprises. A 1000-fold range for angiosperms was first reported in 1982 based on C-values for 993 species (Bennett et al. 1982). This range remained the same for the next 24 years. However, in 2005 the report that a carnivorous plant *Genlisea margaretae* has a 1C-value of only 0.065 pg (Greilhuber et al. 2005) has nearly doubled the range of known C-values in angiosperms (0.065–127.4 pg).

7.5
Understanding the Consequences of Genome Size Variation: Ecological and Evolutionary Implications

In trying to understand the significance of the huge variation in genome size reported, comparative studies in angiosperms have played a leading role in showing that the amount of DNA is correlated with a wide range of different characters and that understanding this relationship has considerable predictive value. At the nuclear level, for example, it has been shown that genome size is correlated with both nuclear (Baetcke et al. 1967) and chromosome volume (Fig. 7.3a; Bennett et al. 1983). Thus, the bigger the genome size or total amount of DNA within the nucleus, the larger the minimum nuclear volume. Similar studies have shown that DNA amount is correlated with a wide range of characters at the nuclear, cellular and tissue level, including the duration of mitosis and meiosis (Fig. 7.3b; Bennett 1977), centromere volume (Bennett et al. 1981), pollen volume (Bennett 1972), stomatal cell size (Masterson 1994), radiation sensitivity (Sparrow and Miksche 1961), and seed size (Fig. 7.3c). These nucleotypic correlations were reviewed recently by Bennett and Leitch (2005a). Clearly, DNA amount correlates closely with many important phenotypic characters.

The value of these nucleotypic correlations is that (i) they apply to all species, irrespective of genome size or chromosome number, and (ii) they have considerable predictive value. For example, the duration of meiosis in *Arabidopsis thaliana* (Brassicaceae) was unknown when Bennett carried out his studies (Fig. 7.3b; Bennett 1971, 1977), yet from the observed correlation between genome size and the duration of meiosis at 20 °C (Fig. 7.3b) it would be predicted to take around 10–20 h. Recently this prediction was confirmed when Armstrong et al. (2003) estimated that the duration of meiosis in *A. thaliana* at 18.5 °C was ca. 24 h (and hence ca. 20 h at 20 °C would be expected assuming a Q_{10} of ca. 2.2; Bennett 1977). (NB The temperature coefficient or Q_{10} represents the increase in reaction rate that results from a temperature rise of 10 °C).

Given the large range of genome sizes observed in plants and the nucleotypic effects that genome size can have on an organism at the nuclear, cellular, and tissue level, it is perhaps not surprising to find that genome size variation can have consequences at the whole plant level, influencing many aspects of a plant's development. One of the recurring themes arising from numerous studies is that large genomes appear to impose constraints on a plant's development, influenc-

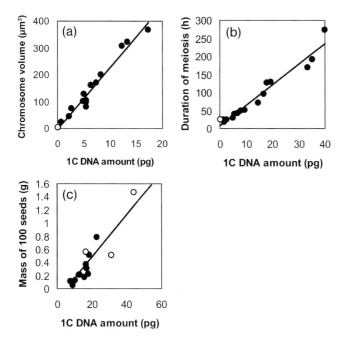

Fig. 7.3 Relationship between DNA amount and (a) total somatic chromosome volume (data taken from Bennett et al. (1983) shown as ●, and from Bennett et al. (2003) as ○); (b) the duration of meiosis in diploid angiosperms (data taken from Bennett (1977) shown as ●, and from Armstrong et al. (2003) as ○); (c) dry seed mass of 16 *Allium* species (data taken from Bennett (1972) shown as ●, and from the Seed Information Database (Flynn et al. 2004) and Plant DNA C-values Database (Bennett and Leitch 2005c) shown as ○).

ing many aspects of when and where a plant may grow. Some examples are outlined below.

7.5.1
Influence of Genome Size on Developmental Lifestyle and Life Strategy

An early study to investigate the consequences of genome size variation on plant development was conducted by Bennett (1972) who showed that the particular developmental lifestyle a plant displays (i.e. whether it can be an ephemeral, annual or perennial) could be influenced by genome size, mainly as a consequence of its relationship with various cellular parameters which play a role in determining growth rate. For example, the duration of meiosis at 20 °C can vary from just under 24 h (Armstrong et al. 2003) in the ephemeral *Arabidopsis thaliana* (1C = 0.16 pg) to nearly 6 weeks in *Fritillaria assyriaca* (1C = 127.4 pg; Bennett 1971). Clearly if a plant is to be an ephemeral (i.e., go from seed to seed in ca. 6–7 weeks), there is no way that it can have a large genome as it simply cannot divide fast enough. These observations led Bennett (1972) to propose a model of the relationship between DNA amount and minimum generation time (= the minimum duration of the period from germination until the production of the first mature seed). He showed that whilst plants with genomes smaller than 1C = ca. 3 pg could adopt any developmental lifestyle under genotypic control, as genome size increased the number of lifestyle options were successively closed. Thus, species with genomes less than 3 pg could be ephemerals (e.g. *Haplopappus gracilis*, Asteraceae; 1C = 2.0 pg), annuals (e.g. *Sinapis arvensis*, Brassicaceae; 1C = 0.4 pg) or perennials (e.g. *Betula populifolia*, Betulaceae; 1C = 0.2 pg). Whereas species with genomes between 3 and 25 pg could be annuals or facultative perennials but not ephemerals, above 25 pg species were restricted to an obligate perennial lifestyle (Fig. 7.4). Interestingly, over 30 years later, no exceptions to these predictions have been found despite an increase of more than 15-fold in available genome size data. Thus knowledge of genome size and its nucleotypic effects have considerable predictive values which can help to unify our understanding of genomes.

Since then it has become increasingly clear that having a large genome precludes other options, not just the type of developmental lifestyle open to a plant. For example, genome size has been shown to play a role in determining what par-

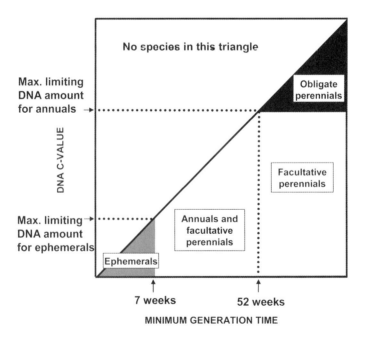

Fig. 7.4 Model showing relationship between DNA amount, minimum generation time and developmental lifestyle options, see Section 7.5.1 (modified from Bennett 1972).

ticular life strategy is adopted, such as whether or not a particular species has the potential to become an important weed.

Many of the key factors which are important for being a successful invasive weed, including rapid establishment and completion of reproductive development, short minimum generation time and fast production of many small seeds, are correlated with low DNA amount and are not possible with large genomes. This raised the question as to whether the option or potential to be a weed was restricted to species with small genomes. In a detailed analysis of DNA amounts for 156 angiosperm weed species (including 97 recognized as important world weeds) Bennett et al. (1998) provided strong evidence that a small genome *is* a requirement for "weediness". By comparing histograms showing the distribution of DNA amounts in 156 weed versus 2685 non-weed species (Fig. 7.5a, b), weeds had a significantly smaller mean C-value (1C = 2.9 pg) than non-weeds (1C = 7.0 pg) and the DNA amounts for weed species (1C = 0.16–25.1 pg) were restricted to the bottom 20% of the range then known for angiosperms (1C = 0.1–127.4 pg). Bennett et al. (1998) also showed that there was a highly significant negative relationship between DNA amount and the proportion of species recognized as weeds (Fig. 7.5c). As DNA amount increased, the proportion of weeds in each sample decreased until at C-values above 25.1 pg no weeds were present. Clearly, weeds appear to be characterized by possessing small genomes and once again it is apparent that having a large genome effectively limits available options. Thus, while all species with small genomes are not weeds, once a genome becomes too big then the option to become one is no longer available.

In a more recent study, Ohri (2005) investigated the consequences of genome size variation on growth form (woody versus herbaceous) in 3874 angiosperms of known DNA content. Generally it was observed that species with a woody growth form were characterized by possessing smaller genomes compared with herbaceous species. However, an analysis of a subset of the data revealed that the mean genome size of tropical woody species was 25% larger than their temperate counterparts. Thus, while the overall results suggest that having a large genome excludes the option to exhibit a woody growth form, clearly the picture is complex and further in-depth analysis is necessary.

7.5.2
Ecological Implications of Genome Size Variation

The impact of genome size in influencing where plants may grow and thus their natural distribution is another topic which has received much attention. It has recently become clear that plants possessing large genomes may be constrained in the range of ecological options available to them (= "the large genome constraint" hypothesis; Knight et al. 2005).

Over the years there have been numerous studies which sought correlations between genome size and various ecological parameters (e.g. latitude, altitude, elevation). Confusingly, many of these seem to give contradictory results when compared (reviewed in Knight and Ackerly 2002; Knight et al. 2005). Much of the

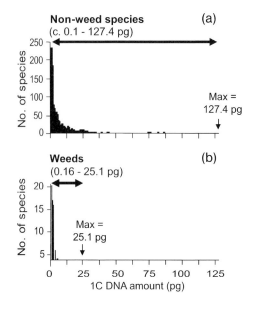

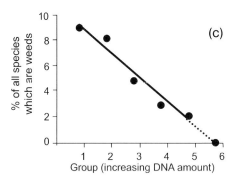

Fig. 7.5 Histograms showing the range of 1C-values in (a) 2685 non-weed and (b) 156 weed species of angiosperms. (c) Relationship between DNA amount and the probability of being a weed. The analysis was based on 2841 species of known 1C-value ranked in order of increasing size including 2719 species with 1C-values of ≤25.1 pg (the maximum for weeds) divided into five groups with 544 (groups 1–4) or 543 (group 5) species and 122 species with >25.1 pg (group 6). The regression line is drawn as a solid line for groups which contain weeds and extended as a broken line for group 6 (see Section 7.5.1). (Reproduced with permission from Bennett et al. 1998).

confusion may have arisen because these studies did not cover the full ecological ranges possible (e.g. elevations from sea level to mountain tops or latitudes from the tropics to the poles), other factors had a large and confounding effect, and/or sample sizes were small. Moreover, data were usually analyzed by linear regres-

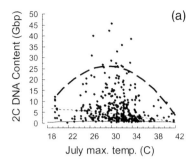 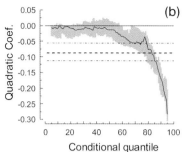

Fig. 7.6 (a) Scatter plot of the mean July maximum temperature inside the range of 421 species in the California flora versus the mean 2C DNA content of those species in gigabase pairs (Gbp). (b) Quantile regression analysis of a, showing a decreasing quadratic coefficient for increasing quantiles. The lines in a correspond to the 5th quantile (thin solid line), the 50th quantile (thin dashed line) and the 95th quantile (thick dashed line). The lines in b correspond to the least squares estimate for the normal mean quadratic function for the relationship depicted in a (dashed line) and the confidence interval of that estimate (double dashed line). The gray area depicts the quantile-dependent confidence interval for the quadratic coefficient. (Reproduced with permission from Knight et al. 2005).

sion analyses which may obscure patterns if the relationships between genome size and ecological variables are non-linear.

To overcome these potential problems, Knight and colleagues (2002, 2005) took a broader approach, looking at genome size and ecological requirements across a broad environmental gradient. They analyzed DNA amounts in 421 species of the Californian flora and looked for relationships between DNA content and various ecological parameters. They also suggested that the relationship between genome size and many environmental factors may be more accurately represented by a unimodal distribution whereby species with low DNA content can exist in any environment but species with larger genomes may be excluded from extreme environments. Thus they used non-linear quantile regression analysis. When genome size versus July maximum temperature (Fig. 7.6a) was plotted, they showed that while species with small genomes occurred throughout the entire range of temperatures reported, species with large genomes tended to be restricted to the middle of the range and declined in frequency at both temperature extremes. Quantile regression analysis confirmed the results. Thus, while the correlation between genome size and July maximum temperature was weak for species with small genomes, it became increasingly significant as the genome size of a species increased (Fig. 7.6b), supporting the idea that species with large genomes are progressively excluded from habitats with extreme July maximum temperatures. Similar results were reported for annual precipitation.

Such studies are in their infancy but illustrate the importance and value of looking for non-linear relationships between genome size across broad and wide-ranging environmental gradients. They also highlight the extension of the trend already observed, namely that species with large genomes are constrained not only in their life cycle and life strategy options but also in the ecological environment that they can occupy.

7.5.3
Implications of Genome Size Variation on Plants' Responses to Environmental Change

Two recent studies further highlight how genome size variation may play a role in determining how plants will respond to environmental change, namely pollution and threat of extinction.

7.5.3.1 Genome Size and Plant Response to Pollution

Vilhar and colleagues (Vidic et al. 2003) investigated the distribution of plants along a steep gradient of heavy metal pollution produced by a former lead smelter in Zerjav, Slovenia. The sites chosen for study were similar in other respects such as geology, aspect, temperature and vegetation type. To study a functionally and taxonomically defined group of plants, the analysis was restricted to 70 herbaceous perennial eudicot species, which comprised more than half of all species at each site. A survey of genome sizes at the different sites revealed that while all sites had species with small genomes, species with large genomes appeared to be excluded from sites with a high metal concentration in the soil. Plotting the results revealed a clear negative correlation between the concentration of contaminating metals in the soil and the proportion of species with large genomes (1C > 5.2 Gbp). These results provide the first direct evidence that plants with large genomes are at a selective disadvantage in extreme environmental conditions caused by pollution. It is currently unclear whether the results reflect functional relationships between the nucleotypic effects of DNA amount and cellular and whole plant physiology or, in this case, DNA amount itself is a direct target of selection pressure in particular because a high frequency of chromosome aberrations were observed in plant material collected in the metal-polluted valley investigated. Clearly further studies are needed but once again the possession of a large genome appears to reduce the options of where a plant may grow.

7.5.3.2 Genome Size and Threat of Extinction

Genome size variation may also play a role in determining how plants will respond to the threat of extinction from environmental change through, for example, habitat loss, invasion of alien species, and pollution. To investigate a possible link between genome size and extinction threat, Vinogradov (2003) extracted genome size data from the Plant DNA C-values database for 3036 diploid species and checked them against the United Nations Environment Programme World Conservation and Monitoring Centre Species database. From this he identified 305 species which were of global concern, 1329 species of local concern, and

1402 species of no conservation interest (i.e. not at risk of extinction). Plotting the range of genome sizes for each of these groups revealed a dramatic relationship between genome size and conservation status. Species at no risk of extinction were characterized by possessing small genomes with a mean genome size of $1C = 7.4$ pg, whereas those at local risk had significantly larger genomes with a mean genome size of $1C = 12.5$ pg and those of global concern had the largest genomes with a mean genome size of $1C = 42.5$ pg. This relationship was shown to hold even when analyses were carried out within families to overcome, to some extent, complicating factors arising from phylogenetic issues. Further analyses showed that the relationship was largely independent of ploidy level (based on C-values for species of known ploidy, not just diploids – 2982 species) and life cycle type (based on C-values for species of known life cycle type – 2510 species). It thus appears that possessing large genomes has consequences for plants relating to their risk of extinction; species with large genomes are at greater risk of extinction than those with small genomes.

7.5.4
Consequences of Genome Size Variation for Survival in a Changing World

Perhaps the above results are not so surprising as many of the consequences of having large genomes may indeed shift the balance of survival towards extinction. In a changing world characterized by habitat destruction, increasing pollution and unstable climate, the ability to colonize and survive in new and possibly extreme environments is essential. Yet many of the characteristics shown to be associated with the possession of a large genome are incompatible with such environments. It has been shown that species with large genomes are slow growing obligate perennials (Bennett 1972) with a restricted ecological distribution (Knight et al. 2005) and sensitive to heavy metal pollution (Vidic et al. 2003). Other studies reinforce this. Sparrow and Miksche (1961) found that species with large genomes were more sensitive to radiation, and Knight et al. (2005) reported that they showed reduced trait variation. For example, a study of seed size showed that while species with small genomes exhibited a wide range of sizes, those with large genomes were generally restricted to producing large seeds with lower dispersal abilities and hence potential to colonize new habitats following environmental change (Knight et al. 2005).

An ability to adapt and evolve in a rapidly changing environment may also be determined in part by genome size, as both Vinogradov (2003) and Knight et al. (2005) reported that species with larger genomes had slower than average rates of diversification. It appears that diversity is constrained in some way in these lineages and this may in turn increase the chances of extinction.

Taken together, such studies highlight the complexity of working with biological systems and reinforce the need to further elucidate the consequences of genome size variation. Perhaps, more importantly, they also stress the need to obtain and incorporate genome size information into computer models and laboratory experiments designed to examine how organisms will respond to the multifaceted environmental changes induced through human activities.

7.6
Methods of Estimating Genome Size in Plants and the Impact of Flow Cytometry

Understanding the biological and evolutionary significance of genome size has only been possible through the development of methods which can provide large numbers of accurate genome size estimates. Yet the path to achieving this has been far from simple and straightforward.

Probably the first plant to have its genome size determined was *Lilium longiflorum* (Liliaceae) in 1951 (Ogur et al. 1951), but this involved a tedious chemical extraction technique for isolating DNA from a known number of cells and staining it colorimetrically. Although such methods were largely accurate, they were complicated and slow and hence rate limiting, thus the generation of data during the early studies of genome size was laborious and slow (Fig. 7.7).

The advent and application of photomicrodensitometry for genome size estimation in the 1960s represented a major step forward and the number of genome size data began to increase more rapidly. Rather than extracting DNA from cells, the method involved staining nuclei fixed on microscope slides and then measuring the amount of light absorbed by the stain. The most commonly used method of staining was the Feulgen reaction and it is still widely practised today.

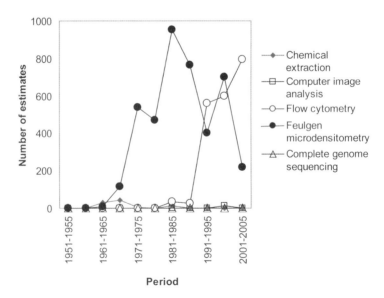

Fig. 7.7 Analysis of the number of genome size estimates made using a range of different techniques. Based on data in the Plant DNA C-values database (release 4.0, October 2005).

7.6.1
The Development of Flow Cytometry for Genome Size Estimation in Angiosperms

While Feulgen microdensitometry still accounts for the majority of genome size estimates reported (e.g. 63% of all estimates in the Plant DNA C-values database (release 4.0, October 2005) were made using this method), the 1980s saw the development of FCM as a viable alternative means.

FCM had, in fact, been used for measuring genome size in the 1970s and early 1980s, but the methods employed were laborious and had to be worked out empirically for each species examined (e.g. Heller 1973; Puite and Tenbroeke 1983). A breakthrough came in 1983 when Galbraith et al. (1983) developed a simple and rapid method of nuclei preparation that appeared to be applicable to a wide range of species. Many researchers started to use this protocol and consequently there was a large increase in the number of genome size estimates published using FCM. By the early 1990s, it had become the method of choice for many researchers (Fig. 7.7) and it continues to be widely used. However, the technique has often been applied without rigorous evaluation of various parameters which have since proved to be critical for obtaining accurate genome size data. Four main areas of concern are listed below (see also Chapters 4, 12 and 18).

7.6.1.1 Choice of Fluorochromes
Following the publication of Galbraith et al.'s paper (1983), a diverse range of fluorochromes was employed for genome size estimation, which is broadly divided into two groups: (i) the DNA intercalating dyes, including propidium iodide and ethidium bromide, and (ii) base-specific fluorochromes, including mithramycin, chromomycin, olivomycin, DAPI and Hoechst. However, in a series of studies Doležel and colleagues (Doležel et al. 1992, 1998) showed that only intercalating dyes were suitable for the accurate estimation of genome size. Further, in a large-scale comparative study involving four laboratories, Doležel et al. (1998) demonstrated that there was excellent agreement between results obtained using either Feulgen microdensitometry or FCM if propidium iodide was used as the fluorochrome. Following discussions at the first and second Plant Genome Size Meetings (1997 and 2003), propidium iodide is now recommended as the fluorochrome of choice for genome size studies (see Key recommendation 5: flow cytometry at http://www.kew.org/genomesize/pgsm/index).

7.6.1.2 Internal Standardization
Another key factor in the generation of accurate genome size data is the use of an internal standard for calibrating the flow histogram, whereby the calibration standard and sample are isolated, stained and analyzed simultaneously. While this was recognized as early as 1991 (Doležel 1991), the importance of internal standardization was not fully appreciated in some early studies, and those using external calibration, especially when reporting intraspecific variation, should be re-evaluated. Certainly, where this has been done, the data have not always been reproducible. For example, Graham et al. (1994) reported a 1.11-fold variation in

genome size for different accessions of soybean (*Glycine max*, Fabaceae) using FCM with external standardization. Yet, a subsequent study by Greilhuber and Obermayer (1997) showed that such variation could be eliminated by internal standardization.

7.6.1.3 The Need for Cytological Data

FCM may give a highly accurate DNA value for a taxon but this value is limited if the chromosome number of the individual plant or tissue analysed is unknown. The significance of differences in DNA amount between different tissues, plants, populations, and species measured using FCM should therefore always be assessed in conjunction with cytological analysis.

Many chromosomal changes and variations (e.g. aneuploidy, chromosome duplications and deletions, sex and supernumerary chromosomes, etc.) are associated with detectable changes in DNA amount. If these are not identified before or after using FCM, then interpretation of the results could be flawed. Thus, although the speed of analysis of FCM is appealing, the essential need for parallel cytological examination, which can be time consuming, should not be understated for a complete evaluation of the results.

7.6.1.4 Awareness of the Possible Interference of DNA Staining

A fourth factor that was largely unappreciated in early studies was the effect that cytosolic compounds, released during nuclei isolation, could have on the fluorochrome staining of DNA. This occurred despite warnings from Greilhuber (1986, 1988a, 1988b) concerning the presence of similar compounds which were observed to severely inhibit the staining of DNA with Feulgen stain. For accurate genome size estimations, using internal standardization, it is assumed that fluorochrome accessibility to DNA is the same in the sample and standard nuclei, yet it is becoming increasingly clear that this is not always the case. One of the first studies to reveal the influence of cytosolic compounds was carried out on sunflower (*Helianthus annuus*, Asteraceae) where the DNA amounts were reported to vary by up to 48% between different leaves of individual sunflowers (Price and Johnston 1996; Price et al. 1998). It was proposed that the intraspecific variation was influenced by the light quantity and/or quality. However, a failure to replicate the data in subsequent experiments led to the realization that the initial results had been caused by the environmentally-induced production of cytosolic inhibitors, which interfered with the fluorescence emission of propidium iodide and/or its binding to DNA (Price et al. 2000).

Since then, it has become increasingly apparent that cytosolic compounds can have a profound effect on genome size estimations leading not only to the false identification of intraspecific variation, as in the case of the sunflowers, but also to erroneous determinations of absolute genome size. Investigations into the diversity and mode of action of such compounds are still in their infancy but their potential presence and effect is something researchers should always be alert to. For example, in an elegant series of experiments which included *Coffea* (Rubiaceae) and *Petunia* (Solanaceae), Noirot and colleagues (Noirot et al. 2000, 2002,

2003, 2005) identified compounds released during co-chopping of the standard and sample that enhanced or inhibited fluorochrome binding to DNA. Their presence resulted in pseudo-intraspecific variation and stoichiometric errors in genome size estimation.

Research has shown that the levels of some of these compounds, which include chlorogenic acids, other phenolics and anthocyanins, are determined in part by environmental factors (e.g. temperature, altitude, light intensity, and drought), and even plants grown in the same environment have the potential to exhibit pseudo-intraspecific variation arising from polymorphisms in the regulatory genes and transcription factors controlling their biosynthesis. Clearly, many studies reporting environmental variation in genome size may need re-evaluation in light of our increased understanding of the role cytosolic chemicals may play in generating artifactual data, and researchers should always be alert to their potential presence.

While there is still much to be learnt about the role cytosolic compounds may play in influencing genome size estimates, ways of minimizing problems associated with these compounds are discussed in Chapter 4 by Greilhuber et al., and by several other authors (e.g. Doležel and Bartoš 2005; Noirot et al. 2005). From the above discussions it should be evident that the initial analysis of genome size for any new plant species requires careful attention to the choice of fluorochrome, the potential presence of interfering compounds, possible environmental and genetic influences determining their concentration, as well as careful cytological analysis. While subsequent studies of the same material may not be so time consuming, the naive view of FCM as a simple and rapid technique for accurate genome size estimation can not longer be taken at face value – care, attention and time are essential if FCM is to be used to its full potential and produce meaningful and accurate genome size data.

7.6.2
Potential for the Application of Flow Cytometry to Other Plant Groups

The use of FCM has not been restricted to angiosperms but has been applied to other plant groups with varying degrees of success.

7.6.2.1 Gymnosperms

In gymnosperms, FCM accounts for 58% (204 out of 348) of all the genome size estimates listed in the Gymnosperm DNA C-values database (Murray et al. 2004). Of these, the majority have used the fluorochrome propidium iodide (193 out of 204 = 94%). Studies of 19 *Pinus* (Pinaceae) species by Wakamiya et al. (1993), which compared genome size estimates obtained using Feulgen microdensitometry and FCM using propidium iodide, showed good agreement. Nevertheless, all the concerns listed above for angiosperms apply to gymnosperms and the numerous claims of intraspecific variation in gymnosperms estimated by FCM (e.g. Bogunic et al. 2003; Hall et al. 2000) should be reinvestigated in light of increased understanding of the technical pitfalls of FCM.

7.6.2.2 Pteridophytes

As noted in Section 7.3.3, representation of genome size data for pteridophytes is poor. Yet out of the 112 estimates currently in the Pteridophyte DNA C-values database (Bennett and Leitch 2004), just over half have been determined by FCM. In a carefully targeted study to increase the amount of genome size data available for pteridophytes, Obermayer et al. (2002) assessed the ease with which measurements could be made by FCM using propidium iodide. While the technique produced distinct peaks in flow histograms for species whose C-values spanned the entire range of genome sizes encountered in pteridophytes so far (i.e. 1C = 0.16–72.67 pg), the ease and speed of obtaining data varied considerable. Some taxa, including members of Equisetales, Psilotales and some "polypodiaceaous" ferns, were found to be relatively easy to measure while others proved more problematic, giving peaks with unacceptably high CVs or in the worst case (i.e. the filmy fern *Trichomanes*, Hymenophyllaceae) no peaks at all. Various approaches were tried to overcome difficulties including increasing the amount of leaf material analyzed or using roots instead of leaves, and in most cases this led to the successful determination of genome size. Further work is clearly needed to optimize the methods (e.g. modification of buffer composition, alteration of stain concentration, etc.) to increase the potential of FCM for estimating C-values in pteridophytes.

7.6.2.3 Bryophytes

There have been several studies estimating bryophyte genome size using FCM, however, those using DAPI as the fluorochrome (e.g. Lamparter et al. 1998; Reski 1998; Reski et al. 1994) are biased because this dye is base-specific (see Section 7.6.1.1). Of the remainder, Voglmayr (2000) reported the most comprehensive survey of genome sizes in 138 moss taxa, selected to increase taxonomic representation of genome size data, using FCM with propidium iodide. No particular problems were mentioned, although small differences in genome size reported for different cytotypes, and suggested to be due to aneuploidy, could not be confirmed as no cytological analyses were conducted. Once again, the importance of cytological analysis to correctly interpret the data is emphasized (see Section 7.6.1.3) and the possibility of cytosolic inhibitors leading to the differences observed cannot be ruled out. Recent studies, again extolling the speed and reliability of FCM, also failed to carry out cytological checks on the material examined (Melosik et al. 2005).

7.7
Recent Developments and the Future of Flow Cytometry in Genome Size Research

FCM has clearly been responsible for the generation of considerable amounts of genome size data in plants. While some values (particularly from the late 1980s and early 1990s) may be flawed and need to be re-measured, the increased understanding and awareness of how various parameters of the technique contribute to generating reliable genome size data means that the future of FCM as an accurate method for genome size determination is assured, provided that appropriate care and attention is taken. Yet it is clear that the initial excitement in the 1980s and early 1990s that the method would be quick and simple should be replaced with a less naive view that the method is reliable but not necessarily quick and cannot be used without due care and attention.

The two Plant Genome Size Workshops at the 1997 and 2003 meetings (see http://www.kew.org/genomesize/pgsm/index) and the workshop held as part of the International Botanical Congress in Vienna 2005 (http://www.kew.org/genomesize/vienna05_report.pdf) have led to recommendations for the accurate determination of genome size (see "The nine key recommendations" at http://www.kew.org/genomesize/pgsm/index). These should be followed together with the key points given in Chapter 4.

Along with increased understanding of the method, there has been the development of new technology and chemicals, which should extend the technique to a wider range of applications. Of particular note is the development of a portable flow cytometer (http://www.partec.de/products/cylab.html) which permits taking the machine to the field rather than bringing the plant to the laboratory. However, while such opportunities potentially increase the range of species which can be analyzed, there are formidable problems to be overcome to ensure accurate genome size data are collected. The researcher still needs to be aware of, and check for, the presence of cytosolic compounds and have a range of plants growing which are suitable as calibration standards. In addition, material should be collected for cytological examination. While a portable flow cytometer has been successfully used for a number of other applications such as the analysis of phytoplankton (Olson et al. 2003; Sosik et al. 2003; Veldhuis et al. 2005; Chapter 13), its use for genome size measurements in the field for land plant groups has yet to be realized.

Other areas of development include the potential of increasing the diversity of plant material which can be analyzed. Despite suggestions that all kinds of plant tissue are suitable for FCM analysis (e.g. Galbraith 1990), fresh leaf material is usually used. However, this can have limitations as the material needs to be analyzed within a reasonable length of time after harvesting, restricting the source of material available for analysis. One alternative to overcome these difficulties has been to collect seeds from the plants of interest, and germinate them in the laboratory to provide fresh, young material suitable for analysis. Suda et al. (2003) demonstrated the applicability of this approach for analyzing 104 Macaronesian angiosperms. More recently, the possibility of measuring genome size directly from specific tissues in ungerminated seed (most often the hypocotyl plus radicle) has been investigated by Sliwinska et al. (2005). This topic is discussed in detail in Chapter 4, but if such an approach turns out to be applicable to a broad range of plants, then it opens up the exciting possibility of obtaining data from the vast diversity of seeds present in numerous seed banks around the world. As these will certainly contain species collected from areas where obtaining fresh leaf material would be extremely difficult, the potential to increase the diversity of genome size data is significant. Of course, this approach has the disadvantage of destroying the seed so that there is no possibility of also carrying out a chromosome count on the same material (see Section 7.6.1.3). Where plenty of seed material is available, this can be overcome by analyzing sufficient numbers of seeds, however for rare seeds such limitation should not be underestimated.

The future importance of FCM for genome size estimation is likely to remain high as the "time bomb" of obsolescence microdensitometers has taken its toll (Bennett and Leitch 2005b). Yet FCM will not be the sole choice available to genome size researchers. The rise of computer-based image analysis systems is likely to continue as this method has already been shown to produce accurate genome size data, comparable with that obtained by FCM and Feulgen microdensitometry (Vilhar and Dermastia 2002; Vilhar et al. 2001; see also Chapter 1). Having the choice of instrumentation will provide researchers with viable alternatives and thus enable genome size data to be collected and fed into holistic genomic studies aimed at increasing the understanding of the biological significance of genome size variation.

References

Adams, M. D., Celniker, S. E., Holt, R. A., Evans, C. A., Gocayne, J. D., Amanatides, P. G., Scherer, S. E., Li, P. W., Hoskins, R. A., Galle, R. F. et al. **2000**, *Science* 287, 2185–2195.

Arabidopsis Genome Initiative. **2000**, *Nature* 408, 796–815.

Armstrong, S. J., Franklin, F. C. H., Jones, G. H. **2003**, *Sexual Plant Reprod.* 16, 141–149.

Baer, R., Bankier, A. T., Biggin, M. D. et al. **1984**, *Nature* 310, 207–211.

Baetcke, K. P., Sparrow, A. H., Nauman, C. H., Schwemmer, S. S. **1967**, *Proc. Natl Acad. Sci. USA* 58, 533–540.

Bedell, J. A., Budiman, M. A., Nunberg, A., Citek, R. W., Robbins, D., Jones, J., Flick, E., Rohlfing, T., Fries, J., Bradford, K. et al. **2005**, *PLoS Biol.* 3, 103–115.

Bennett, M. D. **1971**, *Proc. Roy. Soc. Lond. Series B – Biol. Sci.* 178, 277–299.

Bennett, M. D. **1972**, *Proc. Roy. Soc. Lond. Series B – Biol. Sci.* 181, 109–135.

Bennett, M. D. **1977**, *Phil. Trans. Roy. Soc. Lond. Series B – Biol. Sci.* 277, 201–277.

Bennett, M. D. **1987**, *New Phytologist* 106, 177–200.

Bennett, M. D., Leitch, I. J. **1995**, *Ann. Bot.* 76, 113–176.

Bennett, M. D., Leitch, I. J. **1997**, *Ann. Bot.* 80, 169–196.

Bennett, M. D., Leitch, I. J. **2004**, *Pteridophyte DNA C-values Database (Release 3.0, Dec. 2004)*. http://www.kew.org/genomesize/

Bennett, M. D., Leitch, I. J. **2005a**, Genome size evolution in plants in *The Evolution of the Genome*, ed. T. R. Gregory, Elsevier, San Diego, pp. 90–162.

Bennett, M. D., Leitch, I. J. **2005b**, *Ann. Bot.* 95, 45–90.

Bennett, M. D., Leitch, I. J. **2005c**, *Plant DNA C-values Database (Release 4.0, Oct. 2005)*. http://www.kew.org/genomesize/

Bennett, M. D., Leitch, I. J. **2005d**, *Ann. Bot.* 95, 1–6.

Bennett, M. D., Smith, J. B. **1976**, *Phil. Trans. Roy. Soc. Lond. Series B – Biol. Sci.* 274, 227–274.

Bennett, M. D., Smith, J. B. **1991**, *Phil. Trans. Roy. Soc. Lond. Series B – Biol. Sci.* 334, 309–345.

Bennett, M. D., Smith, J. B., Ward, J., Jenkins, G. **1981**, *J. Cell Sci.* 47, 91–115.

Bennett, M. D., Smith, J. B., Heslop-Harrison, J. S. **1982**, *Proc. Roy. Soc. Lond. Series B – Biol. Sci.* 216, 179–199.

Bennett, M. D., Heslop-Harrison, J. S., Smith, J. B., Ward, J. P. **1983**, *J. Cell Sci.* 63, 173–179.

Bennett, M. D., Leitch, I. J., Hanson, L. **1998**, *Ann. Bot.* 82, 121–134.

Bennett, M. D., Bhandol, P., Leitch, I. J. **2000**, *Ann. Bot.* 86, 859–909.

Bennett, M. D., Leitch, I. J., Price, H. J., Johnston, J. S. **2003**, *Ann. Bot.* 91, 547–557.

Bogunic, F., Muratovic, E., Brown, S. C., Siljak-Yakovlev, S. **2003**, *Plant Cell Rep.* 22, 59–63.

Costa, P., Pot, D., Dubos, C., Frigerio, J. M., Pionneau, C., Bodenes, C., Bertocchi, E., Cervera, M. T., Remington, D. L., Plomion, C. **2000**, *Theor. Appl. Genet.* 100, 39–48.

Doležel, J. **1991**, *Phytochem. Anal.* 2, 143–154.

Doležel, J., Bartoš, J. **2005**, *Ann. Bot.* 95, 99–110.

Doležel, J., Sgorbati, S., Lucretti, S. **1992**, *Physiol. Plant.* 85, 625–631.

Doležel, J., Greilhuber, J., Lucretti, S., Meister, A., Lysák, M. A., Nardi, L., Obermayer, R. **1998**, *Ann. Bot.* 82 (Suppl. A), 17–26.

Fay, M. F., Cowan, R. S., Leitch, I. J. **2005**, *Ann. Bot.* 95, 237–246.

Fleischmann, R. D., Adams, M. D., White, O., Clayton, R. A., Kirkness, E. F., Kerlavage, A. R., Bult, C. J., Tomb, J. F., Dougherty, B. A., Merrick, J. M. et al. **1995**, *Science* 269, 496–512.

Flynn, S., Turner, R. M., Dickie, J. B. **2004**, *Seed Information Database (Release 6.0, October 2004)*. http://www.rbgkew.org.uk/data/sid.

Galbraith, D. W. **1990**, *Methods Cell Biol.* 33, 549–562.

Galbraith, D. W., Harkins, K. R., Maddox, J. M., Ayres, N. M., Sharma, D. P., Firoozabady, E. **1983**, *Science* 220, 1049–1051.

Garner, T. W. J. **2002**, *Genome* 45, 212–215.

Goffeau, A., Barrell, B. G., Bussey, H., Davis, R. W., Dujon, B., Feldmann, H., Galibert, F., Hoheisel, J. D., Jacq, C., Johnston, M. et al. **1996**, *Science* 274, 546–567.

Govaerts, R. **2004**, *Taxon* 53, 144–146.

Graham, M. J., Nickell, C. D., Rayburn, A. L. **1994**, *Theor. Appl. Genet.* 88, 429–432.

Greilhuber, J. **1986**, *Can. J. Genet. Cytol.* 28, 409–415.

Greilhuber, J. **1988a**, Critical reassessment of DNA content variation in plants in *Kew Chromosome Conference III*, ed. P. E. Brandham, HMSO, Kew, pp. 39–50.

Greilhuber, J. **1988b**, *Plant Systemat. Evol.* 158, 87–96.

Greilhuber, J., Obermayer, R. **1997**, *Heredity* 78, 547–551.

Greilhuber, J., Sastad, S. M., Flatberg, K. I. **2003**, *J. Bryol.* 25, 235–239.

Greilhuber, J., Temsch, E. M., Krisai, R. **2004**, Abstract for the 11th Meeting of Austrian Botanists, Vienna, Austria (September 2004).

Greilhuber, J., Borsch, T., Müller, K., Worberg, A., Porembski, S., Barthlott, W. **2005**, *Abstracts of the XVII International Botanical Congress, Vienna, Austria* 138.

Greilhuber, J., Borsch, T., Müller, K., Worberg, A., Porembski, S., Barthlott, W. **2006**, *Plant Biol.* (in press).

Hall, S. E., Dvorak, W. S., Johnston, J. S., Price, H. J., Williams, C. G. **2000**, *Ann. Bot.* 86, 1081–1086.

Hanson, L., Leitch, I. J. **2002**, *Bot. J. Linn. Soc.* 140, 169–173.

Hanson, L., McMahon, K. A., Johnson, M. A. T., Bennett, M. D. **2001a**, *Ann. Bot.* 87, 251–258.

Hanson, L., McMahon, K. A., Johnson, M. A. T., Bennett, M. D. **2001b**, *Ann. Bot.* 88, 851–858.

Hanson, L., Brown, R. L., Boyd, A., Johnson, M. A. T., Bennett, M. D. **2003**, *Ann. Bot.* 91, 1–8.

Hanson, L., Boyd, A., Johnson, M. A. T., Bennett, M. D. **2005**, *Ann. Bot.* 96, 1315–1320.

Heller, F. O. **1973**, *Ber. Deutsch. Botan. Gesell.* 86, 437–441.

Kapraun, D. F. **2005**, *Ann. Bot.* 95, 7–44.

Kapraun, D. F., Leitch, I. J., Bennett, M. D. **2004**, *Algal DNA C-values Database (Release 1.0, Dec. 2004)*. http://www.kew.org/genomesize/homepage.html.

Knight, C. A., Ackerly, D. D. **2002**, *Ecol. Lett.* 5, 66–76.

Knight, C. A., Molinari, N., Petrov, D. A. **2005**, *Ann. Bot.* 95, 177–190.

Lamparter, T., Brucker, G., Esch, H., Hughes, J., Meister, A., Hartmann, E. **1998**, *J. Plant Physiol.* 153, 394–400.

Leitch, I. J., Hanson, L., Winfield, M., Parker, J., Bennett, M. D. **2001**, *Ann. Bot.* 88, 843–849.

Margulies, M., Egholm, M., Altman, W. E., Attiya, S., Bader, J. S., Bemben, L. A., Berka, J., Braverman, M. S., Chen, Y.-J., Chen, Z. et al. **2005**, *Nature* 437, 376–380.

Masterson, J. **1994**, *Science* 264, 421–424.

Melosik, I., Odrzykoski, I. I., Sliwinska, E. **2005**, *Nova Hedwigia* 80, 397–412.

Murray, B. G. **1998**, *Ann. Bot.* 82, 3–15.

Murray, B. G., Bennett, M. D., Leitch, I. J. **2004**, *Gymnosperm DNA C-values Database (Release 3.0, Dec. 2004)*. http://www.kew.org/genomesize/homepage.html.

Noirot, M., Barre, P., Louarn, J., Duperray, C., Hamon, S. **2000**, *Ann. Bot.* 86, 309–316.

Noirot, M., Barre, P., Louarn, J., Duperray, C., Hamon, S. **2002**, *Ann. Bot.* 89, 385–389.

Noirot, M., Barre, P., Duperray, C., Louarn, J., Hamon, S. **2003**, *Ann. Bot.* 92, 259–264.

Noirot, M., Barre, P., Duperray, C., Hamon, S., De Kochko, A. **2005**, *Ann. Bot.* 95, 111–118.

Obermayer, R., Leitch, I. J., Hanson, L., Bennett, M. D. **2002**, *Ann. Bot.* 90, 209–217.

Ogur, M., Erickson, R. O., Rosen, G. U., Sax, K. B., Holden, C. **1951**, *Exp. Cell Res.* 2, 73–89.

Ohri, D. **2005**, *Plant Biol.* 7, 449–458.

Olson, R. J., Shalapyonok, A., Sosik, H. M. **2003**, *Deep-Sea Res. Part I – Oceanogr. Res. Papers* 50, 301–315.

Peer, W. A., Mamoudian, M., Lahner, B., Reeves, R. D., Murphy, A. S., Salt, D. E. **2003**, *New Phytologist* 159, 421–430.

Price, H. J., Johnston, J. S. **1996**, *Proc. Natl Acad. Sci. USA* 93, 11264–11267.

Price, H. J., Morgan, P. W., Johnston, J. S. **1998**, *Ann. Bot.* 82, 95–98.

Price, H. J., Hodnett, G., Johnston, J. S. **2000**, *Ann. Bot.* 86, 929–934.

Pryer, K. M., Schneider, H., Smith, A. R., Cranfill, R., Wolf, P. G., Hunt, J. S., Sipes, S. D. **2001**, *Nature* 409, 618–622.

Puite, K. J., Tenbroeke, W. R. R. **1983**, *Plant Science Letters* 32, 79–88.

Qiu, Y. L., Palmer, J. D. **1999**, *Trends Plant Sci.* 4, 26–30.

Reski, R. **1998**, *Bot. Acta* 111, 1–15.

Reski, R., Faust, M., Wang, X. H., Wehe, M., Abel, W. O. **1994**, *Mol. Gen. Genet.* 244, 352–359.

Robinson, J. G. **2004**, *Science* 304, 53–53.

Sliwinska, E., Zielinska, E., Jedrzejczyk, I. **2005**, *Cytometry* 64A, 72–79.

Sosik, H. M., Olson, R. J., Neubert, M. G., Shalapyonok, A., Solow, A. R. **2003**, *Limnol. Oceanogr.* 48, 1756–1765.

Sparrow, A. H., Miksche, J. P. **1961**, *Science* 134, 282–283.

Suda, J., Kyncl, T., Freiova, R. **2003**, *Ann. Bot.* 92, 153–164.

Temsch, E. M., Greilhuber, J., Krisai, R. **1998**, *Bot. Acta* 111, 325–330.

Veldhuis, M. J. W., Timmermans, K. R., Croot, P., van der Wagt, B. **2005**, *J. Sea Res.* 53, 7–24.

Vidic, T., Greilhuber, J., Vilhar, B. **2003**, *Abstract 31 of the Second Plant Genome Size Meeting held at The Royal Botanic Gardens, Kew, September 2003*. http://www.kew.org/cval/abstracts/cvalAbstract31.html.

Vilhar, B., Dermastia, M. **2002**, *Acta Bot. Croat.* 61, 11–25.

Vilhar, B., Greilhuber, J., Koce, J. D., Temsch, E. M., Dermastia, M. **2001**, *Ann. Bot.* 87, 719–728.

Vinogradov, A. E. **2003**, *Trends Genet.* 19, 609–614.

Voglmayr, H. **2000**, *Ann. Bot.* 85, 531–546.

Wakamiya, I., Newton, R. J., Johnston, J. S., Price, H. J. **1993**, *Am. J. Bot.* 80, 1235–1241.

Watson, J. D., Crick, F. H. C. **1953**, *Nature* 171, 737–738.

Winkler, H. **1920**, *Verbreitung und Ursache der Parthenogenesis im Pflanzen- und Tierreiche*, Gustav Fischer Verlag, Jena.

8
DNA Base Composition of Plant Genomes

Armin Meister and Martin Barow

Overview

The fluorescence of base-specific dyes is correlated with the DNA base composition. However, the intensity of the fluorescence signal is not simply proportional to the total number of AT or GC base pairs. Rather the dyes require a certain (between three and five) number of consecutive bases of the same type to bind to the DNA (= binding length). The relation between base composition and base-specific fluorescence is treated mathematically in this chapter, assuming a random distribution of bases. Moreover, the effect of deviation from randomness is considered. Using the resulting formulae, the base composition of genomic DNA can be determined by flow cytometry. The values for 215 species, which had been estimated in this way, were excerpted from the literature and are presented here in one comprehensive table. Relationships between base composition and genome size and taxonomic position have been claimed in some papers. These correlations are discussed in the light of new representative data. The results obtained by flow cytometry are compared with base frequencies determined by other (mostly older) physico-chemical methods (chromatography, buoyant density centrifugation, thermal denaturation, different reactivity of the bases, UV absorbance) and recent sequencing data. The variability of results obtained by flow cytometry is comparable to those obtained by the other methods. However, flow cytometry requires far less material and time. Finally, possible sources of error of the flow cytometric determination of base composition are discussed.

8.1
Introduction

The genetic information of living organisms is encoded in the sequence of the four bases: adenine (A), thymine (T), guanine (G), and cytosine (C) within the polynucleotide chain of the genomic DNA. The DNA of different species contains different amounts of these bases and the frequencies of the complementary bases adenine + thymine (AT) and guanine + cytosine (GC) are characteristic for each species. If the sequence is known, the AT and GC frequencies are also known. However, sequencing of complex genomes is time consuming and expensive, and at the moment only the sequences of two plant species (*Arabidopsis thaliana*, Brassicaceae (The Arabidopsis Genome Initiative 2000) and *Oryza sativa*, Poaceae (Goff et al. 2002; Yu et al. 2002)) are (almost completely) known. This situation will not change dramatically in the near future. Therefore, physico-chemical properties of the bases were used to determine the base composition of genomic DNA (see Section 8.2.3). Because certain DNA-binding fluorescent dyes have a different affinity for the DNA bases, flow cytometry (FCM) can also be employed for the determination of base frequencies.

8.2
Analysis of Base Composition by Flow Cytometry

While DNA intercalating dyes such as propidium iodide (PI) and ethidium bromide (EB) bind to all bases uniformly, other dyes bind preferentially to one type of DNA base pair (AT or GC): 4′,6-diamidino-2-phenylindole (DAPI), 4′,6-bis (2′-imidazolinyl-4′5′-H)-2-phenylindole (DIPI) and several Hoechst (HO) dyes bind to AT bases within the minor groove of the helix (Larsen et al. 1989; Portugal and Waring 1988), while mithramycin A (MI) and some other antibiotic dyes such as chromomycin A3 (CMA) and olivomycin (OL) bind to GC bases by forming complexes with the helical DNA in the presence of Mg^{2+} ions (Goldberg and Friedman 1971; Van Dyke and Dervan 1983). The binding properties of base-specific dyes have been known for a long time (CMA, MI and OL: Ward et al. 1965; HO 33258: Weisblum and Haenssler 1974; DAPI: Schweizer 1976; DIPI: Schnedl et al. 1977; HO 33342: Brown et al. 1991) and successfully used for visualizing chromosome regions rich in either AT (human C, G and Q bands) or GC (R bands) bases by fluorescence microscopy (Schweizer 1981).

The principle of using dyes with different base preferences in FCM is explained in Fig. 8.1. Leemann and Ruch (1982) were the first to calculate the base composition in plant nuclei stained with DAPI or CMA by measuring the fluorescence intensities of microscopic samples. Their results for *Phaseolus vulgaris* (Fabaceae), *Allium cepa* (Alliaceae) and *Anemone blanda* (Ranunculaceae) agreed well with those obtained by physico-chemical methods (see Section 8.2.3), although a simple linear model was used for the calculation (assuming a binding length of one base pair; see Section 8.2.1). However, the method was rather time consuming because a large number of single nuclei had to be analyzed to achieve a high statistical reliability and precision.

FCM has the important advantage of measuring the fluorescence intensity of a large number of particles (e.g. cell nuclei) within a short time and with high precision (typically 10 000 within a few minutes). Therefore, it has been successfully used for the determination of genome size with DNA-specific dyes (Galbraith et al. 1983). Doležel et al. (1992) were the first to exploit the base specificity of certain dyes for the quantitative determination of base composition in some angio-

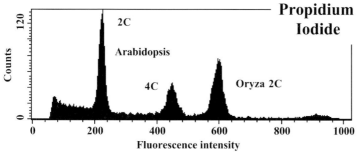
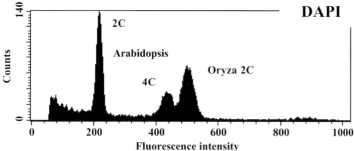
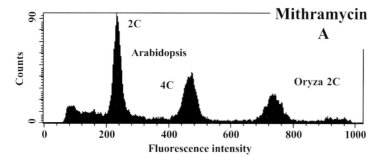

Fig. 8.1 Histograms of relative fluorescence intensities obtained after the analysis of nuclei isolated from *Oryza sativa* (Poaceae) and *Arabidopsis thaliana* (Brassicaceae), and stained with different fluorescent dyes (both species have different base composition). The two left peaks correspond to 2C and 4C nuclei of *A. thaliana*, the right peak corresponds to 2C nuclei of *O. sativa*. Because the base composition of 2C and 4C nuclei of the same species is identical, the ratio between 4C and 2C peaks of *A. thaliana* is (nearly) exactly 2.00 for all dyes, as expected from the doubled DNA content. However, compared to the base-independent propidium iodide (PI), the 2C peak of the relative GC-rich *O. sativa* nuclei is shifted to the right if stained with the GC-binding dye mithramycin A (MI) and shifted to the left if stained with the AT-binding dye DAPI. The 2C peak ratios *Oryza/Arabidopsis* are 2.67 (PI), 2.30 (DAPI), and 3.12 (MI). Reprinted from Meister (2005) with permission from Elsevier.

sperms by FCM. They found that the fluorescence intensity of base-specific dyes was not simply proportional to the absolute amount of the respective bases but that the sum of AT + GC bases calculated by such a simple relationship was always higher than the DNA content determined using a fluorescence dye which was not base-specific, such as propidium iodide.

Godelle et al. (1993) obtained similar results for several species of *Crepis* (Asteraceae). Apparently, there was no linear relation between base content and fluorescence intensity. This was explained by "overspecifity" of dyes, because a certain number of consecutive bases of the same type (AT or GC) were necessary to bind one dye molecule ("binding length"). Based on the calculations of Langlois et al. (1980) who investigated the relation between fluorescence intensity and base composition in metaphase chromosomes for base-specific dyes, Godelle et al. (1993) derived formulae, which took binding length into account. Because the original curvilinear relationship between base composition and fluorescence intensity cannot easily be used for calculations, they proposed a simplified root function, which could be solved with the aid of a pocket calculator. This simplified formula [see Eq. (10)] has been adopted in the majority of base composition calculations since the publication of Godelle's paper (see Table 8.4 summarizing results for 215 species).

8.2.1
Fluorescence of Base-Specific Dyes: Theoretical Considerations

The available data suggest that the known base-specific dyes do not bind to single bases but to groups of consecutive base pairs of the same (GC or AT) type. The length of these groups ("binding length") seems to vary between three and five (Table 8.1). Assuming a random distribution of bases, Langlois et al. (1980) calculated the following relationship between the total frequency p of the respective bases, binding length n, and the frequency f_d of dye molecules bound to groups of bases:

$$f_d = (1 - p) \times p^n / (1 - p^n) \tag{1}$$

Correspondingly, the frequency of bound bases f_n is:

$$f_n = n \times (1 - p) \times p^n / (1 - p^n) \tag{2}$$

This dependence for binding site lengths varying between $n = 3$ and 5 is shown in Fig. 8.2. It is obvious that only a fraction of bases of the same type binds to the base-specific dye. The larger the binding length the greater is the difference. Only for $n = 1$ would all specific bases bind the dye.

The fluorescence intensity F of a nucleus is also proportional to the nuclear DNA content:

$$F = k \times f_n \times \text{Nuclear DNA Content} \tag{3}$$

with k being a proportionality factor. If the base composition of a reference species is known and the intention is to calculate the base composition of a sample

Table 8.1 Binding lengths of some base-specific fluorescent dyes.

Dye	Binding length (bp)	Reference
Hoechst 33258	3	Müller and Gautier 1975
	5	Portugal and Waring 1988
	4	Churchill and Suzuki 1989
	5	Breusegem et al. 2002
Hoechst 33342	5	Brown et al. 1991
DAPI	3	Kapuściński and Szer 1979
	3–4	Portugal and Waring 1988
	4	Barow and Meister 2002
	3	Breusegem et al. 2002
Chromomycin A3	4	Behr et al. 1969
	3	Langlois et al. 1980
Mithramycin A	3	Godelle et al. 1993

species by comparing the fluorescence intensities, the fluorescence ratio has to be compensated for different nuclear DNA contents. This can be done by calculating the "dye factor", DF (Barow and Meister 2002; in a similar form already used by Zonneveld and van Iren 2000). It is defined as:

$$DF = (F_{sample}/F_{reference})/(\text{Nuclear DNA Content}_{sample}/\text{Nuclear DNA Content}_{reference}) \quad (4)$$

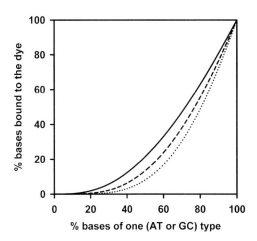

Fig. 8.2 Percentage of bases bound to a base-specific dye as a function of the base frequency for different binding lengths. Solid line, binding length = 3; dashed line, binding length = 4; dotted line, binding length = 5.

DF is the base-specific fluorescence ratio between sample and reference corrected for different genome sizes. If the nuclear DNA content is determined in parallel by comparing the fluorescence intensities of base-unspecific dyes (e.g. PI), the equation becomes (bs = base specific, bu = base unspecific):

$$DF = (F_{sample}(bs)/F_{reference}(bs))/(F_{sample}(bu)/F_{reference}(bu)) \qquad (5)$$

A base-unspecific dye would result in $DF = 1$ by this definition.

The advantage of using this simple dye factor is its independence of any assumption on the binding mechanism, especially on the number of consecutive bases of the same type (binding length), which plays a critical role in the calculations of Langlois et al. (1980). However, it is useful only in comparison to a general reference value (e.g. *Pisum sativum*, Fabaceae with $p = 61.5\%$ AT/38.5% GC; Barow and Meister 2002). If another reference species should be used instead of *P. sativum* (e.g. because of large differences in genome size), the DF of the sample relative to *P. sativum* can be calculated, provided that the DF of the reference standard relative to *Pisum sativum* is known:

$$DF(sample/P.\ sativum) = DF(sample/reference) \times DF(reference/P.\ sativum) \qquad (6)$$

From the Eqs. (2–4), it follows for DF:

$$DF = \{(1 - p_{sample}) \times p_{sample}{}^n/(1 - p_{sample}{}^n)\}/$$
$$\{(1 - p_{reference}) \times p_{reference}{}^n/(1 - p_{reference}{}^n)\} \qquad (7)$$

In this equation, DF (from FCM analysis), $p_{reference}$ and n (from independent experiments) are known; hence the calculation of p_{sample} is in principle possible. However, it is not possible to transform Eq. (7) to a simple formula with p_{sample} on the left side.

Instead, the roots of the following equation in dependence on p_{sample} can be determined:

$$DF - (1 - p_{sample}) \times p_{sample}{}^n/(1 - p_{sample}{}^n)/$$
$$\{(1 - p_{reference}) \times p_{reference}{}^n/(1 - p_{reference}{}^n)\} = 0 \qquad (8)$$

The approximation method of false position (regula falsi; Pollard 1977) may be used for this purpose. The short computer program in BASIC notation is given in Table 8.2.

Godelle et al. (1993) solved this problem by simplifying Eq. (7) to:

$$DF = p_{sample}{}^n/p_{reference}{}^n \qquad (9)$$

This results in the simple formula:

$$p_{sample} = p_{reference} \times DF^{1/n} \qquad (10)$$

Table 8.2 Computer algorithm (in BASIC notation) for exact calculation of base frequency relative to a known reference by the method of false position (regula falsi).

```
F = (1 - BF1) * (BF1 ^ BL)/(1 - BF1 ^ BL)
X1 = .001
X2 = .999
DO
    F1 = (1 - X1) * (X1 ^ BL)/(1 - X1 ^ BL)/F - DF
    F2 = (1 - X2) * (X2 ^ BL)/(1 - X2 ^ BL)/F - DF
    X = X1 - F1 * (X1 - X2)/(F1 - F2)
    DIFF = (1 - X) * (X ^ BL)/(1 - X ^ BL)/F - DF
    IF ABS(DIFF) < .000000001 THEN EXIT DO
    DISCR = SGN(DIFF * F1)
    SELECT CASE DISCR
    CASE 1
        X1 = X
    CASE -1
        X2 = X
    CASE 0
        EXIT DO
    END SELECT
LOOP
BF2 = X
```

BF1, base frequency of the reference; BF2, base frequency of the sample; BL, binding length; DF, dye factor; F, X, X1, X2, DIFF, DISCR, temporary variables used in the course of calculation.

Although this formula has been widely adopted in FCM analyses of base compositions (see Table 8.4), a better approximation would result from the calculation of Langlois et al. (1980) who used a slightly different exponent S instead of n:

$$p_{sample} = p_{reference} \times DF^{1/S} \qquad (11)$$

where S is defined as:

$$S = n/(1 - p_{reference}^{n}) - p_{reference}/(1 - p_{reference}) \qquad (12)$$

Table 8.3 gives the S values for a mean base composition of 40% GC/60% AT, and $n = 1$–5.

Figure 8.3 compares the frequency values relative to a reference value of $p_{reference} = 0.4$ with the binding length $n = 3$ for the exact calculation of Eq. (8), Godelle's simplified equation (10), and the Eqs. (11) and (12) according to Langlois. The graph shows very good agreement between the exact values and those calculated from Eqs. (11) and (12) for p_{sample} varying between 0.3 and 0.5 (corresponding to a DF of between 0.5 and 1.5), while the formula of Godelle shows significant deviation of the resulting curve from the exact curve.

Table 8.3 Corrected exponents S for the simplified calculation of base frequency (assuming $p_{reference} = 0.6$ for AT and $p_{reference} = 0.4$ for GC bases).

Binding length (bp)	S for GC bases	S for AT bases
1	1	1
2	1.71	1.63
3	2.54	2.33
4	3.44	3.10
5	4.39	3.92

Godelle's simplified formula (possibly with the modified exponent S) also allows the calculation of the binding length of an unknown dye by comparison with a dye of known binding length. From Eq. (10), it follows (coefficients 1 and 2 denote the known and unknown dyes, respectively):

$$p_{sample}/p_{reference} = DF(1)^{1/n(1)} = DF(2)^{1/n(2)} \tag{13}$$

$$n(2) = n(1) \times \ln(DF(2))/\ln(DF(1)) \tag{14}$$

The idea is that the fluorescence ratio between two species can be expressed by two different dye factors depending on the dye used.

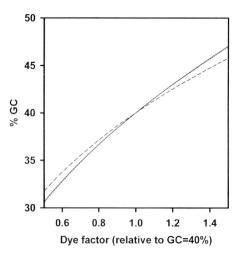

Fig. 8.3 Results of different calculations of DNA base composition (%GC) from the dye factor relative to GC = 40%. Solid line, exact calculation [Eq. (8)]; dashed line, simplified calculation according to Godelle et al. (1993) [Eq. (10)]; dotted line, simplified calculation according to Langlois et al. (1980) [Eqs. (11) and (12)]. While the simplified calculation with a power function according to Godelle et al. (1993) shows small but significant deviations from the exact calculation, the power function with a corrected exponent according to Langlois et al. (1980) is nearly identical to the exact values.

Assuming a binding length of 5 for HO 33342 (Portugal and Waring 1988), Barow and Meister (2002) calculated the binding length of DAPI, on the basis of Eq. (14), by analyzing pairs of species with presumably large differences in dye factor (which is necessary to minimize numerical errors). They obtained values for the DAPI binding length between 2.77 and 4.57 with a mean of 4.22, which corresponds to a binding length of $n = 4$. Large data variations may stem from the difference between (assumed) random distribution of bases and the real base sequence. In addition, accessibility of DNA to dye molecules may also be influenced by cytosolic compounds (Noirot 2002, see also Chapter 4). Such an effect may differ among species (with different amounts and composition of cytosol) and may also depend on the dye used (see Section 8.2.4). Thus, different pairs of species as used in the experiment could have been influenced to a different degree.

Meister (2005) calculated the relative fluorescence intensities, DF, of two nearly completely sequenced genomes of *Arabidopsis thaliana* and *Oryza sativa* (see Fig. 8.1) for different binding lengths by counting the number of groups comprising n (= binding length) consecutive GC or AT bases. Comparison of measured and calculated peak ratios gave the best agreement between experimental and calculated data for a binding length of 1 for all the dyes investigated (DAPI, HO 33258, HO 33342, MI). This is rather surprising as it would indicate a linear relationship between base content and fluorescence intensity, which contradicts previous experimental results (see above). The plausible explanation for this discrepancy may be the incompleteness of the sequenced plant genomes (see Bennett et al. 2003 for *Arabidopsis thaliana*).

An interesting approach for the calculation of the base composition in bacteria was proposed by Sanders et al. (1990). Combining a GC- and an AT-specific dye (CMA/HO 33258), they found a very good linear relationship ($r = 0.99$) between the logarithms of the ratio between CMA/HO 33258 fluorescence and the GC frequency. This is more or less in agreement with the (approximate) power functions of Eqs. (10) and (11). Unfortunately, the method adopted is relatively complicated because it requires two-wavelength excitation and the preparation of a standard curve prior to each set of analyses. This may be the reason why it has not been used outside the authors' laboratory. Moreover, no application in plant science has been reported.

8.2.2
Base Composition of Plant Species Determined by Flow Cytometry and its Relation to Genome Size and Taxonomy

In order to shed light on possible relationships between DNA base composition, genome size and taxonomical classification, we have compiled the results of 257 analyses of DNA base content and genome size in 215 species (Table 8.4). As can be seen from the table, the frequency of AT falls within a narrow range of 47 to 70%. This small difference contrasts with the large variation observed in some other organisms (Fig. 8.4).

Table 8.4 Determination of DNA base content in plants using flow cytometry. If one dye is given in column "Fluorochrome", only %AT or %GC were determined according to the base-specificity of th s dye, using the equations of Godelle et al. (1993). The percentage of the complementary bases is calculated as %AT = 100 − %GC or %GC = 100 − %AT. For other cases, see footnotes for explanation. Means (see column "Species") were calculatec either from different samples/subspecies/locations or with different standards.

Species	Family	2C (pg)	%AT	%GC	Fluorochrome	Reference
Abies concolor	Pinaceae	36.1	60.9	39.1	DAPI	Barow and Meister 2002
Acacia heterophylla	Fabaceae	1.60	61.0	39.0	3 different[a]	Marie and Brown 1993
Actinidia chinensis	Actinidiaceae	4.19	61.3	38.7	3 different[a]	Marie and Brown 1993
Alliaria petiolata	Brassicaceae	2.69	60.9	39.1	DAPI	Barow and Meister 2002
Allium altaicum	Alliaceae	23.3	60.1	39.9	MI	Ricroch et al. 2005
Allium altyncolicum	Alliaceae	27.9	60.0	40.0	MI	Ricroch et al. 2005
Allium ampeloprasum	Alliaceae	33.4	61.5	38.5	MI	Ricroch et al. 2005
Allium angulosum	Alliaceae	27.0	60.0	40.0	MI	Ricroch et al. 2005
Allium cepa	Alliaceae	34.8	69.9	30.1	DAPI/MI[b]	Doležel et al. 1992
Allium cepa	Alliaceae	33.2	59.0	38.7	Hoechst33342/MI[c]	Ricroch and Brown 1997
Allium cepa	Alliaceae	33.7	65.3	34.7	DAPI	Barow and Meister 2002
Allium cepa	Alliaceae	32.4	61.3	38.7	MI	Ricroch et al. 2005
Allium cernuum	Alliaceae	46.5	58.8	41.2	MI	Ricroch et al. 2005
Allium cyaneum	Alliaceae	46.5	59.5	40.5	MI	Ricroch et al. 2005

Allium drummondii	Alliaceae	37.9	59.8	40.2	MI	Ricroch et al. 2005
Allium fistulosum	Alliaceae	23.3	61.0	39.8	Hoechst33342/MI[c]	Ricroch and Brown 1997
Allium fistulosum	Alliaceae	23.5	60.2	39.8	MI	Ricroch et al. 2005
Allium flavum	Alliaceae	30.8	59.5	40.5	MI	Ricroch et al. 2005
Allium galanthum	Alliaceae	28.7	61.1	38.9	MI	Ricroch et al. 2005
Allium heldreichii	Alliaceae	63.3	61.3	38.7	MI	Ricroch et al. 2005
Allium hymenorrhizum	Alliaceae	17.9	60.4	39.6	MI	Ricroch et al. 2005
Allium ledebourianum	Alliaceae	17.4	63.6	36.4	DAPI	Barow and Meister 2002
Allium ledebourianum	Alliaceae	15.6	60.1	39.9	MI	Ricroch et al. 2005
Allium nutans	Alliaceae	41.9	59.8	40.2	MI	Ricroch et al. 2005
Allium obliquum	Alliaceae	23.3	60.6	39.4	MI	Ricroch et al. 2005
Allium porrum	Alliaceae	65.5	63.8	36.2	DAPI	Barow and Meister 2002
Allium porrum	Alliaceae	50.7	60.6	39.4	MI	Ricroch et al. 2005
Allium roylei	Alliaceae	30.7	60.6	39.4	MI	Ricroch et al. 2005
Allium saxatile	Alliaceae	19.8	60.7	39.3	MI	Ricroch et al. 2005
Allium schoenoprasum	Alliaceae	15.4	60.1	39.9	MI	Ricroch et al. 2005
Allium senescens (mean)	Alliaceae	43.4	59.6	40.4	MI	Ricroch et al. 2005
Allium ursinum	Alliaceae	62.1	64.8	35.2	DAPI	Barow and Meister 2002
Allium vavilovii	Alliaceae	32.8	60.7	39.3	MI	Ricroch et al. 2005

Table 8.4 (continued)

Species	Family	2C (pg)	%AT	%GC	Fluorochrome	Reference
Allium zebdanense	Alliaceae	38.4	59.8	40.2	MI	Ricroch et al. 2005
Allocasuarina verticillata	Casuarinaceae	1.90	59.3	41.1	Hoechst33342, MI[c]	Schwencke et al. 1998
Alstroemeria sp.	Alstroemeriaceae	43.5	57.4	42.6	3 different[a]	Marie and Brown 1993
Alstroemeria sp.	Alstroemeriaceae	64.7	57.4	42.6	3 different[a]	Marie and Brown 1993
Alstroemeria sp.	Alstroemeriaceae	90.9	57.2	42.8	3 different[a]	Marie and Brown 1993
Anemone ranunculoides	Ranunculaceae	36.8	59.7	40.3	DAPI	Barow and Meister 2002
Anemone sylvestris	Ranunculaceae	17.0	60.1	39.9	DAPI	Barow and Meister 2002
Aquilegia vulgaris	Ranunculaceae	1.00	60.3	39.7	DAPI	Barow and Meister 2002
Arabidopsis thaliana	Brassicaceae	0.33	59.7	40.3	3 different[a]	Marie and Brown 1993
Arabidopsis thaliana	Brassicaceae	0.42	60.9	39.1	DAPI	Barow and Meister 2002
Armeria maritima	Plumbaginaceae	8.55	58.1	41.9	3 different[a]	Marie and Brown 1993
Asparagus officinalis	Asparagaceae	3.63	59.4	40.6	DAPI	Barow and Meister 2002
Atriplex rosea	Chenopodiaceae/APG: Amaranthaceae	2.11	61.5	38.5	DAPI	Barow and Meister 2002
Begonia socotrana	Begoniaceae	1.23	59.7	40.3	3 different[a]	Marie and Brown 1993
Beta vulgaris	Chenopodiaceae/APG: Amaranthaceae	1.84	62.7	37.3	DAPI	Barow and Meister 2002

Species	Family				Dye	Reference
Brassica napus	Brassicaceae	2.94	61.1	38.9	DAPI	Barow and Meister 2002
Capsicum frutescens	Solanaceae	7.34	64.0	36.0	DAPI	Barow and Meister 2002
Castanea sativa	Fagaceae	1.95	62.7	37.3	DAPI	Barow and Meister 2002
Casuarina glauca	Casuarinaceae	0.70	58.6	40.5	Hoechst33342/MI[c]	Schwencke et al. 1998
Cedrus atlantica (mean)	Pinaceae	32.4	59.3	40.3	Hoechst33342/MI[c]	Bou Dagher-Kharrat et al. 2001
Cedrus brevifolia	Pinaceae	31.8	59.5	40.4	Hoechst33342/MI[c]	Bou Dagher-Kharrat et al. 2001
Cedrus deodara	Pinaceae	32.6	59.1	40.4	Hoechst33342/MI[c]	Bou Dagher-Kharrat et al. 2001
Cedrus libani (mean)	Pinaceae	32.6	59.1	40.4	Hoechst33342/MI[c]	Bou Dagher-Kharrat et al. 2001
Centaurea ragusina (mean)	Asteraceae	3.37	58.7	41.3	Hoechst33342/MI[a]	Siljak-Yakovlev et al. 2005
Chondrilla juncea	Asteraceae	3.12	64.1	35.9	DAPI	Koopman 2002
Chondrus crispus (alga) (Gametophyte, 1C)	Gigartinaceae	0.16	53.7	46.3	3 different[a]	Marie and Brown 1993
Chondrus crispus (alga) (Tetrasporophyte)	Gigartinaceae	0.33	53.7	46.3	3 different[a]	Marie and Brown 1993
Chrysanthemum multicolor	Asteraceae	32.5	63.5	36.5	DAPI	Barow and Meister 2002
Cicerbita plumieri	Asteraceae	6.40	62.4	37.6	DAPI	Koopman 2002
Cichorium intybus	Asteraceae	2.72	64.2	35.8	DAPI	Koopman 2002
Cirsium acaule	Asteraceae	2.62	60.9	39.1	DAPI	Bureš et al. 2004
Cirsium arvense	Asteraceae	2.84	61.9	38.1	DAPI	Bureš et al. 2004
Cirsium brachycephalum	Asteraceae	2.98	59.8	40.2	DAPI	Bureš et al. 2004

Table 8.4 (continued)

Species	Family	2C (pg)	%AT	%GC	Fluorochrome	Reference
Cirsium canum	Asteraceae	2.24	60.6	39.4	DAPI	Bureš et al. 2004
Cirsium eriophorum	Asteraceae	3.60	57.9	42.1	DAPI	Bureš et al. 2004
Cirsium ersithales	Asteraceae	2.33	60.9	39.1	DAPI	Bureš et al. 2004
Cirsium heterophyllum	Asteraceae	2.14	61.8	38.2	DAPI	Bureš et al. 2004
Cirsium oleraceum	Asteraceae	2.32	60.4	39.6	DAPI	Bureš et al. 2004
Cirsium palustre	Asteraceae	2.58	58.8	41.2	DAPI	Bureš et al. 2004
Cirsium pannonicum	Asteraceae	2.44	60.7	39.3	DAPI	Bureš et al. 2004
Cirsium rivulare	Asteraceae	2.40	61.8	38.2	DAPI	Bureš et al. 2004
Cirsium vulgare	Asteraceae	5.54	59.6	40.4	DAPI	Bureš et al. 2004
Crepis dinarica (mean)	Asteraceae	11.7	60.5	39.5	Hoechst33342/MI[b]	Godelle et al. 1993
Crepis dinarica	Asteraceae	11.8	60.8	39.2	3 different[a]	Marie and Brown 1993
Crepis froelichiana (mean)	Asteraceae	11.3	60.3	39.7	Hoechst33342/MI[b]	Godelle et al. 1993
Crepis froelichiana	Asteraceae	11.8	60.7	39.3	3 different[a]	Marie and Brown 1993
Crepis incarnata (mean)	Asteraceae	12.7	60.0	40.0	Hoechst33342/MI[b]	Godelle et al. 1993
Crepis incarnata	Asteraceae	12	60.3	39.7	3 different[a]	Marie and Brown 1993
Crepis praemorsa (mean)	Asteraceae	10.5	59.2	40.2	Hoechst33342/MI[b]	Godelle et al. 1993

Crepis praemorsa	Asteraceae	10.5	59.6	40.4	3 different[a]	Marie and Brown 1993
Crocus cartwrightianus	Iridaceae	7.24	59.5	40.5	DAPI/OL[c]	Brandizzi and Caiola 1998
Crocus sativus	Iridaceae	11.4	59.4	40.7	DAPI/OL[c]	Brandizzi and Caiola 1998
Crocus thomasii	Iridaceae	8.04	58.9	41.3	DAPI/OL[c]	Brandizzi and Caiola 1998
Cucumis sativus	Cucurbitaceae	1.02	62.7	37.3	DAPI	Barow and Meister 2002
Cucurbita moschata	Cucurbitaceae	0.97	61.3	38.7	DAPI	Barow and Meister 2002
Cucurbita pepo	Cucurbitaceae	1.17	61.3	38.7	DAPI	Barow and Meister 2002
Cydonia oblonga	Rosaceae	1.98	59.2	40.8	DAPI	Barow and Meister 2002
Dactylis glomerata	Poaceae	7.13	57.7	42.3	3 different[a]	Marie and Brown 1993
Digitaria setigera	Poaceae	4.57	55.7	44.3	3 different[a]	Marie and Brown 1993
Dryopteris dilatata	Dryopteridaceae	36.2	59.0	41.0	3 different[a]	Marie and Brown 1993
Duchesnea indica	Rosaceae	4.18	60.7	39.3	DAPI	Barow and Meister 2002
Enteromorpha compressa (alga)	Ulvaceae	0.26	53.6	46.4	3 different[a]	Marie and Brown 1993
Eucalyptus globulus	Myrtaceae	1.13	59.9	40.1	3 different[a]	Marie and Brown 1993
Fagus sylvatica	Fagaceae	1.11	60.0	40.0	MI	Gallois et al. 1999
Fagus sylvatica	Fagaceae	1.30	63.3	36.7	DAPI	Barow and Meister 2002
Fagus tortuosa	Fagaceae	1.11	60.2	39.8	MI	Gallois et al. 1999
Fritillaria uva-vulpis	Liliaceae	165.0	64.4	35.6	DAPI	Barow and Meister 2002
Gerbera hybrida	Asteraceae	5.12	59.3	40.7	3 different[a]	Marie and Brown 1993

Table 8.4 (continued)

Species	Family	2C (pg)	%AT	%GC	Fluorochrome	Reference
Gerbera piloselloides	Asteraceae	7.68	59.3	40.7	3 different[a]	Marie and Brown 1993
Gerbera viridifolia	Asteraceae	7.68	59.3	40.7	3 different[a]	Marie and Brown 1993
Ginkgo biloba	Ginkgoaceae	21.6	65.3	34.7	DAPI	Barow and Meister 2002
Glycine max	Fabaceae	2.73	63.6	36.4	DAPI	Barow and Meister 2002
Gymnostoma deplancheanum	Casuarinaceae	0.75	58.7	40.5	Hoechst33342/MI[c]	Schwencke et al. 1998
Haplopappus gracilis	Asteraceae	2.38	61.7	38.3	DAPI	Barow and Meister 2002
Holcus lanatus	Poaceae	3.83	55.0	45.0	3 different[a]	Marie and Brown 1993
Holcus mollis	Poaceae	5.55	55.1	44.9	3 different[a]	Marie and Brown 1993
Holcus mollis	Poaceae	8.18	54.6	45.4	3 different[a]	Marie and Brown 1993
Hordeum vulgare	Poaceae	10.3	55.3	44.7	DAPI	Barow and Meister 2002
Hydrangea anomala (mean)	Hydrangeaceae	2.92	59.2	40.8	Hoechst33342/MI[a]	Cerbah et al. 2001
Hydrangea arborescens	Hydrangeaceae	2.31	59.2	40.8	Hoechst33342/MI[a]	Cerbah et al. 2001
Hydrangea aspera (mean)	Hydrangeaceae	3.59	59.5	40.5	Hoechst33342/MI[a]	Cerbah et al. 2001
Hydrangea heteromalla	Hydrangeaceae	2.95	59.4	40.6	Hoechst33342/MI[a]	Cerbah et al. 2001
Hydrangea involucrata	Hydrangeaceae	5.00	59.3	40.7	Hoechst33342/MI[a]	Cerbah et al. 2001
Hydrangea macrophylla (mean)	Hydrangeaceae	4.08	59.2	40.8	Hoechst33342/MI[a]	Cerbah et al. 2001

Hydrangea paniculata	Hydrangeaceae	3.77	59.6	40.4	Hoechst33342/MI[a]	Cerbah et al. 2001
Hydrangea quercifolia	Hydrangeaceae	1.95	59.3	40.7	Hoechst33342/MI[a]	Cerbah et al. 2001
Hydrangea scandens	Hydrangeaceae	3.75	60.0	40.0	Hoechst33342/MI[a]	Cerbah et al. 2001
Hydrangea seemannii	Hydrangeaceae	2.09	59.7	40.3	Hoechst33342/MI[a]	Cerbah et al. 2001
Hyssopus officinalis	Lamiaceae	1.12	62.5	37.5	DAPI	Barow and Meister 2002
Kappaphycus alvarezii (alga)	Solieriaceae	0.44	56.6	43.4	3 different[a]	Marie and Brown 1993
Lactuca aculeata	Asteraceae	5.96	61.8	38.2	DAPI	Koopman 2002
Lactuca altaica	Asteraceae	6.15	61.6	38.4	DAPI	Koopman 2002
Lactuca dregeana	Asteraceae	6.05	61.7	38.3	DAPI	Koopman 2002
Lactuca indica	Asteraceae	13.0	62.2	37.8	DAPI	Koopman 2002
Lactuca perennis	Asteraceae	4.64	64.2	35.8	DAPI	Koopman 2002
Lactuca quercina	Asteraceae	8.79	61.8	38.2	DAPI	Koopman 2002
Lactuca saligna	Asteraceae	5.04	61.6	38.4	DAPI	Koopman 2002
Lactuca sativa	Asteraceae	6.59	60.3	39.7	DAPI	Barow and Meister 2002
Lactuca sativa	Asteraceae	6.04	61.7	38.3	DAPI	Koopman 2002
Lactuca serriola	Asteraceae	5.92	61.8	38.2	DAPI	Koopman 2002
Lactuca sibirica	Asteraceae	8.44	63.3	36.7	DAPI	Koopman 2002
Lactuca tatarica	Asteraceae	8.63	63.6	36.4	DAPI	Koopman 2002
Lactuca tenerrima	Asteraceae	1.91	64.2	35.8	DAPI	Koopman 2002

Table 8.4 (continued)

Species	Family	2C (pg)	%AT	%GC	Fluorochrome	Reference
Lactuca riminea	Asteraceae	4.30	63.4	36.6	DAPI	Koopman 2002
Lactuca virosa (mean)	Asteraceae	8.53	61.6	38.4	DAPI	Koopman 2002
Laminaria digitata (alga)	Laminariaceae	1.40	54.4	45.6	3 different[a]	Marie and Brown 1993
Laminaria saccharina (alga)	Laminariaceae	1.58	55.3	44.7	3 different[a]	Marie and Brown 1993
Larix decidua	Pinaceae	25.7	60.7	39.3	DAPI	Barow and Meister 2002
Lathyrus annuus	Fabaceae	12.8	61.6	38.4	DAPI	Ali et al. 2000
Lathyrus aphaca	Fabaceae	9.24	61.7	38.3	DAPI	Ali et al. 2000
Lathyrus articulatus	Fabaceae	11.5	62.5	37.5	DAPI	Barow and Meister 2002
Lathyrus cicera	Fabaceae	10.6	60.9	39.1	DAPI	Ali et al. 2000
Lathyrus clymenum	Fabaceae	8.77	61.4	38.6	DAPI	Ali et al. 2000
Lathyrus ochrus	Fabaceae	9.27	61.2	38.8	DAPI	Ali et al. 2000
Lathyrus sativus	Fabaceae	13.4	60.4	39.6	DAPI	Ali et al. 2000
Lathyrus tingitanus	Fabaceae	15.7	59.6	40.4	DAPI	Ali et al. 2000
Lilium bosniacum (mean)	Liliaceae	67.8	65.3	34.7	Hoechst33342/CMA[a]	Muratović et al. 2005
Lilium carniolicum	Liliaceae	67.4	62.1	36.6	Hoechst33342/CMA[a]	Siljak-Yakovlev et al. 2003
Lilium pomponium	Liliaceae	70.3	64.4	35.6	Hoechst33342/CMA[a]	Siljak-Yakovlev et al. 2003

Lilium pyrenaicum	Liliaceae	67.7	62.1	37.9	Hoechst33342/CMA[a]	Siljak-Yakovlev et al. 2003
Lotus corniculatus	Fabaceae	1.38	58.4	41.6	3 different[a]	Marie and Brown 1993
Lycopersicon esculentum	Solanaceae	1.96	50.2[d] (64.5)	49.8[d] (35.5)	DAPI/MI[b]	Doležel et al. 1992
Lycopersicon esculentum	Solanaceae	2.01	60.0	40.0	3 different[a]	Marie and Brown 1993
Lycopersicon esculentum	Solanaceae	2.01	60.0	40.0	Hoechst33342/MI[a]	Blondon et al. 1994
Lycopersicon pimpinellifolium	Solanaceae	2.28	62.9	37.1	DAPI	Barow and Meister 2002
Medicago caerulea	Fabaceae	1.80	61.3	38.7	3 different[a]	Marie and Brown 1993
Medicago quasifalcata	Fabaceae	1.72	61.0	39.0	3 different[a]	Marie and Brown 1993
Medicago sativa	Fabaceae	3.44	61.4	38.6	3 different[a]	Marie and Brown 1993
Medicago sativa subsp. caerulea	Fabaceae	1.80	61.3	38.7	Hoechst33342/MI[a]	Blondon et al. 1994
Medicago sativa subsp. quasifalcata	Fabaceae	1.72	61.0	39.0	Hoechst33342/MI[a]	Blondon et al. 1994
Medicago sativa subsp. sativa (mean)	Fabaceae	3.41	61.4	38.6	Hoechst33342/MI[a]	Blondon et al. 1994
Medicago truncatula	Fabaceae	1.15	61.4	38.6	3 different[a]	Marie and Brown 1993
Medicago truncatula (mean)	Fabaceae	1.09	61.4	38.6	Hoechst33342/MI[a]	Blondon et al. 1994
Melandrium album female	Caryophyllaceae	5.84	58.9	41.1	3 different[a]	Marie and Brown 1993
Melandrium album male	Caryophyllaceae	6.04	59.0	41.0	3 different[a]	Marie and Brown 1993
Momordica charantia	Cucurbitaceae	1.42	63.5	36.5	DAPI	Barow and Meister 2002

Table 8.4 (continued)

Species	Family	2C (pg)	%AT	%GC	Fluorochrome	Reference
Musa acuminata (mean)	Musaceae	1.26	58.1	41.0	Hoechst33342/MI[c]	Kamaté et al. 2001
Musa balbisiana	Musaceae	1.16	58.0	40.3	Hoechst33342/MI[c]	Kamaté et al. 2001
Musa ornata	Musaceae	1.23	58.2	40.5	Hoechst33342/MI[c]	Kamaté et al. 2001
Mycelis muralis	Asteraceae	3.76	63.0	37.0	DAPI	Koopman 2002
Nicotiana tabacum	Solanaceae	9.77	61.9	38.1	DAPI	Barow and Meister 2002
Oryza sativa	Poaceae	1.17	57.0	43.0	DAPI	Barow and Meister 2002
Petunia hybrida	Solanaceae	2.85	59.0	41.0	3 different[a]	Marie and Brown 1993
Phaseolus vulgaris	Fabaceae	1.57	62.5	37.5	DAPI	Barow and Meister 2002
Phragmites australis	Poaceae	2.00	55.4	44.6	3 different[a]	Marie and Brown 1993
Physocarpus opulifolius	Rosaceae	0.71	61.1	38.9	DAPI	Barow and Meister 2002
Picea abies	Pinaceae	38.6	60.1	39.9	DAPI	Barow and Meister 2002
Picea abies (mean)	Pinaceae	37.2	59.4	40.6	Hoechst33342/CMA[a]	Siljak-Yakovlev et al. 2002
Picea omorica (mean)	Pinaceae	33.8	58.3	41.7	Hoechst33342/CMA[a]	Siljak-Yakovlev et al. 2002
Pinus heldreichii	Pinaceae	50.0	60.9	39.1	Hoechst33342	Bogunic et al. 2003
Pinus mugo	Pinaceae	42.8	60.8	39.2	Hoechst33342	Bogunic et al. 2003
Pinus nigra	Pinaceae	46.0	60.9	39.1	Hoechst33342	Bogunic et al. 2003

Species	Family				Method	Reference
Pinus peuce	Pinaceae	54.9	60.3	39.7	Hoechst33342	Bogunic et al. 2003
Pinus sylvestris	Pinaceae	44.2	60.9	39.1	DAPI	Barow and Meister 2002
Pinus sylvestris	Pinaceae	42.5	59.8	40.2	Hoechst33342	Bogunic et al. 2003
Pisum sativum	Fabaceae	9.07	61.4	38.6	DAPI/MI[b]	Doležel et al. 1992
Pisum sativum	Fabaceae	8.37	59.5	40.5	3 different[a]	Marie and Brown 1993
Pisum sativum	Fabaceae	9.07	61.5	38.5	DAPI	Barow and Meister 2002
Polygonum aviculare	Polygonaceae	1.71	59.0	41.0	3 different[a]	Marie and Brown 1993
Polygonum maritimum	Polygonaceae	1.01	59.0	41.0	3 different[a]	Marie and Brown 1993
Porphyra purpurea (alga, 1C)	Bangiaceae	0.3	55.4	44.6	3 different[a]	Marie and Brown 1993
Prenanthes purpurea	Asteraceae	8.43	62.0	38.0	DAPI	Koopman 2002
Psophocarpus tetragonolobus	Fabaceae	1.55	62.0	38.0	3 different[a]	Marie and Brown 1993
Pylaiella littoralis (alga)	Acinetosporaceae	1.10	53.3	46.7	3 different[a]	Marie and Brown 1993
Quercus cerris	Fagaceae	1.91	59.8	40.2	Hoechst33342/MI[a]	Zoldoš et al. 1998
Quercus coccifera	Fagaceae	2.00	59.6	40.4	Hoechst33342/MI[a]	Zoldoš et al. 1998
Quercus ilex	Fagaceae	2.00	60.2	39.8	Hoechst33342/MI[a]	Zoldoš et al. 1998
Quercus petraea	Fagaceae	1.87	58.3	41.7	Hoechst33342	Favre and Brown 1996
Quercus petraea (mean)	Fagaceae	1.90	60.2	39.8	Hoechst33342/MI[a]	Zoldoš et al. 1998
Quercus pubescens	Fagaceae	1.86	57.9	42.1	Hoechst33342	Favre and Brown 1996
Quercus pubescens	Fagaceae	1.91	60.3	39.7	Hoechst33342/MI[a]	Zoldoš et al. 1998

Table 8.4 (continued)

Species	Family	2C (pg)	%AT	%GC	Fluorochrome	Reference
Quercus robur	Fagaceae	1.84	58.3	42.0	Hoechst33342	Favre and Brown 1996
Quercus robur (mean)	Fagaceae	1.88	60.5	39.5	Hoechst33342/MI[a]	Zoldoš et al. 1998
Quercus robur	Fagaceae	2.18	62.7	37.3	DAPI	Barow and Meister 2002
Quercus suber	Fagaceae	1.91	60.3	39.7	Hoechst33342/MI[a]	Zoldoš et al. 1998
Raphanus sativus	Brassicaceae	1.11	47.0[d] (61.4)	53.0[d] (38.6)	DAPI/MI[b]	Doležel et al. 1992
Raphanus sativus	Brassicaceae	1.08	59.6	40.4	3 different[a]	Marie and Brown 1993
Raphanus sativus	Brassicaceae	1.38	60.9	39.1	DAPI	Barow and Meister 2002
Saxifraga granulata	Saxifragaceae	1.35	61.2	38.8	3 different[a]	Marie and Brown 1993
Saxifraga granulata	Saxifragaceae	3.54	61.2	38.8	3 different[a]	Marie and Brown 1993
Saxifraga granulata	Saxifragaceae	3.54	60.6	39.4	DAPI	Redondo et al. 1996
Secale cereale	Poaceae	16.0	55.4	44.6	DAPI	Barow and Meister 2002
Setaria chevalieri	Poaceae	4.46	57.8	42.2	Hoechst33342/MI[a]	Le Thierry d'Ennequin et al. 1998
Setaria holstii	Poaceae	1.70	68.6	41.4	Hoechst33342/MI[a]	Le Thierry d'Ennequin et al. 1998
Setaria incrassata	Poaceae	4.23	58.2	41.8	Hoechst33342/MI[a]	Le Thierry d'Ennequin et al. 1998
Setaria italica (mean)	Poaceae	1.03	56.8	43.2	Hoechst33342/MI[a]	Le Thierry d'Ennequin et al. 1998

Species	Family					Reference
Setaria leiantha	Poaceae	2.40	58.5	41.5	Hoechst33342/MI[a]	Le Thierry d'Ennequin et al. 1998
Setaria macrostachya	Poaceae	3.60	58.8	41.2	Hoechst33342/MI[a]	Le Thierry d'Ennequin et al. 1998
Setaria neglecta	Poaceae	3.50	56.5	43.5	Hoechst33342/MI[a]	Le Thierry d'Ennequin et al. 1998
Setaria palmifolia	Poaceae	3.88	56.6	43.4	Hoechst33342/MI[a]	Le Thierry d'Ennequin et al. 1998
Setaria parviflora	Poaceae	4.82	57.9	42.1	Hoechst33342/MI[a]	Le Thierry d'Ennequin et al. 1998
Setaria pumila (mean)	Poaceae	5.20	57.5	42.5	Hoechst33342/MI[a]	Le Thierry d'Ennequin et al. 1998
Setaria queenslandia	Poaceae	2.76	57.9	42.1	Hoechst33342/MI[a]	Le Thierry d'Ennequin et al. 1998
Setaria sphacelata (mean)	Poaceae	2.69	58.0	42.0	Hoechst33342/MI[a]	Le Thierry d'Ennequin et al. 1998
Setaria viridis (mean)	Poaceae	1.02	58.8	41.2	Hoechst33342/MI[a]	Le Thierry d'Ennequin et al. 1998
Setaria woodii	Poaceae	1.66	59.0	41.0	Hoechst33342/MI[a]	Le Thierry d'Ennequin et al. 1998
Simmondsia chinensis female	Buxaceae/APG: Simmondsiaceae	2.25	61.5	38.5	3 different[a]	Marie and Brown 1993
Simmondsia chinensis male	Buxaceae/APG: Simmondsiaceae	2.24	61.1	38.9	3 different[a]	Marie and Brown 1993
Sinapis arvensis	Brassicaceae	1.35	60.7	39.3	DAPI	Barow and Meister 2002
Solanum tuberosum	Solanaceae	1.77	59.8	40.2	3 different[a]	Marie and Brown 1993
Sonchus asper	Asteraceae	1.45	63.7	36.3	DAPI	Koopman 2002
Sphacelaria sp. (alga)	Sphacelariaceae	3.42	55.2	44.8	3 different[a]	Marie and Brown 1993
Spinacia oleracea	Chenopodiaceae/APG: Amaranthaceae	2.68	61.3	38.7	DAPI	Barow and Meister 2002
Stachys grandiflora	Lamiaceae	12.4	60.3	39.7	DAPI	Barow and Meister 2002

Table 8.4 (continued)

Species	Family	2C (pg)	%AT	%GC	Fluorochrome	Reference
Steptorhamphus tuberosus	Asteraceae	5.61	63.2	36.8	DAPI	Koopman 2002
Taraxacum officinale	Asteraceae	1.78	62.7	37.3	DAPI	Koopman 2002
Teucrium scorodonia	Lamiaceae	2.85	60.5	39.5	DAPI	Barow and Meister 2002
Theobroma cacao	Sterculiaceae/APG: Malvaceae	0.98	61.9	38.1	3 different[a]	Marie and Brown 1993
Trifolium pratense	Fabaceae	1.06	64.8	35.2	DAPI	Barow and Meister 2002
Triticum aestivum	Poaceae	30.9	56.3	43.7	3 different[a]	Marie and Brown 1993
Triticum aestivum	Poaceae	32.8	54.1	45.9	DAPI	Barow and Meister 2002
Ulva rigida (alga)	Ulvaceae	0.33	57.1	42.9	3 different[a]	Marie and Brown 1993
Undaria pinnatifida (alga)	Alariaceae	1.28	56.2	43.8	3 different[a]	Marie and Brown 1993
Urtica dioica	Urticaceae	2.33	61.7	38.3	DAPI	Barow and Meister 2002
Urtica urens	Urticaceae	1.07	63.1	36.9	DAPI	Barow and Meister 2002
Vicia faba	Fabaceae	26.9	62.3	37.7	DAPI/MI[b]	Doležel et al. 1992
Vicia faba	Fabaceae	26.2	61.9	38.1	DAPI	Barow and Meister 2002
Vigna radiata	Fabaceae	1.20	61.4	38.6	3 different[a]	Marie and Brown 1993
Vigna unguiculata	Fabaceae	1.31	61.5	38.5	3 different[a]	Marie and Brown 1993

Species	Family				Method	Reference
Viscum album female	Loranthaceae	155	61.0	39.0	3 different[a]	Marie and Brown 1993
Viscum album male	Loranthaceae	152	60.2	39.8	3 different[a]	Marie and Brown 1993
Zea mays	Poaceae	5.72	44.9	55.1	DAPI/MI[b]	Doležel et al. 1992
Zea mays	Poaceae	5.22–5.35	55.5	44.5	3 different[a]	Marie and Brown 1993
Zea mays	Poaceae	5.90	52.8	47.2	DAPI	Barow and Meister 2002

CMA, chromomycin A3; DAPI, 4′,6-diamidino-2-phenylindole; MI, mithramycin A; OL, olivomycin; "3 different" denotes one AT-specific, one GC-specific, and one base-unspecific dye (S. C. Brown, personal communication).

[a] %AT and %GC were determined separately, but recalculated to add up to 100% as follows: corrected %GC = 100 × original %GC/(original %GC + original %AT); corrected %AT = 100 × original %AT/(original %GC + original %AT).

[b] Absolute AT and GC values (in pg) were calculated separately. The percentage values were calculated as follows: %AT = $AT_{abs}/(AT_{abs} + GC_{abs}) \times 100$; %GC = $GC_{abs}/(AT_{abs} + GC_{abs}) \times 100$.

[c] %AT and %GC were determined separately.

[d] The unusual values for these species suggest that they are possibly not correct. A reanalysis resulted in the values in parentheses (with the agreement of J. Doležel).

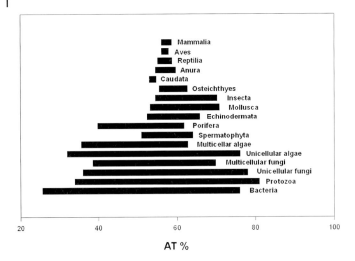

Fig. 8.4 Ranges of AT content of genomic DNA in selected groups of organisms (from Barow 2003).

Although a few authors found significant differences in base composition even among closely related species (Poaceae, King and Ingrouville 1987a, 1987b; *Lactuca*, Koopman 2002), the differences within a genus are mostly small and do not allow species discrimination. The genome size determined with base-unspecific dyes is generally a better way to achieve this goal. Moreover, estimation of base composition by FCM always requires two analyses, one with a base-specific and one with a base-unspecific dye, which increases the effort as well as the experimental error of the measurements.

However, in special cases the determination of base composition can demonstrate unusually high deviation in the base composition of single chromosomes. Ricroch and Brown (1997) compared the GC content of monosomic addition lines of *Allium fistulosum*, carrying a single chromosome from *Allium cepa*, with their parental line and found an unusually low GC content (25%) of *Allium cepa* chromosome 3C compared to the average of *Allium fistulosum* (39.8%).

The relation between base composition and several other biological parameters has also been examined. Vinogradov (1994) found a high correlation ($r = 0.9$) between GC content and genome size in angiosperms (this was, however, based on six species only and data originating from different literature sources). He explained this relationship by the positive effect of a high GC frequency on the physical and chemical stability of large genomes. In the genus *Allium*, a positive correlation was demonstrated between the amount of GC-rich heterochromatin and genome size (Narayan 1988). According to Koopman (2002), there was a negative correlation between AT content and 2C values within the genus *Lactuca*. However, several other reports did not confirm such a relationship. Kirk et al.

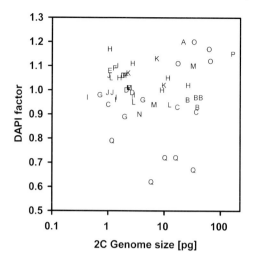

Fig. 8.5 DAPI factor (= dye factor for DAPI) in relation to the 2C genome size for 54 plant species from the 17 families: A, Ginkgoaceae; B, Pinaceae; C, Ranunculaceae; D, Chenopodiaceae/APG: Amaranthaceae; E, Urticaceae; F, Fagaceae; G, Rosaceae; H, Fabaceae; I, Brassicaceae; J, Cucurbitaceae; K, Solanaceae; L, Lamiaceae; M, Asteraceae; N, Asparagaceae; O, Alliaceae; P, Liliaceae; Q, Poaceae. Reprinted from Barow and Meister (2002) with permission from Wiley-Liss, Inc.

(1970) using a chromatographic method for analyzing the AT/GC ratio could not find a significant correlation between GC and genome size among 11 *Allium* species. Cerbah et al. (2001) did not find a correlation between 2C DNA content and the frequency of GC in 16 species and subspecies of *Hydrangea*, and neither did Ricroch et al. (2005) in 24 accessions representing 23 *Allium* species. Furthermore, Ricroch and Brown (1997) failed to confirm a relationship between nuclear DNA amount and base composition in chromosomes of *Allium cepa* using monosomic addition lines of *Allium fistulosum*.

In order to elucidate this issue, Barow and Meister (2002) analyzed a large number (54) of species from 17 different families, including two gymnosperm families with five species (see Fig. 8.5), covering a wide range of 2C genome sizes (0.42–165 pg). However, they did not find a significant correlation between the AT frequency and genome size ($r = 0.18$, $P = 0.19$). The correlation coefficient was even positive instead of negative sign as had been expected from the results of Vinogradov (1994). Moreover, no significant correlation was observed within the investigated families. The overall correlation between the AT frequency and genome size for the data in Table 8.4 results in $r = 0.14$ ($P = 0.03$), which is surprisingly significant, but also in the opposite direction to that anticipated.

Nevertheless, in some cases there are clear differences between families as shown in Fig. 8.5. For example, Poaceae have consistently lower and Alliaceae mostly higher AT frequencies (= DAPI factor) than members of other families.

8.2.3
Comparison of Flow Cytometric Results with Base Composition Determined by other Physico-Chemical Methods

A long time before the advent of FCM, several physico-chemical methods had been used to determine DNA base composition (they are described in detail for example by Johnson 1985). The five most important procedures involve chromatographic methods (Bendich 1957; Kirk 1967; Sulimova et al. 1970; Wyatt 1955), including HPLC (Ko et al. 1977). Chromatography is the most direct means for determination of base composition. Its principle lies in the release of bases from the DNA by acid hydrolysis and their subsequent chromatographic separation. The remaining four indirect methods are based on correlations with results originally obtained by paper chromatography (Wyatt 1955). The first of them, buoyant density centrifugation of DNA in CsCl gradient (Schildkraut et al. 1962), is based on the fact that the density of DNA depends on its base composition. Disadvantages of this procedure are the need for an analytical ultracentrifuge and a long analysis time (typically 44 h for one run). Determination from thermal denaturation of DNA (Marmur and Doty 1962) involves separation of the two DNA strands by heating. This process can be monitored spectrophotometrically and the midpoint temperature of the thermal melting profile (absorbance as a function of temperature) is related to the base composition. Other methods are based on the different reactivity of DNA bases toward bromine reagents (Wang and Hashagen 1964). Bromine reacts with all DNA bases except adenine and this process can be followed spectrophotometrically by measuring absorbance at 270 nm. Following the absorbance change at 270 nm, the base composition is calculated. This procedure has not been widely used, mainly because of its sensitivity to RNA contamination, and therefore has a requirement for pure DNA (in contrast to the majority of other methods). The last of the methods mentioned here is the analysis of UV absorbance of DNA. The UV spectrum of DNA, either intact at different pH values (pH 3.0, Fredericq et al. 1961; pH 7.0, Hirschman and Felsenfeld 1966) or after hydrolysis to free bases (Skidmore and Duggan 1966), depends on base composition. The accuracy of these methods also depends on the purity of the DNA (absence of contaminating RNA). In summary, all physico-chemical methods require large quantities of material, enough time for sample preparation and, in some cases, pure DNA isolates.

As mentioned above, whole genome sequencing (Goff et al. 2002; The Arabidopsis Genome Initiative 2000; Yu et al. 2002) provides exact data on base composition. However, obtaining a complete genome sequence requires large efforts and investments. Until now, none of the plant species has been sequenced to a completion.

In contrast to the physico-chemical methods, FCM needs only a small amount of material (a few milligrams of leaf tissue is sufficient in most cases) and preparation and analysis take only several minutes. Contamination by RNA does not pose any problem because the fluorescent dyes either do not bind to RNA (e.g. DAPI) or RNA can easily be eliminated by RNAse (e.g. in PI staining).

A comparison of GC frequencies determined by FCM with those resulting from physico-chemical methods is given in Table 8.5. As seen from Fig. 8.6, the agreement between both sets of values seems rather poor. However, different GC frequencies for the same species (see e.g. *Allium cepa*) obtained by different physico-chemical methods show that the variation does not necessarily arise from the FCM alone. The correlation between both data-sets (FCM values versus means for other techniques) is $r = 0.684$ ($P < 0.001$), which is also not very acceptable. The relatively narrow range of GC values together with the apparent experimental errors may explain the deviating values.

A review of the literature on base composition shows that about 250 plant species have been analyzed by conventional physico-chemical methods during the last 50 years, which is only slightly more than the number of species (215) investigated by flow cytometry only within the 14 years since the first FCM determination was reported in 1992 (Doležel et al. 1992). To our knowledge, no new analyses of base composition in plants using physico-chemical methods (except the sequencing) have been reported during the last two decades. This indicates a dominance of FCM in this area.

8.2.4
Possible Sources of Error in Determination of Base Composition by Flow Cytometry

Langlois et al. (1980) and Godelle et al. (1993) assumed a random distribution of bases in their equations because the exact base sequence of the analyzed species was mostly not known. However, in cases where the base composition is known (some viruses, bacteria, single chromosomes of higher organisms, and even (nearly) entire genomes of *Arabidopsis thaliana* and *Oryza sativa*), large differences (up to 50%) were found between the dye factors (= relative fluorescence intensity of the bound dye) calculated from the real base distribution and the assumed random distribution, respectively (Barow and Meister 2002). This is not surprising because the repetitiveness of DNA sequences introduces a deviation from randomness, and most nuclear DNA is repetitive. Therefore, to establish the exact fluorescence of base-specific dyes, the exact base sequence must be known. Since this is generally not the case, FCM can only yield the approximate base composition of DNA.

Another weak point in all calculations based on Eq. (1) is the uncertainty of the binding length n (see Table 8.1), which may strongly influence the results. Possibly, the additional specification of the dye factor [Eq. (4)] may be helpful because it is purely an experimental quantity, independent of any theoretical assumption for computing the base composition. To achieve comparability of results, the dye factors should be related to *Pisum sativum* (that is $DF_{Pisum\ sativum} = 1$; Barow and Meister 2002, see Eq. (6)).

There are still other potential sources of error in the determination of base composition by FCM. Breusegem et al. (2002) demonstrated that the results may be influenced by the order of bases of the same type (AT or GC). The affinity of sequences containing four AT bases to HO 33258 and DAPI decreases as follows:

Table 8.5 Comparison of DNA base content determined by flow cytometry and other methods. Mean and standard deviation (SD) of flow cytometric GC values were calculated using data from Table 3.4.

Species	Family	%GC flow cytometry		Alternative method		Reference
		Mean	SD[a]	%GC	Method	
Allium cepa	Alliaceae	34.5	4.3	36	Paper chromatography	Sueoka 1961
				36.8	Bromination	Biswas and Sarkar 1970
				34.6	Ion exchange chromatography	Kirk et al. 1970
				35.6	Ion exchange chromatography	Kirk et al. 1970
				36.6	?[c]	Bonner 1976[b]
				36.6	Paper chromatography	Shapiro 1976[b]
				33.2	Thermal denaturation	Stack and Comings 1979
				31.6	Buoyant density centrifugation	Stack and Comings 1979
				32	Fluorescence of single nuclei	Leemann and Ruch 1982
Allium fistulosum	Alliaceae	39.8	–	37.9	Ion exchange chromatography	Kirk et al. 1970
Allium porrum	Alliaceae	36.2	–	36.8	Ion exchange chromatography	Kirk et al. 1970
Allium ursinum	Alliaceae	35.2	–	35.3	Ion exchange chromatography	Kirk et al. 1970
Arabidopsis thaliana	Brassicaceae	39.7	0.9	34.7	Sequencing	The Arabidopsis Genome Initiative 2000

Species	Family			Method	Reference
Atriplex rosea	Chenopodiaceae/APG: Amaranthaceae	38.5	–	Hydroxyapatite chromatography/ Thermal denaturation	Shapiro 1976[b]
Beta vulgaris	Chenopodiaceae/APG: Amaranthaceae	37.3	–	Bromination	Biswas and Sarkar 1970
				Paper chromatography	Shapiro 1976[b]
Cucumis sativus	Cucurbitaceae	37.3	–	Buoyant density centrifugation	Nagl 1976[d]
				Paper chromatography	Shapiro 1976[b]
Cucurbita pepo	Cucurbitaceae	38.7	–	Paper chromatography	Sueoka 1961
				?[c]	Bonner 1976[b]
				Buoyant density centrifugation	Nagl 1976[d]
				Paper chromatography	Shapiro 1976[b]
				Paper chromatography	Shapiro 1976[b]
Ginkgo biloba	Ginkgoaceae	34.7	–	Paper chromatography	Shapiro 1976[b]
Glycine max	Fabaceae	36.4	–	Thermal denaturation	Biswas and Sarkar 1970
Hordeum vulgare	Poaceae	44.7	–	Paper chromatography	Shapiro 1976[b]
				Paper chromatography	Shapiro 1976[b]
Lathyrus articulatus	Fabaceae	37.5	–	Thermal denaturation	Narayan and Rees 1976
				Buoyant density centrifugation	Narayan and Rees 1976
Lathyrus cicera	Fabaceae	39.1	–	Thermal denaturation	Narayan and Rees 1976
				Buoyant density centrifugation	Narayan and Rees 1976
Lathyrus clymenum	Fabaceae	38.6	–	Thermal denaturation	Narayan and Rees 1976
Lathyrus ochrus	Fabaceae	38.8	–	Buoyant density centrifugation	Narayan and Rees 1976

Values in third numeric column (by row): 35.0; 38.0, 42.2; 38.9, 42.0; 40, 40.8, 37.7, 42.2, 41.0; 34.9; 40.2; 43.1, 41.8; 38.1, 37.8; 39.3, 37.8; 39.3; 40.5

Table 8.5 (continued)

Species	Family	%GC flow cytometry		Alternative method		Reference
		Mean	SD[a]	%GC	Method	
Lathyrus sativus	Fabaceae	39.6	–	41.5	Thermal denaturation	Narayan and Rees 1976
				39.8	Buoyant density centrifugation	Narayan and Rees 1976
Lathyrus tingitanus	Fabaceae	40.4	–	39.3	Thermal denaturation	Narayan and Rees 1976
				37.8	Buoyant density centrifugation	Narayan and Rees 1976
Lycopersicon esculentum	Solanaceae	38.5[e]	2.6	37.3	Thermal denaturation	Biswas and Sarkar 1970
				35.6	Buoyant density centrifugation	Nagl 1976[d]
Medicago sativa	Fabaceae	38.6	–	36.2	Paper chromatography	Shapiro 1976[b]
Momordica charantia	Cucurbitaceae	36.5	–	41.1	Thermal denaturation	Biswas and Sarkar 1970
Nicotiana tabacum	Solanaceae	38.1	–	39.8	?[c]	Bonner 1976[b]
				40.5	Paper chromatography	Shapiro 1976[b]
				40.0	Buoyant density centrifugation	Shapiro 1976[b]
				39.8	Buoyant density centrifugation/Thermal denaturation	Shapiro 1976[b]
Oryza sativa	Poaceae	43.0	–	49.0	Thermal denaturation	Biswas and Sarkar 1970
				43.9	Paper chromatography	Shapiro 1976[b]
				44	Sequencing	Goff et al. 2002
				43.3	Sequencing	Yu et al. 2002

Species	Family				Method	Reference
Phaseolus vulgaris	Fabaceae	37.5	–	40	Paper chromatography	Sueoka 1961
				40.7	?[c]	Bonner 1976[b]
				37.4	Buoyant density centrifugation	Nagl 1976[d]
				35.6	Buoyant density centrifugation	Nagl 1976[d]
				40.7	Paper chromatography	Shapiro 1976[b]
				33	Fluorescence of single nuclei	Leemann and Ruch 1982
Pinus sylvestris	Pinaceae	39.7	0.8	39.4	Paper chromatography	Shapiro 1976[b]
Pisum sativum	Fabaceae	39.2	1.1	38.5	Bromination	Biswas and Sarkar 1970
				37.1	Thermal denaturation	Biswas and Sarkar 1970
				37.7	?[c]	Bonner 1976[b]
				41.9	Paper chromatography	Shapiro 1976[b]
Secale cereale	Poaceae	44.6	–	50.0	Paper chromatography	Shapiro 1976[b]
Solanum tuberosum	Solanaceae	40.2	–	36.2	Buoyant density centrifugation	Nagl 1976[d]
Spinacia oleracea	Chenopodiaceae/APG: Amaranthaceae	38.7	–	37.4	Paper chromatography	Shapiro 1976[b]
Trifolium pratense	Fabaceae	35.2	–	41.2	Paper chromatography	Shapiro 1976[b]
Triticum aestivum	Poaceae	44.8	1.6	48	Paper chromatography	Sueoka 1961
				48.2	Thermal denaturation	Biswas and Sarkar 1970
				47.5	Bromination	Biswas and Sarkar 1970
				48.4	Paper chromatography	Bonner 1976[b]
				48.4	Paper chromatography	Shapiro 1976[b]
				47.5	Paper chromatography	Shapiro 1976[b]
				44.4	Hydroxyapatite chromatography/Thermal denaturation	Shapiro 1976[b]
				48	Fluorescence of single nuclei	Leemann and Ruch 1982

Table 8.5 (continued)

Species	Family	%GC flow cytometry		Alternative method			Reference
		Mean	SD[a]	%GC	Method		
Vicia faba	Fabaceae	37.9	0.3	40.0	Thermal denaturation		Biswas and Sarkar 1970
				37	Fluorescence of single nuclei		Leemann and Ruch 1982
Zea mays	Poaceae	48.9	5.5	47.6	Thermal denaturation		Biswas and Sarkar 1970
				47.8	Bromination		Biswas and Sarkar 1970
				49.1	Paper chromatography		Shapiro 1976[b]
				46.0	Paper chromatography		Shapiro 1976[b]

[a] For single values the standard deviation is indefinite.
[b] If 5-methylcytosine was determined separately, it was included into the cytosine fraction.
[c] "Data from various sources" according to the author.
[d] Recalculated from density values according to Schildkraut et al. (1962).
[e] The corrected value according to footnote d in Table 8.4 was used for calculation of the mean.

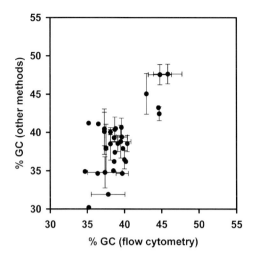

Fig. 8.6 Comparison of GC content as determined by flow cytometry (for comparability, only values calculated with Godelle's equation are included) and other methods. If several values exist, mean and standard deviation are shown.

AATT ≫ TAAT ≈ ATAT > TATA ≈ TTAA. This means that not all sequences containing the same type of DNA bases have identical affinity for a particular fluorochrome.

The binding behavior may also be modified by the accessibility of DNA bases in the chromatin to fluorescent dyes (Darzynkiewicz and Traganos 1988). This effect largely depends on the amount and composition of cytosol and may thus differ between different species as well as between different dyes (Noirot 2002). As a result, the proportionality factor k (in Eq. (4)) considered as constant for all species and therefore eliminated in subsequent calculations, may depend on the specific cytosolic properties of each species. Hence, the actual dye factors, which are determined by the comparison of two different species [Eq. (4)] and are essential for all subsequent equations, may deviate from the theoretical values.

Despite these limitations, FCM is the method of choice in contemporary base composition research. It provides good estimates of base composition within a short time and using a small amount of material. The results are as reliable as those obtained by other more demanding methods (see Fig. 8.6). Only sequence analysis may yield more precise values of base composition provided the whole genome is sequenced, including regions rich in repeated sequences (see Bennett et al. 2003).

8.3 Conclusions

Flow cytometry with base-specific DNA dyes allows the determination of AT and/ or GC base frequencies. The precision is comparable to that of "classical" methods while much less effort is required (i.e. less material and time for preparation and analysis). Therefore, FCM is currently the preferred method for determination of base composition in plant genomes. This becomes evident from the fact that more than 250 FCM analyses (covering 215 plant species) have been published since 1992 (Table 8.4), but virtually none was obtained by the conventional methods. Obviously, sequence analysis of a completely sequenced genome also provides (as a by-product) DNA base composition. However, sequencing is expensive and the genomes of higher plants still likely contain large unsequenced gaps, which may cause an erroneous determination of base composition (Bennett et al. 2003). Owing to these reasons, there is currently no adequate substitute for FCM in DNA base composition research.

Because FCM determination of base composition requires at least two different fluorescent dyes, the analysis is more laborious as compared to the estimation of genome size alone. A practical problem may be the need for UV excitation for the AT-specific dyes (DAPI and Hoechst dyes). New AT-specific dyes, which can be excited in the visible part of the spectrum, or cheaper lasers for UV or extreme (low-wavelength) violet excitation (Shapiro and Perlmutter 2001; Telford 2004) would make the analyses easier and reduce the costs.

Simultaneous staining with dyes of different base specificity could simplify the determination (and possibly enable its automation), but the energy transfer between the dyes (Langlois et al. 1980; A. Meister, unpublished results; for theory see Szöllösi et al. 1998) makes the evaluation difficult, if not impossible. On the other hand, the investigation of the energy transfer could possibly provide deeper insights into the spatial distribution of different bases.

The calculation of base composition by FCM is based on the assumption of random distribution of sequences. However, this is not a realistic premise because nuclear genomes are rich in repetitive DNA sequences. Such deviations from base randomness were confirmed by examination of known genomic sequences. Nevertheless, cytometric analyses provide at least a good approximation of base composition.

Contrary to microorganisms where the base composition is often an important taxonomic criterion, this feature is of limited value in higher plants because of rather small differences between various taxonomic categories (Biswas and Sarkar 1970). Generally, genomes of related species (e.g. within a family) show similar base composition. Consequently, DNA base composition is found to be quite stable within a genus (Ricroch et al. 2005). On the other hand, similar base composition does not exclude differences in base sequences. Therefore, similar base composition may be encountered in unrelated species of different phylogenetic origins. Collectively, these data indicate that base composition has a limited use as a marker for taxa discrimination in plant sciences. Despite some limitations, FCM analysis of base composition provides valuable information which is useful in characterizing plant species.

Acknowledgments

The authors are grateful to Professor Ingo Schubert (IPK Gatersleben, Germany) for critical comments and to Professor S. C. Brown (CNRS, Gif-Sur-Yvette, France) for additional information on his publications.

References

Ali, H. B. M., Meister, A., Schubert, I. **2000**, *Genome* 43, 1027–1032.

Barow, M. **2003**, *Beziehungen zwischen Genomgröße, Basenzusammensetzung und Endopolyploidie bei Samenpflanzen*, PhD Thesis, Martin-Luther-Universität Halle-Wittenberg, Halle.

Barow, M., Meister, A. **2002**, *Cytometry* 47, 1–7.

Behr, W., Honikel, K., Hartmann, G. **1969**, *Eur. J. Biochem.* 9, 82–92.

Bendich, A. **1957**, *Methods Enzymol.* 3, 715–723.

Bennett, M. D., Leitch, I. J., Price, H. J., Johnston, J. S. **2003**, *Ann. Bot.* 91, 547–557.

Biswas, S. B., Sarkar, A. K. **1970**, *Phytochemistry* 9, 2425–2430.

Blondon, F., Marie, D., Brown, S., Kondorosi, A. **1994**, *Genome* 37, 264–270.

Bogunic, F., Muratovic, E., Brown, S. C., Siljak-Yakovlev, S. **2003**, *Plant Cell Rep.* 22, 59–63.

Bonner, J. **1976**, The nucleolus in *Plant Biochemistry*, 3rd edn, ed. J. Bonner, J. E. Varner, Academic Press, New York, pp. 37–64.

Bou Dagher-Kharrat, M., Grenier, G., Bariteau, M., Brown, S., Siljak-Yakovlev, S., Savouré, A. **2001**, *Theor. Appl. Genet.* 103, 846–854.

Brandizzi, F., Caiola, M. G. **1998**, *Plant Systemat. Evol.* 211, 149–154.

Breusegem, S. Y., Clegg, R. M., Loontiens, F. G. **2002**, *J. Mol. Biol.* 315, 1049–1061.

Brown, S. C., Bergounioux, C., Tallet, S., Marie, D. **1991**, Flow cytometry of nuclei for ploidy and cell cycle analysis in *A laboratory guide for cellular and molecular plant biology (BioMethods; 4)*, ed. I. Negrutiu, G. Gharti-Chhetri, Birkhäuser, Basel, pp. 326–345.

Bureš, P., Wang, Y.-F., Horová, L., Suda, J. **2004**, *Ann. Bot.* 94, 353–363.

Cerbah, M., Mortreau, E., Brown, S., Siljak-Yakovlev, S., Bertrand, H., Lambert, C. **2001**, *Theor. Appl. Genet.* 103, 45–51.

Churchill, M. E. A., Suzuki, M. **1989**, *EMBO J.* 8, 4189–4195.

Darzynkiewicz, Z., Traganos, F. **1988**, *Anal. Quant. Cytol. Histol.* 10, 462–466.

Doležel, J., Sgorbati, S., Lucretti, S. **1992**, *Physiol. Plant.* 85, 625–631.

Favre, J. M., Brown, S. **1996**, *Ann. Sci. Forest.* 53, 915–917.

Fredericq, E., Oth, A., Fontaine, F. **1961**, *J. Mol. Biol.* 3, 11–17.

Galbraith, D. W., Harkins, K. R., Maddox, J. M., Ayres, N. M., Sharma, D. P., Firoozabady, E. **1983**, *Science* 220, 1049–1051.

Gallois, A., Burrus, M., Brown, S. **1999**, *Ann. Forest Sci.* 56, 615–618.

Godelle, B., Cartier, D., Marie, D., Brown, S. C., Siljak-Yakovlev, S. **1993**, *Cytometry* 14, 618–626.

Goff, S. A., Ricke, D., Lan, T.-H., and 52 others **2002**, *Science* 296, 92–100.

Goldberg, I. H., Friedman, P. A. **1971**, *Annu. Rev. Biochem.* 40, 775–810.

Hirschman, S. Z., Felsenfeld, G. **1966**, *J. Mol. Biol.* 16, 347–358.

Johnson, J. L. **1985**, *Methods Microbiol.* 18, 1–31.

Kamaté, K., Brown, S., Durand, P., Bureau, J. M., De Nay, D., Trinh, T. H. **2001**, *Genome* 44, 622–627.

Kapuściński, J., Szer, W. **1979**, *Nucleic Acids Res.* 6, 3519–3534.

King, G. J., Ingrouville, M. J. **1987a**, *Genome* 29, 621–626.

King, G. J., Ingrouville, M. J. **1987b**, *New Phytologist* 107, 633–644.

Kirk, J. T. O. **1967**, *Biochem. J.* 105, 673–677.

Kirk, J. T. O., Rees, H., Evans, G. **1970**, *Heredity* 25, 507–512.

Ko, C. Y., Johnson, J. L., Barnett, L. B., McNair, H. M., Vercellotti, J. R. **1977**, *Anal. Biochem.* 80, 183–192.

Koopman, W. J. M. **2002**, *Zooming in on the lettuce genome: species relationships in Lactuca s.l., inferred from chromosomal and molecular characters*, PhD thesis, Wageningen University, Wageningen.

Langlois, R. G., Carrano, A. V., Gray, J. W., van Dilla, M. A. **1980**, *Chromosoma* 77, 229–251.

Larsen, T. A., Goodsell, D. S., Cascio, D., Grzeskowik, K., Dickerson, R. E. **1989**, *J. Biomol. Struct. Dynam.* 7, 477–491.

Leemann, U., Ruch, F. **1982**, *Exp. Cell Res.* 140, 275–282.

Le Thierry d'Ennequin, M., Panaud, O., Brown, S., Siljak-Yakovlev, S., Sarr, A. **1998**, *J. Heredity* 89, 556–559.

Marie, D., Brown, S. C. **1993**, *Biol. Cell* 78, 41–51.

Marmur, J., Doty, P. **1962**, *J. Mol. Biol.* 5, 109–118.

Meister, A. **2005**, *J. Theor. Biol.* 232, 93–97.

Müller, W., Gautier, F. **1975**, *Eur. J. Biochem.* 54, 385–394.

Muratović, E., Bogunić, F., Šoljan, D., Siljak-Yakovlev, S. **2005**, *Plant Systemat. Evol.* 252, 97–109.

Nagl, W. **1976**, *Zellkern und Zellzyklen: Molekularbiologie, Organisation und Entwicklungsphysiologie der Desoxyribonucleinsäure und des Chromatins*, Ulmer, Stuttgart.

Narayan, R. K. J. **1988**, *Theor. Appl. Genet.* 75, 319–329.

Narayan, R. K. J., Rees, H. **1976**, *Chromosoma* 54, 141–154.

Noirot, M. **2002**, *Ann. Bot.* 89, 385–389.

Pollard, J. H. **1977**, The real roots of non-linear equations in *A Handbook of Numerical and Statistical Techniques with Examples Mainly from the Life Sciences*, ed. J. H. Pollard, Cambridge University Press, Cambridge, pp. 18–25.

Portugal, J., Waring, M. J. **1988**, *Biochim. Biophys. Acta* 949, 158–168.

Redondo, N., Horjales, M., Brown, S., Villaverde, C. **1996**, *Bol. Socied. Brot., Séries 2* 67, 287–301.

Ricroch, A., Brown, S. C. **1997**, *Gene* 205, 255–260.

Ricroch, A., Yockteng, R., Brown, S. C., Nadot, S. **2005**, *Genome* 48, 511–520.

Sanders, C. A., Yajko, D. M., Hyun, W., Langlois, R. G., Nasson, P. S., Fulwyler, M. J., Hadley, W. K. **1990**, *J. Gen. Microbiol.* 136, 359–365.

Schildkraut, C. L., Marmur, J., Doty, P. **1962**, *J. Mol. Biol.* 4, 430–443.

Schnedl, W., Mikelsaar, A.-V., Breitenbach, M., Dann, O. **1977**, *Hum. Genet.* 36, 167–172.

Schweizer, D. **1976**, *Chromosoma* 58, 307–324.

Schweizer, D. **1981**, *Hum. Genet.* 57, 1–14.

Schwencke, J., Bureau, J.-M., Crosnier, M.-T., Brown, S. **1998**, *Plant Cell Rep.* 18, 346–349.

Shapiro, H. S. 1976, Distribution of purines and pyrimidines in deoxyribonucleic acids in *Handbook of Biochemistry and Molecular Biology. B: Nucleic Acids, Vol. 2*, ed. G. D. Fasman, CRC Press, Cleveland, pp. 241–311.

Shapiro, H. M., Perlmutter, N. G. **2001**, *Cytometry* 44, 133–136.

Siljak-Yakovlev, S., Cerbah, M., Coulaud, J., Stoian, V., Brown, S. C., Zoldoš, V., Jelenic, S., Papes, D. **2002**, *Theor. Appl. Genet.* 104, 505–512.

Siljak-Yakovlev, S., Peccenini, S., Muratovic, E., Zoldoš, V., Robin, O., Valles, J. **2003**, *Plant Syst. Evol.* 236, 165–173.

Siljak-Yakovlev, S., Solic, M. E., Catrice, O., Brown, S. C., Papes, D. **2005**, *Plant Biol.* 7, 397–404.

Skidmore, W. D., Duggan, E. L. **1966**, *Anal. Biochem.* 14, 223–236.

Stack, S. M., Comings, D. E. **1979**, *Chromosoma* 70, 161–181.

Sueoka, N. **1961**, *J. Mol. Biol.* 3, 31–40.

Sulimova, G. E., Mazin, A. L., Vanyushin, B. F., Beloszerskiy, A. N. **1970**, *Dok. Akad. Nauk SSSR* 193, 1422–1425.

Szöllösi, J., Damjanovich, S., Mátyus, L. **1998**, *Cytometry* 34, 159–179.

Telford, W. G. **2004**, *Cytometry* 61A, 9–17.

The Arabidopsis Genome Initiative **2000**, *Nature* 408, 796–815.

Van Dyke, M. W., Dervan, P. B. **1983**, *Biochemistry* 10, 2373–2377.

Vinogradov, A. E. **1994**, *Cytometry* 16, 34–40.

Wang, S. Y., Hashagen, J. M. **1964**, *J. Mol. Biol.* 8, 333–340.

Ward, D. C., Reich, E., Goldberg, I. H. **1965**, *Science* 149, 1259–1263.

Weisblum, B., Haenssler, E. **1974**, *Chromosoma* 46, 255–260.

Wyatt, G. R. **1955**, Separation of nucleic acid components by chromatography on filter paper in *The Nucleic Acids: Chemistry and Biology. Vol. I*, ed. E. Chargaff, J. N. Davidson, Academic Press, New York, pp. 243–265.

Yu, J., Hu, S., Wang, J., Wong, G.-S., and 96 others **2002**, *Science* 296, 79–92.

Zoldoš, V., Papes, D., Brown, S. C., Panaud, O., Siljak-Yakovlev, S. **1998**, *Genome* 41, 162–168.

Zonneveld, B. J. M., van Iren, F. **2000**, *Euphytica* 111, 105–110.

9
Detection and Viability Assessment of Plant Pathogenic Microorganisms using Flow Cytometry

Jan H. W. Bergervoet, Jan M. van der Wolf, and Jeroen Peters

Overview

Detection of plant pathogenic microorganisms plays a fundamental role in warranty of quality in the food and feed production chains. Early methods used specific host plants and culture media to reveal the presence of pathogens. However, they were laborious and quicker approaches were therefore sought. The advent of immunological techniques, such as Enzyme Linked Immuno Sorbent Assay (ELISA), transformed the phytosanitary testing stations to high-throughput laboratories where large numbers of samples can be analyzed each day. Further demand for rapid tests capable of differentiating between the whole range of pathogens resulted in routine application of multiplexed immunological methods and PCR-based techniques. Flow cytometry is a multi-parameter technique useful for detection of plant pathogenic bacteria and/or assessing their viability. However, its routine application in agriculture is hampered by the large initial investment and requirement for highly trained staff. On the other hand, flow cytometry has a great potential in agricultural research and breeding to identify plant pathogens using specific antibodies, determine their viability, as well as to monitor the effects of crop protective treatments. It is expected that bead-based flow cytometry will become a powerful tool for fast and multiplexed detection of plant pathogens in the near future.

9.1
Introduction

Plant pathogens cause significant reduction in crop yield and quality, and consequently great economic loss worldwide each year. Many efforts have been made to control or prevent their incidence. A combination of strategies has been suggested to achieve this goal, such as the use of disease-free starting material, improved phytosanitary methods, and crop protection systems incorporating chemical treatments. However, if any of the system components fails, an outbreak of

Flow Cytometry with Plant Cells. Edited by Jaroslav Doležel, Johann Greilhuber, and Jan Suda
Copyright © 2007 WILEY-VCH Verlag GmbH & Co. KGaA, Weinheim
ISBN: 978-3-527-31487-4

disease can start rapidly. This three-stage strategy has been successfully implemented for example in the Netherlands to protect potato cultivation. The use of disease-free starting material slows down the spread of pathogens in the field, and a certificate attesting the health status is required for an increasing number of plant propagation materials. Removal of diseased individuals from the field and controlling volunteer plants from the previous season, combined with the use of clean, disinfected tools and materials, and the use of appropriate crop protection agents have all proven efficient in plant disease prevention and control.

For production and selection of disease-free starting material, the use of rapid and reliable methods for detection and identification of pathogens (i.e. viruses, plant pathogenic fungi or bacteria) is of crucial importance. Diagnosis of the disease in an early phase allows the application of a suitable control strategy and may ameliorate possible negative effects. Since the first virus diseases in potato were described (Quanjer et al. 1920; Smith 1931), attempts have been made to develop reliable, robust and inexpensive detection methods. Before serological tests became available, the presence of potato viruses was screened with a callose staining technique (Bokx 1967). However, this method is time consuming and a faster approach for pathogen detection was intensively sought.

Enzyme Linked Immuno Sorbent Assay or ELISA (Clark and Adams 1977) and Immuno Fluorescence microscopy, IF (Schonbeck and Spengler 1979) were the first techniques routinely employed to detect plant pathogens (Maat and Bokx 1978), and they rapidly overtook the plating tests and the bioassays due to their higher speed and specificity. Antibodies for the majority of important plant viruses and bacteria are now available, making ELISA and IF important testing methods. For example, about 60 000 and 4×10^6 potato samples are tested each year with IF and ELISA, respectively, in the Netherlands. Other testing protocols, based on amplification of specific DNA sequences, have been developed in order to further enhance specificity and sensitivity, but they are generally too expensive for routine and large-scale application. Although ELISA and IF provide fast information on the presence or absence of specific antigens produced by the target pathogen, no information on the pathogen viability or virulence is obtained. If this information is needed, culturing bacteria on nutrient media or bioassays using sensitive indicator plants are still very useful.

9.2
Viability Assessment

Viability is defined as the capability of an organism to survive, grow and propagate under favorable conditions. For microorganisms to survive, intact plasma membrane and functioning cell machinery (i.e. DNA transcription, RNA translation, and enzyme activity) are essential. Not surprisingly, methods for viability assessment are mostly based on screening these attributes. Plating assays that involve culturing of microorganisms on selective or semi-selective media were the first procedures developed and are still in use. They allow isolation of pathogens from plant material and their subsequent identification by a range of biochemical or serological tests. This approach is, however, time consuming and requires specific culture media and skilled technicians (Plihon et al. 1995).

A rapid and convenient alternative for providing information on the viability of microorganisms is flow cytometry (FCM). It is easily applicable to a wide range of samples and species (Bunthof et al. 2001; Chitarra and van den Bulk 2003). FCM viability assays rely on the availability of suitable fluorescent probes, which are selected according to their target specificity (e.g. RNA or membrane probes) and optical properties (e.g. fluorescence excitation and emission spectra). Carboxy fluorescein diacetate (cFDA) is an example of a popular enzyme activity probe used as a cell viability indicator (Bunthof et al. 2001). cFDA is a non-fluorescent ester compound that can permeate intact cell membranes. It undergoes hydrolysis inside the cell, after which a green fluorescent compound, carboxy fluorescein, is released and retained inside the cells. Propidium iodide (PI), a red-fluorescent nucleic acid-specific dye, is another probe that has been used to assess microorganism viability. PI does not cross intact cell walls and therefore only enters cells with damaged or compromised membranes. Consequently, the viability of plant pathogenic bacteria and fungi can be assessed following simultaneous staining with PI and cFDA. Another possibility is to use SYTO9, a green-fluorescent nucleic acid dye, which can cross both intact and damaged membranes. As PI can enter cytoplasm and stain double-stranded DNA and RNA only if a plasma membrane is damaged, while SYTO9 can always cross the membrane, the ratio of red PI to green SYTO9 fluorescence is a reliable indicator of the integrity of the plasma membrane. A combination of these dyes has been successfully used to assess the viability of fungal spores and plant pathogenic bacteria.

9.2.1
Viability Tests for Spores and Bacteria

To test the feasibility of FCM pathogen viability assays, we analyzed fungal spores of gray mold (*Botrytis cinerea*) isolated from plate cultures. The samples were stored on ice and aliquots were heated for various periods of time to prepare sub-vital and non-viable cells, respectively. Color compensation (see Chapter 2) was adjusted using viable and non-viable spores stained either with PI or SYTO9, or simultaneously with both dyes. This methodology permitted a clear discrimination of the population of particles stained by both PI and SYTO9 from particles stained by only one of these dyes. Prior to FCM analyses, the stained samples were checked using fluorescence microscopy. The green-fluorescent viable spores were clearly visible as were their red-fluorescent non-viable counterparts (Fig. 9.1). Although the orange-fluorescent spores, representing the intermediate fraction, could be distinguished by fluorescence microscopic observation, the exact ratio between green/red fluorescence signal intensity could only have been determined by FCM (Fig. 9.2).

A similar procedure was employed for the assessment of bacterial viability. Aliquots of overnight-grown culture of *Erwinia carotovora* subsp. *atroseptica* were

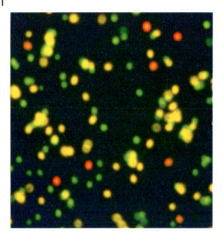

Fig. 9.1 Fluorescent image of viable and non-viable spores of a fungus, *Botrytis cinerea*, that were stained simultaneously with propidium iodide (red fluorescence) and SYTO9 (green fluorescence). Green, viable spores; orange, intermediate (sub-vital) spores; and red, non-viable spores.

exposed to heat for various periods of time to prepare samples differing in the degree of viability. On a scattergram of red fluorescence (x-axis) versus green fluorescence (y-axis), the population of viable green-fluorescent cells appeared as a cluster near the y-axis while the non-viable red-fluorescent cells appeared as a cluster near the x-axis. The cells displaying both green and red fluorescence were located, depending on their status, close to the center of the scattergram. The bacteria in this intermediate fraction were not able to form colonies in the plating assays. This was also observed for the bacterium *Ralstonia solanacearum* but the intermediate population could still cause infection in tomato plants (Van

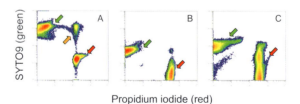

Fig. 9.2 Flow cytometric analysis of fungal spore viability. (a) *Gliocladium roseum*; (b) *Phytophthora infestans*; and (c) *Botrytis cinerea*. The samples were stained simultaneously with propidium iodide and SYTO9. Populations of viable spores are marked with green arrows, intermediate (sub-vital) spores with orange arrows, and non-viable spores with red arrows.

der Wolf et al. 2005). Because it is unclear if this state is caused by DNA damage or other factors, such intermediate populations of bacteria are classified as viable but non culturable (VBNC). It is interesting to note that the VBNC state has also been found in populations of human pathogenic bacteria in food and feed (Baudart et al. 2002; Besnard et al. 2000, 2002; Rahman et al. 2001; Rowan 2004) and it seems that at least some plant pathogenic bacteria may behave in a similar way (Ghezzi and Steck 1999; Manahan and Steck 1997).

In order to confirm viability of subpopulations of fungal cells identified by FCM, mixtures of spores of the fungi *Fusarium culmorum* and *Botrytis cinerea* with different degrees of viability were stained simultaneously with PI and SYTO9, analyzed, and flow-sorted. Isolated viable, intermediate, and non-viable spores were transferred to non-selective media and allowed to germinate and grow. The majority of the green-fluorescent spores germinated rapidly and uniformly at 99%. The performance of the red-fluorescent spores was, as expected, very poor and only a negligible number of spores germinated (4% in *F. culmorum*, 1% in *B. cinerea*). In the intermediate fraction, 57% of *B. cinerea* spores were able to germinate; however, the germination process was much slower when compared to the green-fluorescent counterparts. Such a result indicated that a part of the fungal spore population might have been dead, while the rest had a lower viability but was still able to germinate and grow. This contrasts with the observations made with bacterial cells, where the intermediate fraction showed all the characteristics of cells in a VBNC state (Kell et al. 1998).

Our results showed that although the use of FCM for the assessment of pathogen cell viability is rather straightforward, the optimal amount of nucleic acid dyes and the incubation times need to be experimentally determined for each sample type. In fact, these factors themselves can affect the viability of the microorganisms (Van der Wolf et al. 2005). Once the method is established, the viability of mixtures of different unknown microorganisms can be assessed.

An interesting application of this assay involves the estimation of the total number of bacteria in stem sections of cut flowers. It is known that some bacteria can grow into the vascular bundles of ornamental plants and cause their blockage, resulting in a dramatic decrease in vase life. We have optimized an FCM method for *Gerbera* (Asteraceae), which included homogenization of stem sections about 1 cm in length using a stomacher, passing the homogenate over a Whatman filter, and staining the filtrate simultaneously with PI and SYTO9. A heavy bacterial contamination by microorganisms indicated that dirty containers were used to transport flower cuttings and/or that the lower parts of the stems were not disinfected.

Our experience shows that the FCM viability assay is applicable to a wide range of plant-associated microorganisms, including fungi and bacteria. However, care should be taken with the materials containing a high salt concentration (e.g. sea water) and contaminated with chemical compounds that either interfere with the staining and/or affect the fluorescence of PI and SYTO9, such as commercial bleach. These substances can distort the stoichiometry of the staining and lead to erroneous estimates of microorganism viability.

9.3
Immunodetection

Immunodetection methods (e.g. ELISA and IF) are powerful and widely used techniques for detecting viruses and pathogenic microorganisms. When viruses are assayed, a much-favored option is the Double Antibody Sandwich ELISA (DAS-ELISA). Its principle lies in coating the inner surface of a microtiter plate well with appropriate primary antibody. The sample is then added and incubated for some time. Subsequently, the sample is removed and a secondary antibody conjugated with an enzyme (alkaline phosphatase or peroxidase) is added. Finally, a specific substrate is added and the amount of antigen present in the sample is determined based on the amount of a product formed in the enzymatic reaction. The introduction of robotics allowed the analysis of large numbers of samples in a reliable and cost-effective way. Currently, the main drawback is a rather limited choice of fluorescent labels that can be combined with secondary antibodies. In addition, multiple pathogens often need to be detected simultaneously but, in a standard ELISA, only one parameter (antigen) is assessed per well.

Microscopic immunofluorescent procedures may, at least partially, cope with this problem, as documented by the successful detection of large numbers of bacteria at one time. Bacterial samples are incubated with antibodies tagged with fluorescent dyes and observed using fluorescence microscopy. However, this methodology is time consuming and thus less suitable for large-scale analysis. An additional limitation is that only a few different fluorescent probes (typically three) can be observed using standard microscopic equipment.

More efficient and low-cost methods have recently become available that allow the detection of multiple pathogens by increasing the number of different antibodies per assay. Immuno Flow Cytometry (IFCM), Microsphere Immuno Assays (MIA), and protein microarrays are all examples of this.

IFCM, in which flow cytometry is used to analyze samples incubated with fluorochrome-conjugated antibodies, has been applied for rapid and specific detection of bacteria in a number of research areas including food (Ananta et al. 2004), medical (Pina-Vaz et al. 2005), veterinary (Weiss 2001), environmental (Andrade et al. 2003; Davey and Winson 2003), and plant sciences (Chitarra and van den Bulk 2003). For example, the pathogenic bacterium, *Xanthomonas campestris* pathovar. *campestris* was detected using IFCM in crude seed extracts of cabbage (Chitarra and van den Bulk 2003). Similarly, this method also showed the incidence of *X. campestris* pathovar. *phaseoli* in extracts from field bean seeds (Fig. 9.3).

IFCM can also be combined with the viability assessment, for which the samples are first incubated with a specific antibody tagged with a fluorescent dye (to identify the pathogenic bacteria and fungal spores), and then stained with viability dyes (e.g. SYTO9 and PI). Samples are typically analyzed using a flow cytometer equipped with two lasers, one to detect the labeled antibodies and the other to detect the viability dyes. Because two lasers are employed, there is no need to set up color compensation for the antibody signal. The advantage of IFCM is easy quantification of the microorganisms of interest in a reasonable time (e.g. <1 h).

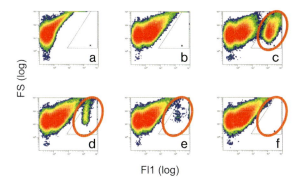

Fig. 9.3 Detection of *Xanthomonas campestris* pathovar. *phaseoli* (Xcp) using immuno flow cytometry in a 10-fold diluted extract from field bean (*Vicia faba*) seeds that were incubated with a specific antibody against Xcp conjugated with fluorescein isothiocyanate (FITC). (a) Bacteria concentration 10^6 ml^{-1}, no antibody added; (b) control bean extract; (c) bacteria concentration 10^6 ml^{-1} + antibody; (d) bacteria concentration 10^5 ml^{-1} + antibody; (e) bacteria concentration 10^4 ml^{-1} + antibody; and (f) bacteria concentration 10^3 ml^{-1} + antibody. The bacterial population is marked by the red ellipse. FS (log), forward angle light scatter (log scale); Fl1 (log), fluorescence of fluorescein (log scale).

However, if a flow cytometer is not equipped with a volumetric sample delivery, enumeration of microorganisms requires the addition of special fluorescent beads at known concentration (Monfort and Baleux 1992). As compared to other serological techniques used for bacterial testing, the reported sensitivity of IFCM of about 10^4 cells ml^{-1} is acceptable. Theoretically, 10^2–10^3 microorganisms ml^{-1} could be detected in an undiluted sample, considering that about 10 μl is a typical volume analyzed per sample.

Both IFCM and FCM viability methods allow for high-throughput screening, including small-scale multiplex detection. Nevertheless, they are rather expensive for routine testing primarily due to the high price of the flow cytometer.

Another system suitable for multiplex pathogen detection are the protein arrays (Templin et al. 2002). They are made on a glass plate surface where proteins (antibodies) are deposited in small spots distributed in a matrix-like pattern. A sample is pipetted onto each of these arrays, incubated, and washed. Secondary antibodies conjugated with a fluorescent reporter molecule are then added, followed by another wash step and final scanning of the slide for presence/absence of the fluorescent signal.

This method is comparable to the universal microsphere array (Joos et al. 2000; Vignali 2000), which theoretically allows simultaneous detection of a large number of pathogens (up to 100). Both the detection level and price per microsphere assay are comparable to standard ELISA, but the instrument (i.e. Luminex 100 ST, a small flow cytometer designed to detect microspheres; Earley 2002) is less expensive than conventional flow cytometers.

9.3.1
Microsphere Immuno Assay

The Microsphere Immuno Assay (MIA) is based on the universal bead array (xMAP) of Luminex (Austin, TX, USA). The microspheres have a diameter of 5.6 μm and are internally stained with two fluorochromes at different ratios. Currently, about 100 different color-coded bead types are available. The beads can also be covalently linked to proteins, peptides, polysaccharides, lipids, and oligonucleotides (Joos 2004). Among other, MIA has proven its value for multiplex detection in human diagnostics (e.g. in cystic fibrosis), and is used in multiplexed assays to study infectious diseases (Kellar 2003) and also in food microbiology (Dunbar et al. 2003). Samples for MIA are transferred to a microtiter plate, incubated first with antibody-coated microspheres and subsequently with secondary antibodies conjugated to a reporter fluorochrome, and analyzed using a Luminex 100 ST (Fig. 9.4). The internal dyes of the microspheres and the reporter fluoro-

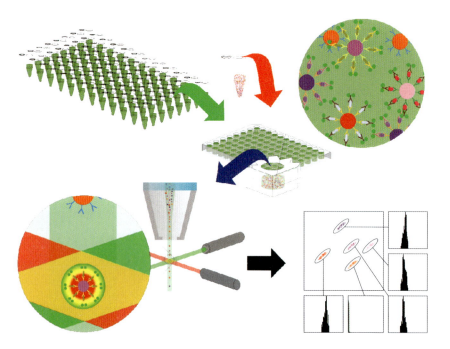

Fig. 9.4 Overview of the Luminex microsphere technology. Sample is prepared and transferred to a 96-well microtiter plate. A mixture of antibodies conjugated with the microsphere sets is added. During the incubation, the pathogens are captured by the antibodies and the secondary antibodies conjugated with a fluorescent reporter are added. Samples are then measured on a Luminex analyzer, and the results are graphically presented. The absence of pathogens results in no signal (no fluorescent reporter will be detected).

chrome are excited with red and green lasers, respectively. When compared to planar antibody arrays, the Luminex technology has superior detection thresholds and a better dynamic range. On the other hand, protein micro arrays seem more suitable for miniaturization (Rao et al. 2004).

9.3.1.1 Detection of Plant Pathogenic Bacteria and Viruses

MIA can successfully be used to detect plant pathogenic bacteria including *Erwinia carotovora* subsp. *carotovora* and *E. chrysanthemi*, which cause soft rot disease. Fundamental components of such assays are the Luminex beads coated with polyclonal antibodies against the targeted microorganisms. Typically, 100 beads of each bead set are measured, which takes only about 10 s. The analysis of all the wells in a 96-well microtiter plate then takes no more than 20 min. The results of this assay are comparable or even better than those obtained by ELISA (Fig. 9.5). Sample enrichment can further enhance the sensitivity. Prior to detection, the sample is transferred to a non-selective medium, in which the microorganism is allowed to propagate, and is then cultured overnight. Such enriched samples can be used in both standard ELISA and MIA (Table 9.1).

Our results suggest that the detection limits of MIA are similar to those of ELISA, and range from 10^7 to 10^6 cells ml^{-1} for *E. carotovora* subsp. *atroseptica* and *E. chrysanthemi*, respectively. However, in the enrichment assays, the MIA proved to be more sensitive and allowed detection of much lower concentrations of both *E. carotovora* subsp. *atroseptica* (10^2 cells ml^{-1}) and *E. chrysanthemi* (10^3 cells ml^{-1}). This is a 100-fold increase in sensitivity when compared to the stan-

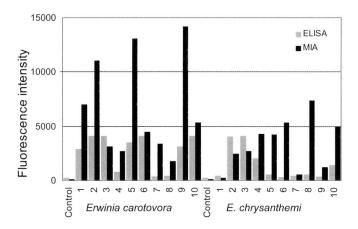

Fig. 9.5 Comparison of flow cytometric Microsphere Immuno Assay (MIA) and Enzyme Linked Immuno Sorbent Assay (ELISA) in the analysis of 10 different samples of potato peel extracts that were infected with *Erwinia carotovora* subsp. *atroseptica* and *E. chrysanthemi*, the causative agents of the black leg disease. Control, the mean value of 10 *Erwinia*-free samples.

Table 9.1 Comparison of Enzyme Linked Immuno Sorbent Assay (ELISA) and the flow cytometric Microsphere Immuno Assay (MIA) in direct and enrichment tests. Known concentrations of *Erwinia chrysanthemi* and *E. carotovora* subsp. *atroseptica* were added to potato peel extracts. The lowest concentration of bacteria per millilitre detected is shown.

Assay[a]	E. chrysanthemi (cfu ml^{-1})		E. carotovora (cfu ml^{-1})	
	ELISA	MIA	ELISA	MIA
Single assay, direct	10^7	10^7	10^6	10^6
Single assay, enrichment	10^5	10^3	10^4	10^2
Duplex, direct		10^7		10^6
Duplex, enrichment		10^3		10^2

cfu, colony forming units.
[a] The duplex assays involved two different populations of beads, each conjugated to a different antibody, to detect both pathogens simultaneously.

dard ELISA. A single-assay MIA (i.e. when only one bead–one pathogen combination is detected per sample) yielded results identical to a modification with simultaneous detection of both pathogens. These results thus show that MIA can be used to carry out a larger number of tests per day without a loss of sensitivity or specificity. We have also found that after a minor modification, the method is suitable for detection of plant viruses, such as potato virus X, potato virus Y, and potato leafroll virus.

9.3.1.2 Paramagnetic Microsphere Immuno Assay

In this procedure, the carboxylated paramagnetic beads from Luminex are used in a similar way as the beads in the MIA no-wash assay. However, the use of magnetic beads permits efficient washing and removal of unbound antibodies by holding the paramagnetic beads at the bottom of the wells using a magnetic support. Although the background in standard no-wash MIA is relatively low, the specific (positive) signal may sometimes approach the background values and thus remains undetected (i.e. results in false negative signals). In the paramagnetic MIA, the non-specific background signal is much lower, giving a better signal to noise ratio and allowing more reliable conclusions (Fig. 9.6).

In some plant organs and tissues, as for example bulbs of *Hyacinthus orientalis* (Hyacinthaceae), it is difficult to prepare samples suitable for ELISA due to the high levels of mucous substances. Nevertheless, we have successfully applied the bead-based array also in this case; paramagnetic beads in 10-fold diluted bulb extracts yielded reproducible results.

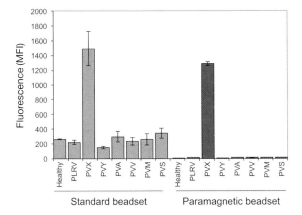

Fig. 9.6 Comparison of standard beads (no-wash procedure) versus paramagnetic beads (wash procedure) in the flow cytometric Microsphere Immuno Assay (MIA) for the detection of potato virus X (PVX) in naturally infected potato leaves. The use of paramagnetic beads permitted discrimination of PVX from other viruses without false positive signals. Error bars represent the standard error of difference (SED); MFI, median fluorescence intensity; PLRV, potato leafroll virus; PVY, potato virus Y; PVA, potato virus A; PVV, potato virus V; PVM, potato virus M; PVS, potato virus S.

9.4
Conclusions and Future Prospects

The flow cytometric techniques described in this chapter, namely the immunofluorescence flow cytometry (IFCM) and the flow cytometric microsphere immuno assay (MIA), are useful for the detection of microorganisms in both agriculture and horticulture. A comparison of their performance with immunofluorescence microscopy (IF) and ELISA is given in Table 9.2. We anticipate that the use of IF in routine diagnostics will decline as it is laborious and the possibility for multiplexing is limited. Automated microscopy with sophisticated software could increase the throughput of IF, but these systems are expensive and require a highly trained staff. Nevertheless, IF will continue playing an important role in pathogen assay development, as it allows direct visualization of non-specific binding and matrix interactions, providing valuable information for further protocol development and fine tuning.

IFCM represents an interesting alternative to the laborious evaluation of stained samples with fluorescence microscopy. It has already shown its potential for the fast and unbiased enumeration of microorganisms in food and vase water. The spread of IFCM, however, may be hampered by the lack of trained staff and the high price of flow cytometers that are equipped with multiple lasers, which are more suitable for multicolor assays.

We envisage that the role of ELISA in routine analysis will change as well. Although many industrial laboratories have built their logistics around ELISA and

Table 9.2 Comparison of Immuno Fluorescence microscopy (IF), Enzyme Linked Immuno Sorbent Assay (ELISA), flow cytometric Microsphere Immuno Assay (MIA), and Immuno Flow Cytometry (IFCM) for the detection of pathogens.

	IF	ELISA	MIA	IFCM
Detection limit (cells per ml)	10^3	10^4	10^4[a]	10^4
Speed	Low	High	Very high	High
Multiplexing	Low	Low	Max. 100	Max. 15
Wash steps	Yes	Yes	Yes/None[b]	None
Sample dilution	Yes	Yes	No	Yes
Viability	No	No	No	Yes

[a] 10^1 after pre-enrichment.
[b] Depending on the assay and bead types; the use of paramagnetic microspheres and/or microtiter filter plates allows washing between the incubations with antibodies and increases the sensitivity.

invested in washing- and pipette robots to reduce manpower (and costs), further costs reduction cannot be expected, because ELISA is not suitable for multiplex detection.

MIA could play an important role in routine diagnostics and could gradually replace ELISA. The transition from ELISA to MIA is relatively straightforward because the nature of both assays and the reagents used are similar. The advantage of MIA over ELISA is the time reduction in sample preparation as smaller amounts of material is needed. This, together with a lower consumption of reagents and consumables, reduces costs without compromising the performance. In fact, the use of the bead-based assay provides another advantage. Nucleic acids can be coupled to the bead surface, allowing for detection of DNA or RNA. Both DNA- and antibody-tagged beads can be analyzed with the same instrument.

Currently, the main bottleneck for rapid implementation of MIA in detecting plant pathogens is the limited number of antibodies available. Development of new antibodies for viruses and bacteria is feasible but time consuming and expensive. Specifically, the use of MIA for the detection of fungi is limited due to difficulties accompanying production of good quality antibodies. The number of antibodies against fungi is still low in the market, which contrasts to the amount of DNA sequence data available. It may be expected that the role of DNA-based methods will increase in time, especially when DNA isolation and purification becomes less difficult and expensive. We believe that flow cytometric microsphere immuno assay, either antibody-based or DNA-based, will soon become an important technology in routine diagnostics of plant pathogenic microorganisms.

References

Ananta, E., Heinz, V., Knorr, D. **2004**, *Food Microbiol.* 21, 567–577.

Andrade, L., Gonzalez, A. M., Araujo, F. V., Paranhos, R. **2003**, *J. Microbiol. Methods* 55, 841–850.

Baudart, J., Coallier, J., Laurent, P., Prevost, M. **2002**, *Appl. Environ. Microbiol.* 68, 5057–5063.

Besnard, V., Federighi, M., Cappelier, J. M. **2000**, *Lett. Appl. Microbiol.* 31, 77–81.

Besnard, V., Federighi, M., Declerq, E., Jugiau, F., Cappelier, J. M. **2002**, *Vet. Res.* 33, 359–370.

Bokx, J. A. **1967**, *Eur. Pot. J.* 10, 221–234.

Bunthof, C. J., Bloemen, K., Breeuwer, P., Rombouts, F. M., Abee, T. **2001**, *Appl. Environ. Microbiol.* 67, 2326–2335.

Chitarra, L. G., van den Bulk, R. W. **2003**, *Eur. J. Plant Pathol.* 109, 407–417.

Clark, M. F., Adams, A. N. **1977**, *J. Gen. Virol.* 34, 475–483.

Davey, H. M., Winson, M. K. **2003**, *Curr. Issues Mol. Biol.* 5, 9–15.

Dunbar, S. A., Zee, C. A., Oliver, K. G., Karem, K. L., Jacobson, J. W. **2003**, *J. Microbiol. Methods* 53, 245–252.

Earley, M. C. **2002**, *Cytometry* 50, 239–242.

Ghezzi, J. I., Steck, T. R. **1999**, *FEMS Microbiol. Ecol.* 30, 203–208.

Joos, T. J. G. **2004**, *Expert Rev. Proteomics* 1, 1–3.

Joos, T. O., Schrenk, M., Hopfl, P., Kroger, K., Chowdhury, U., Stoll, D., Schorner, D., Durr, M., Herick, K., Rupp, S., Sohn, K., Hammerle, H. **2000**, *Electrophoresis* 21, 2641–2650.

Kellar, K. L. **2003**, *J. Clin. Ligand Assay* 26, 76–86.

Maat, D. Z., Bokx, J. A. **1978**, *Netherlands J. Plant Pathol.* 84, 167–173.

Manahan, S. H., Steck, T. R. **1997**, *FEMS Microbiol. Ecol.* 22, 29–37.

Monfort, P., Baleux, B. **1992**, *Cytometry* 13, 188–192.

Pina-Vaz, C., Costa-de-Oliveira, S., Rodrigues, A. G. **2005**, *J. Med. Microbiol.* 54, 77–81.

Plihon, F., Taillandier, P., Strehaiano, P. **1995**, *Biotechnol. Tech.* 9, 451–456.

Quanjer, H. M., Dorst, J. C., Dijt, M. D., van der Haar, A. W. **1920**, *Meded. Landbouwhog.* 17, 1–74.

Rahman, M. H., Suzuki, S., Kawai, K. **2001**, *Microbiol. Res.* 156, 103–106.

Rao, R. S., Visuri, S. R., McBride, M. T., Albala, J. S., Matthews, D. L., Coleman, M. A. **2004**, *J. Proteome Res.* 3, 736–742.

Rowan, N. J. **2004**, *Trends Food Sci. Technol.* 15, 462–467.

Schonbeck, F., Spengler, G. **1979**, *Phytopathol. Zeits. – J. Phytopathol.* 94, 84–86.

Smith, K. M. **1931**, *Proc. Roy. Soc. Lond. Series B, Containing Papers of a Biological Character* 109, 251–267.

Templin, M. F., Stoll, D., Schrenk, M., Traub, P. C., Vohringer, C. F., Joos, T. O. **2002**, *Drug Discov. Today* 7, 815–822.

Van der Wolf, J. M., Sledz, W., Van Elsas, J. D., Van Overbeek, L., Bergervoet, J. H. W. **2005**, Flow cytometry to detect *Ralstonia solanacearum* and to assess viability in *Bacterial Wilt: The Disease and the Ralstonia solanacearum Species Complex*, ed. C. Allen, P. Prior, A. C. Hayward, The American Phytopathological Society Press, St. Paul, Minnesota, USA, pp. 479–484.

Vignali, D. A. A. **2000**, *J. Immunol. Methods* 243, 243–255.

Weiss, D. J. **2001**, *Vet. Pathol.* 38, 512–518.

10
Protoplast Analysis using Flow Cytometry and Sorting

David W. Galbraith

Overview

Flow cytometry and cell sorting requires the availability of single cell suspensions. This chapter describes the preparation and use of protoplasts, plant cells from which the cell wall has been removed, as sources of single cells suitable for flow analysis. It goes on to describe a number of different applications that have been developed for the measurement of specific protoplast properties, and for their selective enrichment via cell sorting. The chapter concludes with an analysis of potential future research directions.

10.1
Introduction

For all species and during most life stages, higher plants comprise multicellular organisms, in which the individual cells are mechanically connected by shared cell walls, and functionally connected in the form of a supracellular symplastic network (Buchanan et al. 2000). Such organization is incompatible with analyses involving flow cytometry (FCM) and sorting, which require single cell, or single particle suspensions (Shapiro 2003). Preparation of wall-less cells (protoplasts) represents a well-established means to convert organized plant tissues into cell suspensions. In this chapter, I focus on the general use of protoplasts in FCM and cell sorting. Applications dealing with transgenic organisms are found in Chapter 17.

10.1.1
Protoplast Preparation

Protoplasts are produced by digestion of the plant cell wall polysaccharides, using lytic enzymes produced from tree-rotting fungi and bacteria. Representative, commercially-available examples are cellulase "Onozuka" from *Trichoderma viride*, and Macerozyme R-10 (a pectinase/hemicellulase mixture isolated from *Rhizopus* spp.). These are incubated with the plant tissues, which are typically peeled or cut into thin strips, in the presence of plasmolytic osmotica such as mannitol or sorbitol. After wall digestion, the released protoplasts are filtered through nylon mesh to eliminate tissue debris, and purified by a combination of low-speed differential and isopycnic gradient centrifugation.

Methods for protoplast production, although well established and empirically optimized for many species and tissue types, display subtle differences, and it is recommended that the eager neophyte search the primary literature (rather than consulting review articles) for methods appropriate for the species and tissues of interest and then contact the authors for technical details. Some species and tissue types remain recalcitrant to protoplast production. However, it is also increasingly obvious that there are sources of variation in protoplast yield and physiological status, relating to growth conditions in general, such as method of production of the plant materials, light regime, temperature, watering schedule, and so forth, as well as the impact of the osmotic and enzymatic treatments required for protoplast release. As our ability to monitor biological and cellular states has become more technologically sophisticated, additional sources of variation have become detectable, including such macroscale examples as prior harvest history and circadian time, and microscale examples such as cell type (Birnbaum et al. 2003, 2005). It will be of considerable interest to partition these sources of variation particularly in terms of their impacts on global gene expression.

Early experiments with protoplasts focused on establishing treatment conditions that yielded large numbers of viable protoplasts from the species and tissues of interest, and were followed by experiments to determine culture conditions, firstly to allow regeneration of the cell wall, and secondly to allow re-initiation of the cell division cycle leading to the production of cell clusters and calli, and finally through induced differentiation to the production of intact plants (for reviews, see Davey et al. 2005; Galbraith 1989). Historically, considerable interest has centered round the concept of employing protoplasts for the production of somatic hybrids, following induced fusion between protoplasts produced from different parental sources. Subsequently, interest shifted towards the use of protoplasts as recipients of genetic material for transformation purposes (Davey et al. 2005), although the development of facile, non protoplast-based methods for plant transformation (e.g. see Clough and Bent 1998) has led to a general lessening of interest in this aspect of protoplast work. The question as to the degree to which, and time course over which, protoplasts preserve patterns of gene expression characteristic of their source tissues, nevertheless remains highly topical (Birnbaum et al. 2003; Galbraith and Birnbaum 2006). It is also clear that protoplasts have considerable potential for providing insight into signal transduction pathways (Sheen 2001), and the development of high throughput FCM methods for analysis of protoplasts should provide insights into signal transduction similar to those emerging for animal and yeast systems (Newman et al. 2006; Sachs et al. 2005).

10.1.2
Adaptation of Flow Cytometric Instrumentation for Analysis of Protoplasts

Plant protoplasts have two general features that directly affect the use of FCM. Firstly, they are, in most cases, larger than mammalian cells (hematopoietic cells: diameters around 10–20 µm; Shapiro 2003) around which the design of flow instrumentation was originally based. Plant cells generally range in diameters from 20 to 100 µm, with notable exceptions that are larger and sometimes much larger than these. Secondly, they are very fragile. The primary function of the cell wall is structural, and removal of this component reveals the plasma membrane which, unlike the situation for most animal cells, lacks extensive reinforcements provided by the cytoskeleton. These features provide two conflicts with conventionally configured flow instruments. The sizes of the standard flow orifices (70 µm) are only slightly larger than the diameters of the protoplasts, and the velocities of the flow stream combined with the abrupt deceleration experienced in collection of the sorted protoplasts adversely affects their integrity.

Adapting flow sorters for the analysis of protoplasts (or more generally "large particles") relies on simple equations describing the physics of droplet formation, but these have complex ramifications in terms of instrument design, performance and operation (Galbraith and Lucretti 2000; Harkins and Galbraith 1987; see also Chapter 2). The flow orifice must be enlarged to accommodate the diameters of the large particles. Although particles that are as large as 68% of the diameter of the orifice can be efficiently analyzed and sorted (Harkins and Galbraith 1987), most groups employ flow tips having diameters from 100 to 200 µm for sorting protoplasts (e.g. see Birnbaum et al. 2005; Galbraith and Lucretti 2000; Hammatt et al. 1990; Harkins and Galbraith 1984, 1987; Puite et al. 1988; Waara et al. 1998; Zilmer et al. 1995). Concomitantly, the system pressure must be lowered. This reduces the flow rate, required to avoid unacceptably large rates of consumption of sheath fluid, and the occurrence of a point of droplet break-off below the flow observation and droplet deflection points. The lowered flow rate decreases the maximal potential rate of droplet formation, driven by the droplet production mechanism (typically a bimorph piezo-electric crystal attached to the flow cell body). The undulation wavelength applied to the flow stream is also a function of the flow cell orifice diameter, in that this wavelength must be greater than π times that diameter for droplet production to occur (Harkins and Galbraith 1987). These factors combine to limit the maximal sort rate; for example, typical droplet drive frequencies must be decreased to around 8 kHz for a 200-µm diameter flow orifice. Given a droplet occupancy of 10%, and a desired positive population representing 1% of the total, positive events can only be collected at a maximal rate of 8 s^{-1}. This low sort rate will not affect downstream analyses that involve protoplast growth in culture, since such growth is possible for amenable species and tissues in microwells containing hundreds to several thousands of protoplasts in small volumes of culture media (<50 µl). It will evidently not affect assays that operate at the level of single protoplasts, including PCR-based methods of molecular biology that permit considerable amplification. However, it will affect protoplast assays that require collection of substantial numbers of protoplasts (including general proteomic assays, and various physiological and enzymatic measurements). The low sort rate will also adversely affect very rare event sorting, for obvious reasons.

10.1.3
Parametric Analyses Available for Protoplasts using Flow Cytometry

Flow cytometric instrumentation is designed to permit analysis of the optical properties of cells. In general, two types of optical signals are collected, those produced by light-scatter, and those produced by fluorescence emission. In all cases, these are time-versus-amplitude signals, which increase from background values to a maximum as the cell completely enters the Gaussian profile of focused laser illumination, subsequently dropping to this background as the cell exits the point of illumination. Processing is done to extract values representing the peaks and integrals of these pulse waveforms, and most instruments are also capable of extracting the pulse-widths. Values are collected, in linear or log-amplified form, in binned histograms typically comprising up to 1024 bins. More sophisticated pattern recognition can be achieved by using flash-digitized pulse waveforms (Godavarti et al. 1996; Murthi et al. 2005; Zilmer et al. 1995) but this approach has not reached commercialization.

Light scatter signals, being produced by all cells, are typically employed to trigger the flow cytometer for detection of all optical signals, scatter and fluorescence, emerging from that particular cell. This observation underscores the importance of providing for flow analysis protoplast suspensions that are as intact as possible. Even minor levels of protoplast disruption produce large numbers of scattering and (sometimes) fluorescent particles. For example, leaf protoplasts contain large numbers of strongly autofluorescent chloroplasts (Galbraith et al. 1988), and the release of these chloroplasts from broken cells can be employed as a measure of protoplast integrity (Fig. 10.1). Excessive numbers of light-scattering particles in samples destined for FCM analysis create difficulties in terms of triggering and in terms of data acquisition. In the first case, the presence of too many particles can overwhelm the triggering process, leading to high levels of signal aborts. In the second, the autoscaling function, if enabled, responds to the rapid accumulation of large numbers of objects having scatter and/or fluorescence signals of low intensities in the first few channels, making it difficult or impossible to identify the peaks of interest in the parametric histogram displays. Protoplast purification provides a simple means to eliminate contributions from subcellular organelles and other cellular debris, and it is recommended this be done prior to implementing flow analysis and sorting. Purification is best done using low-speed isopycnic centrifugation (Galbraith 1990a); in this case, intact protoplasts are of much lower buoyant density, and have a much greater sedimentation value, than subcellular organelles and cellular debris, and can be conveniently purified using various combinations of isotonic sucrose, sugar alcohols (mannitol, sorbitol), and/or salts (KCl).

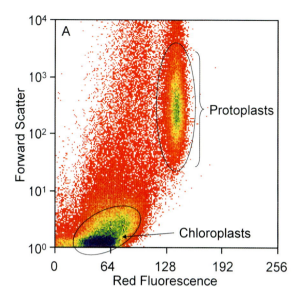
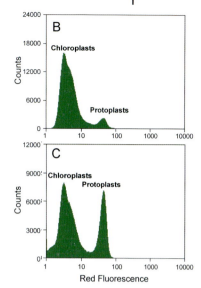

Fig. 10.1 Characterization of protoplasts and chloroplasts using flow cytometry. (a) Biparametric analysis (red fluorescence versus forward scatter) of a freshly-isolated population of *Arabidopsis thaliana* leaf protoplasts reveals two subpopulations, corresponding to protoplasts and free chloroplasts, respectively. (b) Uniparametric analysis (red fluorescence) of the data in Panel A. (c) Uniparametric analysis following protoplast purification by sucrose gradient centrifugation. The proportion of protoplasts is increased by the purification process; if the integrated number of chloroplasts is divided by 120 (the approximate number of chloroplasts per *Arabidopsis* cell; Pyke and Leech 1994, Raser and O'Shea 2005), and compared to the integrated number of protoplasts, the protoplast population is 92.1% intact before and 98.4% after purification.

Light scatter signals are generally collected along two axes, parallel to and orthogonal to the path of laser illumination; these are termed forward scatter (FS) and side scatter (SS) signals respectively. Obscuration bars and neutral density filters are used for signal attenuation, and the finite size of the detection optics means that these signals represent low-angle cones of scattered light. It is particularly unfortunate that the literature has assimilated the technically incorrect notion that the amplitudes of FS signals are proportional to cell size. It has been clearly stated that FS signals, when measured even for simple polystyrene spheres, are complex functions of diameter (d), varying as a function of d^3, d^2, d, and finally $<d$, for different ranges of values of d, and are also affected by the beam shaping geometries that are employed for these measurements (Salzman et al. 1990). For plants, a systematic evaluation of the relationship between FS (log signal) and microspore size strongly suggests a sigmoidal relationship over the range of 18–34 μm (Deslauriers et al. 1991). Interestingly, smaller cells at the tetrad stage produced FS signals that were greater in magnitude than those

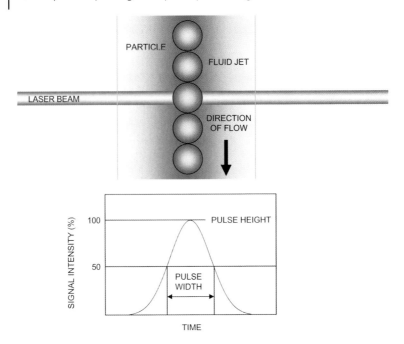

Fig. 10.2 Illustration of the manner in which the pulse waveform is produced during flow cytometric analysis of a cell, and of the measurement of the TOF parameter. The overlapping spheres represent successive temporal stages in the passage of a single particle, centered in the fluid stream, through the laser beam focus.

of the larger microspores, illustrating the complications introduced into FCM analyses by changes in cellular structure.

The concept that side scatter signals provide information about "cellular structure" or "granularity" is less worrisome, since these represent empirical measurements and observations that can be useful for classification of different types of cells. Accurate cell sizing can be carried out based on scatter signals (or on fluorescent signals for that matter, assuming the fluorescence occupies the full volume of the cell), using measurement of the time-of-flight (TOF) of the cells as they pass through the laser illumination point (Fig. 10.2). The pulse-widths of the resultant scatter or fluorescence signals represent a convolution of the cellular diameter with the beam profile. Assuming the cell is large with respect to the beam width, accurate sizing of the cell is possible by measuring the pulse widths (at some defined percentage of the maximal value of the pulse), and subtracting a fixed value corresponding to the beam dimension. This fixed value is determined by measurement of the pulse widths of standard microspheres of different nominal diameters, which will tend to a limiting value (the beam dimension) as the particle diameter decreases (see Galbraith et al. 1988 for an example of this empirical relationship, in which, over the range of 15–55 μm, the TOF signal predicts particle size with a correlation coefficient $(r^2) > 0.99$). TOF measurements have been particularly useful for characterization of phytoplankton, which range in size from <10 to ~1 000 μm (Rutten et al. 2005; Chapter 13).

Flow analysis of fluorescence signals produced by protoplasts can be carried out based on the presence of endogenous fluorophores, representing naturally fluorescent compounds and/or macromolecular complexes found within plant cells, or exogenous fluorophores, representing molecules introduced either by physical addition (i.e. small organic molecules that are fluorescent) or by transgenic expression (i.e. the class of proteins termed Fluorescent Proteins (FPs); the latter will be detailed in Chapter 17). Of the endogenous fluorophores, most attention has been paid to autofluorescent signals produced by chloroplasts, but a number of workers have also employed FCM to examine autofluorescence generated by plant secondary products such as alkaloids (Brown et al. 1984), and to examine fluorescence quenching as a consequence of anthocyanin accumulation (Sakamoto et al. 1994).

Flow analysis can also be undertaken after staining protoplasts using specific fluorochromes. In this situation, it may be necessary to first fix the protoplasts to allow access of the fluorochromes to their binding sites. In contrast, flow analysis can be carried out to monitor the integrity of the protoplast plasma membrane, as reflected by an absence of intracellular (nuclear) fluorescence following addition of propidium iodide. This approach has been particularly valuable in analysis of programmed cell death (Weir 2001).

10.2
Results of Protoplast Analyses using Flow Cytometry and Sorting

10.2.1
Protoplast Size

Measurement of protoplast size is fundamental to a large number of different types of studies, particularly when such measurements are made on individual cells within populations rather than being made as population averages. Many, if not all aspects of cellular biology and physiology scale naturally with cell size, and an ability to correct for, or otherwise deal with variations in cell size is essential for the comprehension of meaningful data extracted from individual cells.

Given the discussion in the previous section, it is unfortunate that very few reports have employed TOF measurements for estimating protoplast size, and for monitoring its changes as a function of genotype, tissue, or imposition of different treatments (Galbraith et al. 1988; Harkins et al. 1990; Meehan et al. 1996). We have described, some time ago, the size distributions of tobacco leaf protoplasts based on TOF measurements (Galbraith et al. 1988). Most of these protoplasts have diameters of between 20 and 50 µm, and these can be very accurately measured for mesophyll protoplasts using TOF values based on chlorophyll autofluorescence. For epidermal protoplasts, which lack chloroplasts, the fluorochromatic signal produced by hydrolysis of fluorescein diacetate (FDA) within living cells can be employed for size measurements of equal accuracy.

10.2.2
Protoplast Light Scattering Properties

FS and SS signals do have general empirical value in the analysis of protoplasts, in much the same way as they do for routine classification of white blood cells into lymphocytes, granulocytes, and monocytes (Shapiro 2003). It has already been mentioned that these signals are routinely employed for triggering in FCM analyses, and in that sense they are indispensable. For example, the FS signal can be employed for discrimination of protoplasts over debris, with intact protoplasts expressing GFP (Green Fluorescent Protein) being defined based on biparametric analysis of SS versus green fluorescence signals (Galbraith et al. 1995; Sheen et al. 1995). Alterations in FS and SS signals also accompany changes in protoplasts from plant cell cultures undergoing apoptosis (O'Brien et al. 1998a). Guzzo et al. (2002) reported an examination of the FS and SS properties of protoplasts prepared from embryogenic carrot cell cultures maintained under a diurnal light/dark regime. In combination with the measurement of red autofluorescence, a signal of plastid differentiation, and of FDA fluorochromasia, identifying viable protoplasts, they were able to distinguish and sort different subpopulations within the protoplasts, which appeared to differ in embryogenic potential.

In recent work, Guzzo et al. (2005) studied the FS and SS properties of protoplasts prepared from transgenic *Arabidopsis thaliana* (Brassicaceae) plants expressing *Glycine max enod40*, a gene rapidly induced in legume root pericycle cells following *Rhizobium* infection and subsequently expressed in dividing cortical cells and in the nodule primordium. They found that the mean FS signals of protoplasts prepared from transgenic plants expressing *enod40* under the regulatory control of the Cauliflower Mosaic Virus 35S were about 6% lower than those of protoplasts from wild-type plants. This statistically significant difference in scatter properties was confirmed by microscope-based measurements of sizes of the cells contained within different tissues. Interestingly, transient expression of *enod40* and treatment of protoplasts with peptides encoded by the *Glycine max enod40* locus and the related gene sequence of tobacco also resulted in lower averages for scatter signals and protoplast size distributions. This effect seemed to be associated with the emergence of bimodality within the protoplast distributions, perhaps implying differential responses of different cell types to *enod40* expression. Clearly, a number of different mechanisms might be responsible for the observed changes, of which the known interactions between *enod40* and sugar metabolism appear to be an attractive possibility. Further experiments will be required to clarify this complex situation.

10.2.3
Protoplast Protein Content

The protein content of protoplasts represents a fundamental cellular descriptor, and the ability to rapidly monitor this parameter using FCM, and correct for its variation, has a number of fundamental and obvious ramifications in plant cell biology and physiology. Naill and Roberts (2005b) have described a FCM method for estimating the protein contents of fixed protoplasts and corresponding individual cells, which involves covalent reaction with fluorescein isothiocyanate. FS and SS signals were employed to define gates representing intact cells, although sorting was available to confirm this assignment. Correlations between the fluorescent signals and conventional (Bradford) measurements of protein were made only indirectly, so the full potential of this flow method remains to be established.

10.2.4
Protoplast Viability and Physiology

Although every effort is made during protoplast production, to reduce as much as possible any losses due to cell disruption, the ability to monitor protoplast viability as well as specific physiological parameters and functions is of obvious importance when employing these protoplast populations for specific biological studies. As indicated previously, protoplast viability can be conveniently measured through use of non-polar fluorochromatic dyes whose hydrolysis, mediated by non-specific cytoplasmic esterases, gives rise to the production of more-polar fluorochromes which differentially accumulate within cells having intact plasma membranes. The specificity for the staining of viable cells therefore relies on two aspects of viability, the retention of active esterases within the cell, and the retention of the fluorescent product of hydrolysis. A number of different fluorochromes are suitable for this purpose, including the prototypical fluorescein diacetate (FDA), and its various derivatives (carboxyfluorescein diacetate, its acetoxymethyl ester, and sulfofluorescein diacetate) although none has been formally tested with protoplasts. Exclusion of propidium iodide from staining the nucleus can also be employed to identify viable protoplasts that are expressing GFP (Halweg et al. 2005); in this case, FDA is unsuitable for viability measurements since the fluorescence emission spectra of fluorescein and GFP extensively overlap.

As a counter-example to viability, apoptosis or programmed cell death (PCD) represents the controlled loss of cellular viability during eukaryotic differentiation and in response to abiotic and biotic stress. Much of what we know about PCD comes from early studies of mammals, and of model organisms such as *Caenorhabditis elegans* (Danial and Korsmeyer 2004; Ellis and Horwitz 1986). It is now known that PCD occurs widely, and most probably ubiquitously, across the eukarya, including plants, and can be classified according to the patterns of cytological changes that are observed (Van Doorn and Woltering 2005; see also Van Doorn and Woltering 2004 and Hörtensteiner 2006 for a discussion of the relationship of plant senescence to PCD and of alterations to chlorophyll occurring as a consequence of senescence). Many of these patterns are conserved across eukaryotes, including changes in chromatin and nuclear structure, nucleosomal fragmentation, loss of mitochondrial membrane potential, and alterations in distribution of phospholipids across the plasma membrane, leading to selective permeabilization of the membrane (Lam 2004; Vanyushin et al. 2004). FCM methods have been extensively developed for analysis of PCD in animal cells (for a recent review, see Darzynkiewicz et al. 2004), and many of these methods have been successfully adapted for plants. For example, O'Brien with co-workers have described FCM methods for analysis of the appearance of phosphatidylserine (PS) at the outer surface of the plasma membrane (annexin V binding) and of nucleosomal fragmentation (O'Brien et al. 1997) and of chromatin compaction and degradation (O'Brien et al. 1998b) in *Nicotiana plumbaginifolia* (Solanaceae) protoplasts induced into apoptosis by treatment with camptothecin (see Weir 2001 for a review). Further studies employing FCM confirm cytological similarities between plant and animal PCD (O'Brien et al. 1998a), with the possible exception of the degree of initial chromatin condensation, which appears greater in plant cells. FCM measurements of loss of nuclear DNA integrity during leaf protoplast culture have also been described for *Brassica napus* (Brassicaceae; Watanabe et al. 2002).

In the context of initiation of PCD, flow cytometric methods for analysis of mitochondrial membrane potential (Yao et al. 2004) have been described using *Arabidopsis thaliana* leaf protoplasts. This work employed FS and SS for the identification of live protoplasts, and used UV autofluorescence as an approximate measure of intracellular NADH levels, and the 488 nm-excited fluorescence of 3,3′dihexyloxacarbocyanine iodide (DiOC6(3)) as a measure of mitochondrial membrane potential (Fig. 10.3). Changes in these parameters were measured following addition to the protoplasts of stimuli known to induce cell death, ceramide, protoporphyrin IX, and the elicitor AvrRpt2, and using protoplasts prepared from wild-type plants, from mutant plants conditional for ectopic cell death, or from transgenic plants overproducing ACD2, a red chlorophyll catabolite reductase (RCCR) located within the plastids. The results were quite complex: subtle differences were sometimes observed in membrane potential in response to these treatments; in other cases, obviously bimodal distributions were generated. This suggests the methods will require further optimization for accurate charting and interpretation of changes in mitochondrial membrane potential. In particular, since the measurements using DiOC6(3) are non-ratiometric, controlling for protoplast size using TOF may turn out to be necessary (ratiometric measurements rely on the correlation between changes in the fluorescence spectrum and the property to be measured, the spectrum typically being analyzed at two different, optimal wavelengths. Such measurements are independent of the amount of loading of the fluorescent dye. Non-ratiometric dyes provide measurements of the property of interest based on the amounts of the dye that are accumulated, therefore being sensitive to the amount of loading, the size/capacity of the object being studied, etc.). A number of different fluorescence probes are

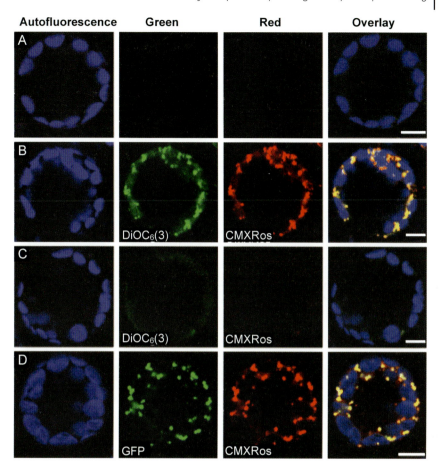

Fig. 10.3 Verification of the mitochondrial location of fluorescence emission from the membrane potential using dye $DiOC_6(3)$. Protoplasts were isolated from 18-day-old wild-type (a–c) and 13Λ9 (GFP labeled mitochondria) (d) leaves of *Arabidopsis thaliana* and observed by laser scanning confocal microscopy. (a) An unstained control cell. Chloroplast autofluorescence was excited at 488 nm and visualized at 738–793 nm. Note that no green and red signals are detectable without staining with $DiOC_6(3)$ and CMXRos, respectively. (b) A cell double-stained with 5 μM $DiOC_6(3)$ and 50 μM CMXRos for 5 min. Note that the localization of the $DiOC_6(3)$ signal matches that of the CMXRos signal. (c) A cell treated with 50 μM CCCP for 1 h and then double-stained with $DiOC_6(3)$ and CMXRos for 5 min. Note the loss of the $DiOC_6(3)$ and CMXRos signals. (d) A protoplast expressing GFP in mitochondria stained with CMXRos. Note that CMXRos co-localizes with mitochondria-targeted GFP. Bars = 8 μm. Slightly modified from Yao et al. (2004). $DiOC_6(3)$, 3,3′-dihexyloxacarbocyanine iodide; CMXRos, MitoTracker® Red (Invitrogen Corpn.); CCCP, carbonyl cyanide m-chlorophenylhydrazone.

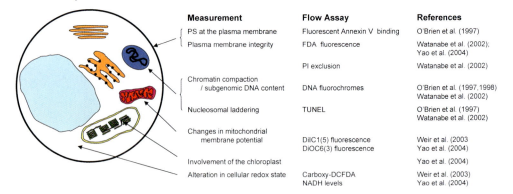

Fig. 10.4 Summary of programmed cell death in protoplasts and flow methods available for analysis of specific cytological and physiological components of this process. Abbreviations: PS, phosphatidylserine; FDA, fluorescein diacetate; PI, propidium iodide; TUNEL, Terminal deoxynucleotidyl Transferase Biotin-dUTP Nick End Labeling; DiIC1(5), 1,1′,3,3,3′,3′-hexamethylindodicarbocyanine iodide; DiOC6(3), 3,3′-dihexyloxacarbocyanine iodide; DCFDA, dichlorofluorescein diacetate.

available that measure membrane potential (see http://probes.invitrogen.com/ for a comprehensive listing). Other FCM methods for measurement of mitochondrial membrane potential and reactive oxygen species have been recently described (Cronje et al. 2004; Weir et al. 2003). These were used to probe the involvement of mitochondrially-generated reactive oxygen species and alterations in mitochondrial membrane potential during apoptosis (Weir et al. 2003), and interactions between various signals (salicylic acid, heat-shock, and pathogen exposure) that are involved in modulating the induction of apoptosis (Cronje et al. 2004). A summary of the various flow assays derived for use in analysis of PCD with protoplasts is given in Fig. 10.4. The interesting interaction reported (Yao et al. 2004) between PCD signals arising from the mitochondria (changes in mitochondrial membrane potential ($\Delta\Psi m$)) and from the plastids (RCCR) will require further study, and implementation of flow assays for chloroplast function (likely involving multiparametric analysis of chloroplast fluorescence properties) appears a logical next step (Fig. 10.4).

In terms of production of fluorescent signals by protoplasts, much interest has centered round the observation that some of the number of commercially important products, such as flavors, colors, and pharmaceuticals (Wink 1999) produced by plants, are autofluorescent. Given that plant cell cultures in general have seen limited use as methods of production of secondary products due to low yields and high variability in productivity levels, the concept of employing protoplast sorting as a potential means to enhance yields and improve homogeneity of production has attracted attention for many years (Brown et al. 1984). This concept relies on the supposition that genetic and/or epigenetic variation emerges in cell culture, resulting in heterogeneity of expression of the desired products. This supposition is most likely correct, based on the body of available knowledge. However, it also requires that sorting of protoplasts containing high product levels results in the establishment of cell cultures which thereafter are stable. This seems unlikely given that our current state of knowledge predicts that tissue culture leads to induction of more, rather than less, variation. Nevertheless, this approach should become increasingly feasible as our knowledge base develops concerning how to manipulate the stability of different states of gene expression. Given these comments, attempts to produce highly-productive cell lines via sorting of protoplasts have been quite limited in number, despite early interest (Brown et al. 1984). Hara et al. (1989) employed FCM to characterize the amounts of berberine fluorescence detected in protoplasts prepared from cell cultures progressively selected for high production of this compound. Sakamoto et al. (1994) described a means for selection of *Aralia cordata* (Araliaceae) protoplasts producing elevated levels of anthocyanins based on quenching of FITC-based protoplast fluorescence. Sorted protoplasts containing high levels of anthocyanins subsequently produced highly-productive cell cultures. Importantly, this method does not require that the secondary product be autofluorescent, which extends the range of possible applications. Further workers have devised flow strategies based on immunofluorescence of antibodies specific for secondary products, which would similarly obviate the requirement for the products themselves to be autofluorescent (Naill and Roberts 2005a).

Publications describing methods of FCM to measure additional physiological parameters have included pH (Brown et al. 1984; Giglioli-Guivarch et al. 1996), as well as the presence on the plasma membrane of molecules recognized by monoclonal antibodies (Desikan et al. 1999) or of receptors for phytohormones (Kitahata et al. 2005; Yamazaki et al. 2003). In the latter two cases, the binding of biotinylated abscisic acid (bioABA) to *Vicia faba* (Fabaceae) guard cell and barley aleurone protoplasts, was monitored following the addition of fluorescently-labeled avidin. Evidently, it is possible that either approach could be modified for the study of additional cellular receptors and surface components of protoplasts.

10.2.5
Protoplast Cell Biology

The ability to employ FCM for rapid and accurate measurement of organelle composition and number provides important information about these crucial cellular parameters. Plastid number and chlorophyll content can be very easily and accurately measured in flow, since chlorophyll autofluorescence provides a signal source that is readily detectable within protoplasts and individual chloroplasts (Galbraith et al. 1988). The signals can be employed for estimation of the chlorophyll contents of individual protoplasts, for the quantification of the numbers of chloroplasts per protoplast, and for estimation of the degree of protoplast integrity within different preparations (see Chapter 11). The presence or absence of chloroplasts can also be employed to distinguish protoplasts of different leaf cell types (epidermal from mesophyll, for example) using FCM, and to subsequently sort these different cell types. Harkins et al. (1990) and Meehan et al. (1996) have employed this approach to quantify gene expression in transgenic plants in a cell type-specific manner (see Chapter 17 for a full discussion). Similar methods for quantifying mitochondrial numbers in isolated populations would be extremely useful.

10.2.6
Construction of Somatic Hybrids

Fusion of protoplasts from somatic cells provides a means to amalgamate genomes that cannot mix via conventional sexual hybridization. A good deal of activity occurred in this area during the period 1965–1995, particularly in terms of optimizing the technologies of protoplast production, manipulation, and fusion, identification of the heterokaryons produced by protoplast fusion, and of isolation and regeneration of the somatic hybrids. Examples of agronomically important somatic hybrids have emerged, notably in *Brassica* and citrus (Davey et al. 2005). In somatic hybridization, a key point is that the proportion of desired heterokaryons in the populations of protoplasts subjected to fusion treatments is typically low. Given the large numbers of different combinations of protoplasts that are employed for somatic hybridization, a general means for identification and isolation of heterokaryons is therefore highly desirable. We demonstrated that protoplasts can be differentially labeled using pairs of exogenous fluorochromes (Galbraith and Galbraith 1979; Galbraith and Mauch 1980; Vankesteren and Tempelaar 1993). Use of fluorescein isothiocyanate and rhodamine isothiocyanate for selective labelling of protoplasts and identification of heterokaryons is illustrated in Fig. 10.5. Selective labeling can also be achieved using various combinations of added and endogenous fluorescence (chlorophyll autofluorescence or, more recently, fluorescence as a consequence of FP expression (Olivares-Fuster et al. 2002)). Flow sorting can then be employed to selectively isolate the heterokaryons produced by induced protoplast fusion, followed by regeneration of somatic hybrid plants (Afonso et al. 1985; Fahleson et al. 1994; Liu et al. 1995).

10.2.7
The Cell Cycle

Protoplasts are not as well suited for cell cycle analysis as are the corresponding nuclei isolated either from protoplasts or intact tissues (Galbraith et al. 1983; Galbraith 1990b; Galbraith et al. 1998; Ulrich et al. 1988). This may be related to the observation that the nuclei represent small objects within a much larger protoplast. Since protoplasts, in general, will be less precisely centered within the fluid stream of the flow cytometer than are the isolated nuclei, they correspondingly will experience a greater degree of variation in fluorescence excitation and this variation will be experienced by a nucleus contained within a fixed protoplast.

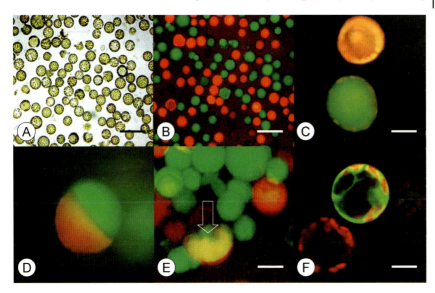

Fig. 10.5 Use of fluorescent labels for marking parental protoplasts and identifying protoplast fusion events. (a) Populations of *Nicotiana tabacum* and *N. nesophila* leaf protoplasts separately labeled with fluorescein isothiocyanate (FITC) and rhodamine isothiocyanate, examined using a light microscope under bright field illumination. Bar represents 150 μm. (b) As for a, except using epifluorescence illumination. (c) As for b, except at higher magnification. Slight chlorophyll autofluorescence is visible in the FITC-labeled protoplasts. Bar represents 20 μm. (d) After induction of fusion, adhesion of different parental protoplasts is observed. Bar represents 20 μm. (e) As for d, but at a later stage of fusion. Merging of the plasma membranes allows formation of heterokaryons, recognized by mingling of the two fluorescent labels (arrow). Bar represents 40 μm. (f) Transfected tobacco protoplast expressing GFP. This method of labeling is also suitable for identification of heterokaryons. Bar represents 20 μm. Panels a–e are modified from Galbraith and Mauch (1980), panel f is modified from Grebenok et al. (1997).

Other factors, including non-specific absorbance or scattering of excitation and emission photons due to protoplast pigmentation and cytoplasmic structure, may also affect the situation. Whatever the explanation, empirically a broadening of the peaks within DNA histograms is observed when stained protoplasts are compared to isolated nuclei (Galbraith 1990b; Ulrich et al. 1988). One obvious advantage in the use of protoplasts over isolated nuclei in studies of the cell cycle is the retention of the cytoplasmic compartment. This permits the potential use of bi-parametric analyses to monitor cellular DNA and total RNA contents simultaneously. Such an approach, which is well established for animal cells, has been successfully employed to quantify DNA and RNA levels within isolated plant nuclei (Bergounioux et al. 1988) but not for protoplasts, as far as I am aware. Nonetheless, the numbers of reports of FCM measurement of the cell cycle using fixed protoplasts remains limited.

10.3
Walled Plant Cells: Special Cases for Flow Analysis and Sorting

If plant cells can be found that comprise natural single cell suspensions, they can be amenable to FCM analysis without the need for protoplast production. A couple of caveats are evident: first, as for protoplasts, the cells must have dimensions that are compatible with the sizes of the flow cell tips. Second, the cells must be reasonably isotropic in terms of their overall dimensions, spherical shapes being ideal. As cells deviate from the spherical shape, their interactions with the fluid stream and the focus of the laser, become increasingly complex. This will generally broaden the distributions of fluorescence and light scatter that are obtained, and in extreme cases can give rise to apparent subpopulations of cells within the flow histograms that are simply artefacts of analysis. This situation can be readily diagnosed through sorting followed by reanalysis.

Examples of walled cells suitable for FCM analysis are provided by pollen, the sperm cells contained within pollen, and microspores, the stage of male gametophyte development immediately prior to the formation of mature pollen. It should be noted that some pollen species are likely to be problematic for flow analysis as a consequence of gross cellular asymmetry, or excessive size, and for some species the means for isolation of sperm and microspores may not be available (see also Chapters 5 and 6). It has already been mentioned that dried pollen provides a suitably indestructible standard for setting up large particle sorting (Harkins and Galbraith 1987). Becker et al. (2003) demonstrated that biologically viable, rehydrated pollen can be isolated by flow sorting based on a combination of pulse-width, forward scatter, and autofluorescence signals, and this approach was taken further by Pina et al. (2005) to allow a description of transcripts that are abundant in this developmental stage.

Sperm cells are found within the cytoplasm of the vegetative pollen grain, being transported along the pollen tube during its growth through the female tissue, and are released to participate in the typical double fertilization leading to production of the zygote and the endosperm. Sperm cells can be released from hydrated pollen by shaking in sucrose solutions, and are purified by Percoll gradient centrifugation (Dupuis et al. 1987). In the case of maize, they are small (7–10 μm diameter), nearly spherical cells, which are therefore entirely suitable for flow analysis in terms of their sizes and optical properties. Zhang et al. (1992) were the first to use FCM to characterize sperm cells isolated from maize. They found the newly-released sperm cells were viable, as indicated by FDA fluorochromasia and exclusion of propidium iodide. Engel et al. (2003) subsequently employed flow sorting to purify maize sperm cells based on FS and Hoechst 33342 fluorescence signals. cDNA libraries were subsequently prepared from these cells, and sequenced. Although the purification appeared effective in eliminating vegetative cell contamination, an unfortunate absence of experimental details regarding the FCM procedures makes this work difficult to assess with complete confidence.

Deslauriers et al. (1991) were amongst the first to apply FCM for the analysis of the optical properties of microspores, in this case isolated from flower buds of *Brassica napus*. They examined the potential of FS and FDA fluorochromasia as a means to recognize differences in microspore embryogenic potential. Schulze and Pauls (1998) extended this to explore the optical properties of cultures of *Brassica napus* microspores shortly following induction of embryogenesis by a heat treatment, and subsequently employed flow sorting to selectively enrich embryogenic cells. In further work (Schulze and Pauls 2002), they employed Calcofluor White as a means to highlight cellulose production by microspores in culture, and indicated that over 4 days in culture, FCM analysis based on FS and blue fluorescence could be used for the identification and sorting of embryogenic cells.

Weir et al. (2005) employed a modified EPICS Elite flow cytometer for the analysis of *Zinnia elegans* (Asteraceae) mesophyll cells isolated following pectolytic hydrolysis of the cell wall. The Zinnia system is unique with respect to the fact that these isolated cells can be induced in culture to differentiate into tracheary elements (TEs) without entering into cell division and with only a moderate increase in size (Fukuda 1997; Fukuda and Komamine 1980). FS versus SS plots of freshly-isolated mesophyll cells show a single tight cluster, and over 48 h in culture this cluster increases in both signal values. After 72 h, a second cluster becomes apparent, having a higher FS signal. From 96 to 168 h in culture, this second cluster becomes increasingly obvious. Flow-based analysis of the fluorescence of cells labeled with fluoroglucinol (identifying lignin) and Calcofluor (identifying cellulose) implies the second cluster comprises completely developed TEs. The flow analyses were further extended to include analysis of intracellular changes in Ca^{2+}, glutathione, pH, and reactive oxygen intermediates. Reciprocal changes in oxidative activity and the levels of reduced glutathione were evident. Further flow analyses of chromatin condensation and nuclear DNA breakdown imply a role for PCD in *Zinnia* TEs differentiation.

10.4
Prospects

The last decade has been remarkable for the rapidity of development of techniques for the study of biological organisms, particularly in the area of genomics and the related 'omics disciplines. Although predicting the future is an uncertain business, we can confidently anticipate further acceleration in the process of techniques and instrument development. At the same time, ancillary components required for the implementation of these techniques (i.e. specific antibodies, recombinant DNA constructions, fluorescent dyes, mutants, etc.) will become increasingly available. Together, this will result in a considerable increase in the rate of general acquisition of new scientific information. More specifically, we can anticipate detailed and comprehensive descriptions of the physiological and cell biological states of different cell types. In combination with techniques of transgenic biology (see Chapter 17), it will become increasingly easy to separate different cell types, particularly using protoplasts, and define the general 'omic status of that cell type (Galbraith and Birnbaum 2006). We will also become more cognizant of the impacts of stochastic processes in single cells (Brandt 2005; Lange 2005), to the extent that multiple analyses of individual cells, rather than analyses of populations, may be necessary to define the full biological capabilities of that particular cell type. Related to stochastics is the question of the role of "noise" within biological systems (Colman-Lerner et al. 2005; Raser and O'Shea 2005), and we can anticipate considerable progress towards answering this and related questions over the next few years. In this respect, multiparametric FCM, coupled to cell sorting, will provide an invaluable tool in these studies, and, in combination with image cytometry, should enable increasing linkage between visible phenotypes and the underlying genomic programs and mechanisms that regulate these phenotypes.

Acknowledgments

Work in the Galbraith laboratory has been supported by grants from the National Science Foundation, and the US Departments of Agriculture and of Energy, and by the University of Arizona Agriculture Experiment Station. I thank Georgina Lambert for providing the data presented in Fig. 10.1, and for reading the manuscript.

References

Afonso, C. L., Harkins, K. R., Thomas-Compton, M., Krejci, A., Galbraith, D. W. **1985**, *Nature Biotechnol.* 3, 811–816.

Becker, J. D., Boavida, L. C., Carneiro, J., Haury, M., Feijo, J. A. **2003**, *Plant Physiol.* 133, 713–725.

Bergounioux, C., Perennes, C., Brown, S. C., Gadal, P. *Cytometry* **1988**, 9, 84–87.

Birnbaum, K., Shasha, D. E., Wang, J. Y., Jung, J. W., Lambert, G. M., Galbraith, D. W., Benfey, P. N. **2003**, *Science* 302, 1956–1960.

Birnbaum, K., Jung, J. W., Wang, J. Y., Lambert, G. M., Hirst, J. A., Galbraith, D. W., Benfey, P. N. **2005**, *Nature Methods* 2, 1–5.

Brandt, S. P. **2005**, *J. Exp. Bot.* 56, 495–505.

Brown, S., Renaudin, J. P., Prevot, C., Guern, J. **1984**, *Physiol. Veg.* 22, 541–554.

Buchanan, B. B., Gruissem, W., Jones, R. L. (eds.) **2000**, *Biochemistry and Molecular Biology of Plants*, American Society of Plant Physiologists, Rockville, USA.

Clough, S. J., Bent, A. F. **1998**, *Plant J.* 16, 735–743.

Colman-Lerner, A., Gordon, A., Serra, E., Chin, T., Resnekov, O., Endy, D., Pesce, C. G., Brent, R. **2005**, *Nature* 437, 699–706.

Cronje, M. J., Weir, I. E., Bornman, L. **2004**, *Cytometry* 61A, 76–87.

Danial, N. N., Korsmeyer, S. J. **2004**, *Cell* 116, 205–219.

Darzynkiewicz, Z., Huang, X., Okafuji, M., King, M. A. **2004**, *Methods Cell Biol.* 75, 307–341.

Davey, M. R., Anthony, P., Power, J. B., Lowe, K. C. **2005**, *Biotechnol. Adv.* 23, 131–171.

Desikan, R., Hagenbeek, D., Neill, S. J., Rock, C. D. **1999**, *FEBS Lett.* 456, 257–262.

Deslauriers, C., Powell, A. D., Fuchs, K., Pauls, P. **1991**, *Biochim. Biophys. Acta* 1091, 165–172.

Dupuis, I., Roeckel, P., Mattys-Rochon, E., Dumas, C. **1987**, *Plant Physiol.* 85, 876–878.

Ellis, H. M., Horvitz, H. R. **1986**, *Cell* 44, 817–829.

Engel, M. L., Chaboud, A., Dumas, C., McCormick, S. **2003**, *Plant J.* 34, 697–707.

Fahleson, J., Eriksson, I., Landgren, M., Stymne, S., Glimelius, K. **1994**, *Theor. Appl. Genet.* 87, 795–804.

Fukuda, H. **1997**, *Plant Cell* 9, 1147–1156.

Fukuda, H., Komamine, A. **1980**, *Plant Physiol.* 65, 57–60.

Galbraith, D. W. **1989**, *Int. Rev. Cytol.* 116, 165–227.

Galbraith, D. W. **1990a**, *Methods in Cell Biol.* 33, 527–547.

Galbraith, D. W. **1990b**, *Methods in Cell Biol.* 33, 549–562.

Galbraith, D. W., Birnbaum, K. **2006**, *Annu. Rev. Plant Biol.* 57, 451–475.

Galbraith, D. W., Galbraith, J. E. C. **1979**, *Zeitsch. Pflanzenphysiol.* 93, 149–158.

Galbraith, D. W., Lucretti, S. **2000**, Large particle sorting in *Flow Cytometry and Cell Sorting*, 2nd edn, ed. A. Radbruch, Springer-Verlag, Berlin, pp. 293–317.

Galbraith, D. W., Mauch, T. J. **1980**, *Zeitsch. Pflanzenphysiol.* 98, 129–140.

Galbraith, D. W., Harkins, K. R., Maddox, J. R., Ayres, N. M., Sharma, D. P., Firoozabady, E. **1983**, *Science* 220, 1049–1051.

Galbraith, D. W., Harkins, K. R., Jefferson, R. A. **1988**, *Cytometry* 9, 75–83.

Galbraith, D. W., Grebenok, R. J., Lambert, G. M., Sheen, J. **1995**, *Methods Cell Biol.* 50, 3–12.

Galbraith, D. W., Doležel, J., Lambert, G., Macas, J. **1998**, Analysis of nuclear DNA content and ploidy in higher plants in *Current Protocols in Cytometry*, ed. J. P. Robinson et al., Wiley, New York, pp. 7.6.1–7.6.22.

Giglioli-Guivarch, N., Pierre, J. N., Vidal, J., Brown, S. **1996**, *Cytometry* 23, 241–249.

Godavarti, M., Rodriguez, J. J., Yopp, T. A., Lambert, G. M., Galbraith, D. W. **1996**, *Cytometry* 24, 330–339.

Grebenok, R. J., Pierson, E. A., Lambert, G. M., Gong, F.-C., Afonso, C. L., Haldeman-Cahill, R., Carrington, J. C., Galbraith, D. W. **1997**, *Plant J.* 11, 573–586.

Guzzo, F., Cantamessa, K., Portaluppi, P., Levi, M. **2002**, *Plant Cell Rep.* 21, 214–219.

Guzzo, F., Portaluppi, P., Grisi, R., Barone, S., Zampieri, S., Franssen, H., Levi, M. **2005**, *J. Exp. Bot.* 56, 507–513.

Halweg, C., Thompson, W. F., Spiker, S. **2005**, *Plant Cell* 17, 418–429.

Hammatt, N., Lister, A., Blackhall, N. W., Gartland, J., Ghose, T. K., Gilmour, D. M., Power, J. B., Davey, M. R., Cocking, E. C. **1990**, *Protoplasma* 154, 34–44.

Hara, Y., Yamagata, H., Morimoto, T., Hiratsuka, J., Yoshioka, T., Fujita, Y., Yamada, Y. **1989**, *Planta Med.* 55, 151–154.

Harkins, K. R., Galbraith, D. W. **1984**, *Physiol. Plant.* 60, 43–52.

Harkins, K. R., Galbraith, D. W. **1987**, *Cytometry* 8, 60–71.

Harkins, K. R., Jefferson, R. A., Kavanagh, T. A., Bevan, M. W., Galbraith, D. W. **1990**, *Proc. Natl Acad. Sci. USA* 87, 816–820.

Hörtensteiner, S. **2006**, *Annu. Rev. Plant Biol.* 57, 55–77.

Kitahata, N., Nakano, T., Kuchitsu, K., Yoshida, S., Asami, T. **2005**, *Bioorgan. Med. Chem.* 13, 3351–3358.

Lam, E. **2004**, *Nature Rev. Mol. Cell Biol.* 5, 305–315.

Lange, B. M. **2005**, *Curr. Opin. Plant Biol.* 8, 236–241.

Liu, J. H., Dixelius, C., Eriksson, I., Glimelius, K. **1995**, *Plant Sci.* 109, 75–86.

Meehan, L., Harkins, K., Chory, J., Rodermel, S. **1996**, *Plant Physiol.* 112, 953–963.

Murthi, S., Sankaranarayanan, S., Xia, B., Lambert, G. M., Rodríguez, J. J., Galbraith, D. W. **2005**, *Cytometry* 66A, 109–118.

Naill, M. C., Roberts, S. C. **2005a**, *Biotechnol. Prog.* 21, 978–983.

Naill, M. C., Roberts, S. C. **2005b**, *Plant Cell Rep.* 23, 528–533.

Newman, J. R. S., Ghaemmaghami, S., Ihmels, J., Breslow, D. K., Noble, M., DeRisi, J. L., Weissman, J. S. **2006**, *Nature* 441, 840–846.

O'Brien, I. E. W., Reutelingsperger, C. P. M., Holdaway, K. M. **1997**, *Cytometry* 29, 28–33.

O'Brien, I. E. W., Baguley, B. C., Murray, B. G., Morris, B. A. M., Ferguson, I. B. **1998a**, *Plant J.* 13, 803–814.

O'Brien, I. E. W., Murray, B. G., Baguley, B. C., Morris, B. A. M., Ferguson, I. B. **1998b**, *Exp. Cell Res.* 241, 46–54.

Olivares-Fuster, O., Pena, L., Duran-Vila, N., Navarro, L. **2002**, *Ann. Bot.* 89, 491–497.

Pina, C., Pinto, F., Feijo, J. A., Becker, J. D. **2005**, *Plant Physiol.* 138, 744–756.

Puite, K., TenBroeke, W., Schaart, J. **1988**, *Plant Sci.* 56, 61–68.

Pyke, K. A., Leech, R. M. **1994**, *Plant Physiol.* 104, 201–207.

Raser, J. M., O'Shea, E. K. **2005**, *Science* 309, 2010–2013.

Rutten, T. P. A., Sandee, B., Hofman, A. R. T. **2005**, *Cytometry* 64A, 16–26.

Sachs, K., Perez, O., Pe'er, D., Lauffenburger, D. A., Nolan, G. P. **2005**, *Science* 308, 523–529.

Sakamoto, K., Iida, K., Koyano, T., Asada, Y., Furuya, T. **1994**, *Planta Med.* 60, 253–259.

Salzman, G. C., Singham, S. B., Johnston, R. G., Bohren, C. F. **1990**, Light scattering and cytometry in *Flow Cytometry and Sorting*, 2nd edn, ed. M. R. Melamed, T. Lindmo, M. L. Mendelsohn, Wiley-Liss, New York, pp. 81–107.

Schulze, D., Pauls, K. P. **1998**, *Plant Cell Physiol.* 39, 226–234.

Schulze, D., Pauls, K. P. **2002**, *New Phytologist* 154, 249–254.

Shapiro, H. M. **2003**, *Practical Flow Cytometry*, 4th edn, Alan R. Liss, Inc., New York.

Sheen, J. **2001**, *Plant Physiol.* 127, 1466–1475.

Sheen, J., Hwang, S., Niwa, Y., Kobayashi, H., Galbraith, D. W. **1995**, *Plant J.* 8, 777–784.

Ulrich, I., Fritz, B., Ulrich, W. **1988**, *Plant Sci.* 55, 151–158.

Van Doorn, W. G., Woltering, E. J. **2004**, *J. Exp. Bot.* 55, 2147–2153.

Van Doorn, W. G., Woltering, E. J. **2005**, *Trends Plant Sci.* 10, 117–122.

Vankesteren, W. J. P., Tempelaar, M. J. **1993**, *Cell Biol. Int.* 17, 235–243.

Vanyushin, B. F., Bakeeva, L. E., Zamyatnina, V. A., Aleksandrushkina, N. I. **2004**, *Int. Rev. Cytol.* 233, 135–179.

Waara, S., Nyman, M., Johannisson, A. **1998**, *Euphytica* 101, 293–299.

Watanabe, M., Setoguchi, D., Uehara, K., Ohtsuka, W., Watanabe, Y. **2002**, *New Phytologist* 156, 417–426.

Weir, I. E. **2001**, *Methods Cell Biol.* 63, 505–526.

Weir, I. E., Pham, N.-A., Hedley, D. W. **2003**, *Cytometry* 54A, 109–117.

Weir, I. E., Maddumage, R., Allan, A. C., Ferguson, I. B. **2005**, *Cytometry* 68A, 81–91.

Wink, M. **1999**, Biochemistry, role, and biotechnology of secondary metabolites in *Functions of Plant Secondary Metabolites and their Exploitation in Biotechnology*, ed. M. Wink, Sheffield Academic, Sheffield, pp. 1–17.

Yamazaki, D., Yoshida, S., Asami, T., Kuchitsu, K. **2003**, *Plant J.* 35, 129–139.

Yao, N., Eisfelder, B. J., Marvin, J., Greenberg, J. T. **2004**, *Plant J.* 40, 596–610.

Zhang, G. C., Campenot, M. K., McGann, L. E., Cass, D. D. **1992**, *Plant Physiol.* 99, 54–59.

Zilmer, N. A., Rodriguez, J. J., Yopp, T. A., Lambert, G. M., Galbraith, D. W. **1995**, *Cytometry* 20, 102–117.

11
Flow Cytometry of Chloroplasts

Erhard Pfündel and Armin Meister

Overview

The present chapter reviews the history and current state of flow cytometry of chloroplasts, which are cellular organelles in higher plants, where photosynthesis occurs. This review includes a general introduction to chloroplast structure and function but, in addition, discusses autofluorescence from chloroplasts in far more detail. We report key advantages of chloroplast analysis by flow cytometry, including identification of specific chloroplast subpopulations by measuring both autofluorescence and light scattering. Further, flow cytometry has been successfully employed to examine binding of fluorescent proteins targeted to the chloroplast, and to study the effects of malfunction of a DNA-unwinding enzyme (gyrase) during chloroplast partitioning by quantification of the chloroplast DNA content. Finally, future prospects for flow cytometry of chloroplasts will also be outlined.

11.1
Introduction

Many prokaryotes and most plants utilize light energy from the sun to synthesize carbohydrates from CO_2. This process is called photosynthesis and, in plants, photosynthesis occurs in specialized cell organelles known as plastids. The present chapter focuses on chloroplasts, the plastid type of higher (vascular) plants and green algae. Chloroplasts are located in the cytosol of photosynthetic cells of higher plants and are flattened or lens-shaped in appearance and their length is in the range of several μm. The many constituent parts of the photosynthetic machinery are located within the thylakoid membranes which are arranged in a highly complex three-dimensional structure inside the chloroplast (Mustárdy and Garab 2003; Staehelin 2003). Chlorophyll molecules are located within thylakoid membranes and give chloroplasts their green color which is easily observed in

Flow Cytometry with Plant Cells. Edited by Jaroslav Doležel, Johann Greilhuber, and Jan Suda
Copyright © 2007 WILEY-VCH Verlag GmbH & Co. KGaA, Weinheim
ISBN: 978-3-527-31487-4

bright-field microscopy but the red fluorescence from chlorophyll molecules results in conspicuous red coloration in fluorescence microscopy. Macroscopically, chlorophyll results in green coloration of leaves and other photosynthetic organs of higher plants.

The structure and composition of chloroplasts is not static but changes in response to the light environment. Studies with whole leaves have demonstrated that chloroplasts acclimate to different light intensities and to different spectral compositions of light by varying the structure of the thylakoid network and the abundance of photosynthetic complexes (Anderson et al. 1988; Evans et al. 2004; Melis 1991). Steep gradients of light intensity and spectral composition also exist inside leaves due to light absorption by chloroplast pigments and light scattering (Smith et al. 1997; Terashima 1989; Vogelmann 1993; Vogelmann and Evans 2002). Gradients of various photosynthetic compounds across leaves (Cui et al. 1991; Evans 1999) support the view that intra-leaf light gradients produce chloroplasts with varying acclimation status. Differentiation of chloroplasts within leaves can also occur in response to compartmentalization of photosynthesis within the leaf as occurs in the so-called C_4 plants (see Section 11.3.2).

Consequently, it becomes evident that information on variations of chloroplast properties within a leaf is an important prerequisite to understand whole-leaf photosynthesis. By analyzing chloroplast suspensions from leaves, flow cytometry (FCM) can provide statistical information on the distribution of the photosynthetic acclimation status. Furthermore, FCM sorting of chloroplasts can yield defined subpopulations for subsequent examination of physical and biochemical factors of acclimation. In addition, FCM permits monitoring of the heterogeneous behavior of individual chloroplasts in both biochemical and molecular biological studies.

In spite of its apparent potential, however, FCM is less frequently employed with cellular organelles, such as chloroplasts, than with whole photosynthetic prokaryotic and eukaryotic cells (Legendre et al. 2001; Toepel et al. 2004; Chapter 13). We believe that the rare use of FCM in chloroplast research results from a lack of awareness of its possibilities rather than from any limitations of the method. To redress this situation and attract more interest in FCM of chloroplasts, this chapter provides information about chloroplasts which is pertinent to flow cytometry. Furthermore, we review the development and current state of the method and, finally, we discuss future prospects of FCM in the study of chloroplasts.

11.1.1
The Chloroplast

This section is mostly confined to chloroplast properties which are relevant for FCM. More detailed information on chloroplast ultrastructure and on the photosynthetic complexes, located in thylakoid membranes, can be found in recent reviews by Mustárdy and Garab (2003), Staehelin (2003), and Dekker and Boekema (2005). Overviews on carbon dioxide fixation reactions in the chloroplast are given by Orsenigo et al. (1997), Schnarrenberger and Martin (1997) and Tolbert (1997).

Chloroplasts are semi-autonomous cell organelles which contain circular DNA that encodes some but not all chloroplast proteins, and chloroplasts have the capacity to synthesize these proteins (Bedbrook and Kolodner 1979; Sugiura 1992). The organelle is surrounded by the "chloroplast envelope" which is a double membrane through which not only nuclear-encoded proteins but also precursors for and products of biochemical reactions within the chloroplasts are transported (Flügge 2000). Chloroplast biochemistry includes, in addition to photosynthesis, many other essential processes such as the synthesis of fatty acids, carotenoids, and amino acids.

The envelope encloses a complex system of thylakoids forming membrane vesicles. Often, chloroplasts of higher plants are "granal" showing two types of vesicles; namely, the grana thylakoids (which are tightly stacked) and the non-stacked stroma thylakoids (which interconnect the grana stacks). While granal chloroplasts occur in the majority of higher plant species, some plants exhibiting C_4 photosynthesis contain granal chloroplasts in mesophyll cells but possess agranal chloroplasts, lacking grana stacks, in bundle sheath cells (Fig. 11.1b).

Within or at the surface of thylakoid membranes, the primary photosynthetic process occurs forming the energy-rich compound, ATP (adenosine triphosphate), and the reductant, NADPH (reduced form of nicotinamide adenine dinucleotide phosphate). These compounds are utilized to drive photosynthetic reactions which reduce carbon dioxide to carbohydrate within the chloroplast stroma, which is the space between thylakoid vesicles and envelope. Formation of ATP and NADPH is linked to the electron transport involving a number of redox compounds located in the thylakoid membranes. The light energy-fueled motors of electron transport are the photosystems (PS) I and II. PS I is located in stroma thylakoid membranes and the outer membranes of grana stacks while PS II is observed in adjoining membranes of neighboring grana thylakoids (Fig. 11.1b). Therefore, reduced abundance of grana is often associated with low PS II/PS I ratios (see Fig. 11.1b).

In higher plant chloroplasts, two groups of pigments, namely, chlorophylls and carotenoids, are involved in light absorption and providing absorbed light energy to the reaction centers of the photosystems (Siefermann-Harms 1985). A number of different carotenoids are present in chloroplasts but the chlorophylls are represented by chlorophyll *a* and chlorophyll *b*. Both are structurally identical except that, in chlorophyll *b*, the B ring-methyl group of chlorophyll *a* (Fig. 11.1c) is replaced by a formyl group. It is generally accepted that nearly all chlorophylls in the thylakoid membrane are non-covalently bound to proteins in PS I and II: only a very minor chlorophyll fraction is bound to other proteins of the photosynthetic electron chain (Pierre et al. 1997). The major part of the carotenoids is also assumed to be non-covalently bound to proteins of the photosystems (Siefermann-Harms 1985) but they may also occur in the free state in thylakoid membranes (Havaux 1998) and in the chlorophyll-free chloroplast envelope (Douce and Joyard 1979).

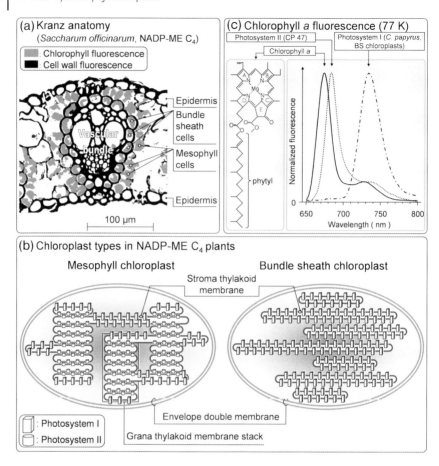

Fig. 11.1 Chloroplasts in NADP-ME C_4 plants. (a) "Kranz" anatomy in C_4 plants which includes the wreath-like arrangement of bundle sheath and mesophyll cells around vascular bundles; fluorescence image from a transverse section of a sugar cane leaf (*Saccharum officinarum*, Poaceae) using confocal laser scanning microscopy. UV-excited blue-green fluorescence from cell walls and blue-excited far-red fluorescence ($\lambda > 710$ nm) from chlorophyll *a* is shown in black and gray, respectively (see Pfündel and Neubohn 1999). Note, that the vascular bundle is surrounded by a ring of thick-walled cells containing chlorophyll, the bundle sheath cells. Chlorophyll fluorescence outside the ring of bundle sheath cells stems from mesophyll cells. (b) Sketch of the situation in C_4 plants with NADP-ME biochemistry (see the text for explanation) in which mesophyll and bundle sheath chloroplasts differ markedly in membrane architecture and occurrence of photosystems I and II (see Section 11.1). (c) Structure and low temperature (77 K) spectra of chlorophyll *a* fluorescence. Pure chlorophyll *a* in acetone emits at shorter wavelengths than chlorophyll *a* in photosystem II but the emission spectrum of chlorophyll *a* in photosystem I is particularly red-shifted. Photosystem II is represented by the isolated pigment–protein complex "CP 47", which is part of the photosystem II core complex, and photosystem I by bundle sheath (BS) chloroplasts from *Cyperus papyrus* (Cyperaceae) purified by flow cytometry, which are virtually devoid of photosystem II (Pfündel et al. 1996). Spectral differences between photosystems are the basis of why the ratio of photosystem I to photosystem II affects fluorescence emission of the entire chloroplast (see Fig. 11.2d).

11.2
Chloroplast Signals in Flow Cytometry

11.2.1
Autofluorescence

Chloroplasts show appreciable levels of autofluorescence in the spectral range from 650 to 800 nm (Fig. 11.2d). This fluorescence originates from chlorophyll *a* because the yield for carotenoid fluorescence is negligibly low (Gillbro and Cogdell 1989) and excited chlorophyll *b*, which fluoresces *in vitro*, is practically non-fluorescent *in situ* because its absorbed light energy is very efficiently forwarded to chlorophyll *a* by fluorescence energy transfer (Govindjee 1995; Karukstis 1992).

Emission spectra of chlorophyll *a* in PS I and II, however, differ because, generally, the spectral behavior of a chlorophyll molecule is affected by its interaction with neighboring molecules and, more specifically, with the different environments that exist for chlorophyll *a* in PS I and in PS II. As a result, the main fluorescence emission peak at low temperature (77 K) of PS I is situated near 730 nm and of PS II near 680 nm (Fig. 11.1c). Although at room temperature, PS I fluorescence exhibits an additional emission shoulder at 690 nm (Croce et al. 1996), autofluorescence at wavelengths longer than 700 nm can be considered to be predominated by PS I emission and at wavelengths shorter than 700 nm by PS II emission. This is also the case under temperatures at which flow cytometers operate.

Furthermore, PS II, but not PS I, exhibits variable fluorescence yields that can range between the minimum "F_0 fluorescence" and the maximum "F_M fluorescence" (Dau 1994; Oxborough 2004). The level of fluorescence depends on the state of PS II reaction centers. In the open state, reaction centers can utilize excitation energy efficiently to drive photosynthetic electron transport and, in this way, their fluorescence is diminished to the F_0 level; but, when reaction centers are closed, more excitation energy is directed towards fluorescence emission so that F_M intensity is observed.

It is important to realize that PS II reaction centers are open in the dark but they are closed by strong light intensities under which the excitation energy arriving at reaction centers exceeds the use of excitation energy by the photochemistry of the photosynthetic process (Govindjee 1995; Schreiber 2004). In the flow cytometer, chloroplasts are normally exposed to low light conditions, which leave most reaction centers in the open state, until they arrive at the interrogation point where they are suddenly illuminated by the high intensity radiation from either an arc lamp or a laser. At this point, light intensity and exposure time determine the velocity and extent of reaction center closure. Hence, chlorophyll fluorescence measured by flow cytometers might be clearly higher than the F_0 and could even reach the F_M level.

The level of chlorophyll fluorescence, F_0, F_M, or intermediate between F_0 and F_M, affects not only the intensity but also the quality of the signal recorded. For example, with a rise from F_0 to F_M fluorescence, the proportion to which PS II

Chloroplasts from *Zea mays* (NADP-ME C$_4$)

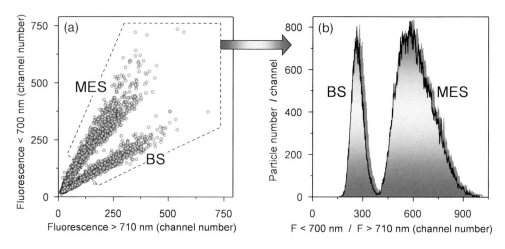

Flow cytometrically purified chloroplasts (*Zea mays*)

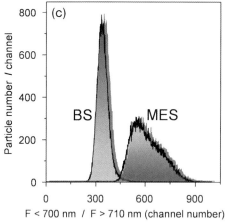
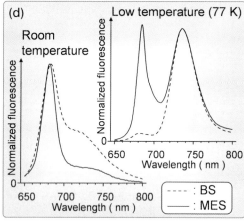

Fig. 11.2 Flow cytometry of chloroplasts from maize leaves. (a, b) Data from flow cytometry of chloroplasts isolated from maize leaves (*Zea mays*, Poaceae). (c, d) Analyses of flow cytometrically-purified bundle sheath (BS) and mesophyll (MES) chloroplasts are shown. (a) Dot plot of chloroplasts resulting from concomitant recording of chlorophyll *a* fluorescence at wavelengths shorter than 700 nm (ordinate) and at wavelengths longer than 710 nm (abscissa). Using the gate drawn as a hatched polygon in panel a, the frequency distribution of the short to long wavelength fluorescence ratios (F < 700 nm/ F > 710 nm) was derived (b). (c) Corresponding fluorescence ratio histograms of BS and MES chloroplasts purified by flow cytometry. (d) Fluorescence emission spectra of these pure chloroplast populations at room and low temperature.

contributes to total fluorescence is expected to rise as has been previously demonstrated with plant leaves (Agati et al. 2000; Pfündel 1998). Also, chloroplasts that suffer from sustained photoinhibition of PS II after exposure to high light stress would be detected much better by F_M rather than by F_0 fluorescence because the latter responds little to sustained photoinhibition (Krause 1994).

Because of its importance, a number of studies have addressed the issue of the level of chlorophyll fluorescence measured in FCM. Ashcroft et al. (1986) employed a dual laser flow cytometer in which intact, that is, envelope-enclosed chloroplasts from spinach leaves (*Spinacia oleracea*; Chenopodiaceae/APG: Amaranthaceae) were first excited by a strong blue laser (476 nm, 250 mW power output) and subsequently analyzed by a much weaker red laser (633 nm, 5 mW power output). By varying the time interval between blue and red excitation from 20 to 50 μs, maximal red-excited fluorescence was measured 25 μs after blue light excitation. As the latter time interval was close to the time required for a complete rise from F_0 to F_M fluorescence (Fig. 11.3), it was concluded that F_M fluorescence is measured 25 μs after the first excitation, and that the signal measured during the first exposure of only a few μs corresponds to F_0 fluorescence.

The above conclusion was questioned by Xu et al. (1990) who analyzed envelope-free spinach chloroplasts using a flow cytometer equipped with a strong blue laser (488 nm, 100 mW power output). It was demonstrated that fluorescence intensity in FCM changed moderately with increasing temperatures from 25 to 70 °C during chloroplasts pretreatment; by comparison, the F_M as recorded with a conventional fluorimeter was similarly temperature-insensitive, but F_0 flu-

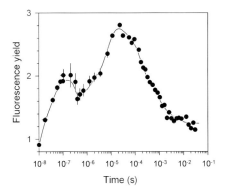

Fig. 11.3 Changes in the yield of chlorophyll fluorescence after a single laser pulse. Dark-adapted green algae (*Chlorella vulgaris*, Chlorellaceae) were illuminated by a saturating 10-ns laser pulse at a wavelength of 337.1 nm and an intensity of 3×10^{14} photons cm^{-2}. The initial rise of the fluorescence curve has been attributed to the reduction of a fluorescence-quenching chlorophyll in the PS II reaction center, and the rise in the μs range has been associated with the disappearance of carotenoid triplets (cf. Govindjee 1995). In FCM, the time interval and intensity of excitation determines whether the initial low fluorescence levels are recorded (comparable to F_0 levels) or whether chlorophyll fluorescence proceeds to the maximum of the fluorescence yield curve (comparable to F_M; see text for further details). Data from Mauzerall (1972).

orescence intensity exhibited a clear maximum at 55 °C. Further, chemically closing PS II reaction centers with DCMU (3-(3,4-dichlorophenyl)-1,1-dimethylurea) to elicit F_M fluorescence did not elevate the fluorescence detected by flow cytometry above that measured in the absence of DCMU. Additionally, the effect of p-benzoquinone, an artificial quencher of chlorophyll a fluorescence, on FCM data was similar to F_M fluorescence measured conventionally and was different from the behavior of F_0 fluorescence. From their data, Xu et al. (1990) concluded that in laser-based flow cytometers, F_M-type fluorescence is measured, and it was suggested that typical time intervals of chloroplast exposure to the flow cytometer's laser beam is considerably longer than that required for the initial fast rise of the PS II fluorescence induction curve (Fig. 11.3). Consistent with Xu et al. (1990), various algae species did not show enhancement of chlorophyll fluorescence after DCMU treatment in laser-based flow cytometers (Furuya and Li 1992; Perry and Porter 1989; Toepel et al. 2005).

Important information on the nature of chlorophyll fluorescence has been provided by Neale et al. (1989) who investigated an algal species (*Chroomonas* sp., Cryptophyta) using FCM. They excited chlorophyll fluorescence of single cells using a blue laser (488 nm, 250 mW output power) and, in parallel, determined the true F_0 and F_M fluorescence utilizing a bench-top fluorimeter. Importantly, an intercalibration procedure permitted the authors to compare fluorescence intensities between the two instruments. Using the latter approach, flow cytometry-measured fluorescence intensities of cells grown under unstressed conditions were found to be intermediate between F_0 and F_M intensities. To explain their data, the authors suggested that the dwelling time of an algal cell in the laser beam was longer than required for the initial fast rise of PS II fluorescence, as was also suggested by Xu et al. (1990), but was shorter than the time required to completely arrive at the F_M level, which agrees with the hypothesis of Ashcroft et al. (1986).

In summary, it appears reasonable to assume that the strong irradiance in laser-based flow cytometers, which is up to 3×10^5 times higher than full sunlight (Neale et al. 1989), induces a fast, sub-microsecond rise in chlorophyll fluorescence so that the fluorescence yield actually measured resides markedly above the F_0 level, but the common exposure intervals of a few microsecond are too short to elevate fluorescence to the true F_M level.

Compared to laser-based flow cytometers, instruments equipped with arc lamps often use lower excitation intensities and employ longer residence times of particles in the beam. Because the behavior of chlorophyll fluorescence is likely affected by the excitation conditions, fluorescence elicited by microsecond exposure to a strong laser beam might well be different from that measured in an arc lamp-based flow cytometer. Indeed, Furuya and Li (1992) demonstrated with algal cells that DCMU treatment significantly enhanced the chlorophyll fluorescence measured in a flow cytometer equipped with an arc lamp but not in a laser-based instrument.

In summary, chlorophyll autofluorescence from chloroplasts needs to be carefully evaluated. Also, additional information is required about the fluorescence properties of chloroplasts before flow cytometric analysis, or of flow-cytometrically sorted chloroplasts, to fully understand the nature of the chlorophyll fluorescence observed in FCM.

11.2.2
Light Scattering

Commonly, the intensity of forward scatter (FSC) is thought to be related to the size of a particle (with questionable reasons, cf. Chapters 2 and 10) and side scatter (SSC) to its internal structure (Phinney and Cucci 1989; Tanke and Van der Keur 1993; see also Chapter 2). However, in the particular case of chloroplasts, an association between FSC and chloroplast size is not always observed. Schröder and Petit (1992) used FCM to analyze two types of chloroplasts from spinach, one with the envelope membrane retained and the other with the envelope removed by hypo-osmotic treatment. Despite similar particle size, as established by microscopic examination, the envelope-free chloroplasts exhibited significantly smaller FSC signals compared to the intact chloroplasts. As the intensity of FSC also depends on the refractive index of the particle (Phinney and Cucci 1989; Tanke and Van der Keur 1993), it appears likely that the low FSC intensities of envelope-free chloroplasts resulted from the removal of the envelope with the subsequent loss of solutes which decreased the refractive index of particles.

We compared the SSC intensity of envelope-free chloroplasts from mesophyll and bundle sheath cells of maize (*Zea mays*, Poaceae) leaves (E. Pfündel and A. Meister, unpublished data). Lower SSC was observed for mesophyll compared to bundle sheath chloroplasts although the former exhibited a more pronounced internal structure due to the presence of many grana stacks than the latter, which possess only a very few grana (Miller et al. 1977). Obviously, our data are not consistent with a simple relationship between SSC intensity and internal structure of chloroplasts. In essence, to interpret scattering signals from chloroplasts, we require complementary information on the particles investigated as has already been suggested above in the case of chloroplast autofluorescence.

11.3
Progress of Research

Flow cytometric signals from chloroplasts are affected by the quality of chloroplast preparations. Basically, two classes of integrity have been analyzed by FCM: envelope-retaining and envelope-free chloroplasts which, subsequently, will be denoted as "intact chloroplasts" and as "thylakoids", respectively. Also, membrane fragments of various sizes can be released from thylakoids and we will refer to these particles as "thylakoid fragments". Usually, non-fixed chloroplasts have been analyzed but we will indicate when chloroplasts have been pretreated with fixation reagents prior to FCM.

11.3.1
Chloroplasts from C_3 Plants

Most higher plants belong to a biochemically-classified group known as C_3 plants. This term refers to the fact that the first stable product of C_3 type photosynthetic fixation of CO_2 is 3-phosphoglycerate, which contains three carbon atoms (Schnarrenberger and Martin 1997). In principle, each C_3 type chloroplast is competent to carry out the entire photosynthetic process and, from this point of view, it might be expected that C_3 chloroplasts isolated from uniform leaf material will result in a single homogeneous population in FCM. In apparent confirmation, frequency distributions for chlorophyll fluorescence intensities did not reveal subpopulations in the case of either intact spinach (a C_3 plant) chloroplasts (Paau et al. 1978; Schröder and Petit 1992) or spinach thylakoids (Xu et al. 1990).

By recording FSC intensities, however, Ashcroft et al. (1986) distinguished two particle populations in preparations of intact spinach chloroplasts which exhibited similar autofluorescence and SSC signals. By experimental means, the authors excluded the possibility that the two populations arose from different orientations of chloroplasts relative to the laser beam, and provided further evidence that both populations represented truly intact chloroplasts. It was hypothesized that the two chloroplasts populations might originate in the upper palisade and lower spongy parenchyma tissue of the leaf. Other researchers also observed two particle populations by measuring FSC intensity in preparations of intact chloroplasts from spinach (Schröder and Petit 1992) and pea (*Pisum sativum*, Fabaceae; see Kausch and Bruce 1994; Subramanian et al. 2001). In contrast to the previous view, these workers suggested that the two populations represented intact chloroplasts and thylakoid contaminants, respectively. To resolve the factors involved in these seemingly contradictory observations, more research is required to characterize the FCM signature of different chloroplast preparations from C_3 plants.

Chloroplasts isolated from *Nicotiana benthamiana* (Solanaceae) leaves and subsequently fixed in 1% formaldehyde showed two populations in SSC versus FSC dot plots (Cho et al. 2004). In this case, light microscopy of the sorted chloroplasts revealed unequivocally that the population with high SSC and FSC intensities consisted of aggregates of two to four chloroplasts while the particles exhibiting low scattering represented single chloroplasts.

Various techniques of labeling and staining have extended the information derived from FCM on chloroplasts beyond that obtained with autofluorescence and scattering alone. Schröder and Petit (1992) bound fluorescein isothiocyanate (FITC)-labeled lectins (i.e. glycoproteins that bind to specific carbohydrates) to chloroplast preparations of different integrity, and measured by FCM the laser-excited green fluorescence from FITC as well as the red autofluorescence from chlorophyll. The studies revealed that galactose-specific lectins bind preferably to intact chloroplasts but to a much lesser degree to thylakoids or thylakoid fragments. The authors speculated that galactolipids of the envelope are responsible for binding of the galactose-specific lectins.

Subramanian et al. (2001) developed a fusion protein that included a peptide, which is normally involved in translocating nucleus-encoded proteins across the chloroplast envelope. Attaching the fluorophore FITC to this protein allowed the authors to employ FCM to study the binding of the protein to intact pea chloroplasts. Flow cytometry provided important data regarding the nature of binding of the peptide; for example, saturation of binding of the fusion protein to chloroplasts and protease-sensitivity of binding sites suggested that specific envelope receptors were mediating the binding process.

Cho et al. (2004) investigated the role of the DNA gyrase in chloroplasts of *Nicotiana benthamiana*. The gyrase enzyme has been suggested to play a central role in topological changes of DNA during transcription and replication. By silencing different gyrase genes in *N. benthamiana*, phenotypes exhibiting yellow or white leaf variegation were observed. The effect of this silencing was analyzed by FCM analysis of propidium iodide-stained chloroplasts. Reduced expression of gyrase resulted in a lower number of chloroplasts per cell, which exhibited much higher DNA content than control chloroplasts. As the abnormal chloroplasts had only one or a few large nucleoids, the results indicated the effect of gyrase on nucleoid partitioning by regulating DNA topology.

11.3.2
Chloroplasts from C_4 Plants

The term C_4 plant refers to the fact that the first product of primary photosynthetic fixation of CO_2 is a four-carbon dicarboxylic acid, oxaloacetate. In C_4 plants, the carbon reactions of sugar formation are constrained to the bundle sheath compartment which corresponds to wreaths of cells situated around vascular bundles (Edwards and Walker 1983; Hatch 1987; Fig. 11.1a). In the so-called mesophyll compartment, which consists of photosynthetic cells outside the bundle sheath compartment, primary CO_2 fixation takes place. The advantage of C_4 over C_3 plants arises from the particularly efficient fixation of CO_2 under stress conditions, which reduce CO_2 concentrations in the leaf. However, C_4 photosynthesis requires more energy to reduce CO_2 to sugar; consequently, C_4 plants need ample light to perform optimally.

There are three biochemical subclasses of C_4 photosynthesis which are named by the enzyme which catalyzes decarboxylation of the C_4 acids in the bundle sheath compartment: NADP-dependent malic enzyme (NADP-ME), NAD-dependent malic enzyme (NAD-ME), or PEP carboxykinase (PEP-CK). In both, NADP-ME- and NAD-ME-type C_4 plants, photosynthetic carbon reactions require ATP and NADPH at considerably different ratios in bundle sheath and mesophyll cells (Edwards and Walker 1983; Hatch 1987). Depending on the actual fraction of ATP/NADPH consumption, PS I/PS II ratios are adjusted because only NADPH production requires the concomitant action of both photosystems while ATP formation can proceed in the absence of PS II (Allen 2003). In fact, PS II/PS I ratios are particularly small in bundle sheath chloroplasts of NADP-ME-type C_4

species while normal ratios exist in the mesophyll chloroplasts; however, NAD-ME-type C_4 species tend to show the opposite pattern (Edwards and Walker 1983; Hatch 1987).

The first FCM study of chloroplasts from a C_4 plant was carried out by Ashcroft et al. (1986) with maize, which is a NADP-ME-type C_4 species. The authors observed that conventionally purified bundle sheath chloroplasts emitted lower chlorophyll fluorescence intensities than mesophyll chloroplasts but a clear FCM separation of the two chloroplast populations was not possible. In any case, the lower emission from bundle sheath chloroplasts concurs with the lower fluorescence yield of PS I relative to that of PS II (Dau 1994) in combination with low PS II/PS I ratios in bundle sheath chloroplasts.

Moreover, two clearly separated populations exhibiting different chlorophyll fluorescence intensities, and probably representing bundle sheath and mesophyll chloroplasts, have been observed in FCM of intact maize chloroplasts by Kausch and Bruce (1994). These data have been obtained with magnetically sorted intact chloroplasts that were practically free of thylakoid contamination. Magnetic sorting involved labeling the outer envelope membrane with antibodies which carried magnetic nano-particles (Fe_3O_4), followed by attachment of the labeled chloroplasts to a magnetized column and elution of non-labeled material; finally, the labeled chloroplasts were harvested after de-magnetizing the column.

In contrast to intact chloroplasts, FCM of thylakoids from maize did not show distinct particle populations emitting different intensities of chlorophyll fluorescence (E. Pfündel and A. Meister, unpublished data). We believe that isolated thylakoids tend to release thylakoid fragments giving rise to particles of quite variable size which, in FCM, appear as a population that extends over a wide range of fluorescence intensities and, hence, blurs the demarcation between populations of bundle sheath and mesophyll thylakoids. This problem has been overcome by simultaneously detecting chlorophyll fluorescence at wavelengths below 700 nm and above 710 nm which are enriched in PS II and PS I fluorescence, respectively (see Section 11.2.1). By plotting short wavelength against long wavelength fluorescence of individual particles, FCM yielded two elongated populations which were clearly separated except for those particles emitting very low fluorescence intensities (Fig. 11.2a; Pfündel and Meister 1996). After appropriate gating, sorting by FCM yielded mesophyll and bundle sheath thylakoids of outstanding purity (Fig. 11.2c).

The probability that particles exhibiting high and low ratios for short/long wavelength fluorescence corresponded to mesophyll and bundle sheath thylakoids, respectively, was confirmed by FCM of conventional preparations of the two types of thylakoids (Pfündel and Meister 1996). Moreover, emission spectra for chlorophyll fluorescence of pure mesophyll and bundle sheath thylakoids sorted by FCM (Fig. 11.2d) corresponded well with the different fluorescence ratios observed in flow cytometry and agreed with the low abundance of PS II in bundle sheath chloroplasts of NADP-ME-type C_4 plants. Also in agreement with different photosystem distributions in the two types of thylakoids, much higher chlorophyll a/chlorophyll b ratios were found in bundle sheath thylakoids than in mesophyll thylakoids after sorting by flow cytometry; PS I is known to possess much higher chlorophyll *a/b* ratios than PS II (Melis 1991).

The FCM method was applied not only to C_4 chloroplasts of other NADP-ME-type species but was also successfully extended to NAD-ME-type species (Pfündel et al. 1996). By investigating thylakoids sorted by FCM, it was shown that within the C_4 sub-types NADP-ME or NAD-ME, conspicuous variations existed in photosystem stoichiometry and PS II-to-PS I energy transfer. Further, fluorescence spectra obtained with thylakoids purified by FCM, in combination with fluorescence spectra from leaves of C_3, C_3-C_4-intermediate and C_4 species of the genus *Flaveria* (Asteraceae), strongly suggested that the functional size of the light-harvesting antenna of PS I in bundle sheath chloroplasts was efficiently increased during evolution of NADP-ME C_4 photosynthesis (Pfündel and Pfeffer 1997). By taking advantage of the fluorescence spectral data obtained with pure bundle sheath and mesophyll thylakoids, a model was derived to estimate the contribution of PS I to total leaf fluorescence (Pfündel 1998). FCM of chloroplasts from C_4 plants, carried out parallel to confocal laser-scanning microscopy, permitted us to observe that variations in PS II concentrations with rather constant PS I concentrations cause the known variations in photosystem stoichiometry in many NADP-ME and NAD-ME C_4 species (Pfündel and Neubohn 1999).

Our FCM analysis, however, was not successful in distinguishing between bundle sheath and mesophyll cells isolated from NADP-ME C_4 species of the genus *Flaveria* (Pfündel and Meister 1998). Probably re-absorption effects, originating in the relatively dense packing of several tens of chloroplasts in each cell, distorted the fluorescence properties of the chloroplasts so that populations of bundle sheath and mesophyll cells overlapped in FCM.

11.4 Conclusion

Considerable information about FCM of chloroplasts has been accumulated during the past 25 years. The particular potential of FCM to analyze and statistically describe chloroplast suspensions has been convincingly exploited to determine interactions between proteins and chloroplasts much better than is possible with simple bulk measurements (Subramanian et al. 2001). Hence, we expect that analogous studies will play an important role in future research. Also, sorting of chloroplasts from C_4 plants by FCM yielded subpopulations of exceptional purity that substantially advanced our understanding of C_4 plants (see Section 11.3.2). Similarly, sorting of chloroplasts by FCM may also provide well-defined starting material for future molecular biological investigations.

As explained in the Introduction, heterogeneous acclimation of chloroplasts is expected in leaves of C_3 plants but FCM data on the variability of C_3 chloroplasts are ambiguous (see Section 11.3.1). We anticipate better progress in this area by applying methods that have already been established for FCM with algae, including the use of different excitation wavelengths (Toepel et al. 2005); this particular approach is well suited to provide information on acclimation of pigment stoichiometries in response to the light micro-climate within the leaf. Moreover, selected combinations of different excitation and emission wavelengths, as has been introduced in principle, for isolated chlorophylls by Meister (1992), can improve the sensitivity of FCM to better inspect chloroplast heterogeneity. In addition, measurements of variable fluorescence (Demers et al. 1991; Furuya and Li 1992; Olson and Zettler 1995) or immuno-labeling of key enzymes of photosynthesis (Orellana et al. 1988) will provide valuable information concerning variability of photosynthetic performance of chloroplasts in the leaf. Certainly, the significance of information derived from FCM will be further increased by parallel examination of leaf sections using modern microspectroscopic methods (Blancaflor and Gilroy 2000; Feijó and Moreno 2004) or isolation of single cells by microdissection for subsequent analyses (Day et al. 2005; Kerk et al. 2003). Ideally, combinations of FCM and microscopic methods will relate statistical data from FCM to spatial arrangement of physiological gradients in a leaf. Such information will lay the basis for a better understanding of whole leaf function, and, eventually, for the function of entire leaf canopies.

Acknowledgments

We thank Dr Robert J. Porra for help in the preparation of the manuscript.

References

Agati, G., Cerovic, Z. G., Moya, I. **2000**, *Photochem. Photobiol.* 72, 75–84.

Allen, J. F. *Trends Plant Sci.* **2003**, 8, 15–19.

Anderson, J. M., Chow, W. S., Goodchild, D. J. **1988**, *Austral. J. Plant Physiol.* 15, 11–26.

Ashcroft, R. G., Preston, C., Cleland, R. E., Critchley, C. **1986**, *Photobiochem. Photobiophys.* 13, 1–14.

Bedbrook, J. R., Kolodner, R. **1979**, *Ann. Rev. Plant Physiol.* 30, 593–620.

Blancaflor, E. B., Gilroy, S. **2000**, *Am. J. Bot.* 87, 1547–1560.

Cho, H. S, Lee, S. S., Kim, K. D., Hwang, I., Lim, J. S., Park, Y. I., Pai, H. S. **2004**, *Plant Cell* 16, 2665–2682.

Croce, R., Zucchelli, G., Garlaschi, F. M., Bassi, R., Jennings, R. C. **1996**, *Biochemistry* 35, 8572–8579.

Cui, M., Vogelmann, T. C., Smith, W. K. **1991**, *Plant Cell Environ.* 14, 493–500.

Dau, H. **1994**, *Photochem. Photobiol.* 60, 1–23.

Day, R. C., Grossniklaus, U., Macknight, R. C. **2005**, *Trends Plant Sci.* 10, 397–406.

Dekker, J. P., Boekema, E. J. **2005**, *Biochim. Biophys. Acta – Bioenergetics* 1706, 12–39.

Demers, S., Roy, S., Gagnon, R., Vignault, C. **1991**, *Marine Ecol. Prog. Series* 76, 185–193.

Douce, R., Joyard, J. **1979**, *Adv. Bot. Res.* 7, 2–116.

Edwards, G., Walker, D. **1983**, C_3, C_4: *Mechanisms, and Cellular and Environmental Regulation, of Photosynthesis*, Blackwell Scientific Publications, Oxford, UK.

Evans, J. R. **1999**, *New Phytologist* 143, 93–104.

Evans, J. R., Vogelmann, T. C., Williams, W. E., Gorton, H. L. **2004**, Chloroplast to leaf in *Photosynthetic Adaptation. Chloroplast to Landscape. Ecological Studies*, (vol. 178), ed. W. K. Smith, T. C. Vogelmann, C. Critchley, Springer-Verlag, Berlin, pp. 15–41.

Feijó, J. A., Moreno, N. **2004**, *Protoplasma* 223, 1–32.

Flügge, U.-I. **2000**, *Trends Plant Sci.* 5, 135–137.
Furuya, K., Li, W. K. W. **1992**, *Marine Ecol. Prog. Series* 88, 279–287.
Gillbro, T., Cogdell, R. J. **1989**, *Chem. Phys. Lett.* 158, 312–316.
Govindjee. **1995**, *Austral. J. Plant Physiol.* 22, 131–160.
Hatch, M. D. **1987**, *Biochim. Biophys. Acta* 895, 81–106.
Havaux, M. **1998**, *Trends Plant Sci.* 3, 147–151.
Karukstis, K. K. **1992**, *J. Photochem. Photobiol. B: Biol.* 15, 63–74.
Kausch, A. P., Bruce, B. D. **1994**, *Plant J.* 6, 6767–779.
Kerk, N. M., Ceserani, T., Tausta, S. L., Sussex, I. M., Nelson, T. M. **2003**, *Plant Physiol.* 132, 27–35.
Krause, G. H. **1994**, Photoinhibition induced by low temperatures in *Photoinhibition of Photosynthesis. From Molecular Mechanisms to the Field*, ed. N. R. Baker, J. R. Bowyer, Bios Scientific Publishers, Oxford, UK, pp. 331–348.
Legendre, L., Courties, C., Troussellier, M. **2001**, *Cytometry* 44, 164–172.
Mauzerall, D. **1972**, *Proc. Natl Acad. Sci. USA* 69, 1358–1362.
Meister, A. **1992**, *Photosynthetica* 26, 533–539.
Melis, A. **1991**, *Biochim. Biophys. Acta* 1058, 87–106.
Miller, K. R., Miller, G. J., McIntyre, K. R. **1977**, *Biochim. Biophys. Acta* 459, 145–156.
Mustárdy, L., Garab, G. **2003**, *Trends Plant Sci.* 8, 117–122.
Neale, P. J., Cullen, J. J., Yentsch, C. M. **1989**, *Limnol. Oceanogr.* 34, 1739–1748.
Olson, R. J., Zettler, E. R. **1995**, *Limnol. Oceanogr.* 40, 816–820.
Orellana, M. V., Perry, M. J., Watson, B. A. **1988**, Probes for assessing single-cell primary production: antibodies against ribulose-1,5-bisphosphate carboxylase (RuBPCase) and peridinin/chlorophyll a protein (PCP) in *Immunochemical Approaches to Coastal, Estuarine and Oceanographic Questions*, ed. C. M. Yentsch, F. C. Mague, P. K. Horan, Springer-Verlag, Berlin, pp. 243–262.
Orsenigo, M., Patrignani, G., Rascio, N. **1997**, Ecophysiology of C_3, C_4, and CAM plants in *Handbook of Photosynthesis*, ed. M. Pessarakli, Marcel Dekker, New York, pp. 1–25.
Oxborough, K. **2004**, *J. Exp. Bot.* 55, 1195–1205.
Paau, A. S., Oro, J., Cowles, J. R. **1978**, *J. Exp. Bot.* 29, 1011–1020.
Perry, M. J., Porter, S. M. **1989**, *Limnol. Oceanogr.* 34, 1727–1738.
Pfündel, E. E. **1998**, *Photosynth. Res.* 56, 185–195.
Pfündel, E. E., Meister, A. **1996**, *Cytometry* 23, 97–105.
Pfündel, E. E., Meister, A. **1998**, Flow cytometry of protoplasts from C_4 plants in *Photosynthesis: Mechanisms and Effects*, (vol. 5), ed. G. Garab, Kluwer Academic Publishers, Dordrecht, pp. 4341–4344.
Pfündel, E. E., Neubohn, B. **1999**, *Plant Cell Environ.* 22, 1569–1577.
Pfündel, E. E., Pfeffer, M. **1997**, *Plant Physiol.* 114, 145–152.
Pfündel, E. E., Nagel, E., Meister, A. **1996**, *Plant Physiol.* 112, 1055–1070.
Phinney, D. A., Cucci, T. **1989**, *Cytometry* 10, 511–521.
Pierre, Y., Breyton, C., Lemoine, Y., Robert, B., Vernotte, C., Popot, J.-L. **1997**, *J. Biol. Chem.* 272, 21901–21908.
Schnarrenberger, C., Martin, W. **1997**, *Photosynthetica* 33, 331–345.
Schreiber, U. **2004**, Pulse-Amplitude (PAM) fluorometry and saturation pulse method in *Chlorophyll Fluorescence: A signature of Photosynthesis*, ed. G. Papageorgiou, Govindjee, Kluwer Academic Publishers, Dordrecht, The Netherlands, pp. 279–319.
Schröder, W. P., Petit, P. X. **1992**, *Plant Physiol.* 100, 1092–1102.
Siefermann-Harms, D. **1985**, *Biochim. Biophys. Acta* 811, 325–355.
Smith, W. K., Vogelmann, T. C., DeLucia, E. H., Bell, D. T., Shepherd, K. A. **1997**, *BioScience* 47, 785–793.
Staehelin, L. A. **2003**, *Photosyn. Res.* 76, 185–196.
Subramanian, C., Ivey, R., Bruce, B. D. **2001**, *Plant J.* 25, 349–363.
Sugiura, M. **1992**, *Plant Mol. Biol.* 19, 149–168.
Tanke, H. J., Van der Keur, M. **1993**, *Trends Biotechnol.* 11, 55–62.
Terashima, I. **1989**, Productive structure of a leaf in *Photosynthesis*, ed. W. R. Briggs, Alan R. Liss, New York, pp. 207–226.

Toepel, J., Wilhelm, C., Meister, A., Becker, A., del Carmen Martinez-Ballesta, M. **2004**, *Methods Cell Biol.* 75, 375–407.

Toepel, J., Langner, U., Wilhelm, C. **2005**, *J. Phycol.* 41, 1099–1110.

Tolbert, N. E. **1997**, *Annu. Rev. Plant Physiol. Plant Mol. Biol.* 48, 1–23.

Vogelmann, T. C. **1993**, *Annu. Rev. Plant Physiol. Plant Mol. Biol.* 44, 231–251.

Vogelmann, T. C., Evans, J. R. **2002**, *Plant Cell Environ.* 25, 1313–1323.

Xu, C., Auger, J., Govindjee. **1990**, *Cytometry* 11, 349–358.

12
DNA Flow Cytometry in Non-vascular Plants

Hermann Voglmayr

Overview

Flow cytometry has not been used extensively in non-vascular plant research apart from aquatic botany, which is dealt with in Chapter 13. Nonetheless, there are also powerful applications of flow cytometry among other groups of non-vascular plants. The most important of these is the estimation of genome size, which is rather challenging due to the comparatively small genomes of mosses and algae. In this chapter, the present knowledge regarding genome size investigations in algae and mosses is summarized, with special reference to the advantages and limitations of flow cytometry as compared to other methods (mainly image analysis). Some other topics, such as cell cycle analysis in algae, are also briefly discussed.

12.1
Introduction

Non-vascular plants are morphologically as well as systematically a very heterogeneous group, comprising eukaryotic algae and bryophytes (see Figs. 12.1 and 12.2 for examples). They include organisms ranging from the unicellular level to the level of morphologically elaborate multicellular organisms consisting of complex tissues. They occupy various environmental niches; eukaryotic algae are an important component of aquatic ecosystems, being the main primary producers and providing important habitats for other organisms. With the advent of molecular phylogenetic methods, the classification of the main non-vascular plant lineages has been in a state of flux and therefore in this chapter mostly informal names are given to particular lineages, following the tree of life project (http://tolweb.org/tree/). Systematically, non-vascular plants do not form a monophyletic group but belong to the four main lineages: stramenopiles (containing diatoms, brown algae, xanthophytes, chrysophytes, raphidophytes, and eustigmatophytes), alveolates (dinoflagellates), rhodophyta (red algae), and green plants (green algae

Flow Cytometry with Plant Cells. Edited by Jaroslav Doležel, Johann Greilhuber, and Jan Suda
Copyright © 2007 WILEY-VCH Verlag GmbH & Co. KGaA, Weinheim
ISBN: 978-3-527-31487-4

Fig. 12.1 Examples of different algal groups (pictures are not to the same scale): (a) *Fucus* sp. (brown algae), sporophyte consisting of real multicellular tissue (haplo-diplont); (b) *Spirogyra* sp. (green algae), a filamentous alga with spirally coiled chloroplasts (haplont); (c) *Pediastrum biradiatum* (green algae), colony made up of 16 independent cells (haplont); (d) *Peridinium* sp. (dinoflagellates), unicellular with conspicuous vertical and horizontal grooves (haplont); (e) *Phacus* sp. (euglenophytes), unicellular with flagellum and without rigid cell wall (haplont); (f) *Volvox* sp. (green algae), with flagellate, interconnected cells forming globose colonies (haplont); (g) *Navicula* sp. (diatoms), cells with silica shells and two brownish chloroplasts (diplont); (h) *Ankistrodesmus* sp. (green algae; asterisk), colony of independent cells (haplont); *Paramecium bursaria* (ciliate) with endosymbiotic *Chlorella* sp. (unicellular green alga; haplont); (i) *Euglena* sp. (euglenophytes), unicellular with flagellum and without rigid cell wall (haplont); (j) *Cosmarium* sp. (green algae), bipartite cell with solid verrucose cell wall and one large chloroplast (haplont).

(a) (b)

Fig. 12.2 Examples of bryophytes, presenting the dominant perennial gametophyte (haploid) and the short-lived sporophyte (diploid). (a) *Marchantia polymorpha* (thallose liverworts); prostrate flattened thallus (gametophyte) showing vegetative reproduction by gemmae (arrowhead) and upright gametangium stands. The plants are dioecious (male and female gametangia on different plants); the male gametangia (antheridia) are located on the upside of the weakly lobed stands (♂; minute dots); the female gametangia (archegonia) on the underside of the strongly lobed stands (♀). The fertilized egg of the archegonium develops into a minute sporophyte (not clearly visible in the picture), which is nourished by the gametophyte. (b) *Polytrichum formosum* (mosses), showing the leafy gametophyte and the sporophyte with beaked sporangium (arrow); on some sporangia the remnants of the archegonium wall (calyptra) are visible (arrowheads).

and bryophytes). In addition to the four main lineages, several other smaller groups contain photoautotrophs (e.g. euglenids, cryptomonads, glaucophytes, and haptophytes), which are traditionally also included in non-vascular plants. Therefore, the term "non-vascular plants" in this chapter is applied in a traditional sense (autotrophic, non-vascular eukaryotes), accepting the fact that they represent a very heterogeneous and polyphyletic group. Some lineages (e.g. dinoflagellates) contain closely related auto- as well as heterotrophic species; in such cases, investigations on heterotrophic species are also considered (e.g. the genus *Pfiesteria*). The *non-vascular plants* merit a specialized chapter because most studies using flow cytometry (FCM) have focused on vascular plants, mainly angiosperms, while very little work has been carried out in the remaining groups of plants.

Several species of algae are of economic importance; some rhodophytes are used for the production of polysaccharides (carrageen, agar). Other rhodophytes (e.g. nori) and brown algae (e.g. wakame, kombu) are cultivated and used as traditional food in Eastern Asia. Some dinoflagellates (e.g. *Pfiesteria*) have recently become known for their negative impacts, as they produce highly poisonous neurotoxins contaminating fish and shellfish, making them unsuitable for human consumption. Other species (e.g. *Caulerpa taxifolia*, a toxic chlorophyte) are highly invasive and have become noxious weeds after their introduction into new areas, representing a threat to whole ecosystems. However, biodiversity and biology have been inadequately researched in many algal lineages and numerous problems remain to be solved, calling for investigations using modern methods.

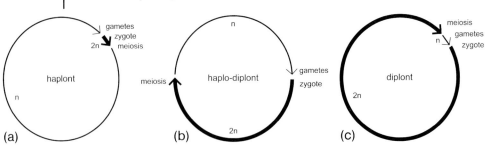

Fig. 12.3 Simplified sketch of life cycles present in non-vascular plants. (a) haplont, with chromosome number n, characterized by meiosis immediately following zygote formation; mitoses only occur in haploid cells; (b) in haplo-diplont, meiosis is separated from zygote formation and vice versa by mitoses of haploid (n) and diploid (2n) cells, respectively; (c) diplont, characterised by meiosis directly producing gametes; mitoses only occur in diploid (2n) cells.

Bryophytes are important constituents of mainly terrestrial ecosystems. Many species are highly resistant to drought and can inhabit sites under extreme conditions. Concerning biomass, they have an important role especially in temperate wetlands. In bogs, *Sphagnum* (Sphagnaceae) and some other mosses are responsible for peat accumulation.

In understanding the biology of non-vascular plants, their life cycle is of central importance. Life cycles of non-vascular plants are very diverse, and it would be beyond the scope of this chapter to deal extensively with this topic; for details on a specific systematic group botany textbooks should be consulted. However, some general features concerning life cycles of non-vascular plants relevant to this chapter are briefly presented below.

Figure 12.3 illustrates generalized schemes of life cycles found in non-vascular plants. The features determining the life cycle are the chronology of mitoses, meiosis and zygote formation. When zygote formation is immediately followed by meiosis, all cells (except for the zygote) are haploid and the organism is consequently termed a haplont (Fig. 12.3a). Conversely, when meiosis directly results in the production of gametes (which immediately fuse to form a zygote), all cells (except for the gametes) are diploid and the organism is termed a diplont (Fig. 12.3c). However, in most non-vascular plants, the stages of zygote and meiosis are separated by mitoses of haploid and diploid cells, respectively, resulting in the longer duration of haploid and diploid stages. The organism is then termed a haplo-diplont (Fig. 12.3b). If the organism is multicellular and the cells are differentiated into vegetative and generative cells (gametes), the haploid stage of the haplo-diplont is termed gametophyte, whereas the diploid stage is termed sporophyte. The gametophyte produces the gametes and the sporophyte the spores (usually meiospores, as a result of meiosis). Depending on the systematic group, sporophytes and gametophytes may be morphologically very similar (= isomorphic; e.g. some green and brown algae) or very different (= heteromorphic; e.g. bryophytes); in the latter case one stage is usually long-lived and dominant, whereas the other is ephemeral. The haplo-diplont life cycle can be complex through the presence of two subsequent sporophyte generations, of which only the second produces the meiospores (red algae). In mosses, the gametophyte is highly differentiated; the germinating spore produces a filamentous protonema, which is similar to filamentous algae. The protonema cells initially contain numerous chloroplasts (termed chloronema), but later produce cells with oblique cell walls and fewer chloroplasts (termed caulonema), which finally give rise to the typical moss shoots.

In non-vascular plants, flow cytometry (FCM) has been widely applied in environmental, ecological and physiological studies of aquatic algae, covering issues such as identification and enumeration of phytoplankton populations (Olson et al. 1988, 1989; Toepel et al. 2005), phytoplankton monitoring (Becker et al. 2002; Rutten et al. 2005), detection of new oceanic taxa (Chisholm et al. 1988; Courties et al. 1994), detection and enumeration of living and dead cells (Van de Poll et al. 2005), photosynthetic activity of phytoplankton as an indication of the presence of toxic substances in limnic and marine ecosystems, and so forth. A comprehensive review on the use of FCM in unicellular algae is given by Collier (2000), and a methodological compendium of DNA and RNA analysis in phytoplankton by Marie et al. (2000). Investigations concerning phytoplankton have in common the fact that they measured various properties of the cells such as "cell size" (by scatter, which however, also depends on particle characters other than size; see Chapter 2) and autofluorescence of photosynthetic pigments, the intensity of which was then linked to specific ecological aspects. Such applications are dealt with in detail in Chapter 13 and will thus not be considered here.

Apart from the above-mentioned environmental issues, FCM has not been widely used in non-vascular plant research despite its obvious potential to generate extensive data sets within a reasonable time (Kapraun et al. 2004). This may in part be due to the fact that, compared to vascular plants, fewer researchers are working with non-vascular plants, which, in addition, are much more challenging with respect to material acquisition and species determination. Moreover, there are some specific methodological issues in non-vascular plant research as described in detail below.

12.2
Nuclear DNA Content and Genome Size Analysis

Investigations using flow cytometry in non-vascular plants other than those aimed at plankton analysis, largely focused on DNA content. The nuclear DNA amount or genome size may be determined either in relative or in absolute units. The former approach does not require absolute quantification and has mainly been used in cell cycle studies (e.g. in the moss *Physcomitrella patens*, Funariaceae; Schween et al. 2003) and for the estimation of DNA ploidy levels within the life cycle (e.g. in the brown algae *Laminaria* and *Macrocystis*; Gall et al. 1996) or among different individuals (e.g. in *Sphagnum*; Śliwińska et al. 2000).

Measurement of genome size in absolute units is a more challenging task. To achieve reliable results, involvement of an appropriate internal standard with known genome size is essential. Absolute genome size of the unknown sample can then be calculated by comparing standard/sample peak positions (see Chapter 4 for explanation). In addition, fluorochromes should be selected which bind to nuclear DNA independently of its base composition.

12.2.1
General Methodological Considerations

Various nucleus isolation protocols, reference standards (such as chicken red blood cells, trout red blood cells, the angiosperms *Arabidopsis* (Brassicaceae), soybean, and tobacco, the red alga *Chondrus crispus* (Gigartinaceae), and others) and fluorescent dyes have been used for the determination of absolute genome size. The following sections therefore summarize methodological issues affecting the quality of the results. Whereas some recommendations are of general concern for plant DNA flow cytometry (see Chapter 4), others are more relevant to non-vascular plants.

12.2.1.1 Isolation and Fixation of Nuclei

Generally, measurements of nuclear DNA content have been carried out using isolated nuclei, because the cells contain also other organelles with their own DNA. Autofluorescent particles such as chromatophores should be completely removed. If enough tissue is available, chopping with a sharp razor blade (Galbraith et al. 1983) is a simple but very efficient method of isolation, and is applied in brown and green macroalgae (Asensi et al. 2001; Gall et al. 1996; Peters et al. 2004) and bryophytes (Voglmayr 2000). Buffers are described in Chapter 4 and also commented on in Chapter 18.

Whereas chopping appeared to be the method of choice in Laminariales (Asensi et al. 2001; Gall et al. 1996), it largely failed in Ectocarpales (Peters et al. 2004). Isolation of protoplasts by enzymatic digestion of cell walls is an option if chopping fails. This method was applied to vegetative thallus tissue by Le Gall et al. (1993; Gall et al. 1996) in their investigations on genome size of multicellular red, brown and green algae. The vegetative thalli were digested with cellulolytic enzymes to produce protoplasts, which were then subjected to lysis in a buffer to release the nuclei. This approach is more time consuming and requires optimization of the cell-wall digestion protocol for the particular group under investigation. Neither this method worked with Ectocarpales (Peters et al. 2004).

Whitley et al. (1993) lysed the predatory marine dinoflagellate *Oxyrrhis marina* in phosphate buffered saline plus detergent and succeeded in separating the nuclei of the flagellate from the prey cells. They used the fluorescent dyes Hoechst

33258 and DAPI to stain DNA. However, the same method applied to intact methanol-fixed cells was not successful. Cell cycle analysis showed that in this taxon, unlike most phytoplankton species, cells spent more time, ca. 50% of the cell cycle duration, in $G_2 + M$ phases, when actively dividing.

Zooids or zoospores from laboratory cultures were collected live by Peters et al. (2004) in brown algae (Ectocarpales) and by Parrow and Burkholder (2002) in the dinoflagellate genus *Pfiesteria*. The cells were lysed in a buffer to release the nuclei, which were stained with SYBR Green I, or propidium iodide (PI) and SY-TOX Green for FCM.

Although measurements of unicellular algae may seem to be trivial, several problems can be encountered. LaJeunesse et al. (2005) reported that staining whole cells of Dinophyceae increased background fluorescence. Therefore, lysis of cells may be required. Pigmentation of the cell wall may decrease the intensity of nuclear fluorescence and increase non-specific background noise. In addition, a thick cell wall may reduce accessibility of the fluorochrome, while irregular cell shapes, commonly found in algae (see Fig. 12.1), may disturb the laminar flow of the instrument.

Unicellular algae were used directly for FCM analyses either fresh (Vaulot et al. 1994, in Haptophyceae/Prymnesiophyceae) or after fixation (LaJeunesse et al. 2005, in Dinophyceae). While ideally the cells should be subjected to lysis, after which the nuclei are released, fixation with methanol–acetic acid (LaJeunesse et al. 2005) has the advantage that chlorophyll and other autofluorescent pigments are removed, which can otherwise result in prominent background noise. Fixation in diluted methanol (up to 35% v/v) as reported by Eschbach et al. (2001) is probably not appropriate. In order to precipitate DNA, alcohols should be added at final concentrations of 70% at least. LaJeunesse et al. (2005) pelleted dinoflagellate cells by centrifugation, fixed them using the classical methanol:acetic acid (3:1) fixation solution, stored them in 90% methanol, and measured the DNA content using FCM after staining with DAPI in MOPS buffer (cf. Chapter 4). Eschbach et al. (2001) studied fixation methods for FCM in *Chrysochromulina polylepis*, a toxic marine member of the Prymnesiophyta (or Haptophyta). They compared freshly dissolved paraformaldehyde in seawater, formaldehyde, glutaraldehyde (GA) and methanol (see above) in combination with a DNA fluorochrome, SYTOX Green. It should be noted that formaldehyde and glutaraldehyde, to an even greater extent, are protein-crosslinking fixatives, while methanol at higher concentrations precipitates nucleoproteins. The mechanisms of fixation are thus quite different. The aldehydes were used at concentrations ranging from 0.25 to 4%, and methanol at 15–35% (the latter concentrations being hardly comparable to the higher methanol concentrations commonly used in cytological studies). Glutaraldehyde at low concentration was found to be the most efficient fixative while methanol was the most unsuitable, as it inhibited nuclear fluorescence. In this study, retention of red chlorophyll fluorescence after GA fixation was deemed advantageous as a further parameter to discriminate cells.

Fixation can certainly influence the staining properties of the nuclei. From biomedical studies it is known that prolonged formaldehyde fixation leads to reduced nuclear PI fluorescence (Leers et al. 1999), and this certainly also applies to fixation with GA and staining with other fluorochromes. Corresponding studies in plants have not been carried out. Here it should be mentioned that formaldehyde as a fixative plays a positive role in Feulgen densitometry: this monovalent aldehyde inactivates condensable tannins by polymerization *in situ* and is compatible with the Feulgen reaction, because the non-covalently bound aldehyde can be completely washed out and no reactive aldehyde groups remain in the tissue. This is not the case with glutaraldehyde (Greilhuber and Temsch 2001). It remains to be demonstrated whether formaldehyde fixation is useful in DNA flow cytometry of brown macroalgae (see below).

12.2.1.2 Standardization

The term *standard* in the present context means a biological material included in an experiment or set of tests to serve as a reference, to which all measurement data can be related and so become generally comparable. An *internal standard* is included in the very same test to guarantee identical experimental conditions for the unknown sample and the standard. Its application is generally considered superior to external standardization (see Chapter 4 for discussion on standardization).

For estimation of genome size in absolute units, an internal standard of *known* genome size is co-processed with the specimen. In the resulting flow histogram, the ratio of specimen/standard peak means (or modal values) is calculated, which is then multiplied by the absolute genome size of the reference standard to give the absolute genome size of the specimen. Absolute genome size may be expressed in DNA picograms (pg) or basepairs (bp); for the conversion formula of these, see Doležel et al. (2003) and Chapter 4. Note also that there is a distinction between C-value (holoploid genome size) and Cx-value (monoploid genome size); for definitions and discussion see Chapter 4.

Selection of a suitable internal standard species is a crucial step in achieving reliable data not only in absolute but also in relative genome size measurements. An internal standard should be easily accessible, well characterized taxonomically, and its genome size should be reliably established within narrow limits. In addition, its genome size should be close to that of the sample to be measured (say, 0.4-fold to 2.5-fold of the unknown sample) to avoid errors due to instrument nonlinearity and to enable the use of the linear amplification scale. Actually, only a few of the published investigations on absolute genome size in non-vascular plants meet these criteria. In many studies, fixed chicken red blood cells (CRBCs) were used (LaJeunesse et al. 2005; Parrow and Burkholder 2002; Vaulot et al. 1994). However, even at a suitable genome size ratio to the unknown sample, CRBCs are not considered to be an ideal standard due to their strong DNA compaction (Hardie et al. 2002) and, more importantly, because of (usually) a different history of fixation and storage. In addition, the species is possibly not stable in genome size. Male and female chicken differ by 2.7% because of sex chromosomes (Nakamura et al. 1990), which may introduce additional but minor experimental errors. Therefore, the use of CRBCs as an internal standard is not fully reliable (cf. Chapter 4) and should be strongly discouraged in high-precision estimations of genome size in absolute units. However, CRBCs are useful as a landmark and approximate reference for relative genome size measurements (but note that different samples of CRBCs can differ; cf. Chapter 4).

For plant samples, a plant reference standard is highly preferable. For monadoid protists of low genome size, a strain of *Chlamydomonas* sp. may be appropriate as the primary standard as may be other monadal taxa, which are easy to cultivate, and whose genomes may be sequenced in the near future. Although sequencing does not usually include repetitive and heterochromatic DNA, this type of DNA should constitute a minor component in small genomes of protists. The selection of standard species with suitable genome size remains a problem that has not been solved until now in algal DNA flow cytometry research.

12.2.1.3 Fluorochromes for Estimation of Nuclear DNA Content

The most commonly used DNA fluorochromes have been 4'-6-diamidino-2-phenylindole (DAPI), ethidium bromide (EB) and propidium iodide (PI). DAPI is known to have a preference for AT-rich regions of DNA (Doležel et al. 1992; see also Chapters 4, 8, and 18), but does not stain RNA, whereas ethidium bromide (EB) and propidium iodide (PI) show no base preference, but give a somewhat lower fluorescence yield. The advent of green (532 nm) solid state lasers, which are more efficient for PI excitation than the turquoise line (488 nm) of conventional argon ion lasers may lead to practical improvements. PI and EB stain also RNA and thus require RNase treatment to be specific for DNA (compare Chapters 4 and 8). Apart from these, other DNA binding fluorochromes were applied, such as SYBR Green 1 and SYTOX Green, which have less defined binding specificities. The former is commonly used for DNA quantification in Real-Time PCR (e.g. Bustin 2005), whereas it has only rarely been utilized in the FCM of non-vascular plants (e.g. Peters et al. 2004). It is an intercalating dye which binds to double-stranded DNA, but also shows base-specific preferences under certain conditions, which is considered to be due to minor groove binding (Zipper et al. 2004). Binding properties and base specificity vary greatly under different dye/base pair ratios and salt concentrations; under high dye/base pair ratios preferred binding to AT-rich sequences was observed (Zipper et al. 2004). As a result, SYBR Green 1 may be useful for ploidy and cell cycle investigations, but should not be used uncritically for the estimation of DNA content in absolute units.

SYTOX Green is a DNA-binding fluorochrome, which only penetrates dead cells and has thus been mainly used for discrimination of dead versus living cells especially in prokaryotes (e.g. Roth et al. 1997), but also in phytoplankton (e.g. Veldhuis et al. 2001). Little is known about the exact binding properties of this stain, which presently limits its applicability in genome size investigations.

TOTO and YOYO are analogous cyanine dyes which bind to double-stranded and single-stranded DNA, to RNA, and even seem to exhibit some specificity for certain base sequences (Jacobsen et al. 1995; Rye and Glazer 1995; Rye et al. 1992, 1993).

The impression from a perusal of the literature is, that there is a need for comparative studies, applying static fluorescence microscopy and FCM, using higher plants with well-known genome sizes such as *Arabidopsis thaliana* (Brassicaceae), *Pisum sativum* and *Vicia faba* (Fabaceae), *Allium cepa* (Alliaceae), and others (Doležel et al. 1998), to compare results obtained with SYTOX Green, SYBR Green, PicoGreen, YOYO-1, and YOPRO-1 (applied for instance by Marie et al. 1996, 1997) with those obtained with PI, to obtain empirical estimates of their performance in DNA content measurement. According to Marie et al. (1996), who investigated marine prokaryotes and *Escherichia coli*, YOYO-1, YO-PRO-1 and Pico Green can be used only on aldehyde-fixed cells, require co-factors (EDTA, potassium, citrate), and are highly sensitive to ionic strength. As a result of their fluorescence yield they were considered to be suitable for counting cells in natural samples but unsuitable for conducting cell cycle analyses. It may be concluded that these fluorochromes should not be used uncritically for DNA content measurements.

12.2.1.4 Secondary Metabolites as DNA Staining Inhibitors

Non-stoichiometric DNA staining due to the interference of secondary metabolites has only recently received wider attention in FCM. Whereas staining errors are generally accepted to be of major concern in quantitative DNA staining with the Feulgen method, their possible effects in FCM have long been neglected (cf. Chapter 4). Nevertheless, there are indications that cytoplasmic compounds also interfere with fluorescence staining in liverworts and algae such as *Spirogyra* (Chlorophyta) and other Conjugatae. These are known to contain gallotannin in the vacuole (Cannell et al. 1988; Czapek 1910), which acts as an inhibitor of the Feulgen reaction and fluorochrome staining (E. M. Temsch, unpublished data; cf. also Greilhuber 1997). Phlorotannins are well-known phenolics of brown algae (Ragan and Glombitza 1986), and they would likely interfere with fluorochromes in FCM, if present. Therefore, the potential disturbing effects of secondary metabolites should also be considered carefully in work with algae.

12.2.2
DNA Content and Genome Size Studies

The most frequent application of FCM in vascular plants is for the estimation of DNA ploidy level and genome size (see Chapters 4 and 5). In contrast, FCM has not been widely applied to genome size estimation in non-vascular plants. According to Kapraun (2005) and Kapraun et al. (2004) in Bennett and Leitch (2005), out of about 253 measurements of genome size in eukaryotic algae, only 11 were obtained using FCM (Le Gall et al. 1993; but see additional work by Peters et al. 2004). Spring et al. (1978) determined, in a pioneering FCM experiment, genome sizes of *Acetabularia acetabulum* (1C = 0.92 pg) from gametes and zygotes, and of a haploid strain of *Chlamydomonas reinhardi* (1C = 0.28 pg), using RBCs of the fish *Betta splendens* (2C = 1.3 pg) as internal standard.

In bryophytes, of the 176 measurements reported by Bennett and Leitch (2005), 133 were obtained with FCM, which is mainly due to the extensive investigations of Voglmayr (2000), who exclusively studied mosses.

Published and listed genome sizes (1C-values) in algae range from 0.01 to 19.6 pg (Bennett and Leitch 2005) and in bryophytes from 0.17 to 4.05 pg (Renzaglia et al. 1995; Voglmayr 2000). Such DNA amounts generally lie within the range measurable using FCM, although the lowest category in known eukaryotic algal values, comparable to that of the yeast, *Saccharomyces cerevisiae* (1C = 0.012 pg), will probably require flow cytometers capable of precisely measuring very weak fluorescence signals.

12.2.2.1 **Algae**

Algae are methodologically demanding organisms with respect to FCM investigation. Therefore, it is not surprising that most genome size estimates have been made using other methods (Bennett and Leitch 2005; Kapraun 2005). As can be seen from Fig. 12.1, algal cells often have solid cell walls, which may pose problems during staining and measurement. Separation of cells may not be easy in colony-forming algae. In filamentous algae, it is often difficult to release the cytoplasmic contents. Most species do not have massive amounts of tissue, which could be easily homogenized by chopping to obtain nuclear suspensions. Therefore, sample preparation (see above) is usually more elaborate than in higher plants, depending both on the systematic group and on organization level (unicellular, filamentous or thalli consisting of complex tissues).

The taxonomic spectrum of eukaryotic algae is by far the least covered topic in the FCM literature. The major taxa are Haptophyta, Phaeophyta, Dinophyta, Rhodophyta, and Chlorophyta. The remarkable work of Le Gall et al. (1993), on marine macroalgae, contains data on two green algae (*Enteromorpha compressa*, 1C = 0.13 pg; *Ulva rigida*, 1C = 0.15 pg), five brown algae (*Laminaria digitata*, 1C = 0.70 pg; *L. saccharina*, 1C = 0.79 pg; *Pilayella littoralis*, 1C = 0.55 pg; *Sphacelaria* sp., 1C = 1.70 pg; and *Undaria pinnatifida*, 1C = 0.64 pg), and three red algae (*Chondrus crispus*, 1C = 0.16 pg; *Kappaphycus alvarezii*, 1C = 0.22 pg; and *Porphyra purpurea*, 1C = 0.30 pg).

Other studies are restricted to certain taxonomic groups and concern the Haptophyta or Prymnesiophyta, the Phaeophyta, and the Dinophyta.

Haptophyta (Prymnesiophyta)   Vaulot et al. (1994) studied *Phaeocystis* (Prymnesiophyceae) using cells fixed with formaldehyde. Fluorochromes used were Hoechst 33342, chromomycin A_3 (CMA_3) and propidium iodide. CRBCs were used as the reference material for determination of DNA content (pg) and base pair ratio (for detailed discussion on the base pair ratio determination see Chapter 8). Sixteen strains assigned to *Phaeocystis globosa* and *P. antarctica* (mostly monadal marine flagellates) from the Atlantic, Gulf of Mexico, Mediterranean (Naples), and the Antarctic were studied. For sensitive discrimination of strains one strain was used as the standard and CMA_3 as a DNA stain. Haploid and diploid strains were identified, and both levels occurred in some strains according to flagellate (haploid) or colony (diploid) stage. In one strain without colonies haploid and diploid flagellates were found to co-exist. Co-processing strains in a one plus one fashion would permit strains of distinct genome size to be differentiated based on bimodal versus unimodal histogram peaks. Five groups of strains were identified in this way, and the groups exhibited a geographical pattern. Monoploid genome sizes (1Cx-values: see Chapter 4) varied 1.73-fold, between 0.12 and 0.22 pg. 1C-values varied between 0.12 and 0.44 pg. Base pair ratios estimated with Hoechst 33342/PI and CMA_3/PI diverged widely, ranging from 51 to 57% GC, and were considered unreliable, probably because of the large difference in GC content between *Phaeocystis* and CRBCs (with 42.7% GC).

Houdan et al. (2004) studied the heteromorphic life cycles of four marine plankton coccolithophore genera of Prymnesiophyceae. Coccoliths are calcified organic scales covering the cell which are characteristic of this class. Cultures of *Emiliana huxleyi*, *Coccolithus pelagicus*, *Calcidiscus leptoporus* and *Coronosphaera mediterranea* were investigated, and two monadal species, *Micromonas pusilla* (green algae, Prasinophyceae) and *Isochrysis galbana* (Haptophyceae), both of lower DNA content, were used as co-processed internal references. FCM using SYBR Green as the DNA fluorochrome, applied to isolated nuclei after cell lysis, allowed the identification of two DNA ploidy levels in these Prymnesiophytes, the haploid level characterized by holococcolith-bearing or non-calcifying cells, and the diploid level characterized by heterococcolith-bearing cells.

Phaeophyta Gall et al. (1996) reported the formation of polyploid, mostly abnormal sporophytes after regeneration from tissue cultures obtained from excised sporophyte tissues, and from unisexual cultures obtained by induced parthenogenesis of the Laminariales *Laminaria saccharina*, *L. digitata* and *Macrocystis pyrifera*. DNA ploidy levels were determined by FCM, as chromosome counting was difficult due to the high number and small size of the chromosomes. Nuclear suspensions were obtained by chopping the tissue in isolation buffer, and subsequently staining with Hoechst 33258 or DAPI. Chicken red blood cells (CRBCs with $2C = 2.33$ pg) or occasionally the diplophasic tetrasporophytes of the red alga *Chondrus crispus* with $2C = 0.33$ pg (Le Gall et al. 1993) were used as the internal reference. From the relative data in Gall et al. (1996), it can be estimated that *L. digitata* and *L. saccharina* genomes were in the range of 0.58 to 0.70 pg (1C), respectively, which is compatible with the values from Le Gall et al. (1993) mentioned above. For *Macrocystis pyrifera* the estimated genome size is about 0.57 pg (1C).

Asensi et al. (2001) investigated vegetative regeneration of cells isolated from sporophytes of *Laminaria digitata*. Whereas chromosome counts proved that all cells were diploid ($2n =$ ca. 60–61), the cells of the early filamentous stage uniformly had a 2C DNA level, and normal sporophytes developing from these cell-filaments had a 4C level. This was supposed to be an indication of "polyteny" involved in the regeneration process. Specifically, it was suggested that in the sporophytes with 4C level instead of four sets of chromosomes (each chromosome with a single chromatid) there were two sets with chromosomes having two chromatids instead of one each (Asensi et al. 2001). The premise of this assumption is that the majority of nuclei in both developmental stages of the alga were unreplicated. However, a note on terminology is in order here. The term *polyteny* describes a condition of somatic polyploidy characterized by polytene or giant chromosomes (cf. Chapter 15). Mitotically active chromosomes having an unreplicated internal bi-stranded structure would be better termed *bineme* (Rieger et al. 1976). However, bineme or polyneme chromosome models are out of date. The current view is that the DNA in anaphase chromosomes (i.e. a chromatid) consists of one single strand of double-helical DNA. Thus, the case of *Laminaria digitata* deserves further investigation. It would appear more plausible to assume, that the cells in the regenerated sporophytes transgress S-phase quickly and accumulate in 4C, so that nuclei in 2C are not detected by FCM. An example from the literature is the gametophyte of the moss *Physcomitrella patens*, in which cells in 1C are so few in number that they are not detected in FCM histograms (see below; Schween et al. 2003; Zoriniants et al. 2005).

Nevertheless, it should be borne in mind, that algal chromosomes are apparently not always bound to orthodox models that are valid for higher plants or animals. According to Goff and Coleman (1986) in some red algae and according to Kapraun (1994) in some green macroalgae there is during growth an "incremental size decrease associated with a cascading down of DNA contents" (Kapraun 2005). This means, that the tip cell of such algae, from which the other cells are derived by mitoses, has significantly more DNA (64–128 times more in the red alga *Polysiphonia*) in the nucleus than mature somatic or gametic nuclei. This amount of DNA and also the number of chromosomes are progressively reduced while mitoses without DNA synthesis occur until the typical (gametic) genome size is attained. These data have been obtained with static cytofluorometry after staining with DAPI. Such results add another bizarre mechanism to the repertoire of the eukaryotic genome behavior, which should be further supported by other techniques of investigation, for instance, flow cytometry. However, it is debatable whether cytofluorometric DNA content studies in such algae are always carried out under conditions which guarantee stoichiometric DNA staining with DAPI (e.g. Bleckwenn et al. 2003).

Peters et al. (2004) studied genome size in Phaeophyceae with the intention of finding a brown alga of optimal characteristics for genome sequencing. They used SYBR Green to stain nuclei from lysed zoids or gametes. Nuclei released from chopped tissue of the red alga *Chondrus crispus* (1C = 150 Mbp) and gamete nuclei of *Ectocarpus siliculosus* (1C = 240 Mbp in European accessions) served as reference materials. Genome sizes were lower in Ectocarpales (127–500 Mbp) than in Laminariales (580–720 Mbp) and Fucales (1095–1271 Mbp). The Ectocarpales were: *Striaria attenuata* (127 Mbp), *Scytosiphon lomentaria* (222 Mbp), *Ectocarpus siliculosus* from Peru and Europe (214 and 240 Mbp, respectively), and *E. fasciculatus* (290 Mbp). The Fucales were: *Fucus serratus* (1095 Mbp), *F. vesiculosus* (1140 Mbp), and *Ascophyllum nodosum* (1271 Mbp). The genome sizes determined earlier by Le Gall et al. (1993) are given by Peters et al. (2004) in Mbp as follows. Ectocarpales: *Undaria pinnatifida* (580 Mbp); Laminariales: *Laminaria digitata*

(640 Mbp), and *L. saccharina* (720 Mbp); Sphacelariales: *Sphacelaria* sp. (1550 Mbp). *Ectocarpus siliculosus* (Ectocarpales) was proposed by Peters et al. (2004) as a model species for genomic studies in brown algae, mainly because of the relatively small genome size, the ease of cultivation, morphological differentiation, sexual reproduction and the large amount of genetic knowledge about this species.

Dinophyta LaJeunesse et al. (2005) investigated 11 genera of Dinophyceae (18 species) using DAPI as the DNA stain and CRBCs as a standard with 2C amount assumed to be 3.0 pg (cf. Chapter 4). Dinophyceae have comparatively very large genomes. The smallest 1C-values were found in the symbiont of marine evertebrates, *Symbiodinium*, ranging from 1.9 to 4.8 pg in 29 strains. In free-living taxa, however, 1C amounts ranged from 3.6 pg in *Katodinium rotundatum* to 115 and 225 pg in *Prorocentrum micans*, the variation in the latter species possibly caused by a polyploidization event. When more than one strain per species was available, the C-values were fairly similar. Not unexpectedly, the DNA amounts were positively and significantly correlated with chromosome volume and cell size. As the authors discuss, the small genome in *Symbiodinium* may be related to the small cell size required by a symbiontic cellular organism.

Kremp and Parrow (2006) studied the formation of cysts in *Scrippsiella hangoei*, a peridinoid flagellate of the Baltic Sea. The supposed sexual origin of these cysts could be largely disproved, because most of these had a 1C DNA content, on basis of the flagellate cell population, which cycled between 1C and 2C. SYTOX Green was the DNA stain applied to lysed cells, and triploid trout RBCs ($2C = 3Cx = 8.25$ pg) were the standard. 1C of *Scrippsiella hangoei* was determined as 23.26 pg DNA.

Cell Cycle Studies in Algae In algae, cell cycle analyses have frequently been performed with FCM, such as in the green alga *Chlorella* (Gerashchenko et al. 2000, 2001; Kadano et al. 2004; McAuley and Muscatine 1986), the euglenophyte *Euglena gracilis* (Carré and Edmunds 1993), and the dinoflagellates *Pfiesteria* (e.g. Lin et al. 2004; Parrow and Burkholder 2002) and *Scrippsiella* (Kremp and Parrow 2006). The aims of these analyses were to reveal stage-specific C-levels or ploidies in the life cycle (e.g. cysts in *Scrippsiella*, see above, and zoospores in *Pfiesteria*) or to elucidate the influence of cyclic adenosine monophosphate (AMP) on circadian cell division rhythmicity (*Euglena*), and the influence of the host cell cycle on the cell cycle of endosymbionts. In green paramecium (*Paramecium bursaria*), the FCM results indicated that the cell cycle of the symbiontic *Chlorella* might be controlled by the host (Kadano et al. 2004).

12.2.2.2 Bryophytes

As has been demonstrated by Voglmayr (2000), FCM offers a rapid means of generating extensive data on absolute genome size. However, FCM has so far been successfully applied only to mosses (Reski et al. 1994; Lamparter et al. 1998; Temsch et al. 1999; Voglmayr 2000; Schween et al. 2003; Melosik et al. 2005) while there is still no published data for hornworts and liverworts. In mosses, obtaining suspensions of nuclei proved to be very simple (Voglmayr 2000); moss shoots were thoroughly chopped with a sharp razor blade together with an internal standard in a citrate buffer. In general, FCM worked well with all species of mosses examined. A general point to be considered here is the fact that bryophytes, especially mosses and in particular *Sphagnum*, house fauna and flora, whose nuclei, if present, must not be mistaken for the target nuclei.

To calculate absolute genome sizes in bryophytes, different internal standards have been employed. Reski et al. (1994), investigating *Physcomitrella patens* (Funariaceae), and Lamparter et al. (1998), investigating *Ceratodon purpureus* (Ditrichaceae), used *Arabidopsis thaliana*, which has a genome size similar to that of mosses. The internal standard used for the extensive investigations of Voglmayr (2000) was soybean (*Glycine max*, Fabaceae; Fig. 12.4), while Melosik et al. (2005) used *Petunia hybrida* (Solanaceae) for the genus *Sphagnum*.

Different nuclear stains have been applied. Reski et al. (1994) and Lamparter et al. (1998) used DAPI, which binds preferentially to AT-rich regions of DNA (Doležel et al. 1992) and is therefore not recommended for absolute genome size determination. Temsch et al. (1999), Voglmayr (2000) and Melosik et al. (2005) used propidium iodide, an intercalating dye without base preference. There is a wide consensus that this fluorochrome yields reliable data with regard to absolute genome size. Temsch et al. (1999) and Voglmayr (2000) demonstrated that the

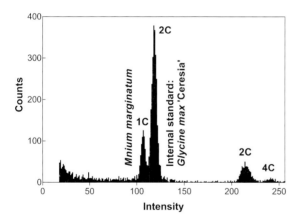

Fig. 12.4 Estimation of nuclear genome size in a moss species *Mnium marginatum* (see also Voglmayr 2000). Histogram of relative fluorescence intensities was obtained after FCM analysis of propidium iodide-stained nuclei. The y-axis represents the number of fluorescent events, the x-axis the fluorescence intensity of nuclei, which corresponds to DNA content, and is given on a linear scale. As an internal standard, soybean (*Glycine max*) 'Ceresia' was used, with an absolute 2C-value of 2.268 pg. Relative ratio 1C peak of *Mnium marginatum*/2C peak of soybean is 0.903, resulting in an absolute genome size of 2.05 pg (1C) for *Mnium marginatum*. In addition to absolute genome size calculation, information about cell cycle parameters is also obtained; most of the moss nuclei (ca. 63%) are unreplicated (1C), whereas ca. 37% are replicated with a 2C DNA amount.

results of propidium iodide FCM are highly comparable with those of Feulgen DNA image analysis.

As most data on absolute genome size in mosses have been obtained by Voglmayr (2000), this study is briefly described here. From a total of 289 accessions, the majority (274) were measured using propidium iodide FCM and 53 accessions using Feulgen DNA image analysis. All major groups of mosses were sampled except for *Sphagnum* (for FCM of this genus, see Temsch et al. (1999) and Melosik et al. (2005)). Genome sizes (1C) were low, although higher than those in *Arabidopsis thaliana* (1C = 0.16 pg); they ranged from 0.174 to 2.16 pg, resulting in only a 12-fold variation, which is remarkably low compared to the approximately 2000-fold variation in angiosperms (Bennett and Leitch 2005; Chapter 7). Within this narrow range, the vast majority of moss species had an absolute genome size (1C-values) between 0.3 and 0.6 pg, which indicates strong selection pressure for small genomes. If more than one accession was measured from the same species, genome sizes usually did not differ significantly from each other, except for some incidences of polyploidy. Interestingly, absolute genome sizes of some species differed remarkably between the studies of Renzaglia et al. (1995) and Voglmayr (2000). In the former study, Feulgen densitometry of bryophyte sperm nuclei was used, with chicken erythrocytes as the internal standard. The reasons for these discrepancies are unknown, but may involve methodological issues such as differences in the level of DNA compaction between standard and specimen nuclei (bryophyte sperm nuclei are highly compacted) and/or the use of chicken erythrocytes as the internal standard. The results of Feulgen staining are rather sensitive to methodological factors but Voglmayr (2000) obtained almost identical genome sizes in all species investigated using both FCM and Feulgen DNA image analysis. The approach used demonstrated the efficient use of different methods when values were to be corroborated. For this purpose, FCM and Feulgen image analysis complement each other ideally.

A disadvantage of propidium iodide compared to DAPI is that the small genome size of bryophytes (see above) often approaches the resolution limits of bench-top flow cytometers. Debris can become troublesome when a considerable number of replicated and/or endopolyploid cells are present and G_1 cells are almost completely restricted to mitotically cycling cells, as in the moss *Physcomitrella patens* (Schween et al. 2003; Zoriniants et al. 2005), in which the G_1 peak may be overlooked. In such species, it is advisable to corroborate the FCM data by independent microscopical methods. Low degrees of endopolyploidy occur often in moss shoots (H. Vogmayr, unpublished observations).

Measuring relative DNA contents, Śliwińska et al. (2000) determined DNA ploidy levels in different *Sphagnum* species (peat moss). They obtained nuclear suspensions by chopping the vegetative tissue in isolation buffer and used DAPI to stain the nuclei. For ploidy determination, they usually compared the peak channels among separate runs of individual specimens. To corroborate these results, Śliwińska et al. (2000) also co-processed both haploid and diploid specimens and therefore measured DNA contents simultaneously within the same run. This enabled the calculation of an exact DNA content ratio between the haploid and diploid accessions, which was close to 1:2, in agreement with observations by Temsch et al. (1998) using Feulgen image densitometry.

Cell cycle analysis has been performed in the moss *Physcomitrella patens* (Schween et al. 2003). The authors investigated the chloronema and caulonema stages of the protonema of the gametophyte (see Section 12.1 for explanation). Cell cycle parameters, including the influence of plant growth regulators and nutrients on cell cycle and cell differentiation, were studied during a 22-day culture period. The nuclear suspension, obtained by chopping the plant material grown in axenic liquid cultures, was stained by DAPI. Schween et al. (2003) concluded that the chloronema cells were mainly in the G_2 stage whereas the caulonema cells were in the G_1 stage. They also reported the presence of endopolyploid cells (4C).

12.3
Future Perspectives

FCM has great potential in numerous areas of non-vascular plant research, including developmental biology, taxonomy, ecology, and population biology. The high throughput opens new prospects for large-scale population studies. Numerous morphologically similar bryophyte species are known to have different ploidy levels (Fritsch 1991), and even within the same species different cytotypes have been found. FCM makes possible DNA ploidy screening in large population samples and could help in elucidating the extent of ploidy variation at different spatial scales (cf. for angiosperms Chapter 5). Genome size can be a reliable species-specific marker in taxonomically complex alliances. For example, the genus *Sphagnum* has numerous species which are difficult to discriminate morphologically but which are often well characterized by their C-value (i.e. ploidy level), and FCM may be used for taxa identification (Melosik et al. 2005). Recent investigations show that herbarium specimens can be used for DNA ploidy level screening in angiosperms (see Chapter 5), which should also be considered for some non-vascular plants. Bryophytes seem to be particularly promising due to their high resistance to drought, which by experience is found to be paralleled by the amenability of herbarium specimens to FCM. Voglmayr (1998) found that specimens stored for up to 5 months did not show any shift in genome size as compared to fresh individuals. Using Feulgen DNA image analysis, Temsch et al. (2004) demonstrated that DNA ploidy level can be estimated in herbarium specimens up to 30 years old, which may also be feasible using FCM provided a proper methodology is developed. It would be worthwhile to test herbarium specimens of algae. For instance, vouchers of the monostromate green alga *Monostroma* (Chlorophyceae) with $1C = 255$ Mbp (Kapraun 2005) could be a promising material to begin with, because a relatively fast drying process can be assumed in such vouchers, which may keep the nuclei in a good condition.

Evolutionary changes in genome size among closely related taxa can be documented by FCM, which can be linked to ecological parameters which act as selection pressure on genome sizes. If combined with molecular genetic and phylogenetic analyses, new insights into the evolution of genome size can be gained, which has, for example, been recently done by LaJeunesse et al. (2005) for the genus *Symbiodinium* (Dinophyceae). Hypotheses such as a decrease in genome size in endosymbiontic organisms can be evaluated using FCM; there are numerous candidates in several algal lineages (e.g. green algae, dinoflagellates). Correlations of genome size with cell size, cell morphology, cell cycle duration and life forms can also be investigated and may reveal novel insights into internal evolutionary constraints. FCM is also a powerful tool for detailed investigations of the life cycle of algae, and the effects of different environmental conditions and chemicals on life cycle parameters.

12.4
Conclusion

FCM is increasingly applied to the elucidation of numerous challenging questions of contemporary non-vascular plant research. Apart from issues related to aquatic FCM, which are discussed in Chapter 13, flow cytometry has hitherto been used mainly for the estimation of nuclear DNA content. Using fluorescent dyes which bind to nucleic acids, FCM becomes a convenient method for cell cycle analyses and the estimation of DNA ploidy and/or genome size in absolute units. However, there are several potential pitfalls, and it is necessary to meet high methodological standards to guarantee reliable data (see also Chapters 4 and 5). New and improved flow cytometers offer promising possibilities to obtain rapidly more data on absolute genome size in non-vascular plants, for which still very incomplete information exists compared to vascular plants. Nevertheless, FCM data may need corroboration by other cytological methods. This is especially important if new taxonomic lineages, for which there is little expertise, are to be studied. Non-vascular plants have been very unevenly sampled for genome size. The bryophytes serve as an impressive example for a deplorably insufficient knowledge in liverworts, whereas the data basis for mosses is quite good (Voglmayr 2000). FCM as a method can, in combination with other techniques (e.g. molecular phylogenetics), provide novel insights into systematics, phylogeny, ecology, evolution, and population biology of non-vascular plants, and is indispensable for identifying the most appropriate candidates to become novel genetic standard organisms among the representatives of the major taxonomic algal groups.

Acknowledgments

I thank J. Greilhuber for useful comments and suggestions.

References

Asensi, A., Gall, E. Ar, Billot, D. M. C., Kloareg, P. D. B. **2001**, *J. Phycol.* 37, 411–417.

Becker, A., Meister, A., Wilhelm, C. **2002**, *Cytometry* 48, 45–57.

Bennett, M. D., Leitch, I. J. **2005**, *Plant DNA C-values Database (release 4.0, October 2005)*, http://www.rbgkew.org.uk/cval/homepage.html.

Bleckwenn, A., Gil-Rodriguez, M. C., Medina, M., Schnetter, R. **2003**, *Eur. J. Phycol.* 38, 307–314.

Bustin, S. A. **2005**, Real-Time PCR in *Encyclopedia of Medical Genomics and Proteomics*, ed. J. Fuchs, M. Podda, Marcel Dekker, New York, pp. 1117–1125.

Cannell, R. J. P., Farmer, P., Walker, J. M. **1988**, *Biochem. J.* 255, 937–941.

Carré, I. A., Edmunds, L. N. **1993**, *J. Cell Sci.* 104, 1163–1173.

Chisholm, S. W., Olson, R. J., Zettler, E. R., Goerick, R., Waterbury, J. B., Welschmeyer, N. A. **1988**, *Nature* 334, 340–343.

Collier, J. L. **2000**, *J. Phycol.* 36, 628–644.

Courties, C., Vaquer, A., Troussellier, M., Lautier, J., Chrétiennot-Dinet, M. J., Neveaux, J., Machado, C., Claustre, H. **1994**, *Nature* 370, 255.

Czapek, F. **1910**, *Ber. Deutsch. Bot. Gesellsch.* 28, 147–159.

Doležel, J., Sgorbati, S., Lucretti, S. **1992**, *Physiol. Plant.* 85, 625–631.

Doležel, J., Greilhuber, J., Lucretti, S., Meister, A., Lysák, M. A., Nardi, L., Obermayer, R. **1998**, *Ann. Bot. (Suppl. A)*, 82, 17–26.

Doležel, J., Bartoš, J., Voglmayr, H., Greilhuber, J. *Cytometry* **2003**, 51A, 127–128.

Eschbach, E., Reckermann, M., John, U., Medlin, L. K. **2001**, *Cytometry* 44, 126–132.

Fritsch, **1991**, *Bryophyt. Biblioth.* 40, 1–352.

Galbraith, D. W., Harkins, K. R., Maddox, J. M., Ayres, N. M., Sharma, D. P., Firoozabady, E. **1983**, *Science* 220, 1049–1051.

Gall, E. Ar (Le Gall, Y.), Asensi, A., Marie, D., Kloareg, B. **1996**, *Eur. J. Phycol.* 31, 369–380.

Gerashchenko, B. I., Nishihara, N., Ohara, T., Tosuji, H., Kosaka, T., Hosoya, H. **2000**, *Cytometry* 41, 209–215.

Gerashchenko, B. I., Kosaka, T., Hosoya, H. **2001**, *Cytometry* 44, 257–263.

Goff, L. J., Coleman, A. W. **1986**, *Amer. J. Bot.* 73, 1109–1130.

Greilhuber, J. **1997**, The problem of variable genome size in plants (with special reference to woody plants) in *Cytogenetic Studies of Forest Trees and Shrub Species*, ed. Z. Borzan, S. E. Schlarbaum, Croatian Forests, Inc., and Faculty of Forestry, University of Zagreb, Zagreb, pp. 13–34.

Greilhuber, J., Temsch, E. M. **2001**, *Acta Bot. Croat.* 60, 285–298.

Hardie, D. C., Gregory, T. R., Hebert, P. D. N. **2002**, *J. Histochem. Cytochem.* 50, 735–749.

Houdan, A., Billard, C., Marie, D., Not, F., Sáez, A. G., Young, J. R., Probert, I. **2004**, *Systemat. Biodiversity* 1, 453–465.

Jacobsen, J. P., Pedersen, J. B., Hansen, L. F., Wemmer, D. E. **1995**, *Nucleic Acids Res.* 23, 753–760.

Kadano, T., Kawano, T., Hosoya, H., Kosaka, T. **2004**, *Protoplasma* 223, 133–141.

Kapraun, D. F. **1994**, *Cryptogamic Bot.* 4, 410–418.

Kapraun, D. F. **2005**, *Ann. Bot.* 95, 7–44.

Kapraun, D. F., Leitch, I. J., Bennett, M. D. **2004**, *Algal DNA C-values database (release 1.0)*, http://www.kew.org/cval/homepage.html.

Kremp, A., Parrow, M. W. **2006**, *J. Phycol.* 42, 400–409.

LaJeunesse, T. C., Lambert, G., Andersen, R. A., Coffroth, M. A., Galbraith, D. W. **2005**, *J. Phycol.* 41, 880–886.

Lamparter, T., Brücker, G., Esch, H., Hughes, J., Meister, A., Hartmann, E. **1998**, *J. Plant Physiol.* 153, 394–400.

Leers, M. P. G., Schutte, B., Theunissen, P. H. M. H., Ramaekers, F. C. S., Nap, M. **1999**, *Cytometry*, 35, 260–266.

Le Gall, Y., Brown, S., Marie, D., Mejjad, M., Kloareg, B. **1993**, *Protoplasma* 173, 123–132.

Lin, S., Mulholland, M. R., Zhang, H., Feinstein, T. N., Jochem, F. J., Carpenter, E. J. **2004**, *J. Phycol.* 40, 1062–1073.

Marie, D., Vaulot, D., Partensky, F. **1996**, *Appl. Environ. Microbiol.* 62, 1649–1655.

Marie, D., Partensky, F., Jacquet, S., Vaulot, D. **1997**, *Appl. Environ. Microbiol.* 63, 186–193.

Marie, D., Simon, N., Guillou, L., Partensky, F., Vaulot, D. **2000**, *Curr. Protocols Cytometry* 11.12.1–11.12.14.

McAuley, P. J., Muscatine, L. **1986**, *J. Cell Sci.* 85, 73–84.

Melosik, I., Odrzykoski, I. I., Śliwińska, E. **2005**, *Nova Hedwigia* 80, 397–412.

Nakamura, D., Tiersch, T. R., Douglas, M., Chandler, R. W. **1990**, *Cytogenet. Cell Genet.* 53, 201–205.

Olson, R. J., Chisholm, S. W., Uettler, E. R., Armbrust, E. V. **1988**, *Deep Sea Res.* 35, 425–440.

Olson, R. J., Zettler, E. R., Anderson, O. K. **1989**, *Cytometry* 10, 636–643.

Parrow, M. W., Burkholder, J. M. **2002**, *J. Exp. Marine Biol. Ecol.* 267, 35–51.

Peters, A. F., Marie, D., Scornet, D., Kloareg, B., Cock, J. M. **2004**, *J. Phycol.* 40, 1079–1088.

Ragan, M. A., Glombitza, K. W. **1986**, *Prog. Phycol. Res.* 4, 130–241.

Renzaglia, K. S., Rasch, E. M., Pike, L. M. **1995**, *Am. J. Bot.* 82, 18–25.

Reski, R., Faust, M., Wang, X.-H., Wehe, M., Abel, W. O. **1994**, *Mol. Gen. Genet.* 244, 352–359.

Rieger, R., Michaelis, A., Green, M. M. **1976**, *Glossary of Genetics and Cytogenetics*, 4th edn, Springer-Verlag, Berlin, Heidelberg, New York.

Roth, B. L., Poot, M., Yue, S. T., Millard, P. J. **1997**, *Appl. Environ. Microbiol.* 63, 2421–2431.

Rutten, T. P. A., Sandee, B., Hofman, A. R. T. **2005**, *Cytometry* 64A, 16–26.

Rye, H. S., Glazer, A. N. **1995**, *Nucleic Acids Res.* 23, 1215–1222.

Rye, H. S., Yue, S., Wemmer, D. E., Quesada, M. A., Haugland, R. P., Mathies, R. A., Glater, A. N. **1992**, *Nucleic Acids Res.* 20, 2803–2812.

Rye, H. S., Dabora, J. M., Quesada, M. A., Mathies, R. A., Glazer, A. N. **1993**, *Anal. Biochem.* 208, 144–150.

Schween, G., Gorr, G., Hohe, A., Reski, R. **2003**, *Plant Biol.* 5, 1–9.

Śliwińska, E., Krzakowa, M., Melosik, I. **2000**, Estimation of ploidy level in four *Sphagnum* species (Subsecunda section) by flow cytometry in *The Variability in Polish Populations of Sphagnum taxa (Subsecunda section) according to Morphological, Anatomical and Biochemical Traits*, ed. M. Krzakowa, I. Melosik, Bogucki Science Publishers, Poznań, pp. 137–144.

Spring, H., Grierson, D., Hemleben, V., Stöhr, M., Krohne, G., Stadler, J., Franke, W. W. **1978**, *Exp. Cell Res.* 114, 203–215.

Temsch, E. M., Greilhuber, J., Krisai, R. **1998**, *Bot. Acta* 111, 325–330.

Temsch, E. M., Greilhuber, J., Voglmayr, H., Krisai, R. **1999**, *Abhand. Zool.-Bot. Gesellsch. Österreich* 30, 159–167.

Temsch, E. M., Greilhuber, J., Krisai, R. **2004**, Polyploidiegradbestimmung an Herbarbelegen bei Moospflanzen in *11. Österreichische Botanikertreffen in Wien, 3–5 September 2004. Kurzfassung der Beiträge, Vorträge und Poster*, ed. C. König, M. A. Fischer, Institut für Botanik, Wien, pp. 71–72.

Toepel, J., Langner, U., Wilhelm, C. **2005**, *J. Phycol.* 41, 1099–1109.

Van de Poll, H., van Leeuwe, M. A., Roggeveld, J., Buma, A. G. J. **2005**, *J. Phycol.* 41, 840–850.

Vaulot, D., Birrien, J.-L., Marie, D., Casotti, R., Veldhuis, M. J. W., Kraay, G. W., Chrétiennot-Dinet, M.-J. **1994**, *J. Phycol.* 30, 1022–1035.

Veldhuis, M. J. W., Kraay, G. W., Timmermans, K. R. **2001**, *Eur. J. Phycol.* 36, 167–177.

Voglmayr, H. **1998**, *Genome size analysis in mosses (Musci) and downy mildews (Peronosporales)*, PhD Thesis, University of Vienna, Austria.

Voglmayr, H. **2000**, *Ann. Bot.* 85, 531–546.

Whiteley, A. S., Burkill, P. H., Sleigh, M. A. **1993**, *Cytometry* 14, 909–915.

Zipper, H., Brunner, H., Bernhaben, J., Vitzthum, F. **2004**, *Nucleic Acids Res.* 32, e103.

Zoriniants, S., Temsch, E. M., Greilhuber, J. **2005**, DNA cytophotometrivc study of the *Physcomitrella patens* gametophyte in *Moss 2005, Brno, July 23rd–26th, 2005*, Abstracts, p. 25.

13
Phytoplankton and their Analysis by Flow Cytometry

George B. J. Dubelaar, Raffaella Casotti, Glen A. Tarran, and Isabelle C. Biegala

Overview

This chapter outlines how flow cytometry can be used for the analysis of phytoplankton: from basic and straightforward analysis to more challenging applications. Whereas most applications of flow cytometry in aquatic science are still laboratory-based, the emphasis of this chapter is on its potential to be used *in situ*, operated on a high frequency basis. This permits high-resolution sampling in time and space, which is crucial for our understanding of aquatic microbial ecosystems. Beginning with a basic description of the target particles, a sketch is made of the aquatic environment of these microorganisms and specific properties of the environment that may be very different from typical biomedical or plant research conditions. Some phytoplankton-related limitations and pitfalls are discussed as well as special instruments and instrument modification for phytoplankton analysis. Phytoplankton sampling is described with consideration to "critical scales", including the use of platforms such as research ships and ships of opportunity, submerged use in vertical casts or on autonomous underwater vehicles and moored platforms. Various applications are presented, including the screening of species, probing phytoplankton biodiversity, monitoring "harmful algal blooms", studying population-related processes, and cell-related processes and functioning. Plankton abundance patterns in the sea are assessed and the coupling of flow cytometry data to ocean optics and physics is advocated. Finally, the potential use of flow cytometry in protection and warning is discussed, including the monitoring of harmful algal blooms, water quality in the aquaculture industry, bathing waters and the drinking water industry, as well as with bio-indicatiors and the control of ship's ballast water treatment systems.

Flow Cytometry with Plant Cells. Edited by Jaroslav Doležel, Johann Greilhuber, and Jan Suda
Copyright © 2007 WILEY-VCH Verlag GmbH & Co. KGaA, Weinheim
ISBN: 978-3-527-31487-4

13.1
Introduction

In as much as phytoplankton can be regarded as a "special type" of plant, this chapter should be regarded as a "special" addition to this book on plant flow cytometry. Its purpose is to introduce readers to the arena of aquatic flow cytometry, describing both the various applications and their biological significance as well as the unique features of flow instruments and methodology.

13.2
Plankton and their Importance

13.2.1
Particles in Surface Water

Surface waters in lakes, rivers, seas and the oceans contain a wealth of microscopic particles, both living and non-living. The non-living component consists of mineral particles and particles of biological origin such as dead cells, debris and organic matter. River water and water in coastal zones in particular, are often heavily loaded with mineral particles such as clay, silt and small sand grains. The living component is generally referred to as "plankton": microscopic organisms suspended in the water column. These living particles comprise viruses, bacteria, archaea and eukaryotic phytoplankton and zooplankton. Viruses are the most abundant biological agents in seawater. They infect bacteria, phytoplankton and zooplankton and may be important in controlling the abundance and composition of microbial communities. Bacteria are vital components of the aquatic microbial community. Many break down particulate matter, such as cells and detritus, and take up dissolved organic matter, converting it into cell mass, whereas some contain chlorophyll and other pigments and are photosynthetic. Despite being classified among the Bacteria, because of their photosynthetic activity cyanobacteria are considered to belong to the *phyto*plankton. The use of flow cytometry (FCM) for counting natural planktonic bacteria and understanding the structure of planktonic bacterial communities has been discussed by Gasol and Del Giorgio (2000).

Archaea are a newly emerging group of planktonic organisms (Giovannoni and Stingl 2005). They are prokaryotes, similar in size and shape to bacteria, being generally less than 1 µm in size. They are being found in many aquatic environments using molecular techniques, although their role in the plankton is still under investigation. Eukaryotic plankton can be divided into zooplankton and phytoplankton, with some species being mixotrophic. Many zooplankton are herbivores and graze on phytoplankton. Zooplankton are divided into protozoa and metazoa. The protozoa are made up of single celled flagellates, ciliates and sarcodines (amoebae and radiolarians), whereas the metazoans include rotifers, cladocera, copepoda, salps, and others.

13.2.2
Phytoplankton

The phytoplankton (algae s.l.) are planktonic plants occurring as single cells, colonies or filaments. Like all plants, they use carbon dioxide and light to produce biomass in the process of photosynthesis. Overall, phytoplankton are responsible for 45% of primary production on earth and as such are considered to be a massive CO_2 pump owing to their photosynthetic activity. Phytoplankton and the rainforests are the two "lungs" of our planet, playing a key role in gas exchange with the atmosphere. Central parts of the oceans, which are deprived of nutrients, are dominated by *Prochlorococcus*, a cyanobacterium about 0.6 µm in size, amongst the smallest phytoplankton and the most abundant photosynthetic organisms on earth. Due to their photosynthetic activity, the ancestors of this organism contributed to the origin of our oxygenated atmosphere. These discoveries, which contributed to the understanding of the functioning of our planet, could not have been possible without the help of FCM (Chishdm et al. 1988, Li 1995). On a smaller scale, the mineral enrichment of inland and coastal waters due to human activity leads not only to increased phytoplankton growth, but also to the occurrence of population explosions of nuisance species. The aquaculture industry is particularly vulnerable to phytoplankton related damage.

Phytoplankton range in size from abundant sub-micron cyanobacteria to colonial and filamentous species that can reach several millimeters in length and are visible to the naked eye. Figure 13.1 gives an impression of their enormous size range. Phytoplankton are commonly subdivided into three size classes: picophytoplankton (<2 µm), nanophytoplankton (2–20 µm), and microphytoplankton (>20–200 µm).

The diversity of phytoplankton is very high and representatives of most algal divisions may be found among these aquatic organisms. The golden-brown algal line (containing chlorophylls *a* and *c*) includes diatoms and dinoflagellates, which belong to the best known and species-rich groups in the plankton. The green algal line (with chlorophylls *a* and *b*) includes green algae and euglenophytes, which are also very common in marine coastal and offshore waters. The most numerous phytoplankton, however, belong to the cyanobacteria (with chlorophyll *a* and biliproteins), which include free-living prochlorophytes (Urbach et al. 1992). These are ubiquitous and are able to thrive in very dim light at the base of the euphotic zone (Chisholm et al. 1988; Waterbury et al. 1979). The red algal line (with chlorophyll *a* and biliproteins) is common in benthic habitats of tropical reef waters and in Norwegian coastal waters (Paasche and Throndsen 1970), but is not generally detected in oceanic plankton.

13.2.3
Distributions in the Aquatic Environment

All the above-mentioned organisms as well as the non-living particles may occur in greatly varying abundances and compositions in the water column as compo-

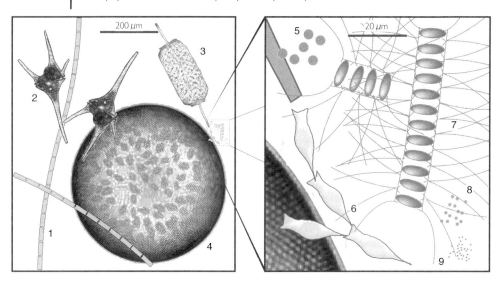

Fig. 13.1 Schematic drawings of some phytoplankton taxa illustrating their size differences. Bacillariophyta: (1) *Melosira*, (3) *Ditylum*, (4) *Coscinodiscus*, (7) *Chaetoceros*; Pyrrhophyta: (2) *Ceratium*; Chlorophyta: (5) *Chlorella*; Chrysophyta: (6) *Dinobryon*; Cyanobacteria: (8) *Synechococcus*, (9) *Prochlorococcus*.

nents of the microbial community, affecting other organisms by their presence and vice versa (predation) as well as being affected by physical and environmental factors and nutrients. Of crucial importance for the growth and succession of phytoplankton is light penetration into the water column, which is in turn reduced by the living and non-living particles. Temperature is also an important factor with large variations in temperature tolerance among the algae. Turbulence plays an important role as well. These factors lead to survival strategies such as colony formation, aggregation and buoyancy regulation. Throughout the annual cycle there is a temporal pattern in the development of communities of particular taxonomic composition and dominance hierarchy. Small microalgae are typically grazed by small microzooplankton with comparable growth rates, rapidly responding to changes in prey density. Larger microalgae are grazed by copepods with growth rates an order of magnitude lower, hence leading to temporally separated blooms. The composition of surface waters is intrinsically "patchy" and dynamic. Overall production by phytoplankton over an entire ocean can vary by 25% between years and up to a factor of 10 at a single location. Occurrence of single species may show much larger fluctuations. Patchiness is defined as variability in the range of 10 m to 100 km horizontally and/or 0.1 m to 50 m vertically in marine systems. Vertical variability such as "thin layers" (e.g. the "Deep Chlorophyll Maximum", a temporally persistent and highly productive region near the bottom of the euphotic zone) are typical for stratified water columns. Storm cycles may in turn account for mixing, variance in water movement and nutrient input. The persistence of "patchiness" is based on the interplay of patch-formation processes like population growth and patch-dissipation due to turbulence (Mackas et al. 1985).

13.3
Considerations for using Flow Cytometry

According to Peter Burkill from the National Oceanographic Centre, Southampton, UK, "Natural marine waters are complex with components that are often finely balanced. As man's activities influence marine waters at an ever increasing rate, so it becomes increasingly important to understand their complexity and fine balance. Rapid, precise and objective techniques for the analysis of phytoplankton in marine waters are increasingly required. The stepping stone towards this lies with the development and deployment of suitable equipment. Flow cytometry clearly has much to offer here, although most instruments are suboptimal for the analysis of phytoplankton".

13.3.1
Analytical Approach

The traditional method of analyzing phytoplankton populations with the use of microscopic techniques so that a trained scientist can recognize the morphology and size of different phytoplankton taxa (Utermohl 1958), is very time consuming and expensive. The sampling frequency and spatial coverage are generally low, resulting in limited data interpretation (Baretta et al. 1998; Smayda 1998). On the other hand, conventional quantitative oceanographic measurements only permit quantification of bulk properties of the water such as the concentration of chlorophyll a per m^3. This provides at best only average properties of a hypothetical "typical" cell (Platt 1989). Our progress in understanding the functioning of microbial ecosystems and their response to external factors is being held back by these approaches. To analyze aquatic ecosystems in detail, it is crucial to collect information at the level of the single cell to account for biodiversity, viability, specific functions, and activities. This can be achieved using FCM due to its wide analysis range capability (from small zooplankton, phytoplankton and bacteria, to viruses) and its high count rate (up to thousands of cells per second).

Flow cytometer-based counting is accurate, direct and reliable, especially with instruments equipped with a volumetric sample delivery system (Rutten et al. 2005). When discrepancies occur with traditional methods, critical evaluation should follow. For example, validation trials were performed in the early 1990s in the Netherlands, comparing standard microscopic counting of the relatively easy recognizable *Rhodomonas* sp. (Cryptomonadaceae) cells in natural North Sea samples and FCM analysis of the same samples using the aquatic Optical Plankton Analyzer flow cytometer (Dubelaar et al. 1989). Large discrepancies between the FCM and microscope results were observed in the first year but they largely disappeared in the second year, when transmission light microscopy (Utermohl 1958) was replaced by fluorescence microscopy (Dubelaar et al. 2004).

13.3.2
Limitations and Pitfalls of using Biomedical Instruments

Aquatic scientists have achieved many great discoveries using FCM. However, this technology was designed and commercialized for biomedical applications. There are, therefore, a number of compromises that aquatic scientists have had to make whilst using such instruments, as they were not designed with aquatic applications in mind. Some of these compromises are briefly mentioned below.

(i) Laser emission lines and filter set ups for fluorescence detection are not optimal for auto-fluorescent phytoplankton. Chlorophyll has a much larger Stoke's shift as compared to the fluorochromes normally used in FCM, with optimal excitation at 430–440 nm and emission at 680–690 nm.

(ii) Accurate cell counts and cell size determinations are necessary for many ecological studies, but instruments are usually not very good at providing these (most instruments use a differential air pressure system for sample and sheath fluid delivery instead of a volumetric delivery system).

(iii) Particles in water span a very wide size range compared to mammalian cells commonly analyzed using FCM. Many instruments have sufficient sensitivity to detect the particles at the smaller end of the size spectrum (sub-micron) but only a few are able to cope with the particle sizes at the opposite end of the spectrum (1 mm and larger). Particle size-related problems are due to less efficient uptake from the sample flask into narrow sample tubing, maintaining them in suspension throughout the complete fluidics trajectory in the instrument, the risk of partial or complete clogging of an instrument as well as problems with sorting devices. Data deterioration and deviation from linearity is caused by incomplete illumination by the laser light and detector saturation with cells and colonies that are too large.

(iv) Particle concentrations encountered in aquatic samples also vary greatly. In eutrophic conditions phytoplankton blooms can reach densities of 10^6 cells ml^{-1}, which is similar to cell suspensions used in biomedical applications. High concentrations in the order of 10^7 particles ml^{-1} are typically reached by the smallest particles (i.e. viruses and sediment). In low nutrient oligotrophic waters, particle concentrations decrease, however, down to less than one individual per millilitre for the larger species. These low natural concentrations require high sample analysis rates, which are a problem at typical FCM sample rates of 0.2–2 µl s^{-1}.

(v) Instruments are usually set up to measure discrete samples, while continuous sampling in a flow stream is useful for many aquatic applications.

(vi) The passage through a flow cytometer may damage vulnerable cells and colonies. High intensity laser light may cause photosynthetic shock and electrical shock may be caused by sorting. Large fragile particles may be broken by the high fluid shear in small nozzles.

(vii) The length of many phytoplankton species may also cause artifacts. For instance, the duration of signals of long cells and colonies present in phytoplankton samples is often truncated by the design properties of the instrument's electronics.

(viii) Another artifact can be due to non-homogeneously distributed chlorophyll. When a distance between adjacent pigmented regions within a single particle (a large cell or a colony) exceeds the width of the laser focus, the fluorescence signal may drop below the triggering level, ending the processing of this particular particle prematurely. Large intercellular spaces within colonies occur in various species, for instance *Thalassiosira, Skeletonema* (both Bacillariophyta), and others; low pigmented heterocysts in filamentous cyanobacteria such as *Anabaena* may cause the same effect.

13.3.3
Instrument Modification and Specialized Cytometers

Various scientists have modified existing flow cytometers for use with phytoplankton samples, for instance to increase the sample flow rate and/or the dynamic range. Others have designed their own instruments. In 1983, Olson et al. (1983) described a flow cytometer for the analysis of fluorescence signals in phytoplankton consisting of an epifluorescence microscope, a photomultiplier, a "Coulter" size analyzer and an inexpensive quartz capillary flow system as the basic components. A low-cost portable flow cytometer with a wide (300 µm) nozzle was designed and built by Cunningham (1990a). Frankel et al. (1990) developed a high-sensitivity flow cytometer for studying picoplankton, suitable for shipboard FCM, and Hüller et al. (1991) reported on a macro flow planktometer for analysis of large marine plankton organisms (>100 µm). The analyses included electrical impedance volume, fluorescence and beam attenuation. A special instrument for measurements of field samples with a large particle size range (Optical Plankton Analyser, OPA) was reported by Dubelaar et al. (1989), Peeters et al. (1989), and Balfoort et al. (1992). A very wide range of particle sizes could be analyzed in one run, with a very low level of fluid shear, and a large sample flow.

Based on this instrument the "EurOPA" flow cytometer was developed (Dubelaar et al. 1995) with add-on features such as diffracted light pattern detection and analysis (Cunningham and Buonaccorsi 1992), pulse shape acquisition and analysis, imaging-in-flow (Wietzorrek et al. 1994), and more automated data analysis (Carr et al. 1994, Wilkins et al. 1996), particularly by artificial neural networks (Boddy 1994). Changing the approach to a compact and transportable design (Dubelaar and Gerritzen 2000; Dubelaar et al. 1999), the core technology was redesigned as a (commercially available) series based on a very small footprint instrument, available for various platforms such as a bench-top version (CytoSense) for static or mobile laboratories (ship) as well as a submersible version (CytoSub) for operation at depths of 200 m, and a moored (floating) version (CytoBuoy) placed inside a small spherical buoy with radio-transmission of data over line of sight distances (Fig. 13.2b–d). The particle size range extends from submicron pico-

Fig. 13.2 Instruments designed for aquatic flow cytometry. (a) Submersible Flow Cytobot (H. Sosik and R. Olson, Woods Hole Oceanographic Institution, Woods Hole, MA, USA). (b) Submersible CytoSub (CytoBuoy b.v., Nieuwerbrug, Netherlands) mounted on a rosette sampling frame. (c) Floating CytoBuoy (CytoBuoy b.v.), a flow cytometer inside a 90-cm spherical buoy for operation within line of sight (approximately 10 miles) from a land station. (d) Testing radio data transfer.

plankton to the large diatoms and their chains and filaments. The data format includes the digitized pulses (one-dimensional scanned particle shapes) for morphological analysis of these particles. Another submersible instrument was developed by Olson et al. (2003). This automated submersible flow cytometer called Flow Cytobot (Fig. 13.2a) has been deployed at coastal cabled observatory sites for analyzing pico- and nanophytoplankton at relatively shallow depths. A hybrid instrument, FlowCAM of Fluid Imaging Inc. (Edgecomb, ME, USA) acquires images of (20 μm and larger) cells in a sample stream passing a fan of laser light generating also light scatter and fluorescence signals (Sieracki et al.

1998). Different objectives can be used for different magnifications; the camera image is digitized and a sub-image containing the cell is stored in real time (Fig. 13.3a).

13.3.4
Sizing and Discrimination of Cells

Flow cytometers may provide several types of signal that can be used for sizing and characterization of measured particles. The most commonly used are forward light scatter, perpendicular (side) light scatter and fluorescence. Less often used are excitation light beam attenuation and the "time of flight". The time of flight is the time a particle needs to cross the laser beam, which is proportional to the particle length (see also Chapter 2). The beam attenuation results from the removal of energy from the laser beam during passage of a particle. Although not a standard parameter, it can nonetheless be measured. Forward and side light scatter relate to size in different ways. The side scatter may yield the most straightforward relationship (proportional to particle cross-section) for particles > 1 µm upward, with low refractive index (Morel 1991). However, it is known to be very sensitive to internal cellular structures. Forward light scatter is proportional to cellular cross-section only for very large cells (>50 µm), and highly absorbing cells, and shows variable behavior with decreasing size. The emitted fluorescence is proportional to the number of absorbed photons and the fluorescence quantum yield of a fluorophore. The absorption efficiency may vary from linearity with small particle size to a constant value for large particles; therefore calibration is required in order to use fluorescence as a size indicator.

The ratio between various fluorescence signals may be a useful discrimination tool. The presence of specific accessory antenna pigments varies between the main taxonomic phytoplankton groups. Fluorescence emission and excitation characteristics are influenced by these pigments and can be used to classify phytoplankton populations. Whereas high-resolution spectrophotometric spectra refer to the bulk optical properties of a sample, FCM probes spectral properties of individual cells at high speed. Owing to the physical organization of the energy transfer between pigments, it is more efficient in terms of discriminatory power to analyze excitation spectra instead of emission spectra (Owens 1991). Unfortunately, there are seldom more than one or two laser wavelengths available to serve as an excitation source. On the emission side things are slightly better. The low duration and amplitude of the fluorescence signals of the individual particles allow only a crude spectral decomposition of the emitted fluorescence, and typically, three to five successive dichroic mirrors feeding different photomultiplier tubes (PMTs) are used. These excitation/emission data can be used to determine pigment ratios for each individual particle in a sample. These ratios are surprisingly narrow and stable for a healthy phytoplankton population and can be used to classify each individual into different taxonomic groups. Using three lasers as excitation light sources, Hofstraat et al. (1991) recognized cyanobacteria, cryptophytes, chlorophytes, and prasinophytes. Olson et al. (1989) used 488 and 515-nm

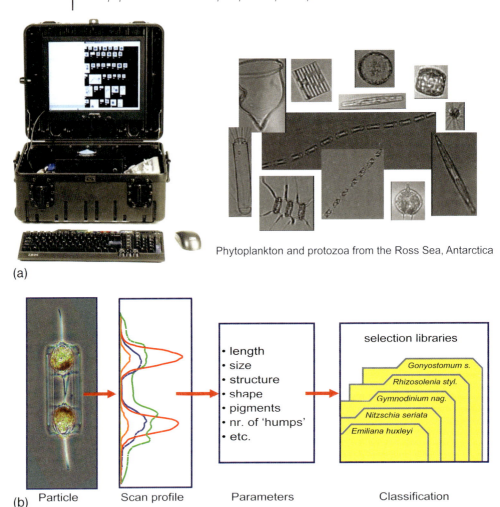

Fig. 13.3 (a) Example of data output of FlowCAM (from Kent Peterson, FlowCAM Technology: Digital Imaging Flow Cytometry for Oceanographic Research. Presentation at the ASLO/TOS Ocean Research 2004 Conference, February 2004, Honolulu, HI, USA; unpublished). (b) Principle of scanning flow cytometry (cytosense). The sample is injected into a particle-free sheath fluid that narrows down the suspension into a very thin line of fluid in which the particles are gently stretched out into a single file, moving at a fixed high speed exactly through the middle of a focused laser beam. The amplitude, length and shape of the scatter and fluorescence signals represent the size, length and shape of the particles, as well as the distribution of its "body parts" and chloroplasts along its length axis. With chain-forming diatom colonies, the single cells show up as "humps" in the scan. After parameterization, this morphological information is analyzed by standard cluster analysis.

excitation wavelengths and florescence detection at phycoerythrin and chlorophyll emission bands to distinguish cryptophytes, a rhodophyte, coccolithophorids, and chlorophytes in a mixture of 26 cultures. Similarly, a high side scatter to forward scatter ratio can be used to discriminate cells with high refractive index, such as the calcite coccoliths surrounding coccolithophores or the intracellular gas vacuoles in certain species.

13.3.5
More Information per Particle: From Single Properties to (Silico-) Imaging

Data analysis is a bottle neck in many applications involving FCM in aquatic sciences. In the biomedical field, the cytometrist is usually faced with only a few cell types to differentiate, whereas in aquatic situations the number of cell types is typically up to an order of magnitude greater. In addition, all these species appear in greatly varying concentrations, in different life cycle stages, in different states of aggregation and are typically accompanied by large numbers of debris and non-biological particles. With regard to assessing the microbial composition, it is necessary to count and identify as many groups and species as possible. As diversity increases, the number of measured particles per sample has to increase accordingly, to allow for clustering (requirement for subpopulation detection and analysis) of the less abundant species. This implies the generation of sets of measured data for tens of thousands of particles per sample. At the moment, autofluorescence at various excitation/emission wavelengths, together with measured light scattering properties are used to differentiate between species, keeping the data set per measured cell limited to only three to six numbers. With more detectors the amount of data per particle increases and combined with high numbers of particles to be analyzed, data overload may become a problem. One method to cope with this is artificial neural networks (ANNs). Boddy et al. (2000) presented an ANN-based analysis of 72 cultured species in a mixed sample with an overall identification rate of at least 70% using standard FCM data based on seven optical parameters. Wilkins et al. (1999) have shown a 92% success rate for 34 somewhat more distinct species measured with an 11-parameter flow cytometer.

FCM data provides less complex information compared to microscopy. However, cytometry can offer more detailed analysis of phytoplankton by deploying morphological and/or physiological analysis techniques such as forward light diffraction analysis, photosystem probing, pulse shape analysis (can be regarded as one-dimensional scanning or "silico-imaging"), and traditional imaging. The discriminatory potential depends on the number of independent properties measured for each particle, as well as the parameterization of these properties. The latter is also crucial for the overall applicability of the technology in terms of automated data analysis and the processing of large data sets.

An interesting development has been the "Pump-During-Probe" flow cytometer (Olson et al. 1996) in which the time course of chlorophyll fluorescence yield is measured during a 150-ms excitation flash provided by an argon ion laser. This provides estimates of the potential quantum yield of photochemistry and the

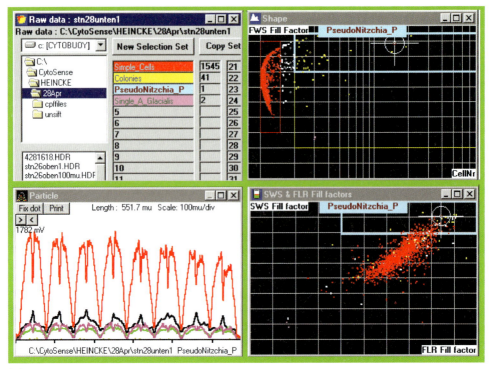

(a)

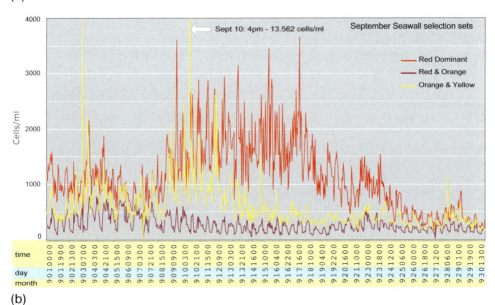

(b)

functional absorption cross-section for photosystem II, for either individuals (in large cells) or populations (in small cells).

Another approach, which is similar in data load but more straightforward in terms of instrumentation, is one-dimensional scanning FCM (also called "silico-imaging"). In FCM, signals coming from light detectors are normally quantified only by their maximum height and area. However, it is relatively easy to digitize and store the complete signal profiles in real time for morphological analysis (Fig. 13.3b). The information thus obtained relates to the shape of the particles in a format that allows efficient processing of large data sets (Cunningham 1990b; Dubelaar et al. 2000; Rutten et al. 2005). For example, the data cluster analysis application CytoClus operating with the five-parameter flow scanning cytometer CytoSense (CytoBuoy b.v., Netherlands), reduces each of the five digitized pulse profiles per particle to six parameters: the pulse length and average amplitude, the fill factor, the asymmetry, the inertia, and the number of "humps" in the profile, which is roughly the number of cells in a filamentous/chain forming colony. From these 30 (five profiles × six parameters) available parameterized values for each particle, only 14 have been used so far in the data processing software, allowing typically between 30 and 50 groups of particles in natural samples of fresh and marine waters to be distinguished.

The discrimination power is shown with data from a cruise on the North Sea, on the R/V Heincke (Alfred Wegener Institute (AWI), Germany) in April/May 2003. Samples were collected at various depths at successive stations. During night time, samples were analyzed at 10-min intervals from a sea water supply hose. Based upon selection criteria obtained from a data file of a culture of *Pseudo-nitzschia pungens* (Bacillariophyta), a set of measured data files was screened for the presence of *P. pungens*. This species was occasionally found in low numbers. An example of a single "matching result" is shown in Fig. 13.4a. Table 13.1 shows how the pulse profile information increases the discrimination power for the detection of *Pseudo-nitzschia* in 10 samples. Adding the length to the "classical" parameters (not shown separately) resulted in a reduced number of false positives, but only after screening with the full pulse shape parameter

Fig. 13.4 (a) Example of the CytoClus (data particle classification software, CytoBuoy b.v.) user interface. Only two of the seven dot-plots are shown (top and bottom right). Top left: the file directory structure. Bottom left: pulse data (one-dimensional "image") of the first particle found inside the selected category (*Pseudo-nitzschia pungens*). The toggle arrows allow browsing through the other particle profiles in the category, if present. FWS, forward scatter; SWS, sideward scatter; FLR, red fluorescence intensity. The fill factor characterizes the pulse profile generated by each particle (see Section 13.3.5 for details). (b) Flow cytometric analyses (with CytoSense, a bench-top flow cytometer particularly suitable for phytoplankton, CytoBuoy b.v.) on the seawall of San Francisco Bay, California, during September 2004. Counts of phytoplankton cells in the seawater are given for three groups with different pigment composition. (Courtesy of Professor R. Dugdale, Romberg Tiburon Centers of the San Francisco State University, CA, USA).

Table 13.1 CytoSense (CytoBuoy b.v., Nieuwerbrug, Netherlands) data from 10 stations on a North Sea cruise (on the Heincke, Alfred Wegener Institut, Bremerhaven, Germany, April/May 2003). *Pseudo-nitzschia* colonies were identified from the total number of measured fluorescent particles by using a selection set based on classical flow cytometry parameters first (pulse height and area), followed by adding the pulse length parameter only and then by adding more pulse shape parameters (see Section 13.3.5 for details).

Heincke cruise	Number of individuals		
	Fluorescent particles	*Pseudo-nitzschia* set	
File name		Classic properties plus length	Same set with pulse shape
Station 26	3396	74	1
Station 27 top	3877	28	0
Station 27 bottom	3187	120	1
Station 28 top	3243	53	0
Station 28 bottom	3148	41	1
Station 29 top	5321	83	4
Station 29 bottom	4480	60	2
Station 30 top	5181	29	0
Station 30 bottom	3935	30	0
Station 31 bottom	6766	8	0
Totals	42534	526	9

set did the number of positively identified *Pseudo-nitzschia* colonies match with reality. "Silico-imaging" (in contrast to video-imaging) runs at full flow cytometer particle throughput rate (up to 1000 or more particles per second) and full particle size range (submicron to millimetre), which allows accurate determination of particle assemblage composition and concentrations at high throughput rates.

To acquire video images of cells and store these alongside the more usual light scattering and fluorescence measurements is very helpful for identification of measured particles: clusters in data space are revealed. This was achieved with the EurOPA instrument (Dubelaar et al. 1995), which featured "imaging-in-flow" of pre-selected cells for identification of clusters found in data space. This allowed relatively fast identification of clusters, but it remains an interactive procedure in which the user examines a series of images (Rutten et al. 2005). A similar development is the FlowCAM (Fluid Imaging Inc.), which acquires video images of (20 μm and larger) cells in a sample stream passing a fan of laser light yielding light scatter and/or fluorescence signals serving as trigger source (Sieracki et al. 1998). The software allows grouping of the acquired images on the basis of basic geometrical properties with cell sizes being measured directly from the images.

13.4
Sampling: How, Where and When

13.4.1
Sample Preparation

The size of a sample and the number of steps involved in sub-sampling affect the significance of the results in a statistical manner. Whereas the typical FCM sample volume is only tens to hundreds of microlitres, the analyzed number of particles can be up to hundreds of thousands. The design of a sampling strategy therefore depends on the type of environment the samples are taken from and the relative concentration of the target particles. FCM samples in the biomedical laboratory typically have concentrations of about 10^6 particles ml^{-1}. These concentrations can often be attained by bacteria, particularly in freshwater systems and coastal waters, and are the norm for viruses in aquatic systems. On the other hand, phytoplankton only reach such high numbers in exceptional cases such as dense blooms. Most natural phytoplankton samples have lower concentrations of cells by two to three orders of magnitude. Small cells typically occur at higher concentrations, whereas the larger species may bloom at cell densities of only a few thousands per litre. One way to increase the number of cells counted is to enrich samples before the analysis. However, enrichment of samples by filtration or centrifugation inevitably leads to a change in the relative frequency of particles. For example, Hofstraat et al. (1990) examined the effect of simple filtration, tangential filtration and centrifugation on the length of *Skeletonema costatum* colonies. It was shown that damage to fragile particles, especially the largest colonies, was unavoidable. Centrifugation caused the least damage but was still not perfect.

In many aquatic applications, the analysis is carried out in the laboratory. Live samples quickly degrade, altering community composition, so it is generally necessary to preserve samples and transport them from the study site back to the laboratory. Many methods for the preservation of phytoplankton samples have been developed in the past, but none have been ideal or generally applicable. Formaldehyde and Lugol's iodine fixation modify cell shape and drastically affect fluorescence, respectively; ethanol fixation results in photosynthetic pigment extraction and therefore loss of cell autofluorescence. Low concentrations of glutaraldehyde and formaldehyde perform well for analysis of freshwater samples within about 7 days. However, this is not sufficient for many marine applications as research cruises often last for many weeks. Vaulot et al. (1989) described a method consisting of 1% (v/v) glutaraldehyde fixation followed by storage of the samples in liquid nitrogen. This method works well with picoplanktonic populations. However, a significant proportion of larger and more fragile cells are usually lost. A protocol, consisting of immediate fixation with 0.1 to 0.5% (w/v) formaldehyde and storage at 4 °C, was reported by Premazzi (1992) and tested using cultures of the dinoflagellate *Gymnodinium corii*. Size and chlorophyll autofluorescence of the cultures were preserved very well for up to 4 months. This study concerned only a single species and is not representative of heterogeneous phytoplankton communities. Generally, the ideal situation is to analyze fresh samples soon after collection without pre-concentration. Where this is not possible, the data obtained from stored samples should be interpreted with caution. Recently, a very simple protocol involving a surfactant Pluronic F-68 has been developed, which allows picoplankton sample enrichment for FCM. This protocol can be used for the least represented group of picoeukaryotes (down to 10^2 cell ml^{-1}) avoiding any cell loss or cell damage (Biegala et al. 2003).

13.4.2
Critical Scales and Sampling Frequency

"Critical scales" are the temporal and/or spatial scales at which data must be collected in order to resolve patterns and processes. If the sampling is not done at critical scales, the fundamental patterns of distribution of organisms in the water environment remain obscure as well as the processes which control the distributions and the dynamics over a broad range of temporal and spatial scales (Donaghay 2004). Taking bottle samples for plankton counting and taxonomy can be considered as taking a snapshot in time/space of the continually fluctuating state of the ecosystem. It is tempting to interpret the resulting data point as if it represents a realistic (average) value for that time period/area. The reliability of such an approach is inversely proportional to the "under-sampling" of the actual ecosystem variability. Therefore, ideally the relevant ecosystem variability (the critical temporal scale) should be known before deciding on a sampling regime.

In aquatic ecosystems, water movement and tidal currents may generate significant fluctuations with time constants of minutes to hours. It is well known that small microbes may double their numbers in a few hours. Consequently, in both stable and/or well-mixed situations significant changes may appear within one to a few days, depending on environmental factors such as sunlight, temperature and ecological factors such as grazing and viral attack. Sampling frequency, therefore, is an important consideration when undertaking any form of temporal or spatial study. For example, daily FCM analysis of over 30 phytoplankton groups and taxa in surface water (Oude Rijn canal, The Netherlands) showed that certain species can bloom and begin to disappear again within a week, whereas others remained constant (Dubelaar et al. 2004). Higher frequency sampling can reveal even finer scale changes in plankton communities. Hourly FCM analysis of phytoplankton abundance of some major groups of phytoplankton was carried out from the seawall at the mouth of the San Francisco Bay during September 2004 (R. Dugdale et al., personal communication). Sampling was achieved by pumping seawater samples up to the flow cytometer. The data show strong hourly fluctuations governed by the tidal movement (Fig. 13.4b).

In water quality monitoring applications, conducted by government agencies and other regulatory bodies, sampling frequencies rarely exceed two per month. Time series of phytoplankton counts at higher sampling frequency are scarce. Li and Dickie (2001) reported a multi-year series of phytoplankton counts of the Bedford Basin near Dartmouth, Nova Scotia, with a weekly sampling frequency. This series shows considerable ecosystem fluctuations, which would have been missed almost entirely if the sampling frequency had been reduced to once or twice a month. Figure 13.5a shows the number of phytoplankton cells in the Bedford basin during 2000 (Li and Dickie 2001). Monthly sampling was simulated by phytoplankton analysis every first and every second week of the month respectively. The aliasing caused by under-sampling results in two significantly differing graphs (Fig. 13.5b) both representing the same data. On many occasions the critical scales are not met as the sampling frequency is a compromise between a need to obtain sufficient resolution to understand what is going on, and the feasibility and costs involved to conduct the study. FCM may play an important role in finding the critical scales and to aid in designing adequate sampling and measurement programs for research and monitoring of surface waters with regard to analysis of microbial assemblages.

13.4.3
Platforms for Aquatic Flow Cytometry

Potential applications of FCM are to analyze phytoplankton over wide spatial scales on moving platforms such as ferries and research vessels. There are two approaches for operation onboard ships: rugged bench-top instruments for operation onboard and submersible instruments for operation on the ship's winch. The bench-top instruments can be used interactively, measuring discrete samples taken at stations at various depths using a rosette water bottle sampler. Online autonomous sampling is the method of choice if a continuous seawater supply is available and the ship is on passage, effectively conducting transects, such as is usual with ferries and other ships of opportunity. A submersible flow cytometer can be lowered down the water column for real *in situ* analyses, for instance directly probing the deep chlorophyll maximum found at the bottom of the mixed layer in the open ocean. These "Thin Layers" – structures of the water column – were discovered by sampling coastal marine waters at "critical scales" and may only be a few centimeters to a meter in vertical depth, but in the order of kilometers in the horizontal extent. They have important implications for marine ecology, and for ocean optics and acoustics. In comparison to ship-operated profilers, bottom-up profilers are better suited for precision measurements (Donaghay 2004), such as those needed to locate layers of high abundance of certain species that may be as thin as 10 cm while spatially coherent over hundreds of meters. The bottom-up profilers consist of an anchored bottom station from which a slightly buoyant instrument package is reeled upwards while taking (semi-)continuous measurements.

Autonomous Underwater Vehicles (AUVs) are an emerging platform for analytical instruments to observe the marine environment over wide spatial scales. This will be of great value to researchers interested in fields such as oceanography, macro-ecology, and marine ecosystems. An example is AutoSub, a UK Government-funded AUV capable of collecting physical, biological, chemical, and geophysical data to depths of 6000 m and over transects of up to 8000 km (see insert in Fig. 13.8b). A CytoSub submersible flow cytometer was tested in the Autosub, cruising through a coccolithophore bloom SE of the Isles of Scilly

304 *13 Phytoplankton and their Analysis by Flow Cytometry*

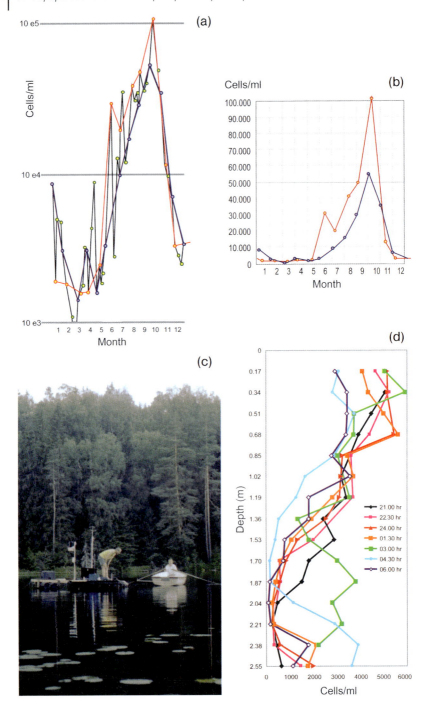

(white "cloud" in the satellite image). The coccolithophore cells of *Emiliana huxleyi*, although outnumbered by other smaller species dominate the water leaving back scattered light as seen by the satellite, owing to their strongly scattering extracellular calcite plates (liths). Because the various groups/species can be distinguished from each other by FCM, this enables the determination of not only the cell numbers but also the fluorescence (pigment) and light scattering properties per group. The data from the flow cytometer onboard Autosub confirmed that the coccolithophore cells were mostly causing the light reflection. The coccolithophore cell counts corresponded well with the turbidity and backscatter measurements (Cunningham et al. 2003).

Another mode in which specialized flow cytometers may be used is a real-time monitoring tool in a fixed *in situ* location. Special flow cytometers, configured for autonomous operation, may be placed on floating moored platforms provided there is electrical power, data transmission capability and instrumental options. Data transmission by satellite, whilst feasible, is very expensive due to the high data load. However, most applications are near-shore. This means large data sets can be easily transferred by line of sight radio communication. Besides the Cyto-Buoy (Fig. 13.2c), small scale experiments may also be carried out with a waterproof or even a splash-proof bench-top instrument mounted on a floating platform. An experiment was performed with a CytoSense instrument on a small raft in a shallow stratified Finnish lake (Fig. 13.5c, d). A series of depth profiles of 15 depths at 17-cm intervals was taken automatically using a hose system lowered with a fishing rod reel with the FCM measuring in an on-line mode. Each 1.5 h a complete cycle was performed over all depths. In this way it was possible to study the fine-scale phytoplankton composition as a function of the water depth at high frequency.

13.5
Monitoring Applications

13.5.1
Species Screening: Cultures

Many institutes and university departments in the field of marine, coastal and limnological research maintain culture collections of phytoplankton species and

---

Fig. 13.5 Weekly phytoplankton cell counts obtained using flow cytometry in the Bedford Basin, Canada, Atlantic Coast, from January to December 2001 (based on data of Li, 2001). (a) Black line represents original data, the red and blue lines represent sampling every first and second week of the month, respectively. (b) The resulting graphs on a linear scale. (c) CytoSense (CytoBuoy b.v.) being transported to the platform, Lake Lammi, Finland, July 2002; and (d) Series of successive depth profiles for one of the ca. 20 groups of particles/organisms that were identifiable in this particular data set.

strains. The conditions of the cultures are often not well known. FCM is a powerful diagnostic tool for rapid screening of the status of these cultures. Information on cell concentration, cellular properties and their statistics such as the mean and variance of fluorescence (pigment status) and light scatter (size, shape, physiology) can be directly obtained. Contamination with cells of a different species or bacteria can easily be detected and quantified. Figure 13.6 shows the result of an analysis of a sample containing *Pyramimonas grossii* (Prasinophytae). Various scat-

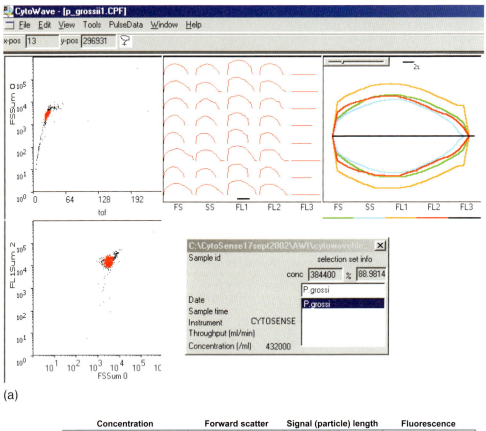

(a)

(b)

	Concentration			Forward scatter		Signal (particle) length		Fluorescence	
SelectionSet name:	all particles #/ml	selection set %	#/ml	mean value (a.u.)	Coef. Var. %	mean value µm	Coef. Var. %	mean value (a.u.)	Coef. Var. %
P. grossi	432000	89	384400	3018	18	7	9	16030	26

Fig. 13.6 Flow cytometry data from a culture of *Pyramimonas grossii*. (a) This example (software: CytoWave, CytoBuoy b.v.) shows two scatter plots in which the *P. grossii* cluster is highlighted in red. The insert shows the concentration. The pulses overview (top middle) shows that the culture is very uniform, containing only single cells. In the pulse detail view (top right) it is possible to scroll through pulses from individual cells. (b) Statistical output for the data.

ter plots can be chosen using available measured entities such as scatter or fluorescence. The corresponding statistics can be directly presented using a standard spreadsheet program.

Phytoplankton species forming large colonies pose problems to conventional microscopic and automatic particle analysis techniques and most flow cytometers also have strict upper particle size limits that are too small for many phytoplankton colony formers. Instruments for large size ranges that collect images or scans of particles may be used to study the population dynamics of colony formers, filamentous algae and chain-forming diatoms. The length of particle pulse profiles as measured with a scanning cytometer is directly related to the particle's length. The number of cells in a chain can be obtained by Fourier analysis of the pulse profiles. Scanning flow cytometers are also useful for the analysis of species forming non-linear colonies of more or less identical cells. The size of fluorescence distributions can be normalized to that of a mean single cell even for more or less amorphous aggregations such as the cyanobacterium *Microcystis aeruginosa* (Dubelaar et al. 1995). The kinetics of (dis)aggregation processes can also be monitored, even rapid changes, if the sampling frequency is sufficient.

13.5.2
Phytoplankton Species Biodiversity

Screening for specific target species in natural waters can be done well with FCM if these species are distinct enough to distinguish them from the non-target counterparts and their concentration is sufficient to have a significant count after a reasonable time of sampling. Although most commercially available flow cytometers are not suited to measuring large volumes of sample fluid, some instruments were designed for a high flow rate. With appropriate automation and real-time data reduction, such instruments may operate over longer periods of time without operator intervention to process larger sample volumes while looking for "target" species, either in the laboratory or *in situ*. What remains is the uniqueness of the target particle as measured by the flow cytometer.

Currently the most promising tools to assess both aspects of biodiversity, i.e. species richness and abundance of species employ molecular methods (Biegala 2003; Collier 2000; Not et al. 2004; Romari and Vaulot 2004; Simon et al. 1994; Vaulot et al. 2004). The most direct approach is to develop a species-specific agent to which a fluorescent label can be attached. Thus, Vrieling and Anderson (1996) and Vrieling et al. (1995, 1996, 1997) showed that antisera against purified cell walls and against extruded trichocystal cores of the organism allowed immunofluorescent detection of the dinoflagellates *Prorocentrum micans*, *Gyrodinium aureolum* and *Gymnodinium nagasakiense*. An even more promising strategy involves the use specific oligonucleotides (Jonker et al. 2000; Simon et al. 1997). These can be used as primers for quantitative PCR (polymerase chain reaction), dot blot hybridization, and whole cell (i.e. *in situ*) hybridization. Each technique possesses its advantages and drawbacks, but among them fluorescence *in situ* hybridization of oligonucleotide probes (FISH) is the most straightforward and easiest to use. Once fluorescently tagged, cells can be enumerated by different methods.

Epifluorescence microscopy is the most commonly used method, confocal microscopy allows detection of cells associated with particles, and FCM allows a rapid enumeration of a large number of cells. The sorting capacities of new flow cytometers have led to the acquisition of more phylogenetic information on a fluorescently-tagged population to assess their detailed species richness.

13.5.3
Harmful Algal Blooms (HABs)

Blooms of toxic algae have been commonly called "red tides", due to the fact that blooms are often caused by dinoflagellates whose pigments tint the water with a reddish color. The scientific community now uses the term "harmful algal bloom" or HAB. HABs are caused by a diverse group of organisms with serious impacts for humans and coastal ecosystems, including the dinoflagellates *Alexandrium tamarense* (paralytic shellfish poisoning), *Dinophysis* (diarrhetic shellfish poisoning), *Pfiesteria piscicida* (kills fish at mid-Atlantic latitudes), *Karenia brevis* (= *Gymnodinium breve*; neurotoxic shellfish poisoning), diatoms such as *Pseudo-nitzschia* sp. (amnesiac shellfish poisoning) and *Chaetoceros* sp. (kills fish), pelagophytes such as *Aureococcus anophageffarens* and *Aureoumbra lagunensis* (brown tides), and cyanobacteria such as *Anabaena*, *Aphanizomenon*, *Microcystis*, and *Synechococcus elongatus* (harmful cyanobacterial blooms). Prevention of HABs is unlikely, although "control" seems feasible in the future. Mitigation as an effective strategy is well established. A rapid and reliable method for the specific detection of harmful algal strains is still badly needed. Molecular probes to external and internal cell features (as previously described in Section 13.5.2) are one way forward. Currently, probes can be used in conjunction with FCM in laboratory-based operations. Whereas standard instruments can be used as interactive discrete sampling devices, some "aquatic" cytometers can be deployed as continuous, automated *in situ* systems. These instruments discriminate the target species on the basis of their morphology, which may be specific for certain species (such as *Pseudo-nitzschia*) but less specific for others. A CytoSense analysis of a series of cultures from the collection of the AWI showed that the recognition of several types of Dinophyceae was very specific and almost no false positives were generated by the other cultures (Table 13.2).

13.6
Ecological Applications

13.6.1
Population-related Processes

Whilst monitoring and surveying programs involving quantification and classification of phytoplankton is generally targeted at specific species of interest, fundamental research in aquatic systems also employs flow cytometers to classify and

Table 13.2 The results of analysis of some cultures from the AWI Bremerhaven measured with a CytoSense flow cytometer. Narrow selection sets were determined for *Alexandrium minutum, A. catanella* and *Gymnodinium nagasakiense*. The vertical columns show how many individuals from other cultures match those in the selection sets (false positives).

Cultures from AWI		Sample volume (ml)	Total number of particles	Number of particles found matching the selection criteria		
Taxon designation	Code			*Alexandrium minutum* AL1T	*A. catanella* BAHME255	*Gymnodinium nagasakiense* PLY561
Synura uvella	CCMP870	0.015	1393	0	0	4
Cyclotella caspia		0.014	1042	0	0	0
Cyclotella cryptica		0.013	1316	0	0	0
Cyclotella meneghiniana		0.016	1234	0	0	2
Cyclotella sp.	1435	0.024	1585	0	0	0
Alexandrium catenella	BAHME255	0.276	675	9	97	0
Alexandrium minutum	AL1T	0.095	607	325	0	0
Alexandrium ostenfeldii	LF37	0.89	475	0	1	0
Alexandrium tamarense	GTPP01	0.18	938	3	0	7
Alexandrium tamarense	SNZB01	0.644	667	19	19	0
Alexandrium taylori	AY2T	0.854	1061	0	0	0
Gymnodinium fuscum	CCMP1677	0.463	338	0	0	0
Gymnodinium nagasakiense	PLY561	0.131	545	0	0	315

Table 13.2 (continued)

Cultures from AWI		Sample volume (ml)	Total number of particles	Number of particles found matching the selection criteria		
Taxon designation	Code			*Alexandrium minutum* AL1T	*A. catanella* BAHME255	*Gymnodinium nagasakiense* PLY561
Gymnodinium nagasakiense	GymNagas0204	0.051	695	0	0	219
Gymnodinium nagasakiense	GymNagas0403	0.249	834	0	0	86
Gymnodinium nagasakiense	GymNagas0204	0.085	667	1	0	246
Gymnodinium varians	CCMP421	0.165	1393	0	0	0
Gyrodinium aureolum	K0303	0.436	891	5	0	2
Heterocapsa triquetra	CCMP448	0.097	1093	0	0	0
Synedra sp.		0.01	1222	0	0	1

quantify phytoplankton, as well as bacteria and, more recently, viruses. These basic abundance data are used to quantify the standing stocks of phytoplankton in aquatic ecosystems from lakes and rivers to ocean basins (Tarran et al. 2006). FCM can also be used to study bloom development and succession by monitoring changes in phytoplankton standing stocks (Burkill et al. 2002; Tarran et al. 2001), and to study controlling factors in ecosystem function, such as grazing and virus infection.

For example, phytoplankton can, through their autofluorescent characteristics, be thought of as tracer particles in much the same way as fluorescent microspheres when used in particle uptake experiments. Phytoplankton assemblages have been used to assess grazing rates, particle selectivity, and endocytotic abilities in various marine species, from single-celled organisms to higher invertebrates (Cucci et al. 1989). For instance, Christaki et al. (1999) compared the consumption of two picoplankters *Synechococcus* and *Prochlorococcus* by an algivorous ciliate, *Strombidium sulcatum*, and a bactivorous ciliate, *Uronema* sp., using FCM. Jonker et al. (1995) studied grazing by *Daphnia* on size classes of phytoplankton in a freshwater lake. If the grazing organisms are not too big they themselves can be analyzed by FCM to quantify the total number of ingested particles per individual directly, the amount of ingested fluorescence being proportional to the number of prey items grazed, at least during the initial ingestion phases.

FCM can be used to study other aquatic microbial processes thanks to the capability of certain instruments to sort populations from samples and conduct rate-related studies (Li 1994). The major processes associated with plankton involve the cycling of elements such as carbon and sulfur, nitrogen, phosphorus, and iron (Mills et al. 2004). Abundance data can be used to determine the contribution of particular groups to total primary production. FCM sorting can also be used to directly measure group-specific rates of primary production and nutrient uptake in picoplanktonic populations (Li 1994; Zubkov et al. 2003). Certain phytoplankton groups are responsible for the emission of approximately 20–50 million tonnes of sulfur into the atmosphere annually through their production of sulfur compounds like dimethylsulphoniopropionate (DMSP). FCM can be used to quantify the cellular concentrations of DMSP in plankton and can then help in our understanding of the passage and transformation of these compounds through the marine food-web (Archer et al. 2001; Burkill et al. 2002). Nutrient limitation and utilization by phytoplankton and their effect on other processes can also be better understood using FCM to quantify specific components of the plankton community and to investigate their response to different nutrient regimes, both in terms of abundance and changes in optical properties.

13.6.2
Cell-related Processes and Functioning

There are various ways of addressing the internal physiology and the health status of phytoplankton cells using FCM. In the case of mixed communities, differences in the physiology of the co-occurring species (inter-specific variations) can be used to explain species dominance and/or succession. However, since FCM measures each single cell separately, it also offers the possibility of studying differences within a single population (intra-specific variation). In particular, the last point addresses a classical concept that a population of cells would be uniform in their response.

Insight into the cell physiology assists us in understanding species and ecosystem dynamics. The most simple of all physiological responses is the autofluorescence signal of chlorophyll *a* in phytoplankton. The pigment properties respond not only to changes in light conditions but also to nutrients as well as to trace metals such as iron. Cell cycle analysis after the measurement of DNA content (see also Chapter 14) has been successfully used to assess *in situ* growth rates of a single species. Recently, Veldhuis and Wassman (2005) applied the method to different subpopulations of the colony forming phytoplankter *Phaeocystis* (Prymnesiophyceae).

Besides probing the cell cycle by measuring cellular DNA content with fluorescent DNA stains, the mean cell size and diel variations in light scattering proper-

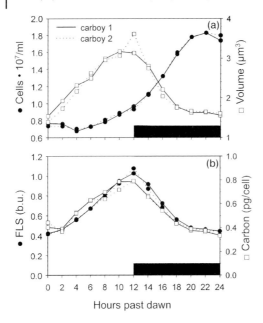

Fig. 13.7 Time series of (a) cell concentration and cell volume, and (b) flow cytometric forward light scatter (FLS, bead units) and carbon per cell for replicate carboys of *Micromonas pusilla* cultures. The black bar denotes when the lights were off in the incubator (12–24 h past dawn).

ties can be followed (for the analysis of DNA content see also Chapter 12). Du-Rand et al. (2002) conducted experiments with mono-cultures of *Micromonas* (Prasinophytae) and found that the cells increased in size and carbon content during the light period and then decreased in size with cell division during the dark period. FCM forward and side light scattering followed the same diel pattern, as did cross sections for attenuation, scattering and absorption (Fig. 13.7). The refractive index, calculated using the anomalous diffraction approximation, did not show any significant trend with the light/dark cycle. Since the single-cell measurements of forward light scattering were strongly correlated with independent measurements of cell volume and attenuation cross section, this data set could help provide calibrations for the use of FCM measurements of similar phytoplankton made at sea, to estimate cell size and growth rates and contributions to bulk optical properties such as beam attenuation.

The use of fluorescent oligonucleotide probes targeting the small subunit of ribosomal RNA (18S or 16S rRNA) could provide population-specific proxies of growth rate in the natural environment. Biegala et al. (2003) demonstrated on different picoeukaryote cultures that the fluorescence of cells tagged with probes changed significantly from exponential to stationary phase. In the natural environment similar changes of fluorescence were observed within specifically tagged picoplanktonic populations. More research needs to be done in that area to develop such species-specific growth rate proxies.

Another approach is to address the general physiological status of cells by loading them with a non-fluorescent substrate which, after enzymatic hydrolysis, results in a bright fluorescent product. It is, however, crucial that the end-products after cleavage be retained inside the cell, which is not always the case. Also, the status of cell membranes appears to be a useful tool to determine the viability of cells. Veldhuis et al. (2001) determined the viability of different phytoplankton species based on a staining procedure using the nucleic acid dye SYTOX Green. The assay is based on the particular characteristic of this stain that it can only penetrate cells with a compromised plasma membrane but cannot penetrate membranes of living cells (Roth et al. 1997). Thus the DNA of viable phytoplankton cells will not stain whereas the DNA of non-viable cells will show a bright green fluorescence.

Many cellular process studies have investigated phytoplankton cultures as well as field samples. During the past few years several of these FCM–cell physiology assays have been used in nutrient limitation studies, for toxicity tests, to determine the role of viruses and even in ships' ballast water research. These studies have altered our perceptions about the apparent homogeneity of populations, not only in the field but even in algal mono-cultures.

FCM systems may play an important role in the assessment of phenomena studied in the field of biological oceanography as well as ecophysiology of plankton. Ecosystem dynamics are driven by the interplay between the dynamics of the surface ocean mixed layer and the depth of light penetration. A "cascade of turbulence phenomena" leads to a range of mixing scales with photosynthesis, depending on irradiance, photo-inhibition and photo-adaptation, temperature, nutrients, phenotypic, and genotypic variance. FCM data can be included in a larger data set alongside environmental parameters in order to estimate eco-physiological key processes in the upper water column. A continuous and relevant process with respect to photosynthesis is light acclimation of phytoplankton. FCM can be used to study photo-acclimation with respect to light variations at different scales of time and space. *In situ* studies, as well as controlled environments (laboratory and *in situ*-simulated environments), were reported by Brunet et al. (2003), in relation to vertical water movements, and by Dusenberry et al. (2001). Other topics studied are mesoscale features of phytoplankton and planktonic bacteria in coastal areas induced by external water masses (Casotti et al. 2000). Dignum et al. (2004) developed a straightforward method to determine phosphate availability for individual algae. This technique was successfully applied in 2001 in the Loosdrecht Lakes (the Netherlands) and uses the response that phytoplankton cells themselves show when there is a lack of phosphate, by conversion of specific enzyme activity into a localized fluorescent signal. The fluorescence of individual algae is measured on a flow cytometer. The full benefit of such diagnostic tools requires integration of field data with experimental assessment of relevant population characteristics, and prediction of the complex interactions between algae and their environment. Different types of lakes are expected to show seasonal and local variations in the nutrient status.

13.6.3
Plankton Abundance Patterns in the Sea: Indicators of Change

FCM is well suited to the examination of multiscale patterns in plankton ecology. Since measurements are easily taken at short time scales and at small spatial scales, and since environmental monitoring programs can be sustained for long time periods over large geographic regions, possibilities exist to scale microbial interactions to regional and global phenomena (Li et al. 2006a, 2006b). Long-term studies of phytoplankton abundance and diversity indicate that directional changes have occurred on both large (Richardson and Schoeman 2004) and small spatial scales (Ribera d'Alcalà et al. 2004) in many places. Inherently, biological populations undergo strong natural fluctuations. The ability to discern real change from natural variability depends on a time series of relevant measurements made at an appropriate frequency for an extended period of time over a wide area. Because FCM is designed for rapid quantitative screening of fluorescence and size-related characteristics of single cells, it becomes feasible to map selected aquatic microbes at high resolution in both space and time, limited only by the number of water samples recovered in hydrographic surveys. Yet it remains true that no map of a plankton variable derived from shipboard surveys can capture all the variance at relevant space and time scales. However, statistical examination of extremely large data sets provided by FCM can reveal general phenomenological patterns that indicate intrinsic, evolutionary or extrinsic constraints on variation. In large data sets, thousands of individual contingent case histories can be subsumed to provide a detail-free holistic view (Li 2002). Patterns of abundance for cells of different sizes examined in relation to attributes such as autotrophic biomass, primary production and water column stability indicate the possible reactions that phytoplankton communities may adopt in the face of environmental change.

13.7
Marine Optics and Flow Cytometry

In situ cytometer systems open up new possibilities for establishing a link between measurements of single particle optics and bulk inherent optical properties, based on assessing populations of marine particles. When flow cytometers were first introduced to the marine sciences, instruments were taken to sea in dedicated container laboratories with separate industrial cooling sections (Tarran and Burkill 1992). However, the introduction of air-cooled argon lasers soon led to compact bench-top designs, and it was anticipated that FCM would make a major contribution to our understanding of the nature and dynamics of particle suspensions in the sea (Demers 1991). The potential value of the technique was demonstrated by the detection of ubiquitous and previously unknown populations of marine prochlorophytes (Chisolm et al. 1988). Early studies of single-cell phytoplankton optics included estimates of absorption and cross-sections (Perry and Porter 1989), refractive indices (Spinrad and Brown 1986), forward scattering patterns (Cunningham and Buonaccorsi 1992), and anomalous scattering from cyanobacteria (Dubelaar et al. 1987). However, from the point of view of marine optics, much of the research momentum seems to have been lost in recent years.

The few research groups currently carrying out FCM at sea are concentrating on problems in microbiology (Sieracki et al. 1995), picoplankton physiology (Boelen et al. 2000), and grazing by microheterotrophs (Zubkov et al. 2000). Recent attempts to draw up a database for the optical properties of marine particles (Stramski and Mobley 1997) show that there is an urgent need for information on natural particle suspensions. The importance of establishing a link between measurements of single particle optics and bulk inherent optical properties was first discussed by Morel (1991), but remarkably little progress has been made since. We anticipate that the availability of *in situ* FCM systems will open up new possibilities for studying populations of marine particles *in situ*, and there are a number of identifiable topics of current ecological interest which would benefit from the technique. These include detecting thin layers of heterotrophic activity which are believed to exist close to major pycnoclines, monitoring the physiological status of sinking algal blooms, investigating variations in diatom chain length under different physical conditions (e.g. turbulence regimes), and counting phytoplankton cells in waters which bear a heavy burden of other suspended particles. An example of the potential use of *in situ* flow cytometers has already been described earlier in this chapter and involved a CytoSub flow cytometer, with additional marine optics instrumentation onboard an AUV, passing through a coccolithophore bloom. The data generated by CytoSub along the transect corresponded well with reference measurements from other marine optics instrumentation (Fig. 13.8).

The main potential contribution to ocean optics is the possibility of accurately characterizing mixed particle assemblages in terms of the numerical proportions of different classes of the materials and of the distribution of sizes within each class. This information should provide new insight into variations in volume scattering functions, particularly in coastal waters. Wider application of FCM to the determination of single particle optical characteristics may help advance our understanding of the characteristics of mixed particle suspensions and hence, in an interesting contrast of physical scales, aid in the interpretation of satellite measurements of remote sensing reflectance.

13.8
Future Perspectives

The hydrological and ecological composition of our surface waters is intrinsically patchy and dynamic. Therefore the determination of "critical scales", and deploy-

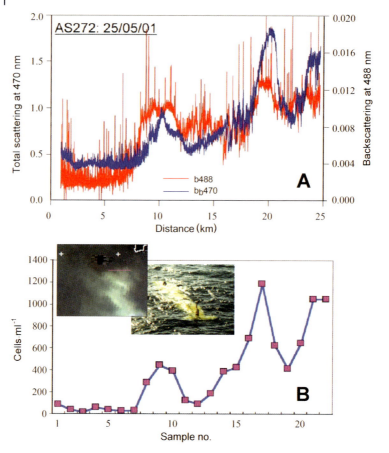

Fig. 13.8 Analysis of coccolithophore (Haptophyta) bloom SE of Isles of Scilly, SW England, 24 May 2001. (a) Ac9 (flow-through absorption attenuation meter, WET labs, Philomath, OR, USA) total scattering (left scale) and Hydroscat (multi-wavelength optical backscattering sensor, HOBI Labs, Tucson, AZ, USA) back-scattering (right scale) transects from Autosub (autonomous underwater vehicle, Southampton Oceanography Center and NERC). (b) *Emiliana huxleyi* cell counts from CytoSub (submersible flow cytometer, CytoBuoy b.v.) traveling in the Autosub submarine. Left insert, SeaWiFS (Sea-viewing Wide Field-of-view, NASA) RGB image of coccolithophore bloom (Remote Sensing Data Analysis Service, PML, UK), the purple line marking the Autosub transect. Right insert, the Autosub on a mission.

ing the technology to sample at these required scales often leads to the discovery of new kinds of patterns or phenomena. This information may lead us to abandon old ideas, ask new questions, and revolutionize scientific paradigms (Donaghay and Osborn 1997). Perhaps the most exciting perspective of the advent of modern, easy-to-use and especially the aquatic flow cytometers is our increasing ability to analyze microbial assemblages at their relevant critical scales in time and space. Integration with other technologies seems to be the way forward

(Babin et al. 2005). Further development of dedicated combinations of instrumentation, pre-processing protocols and data analysis algorithms as well as merging data from different scales from microscope to satellite will allow a shift in applications towards powerful solutions for regular monitoring and control situations as well as *protection and early warning* applications.

The potential applications are numerous, in which FCM can be used to detect phytoplankton and other organisms, such as bacteria in industrial and regulatory situations. Harmful algal blooms and pathogenic bacteria are a constant threat in the aquaculture industry, drinking water industry and for bathing and recreational waters. Regulations are becoming more stringent (e.g. EU Bathing Waters Directive, EU Shellfish Waters Directive) and regulatory authorities are looking for more rapid techniques to provide early warning of potential pollution incidents and also to monitor the disappearance of pollution so that the situation can be declared safe again. Flow cytometry, either as a stand-alone technique or in conjunction with other techniques, such as fluorescent species-specific probes, is well placed to provide rapid solutions for regulatory authorities. Autonomous and high frequency sampling may provide detailed information about community composition and abundance. This would make it easy to detect harmful species at pre-bloom concentrations and also to assess physiological states and viability. This information is crucial for making rapid decisions regarding harvesting aquaculture stock, take other appropriate measures to protect production or, in the case of bathing waters, informing the public about safety issues on beaches and around recreational lakes.

FCM could also be used to detect pollution by herbicides and other toxins (Readman et al. 2004) by analyzing their effects on phytoplankton abundance and cellular characteristics (Fig. 13.9). Using a sufficiently high sampling frequency enables the detection of sudden changes. For example, the fluorescence increase shown in Fig. 13.9b indicates an initial blocking of the photosystem followed by cell death. FCM combined with phytoplankton as bio-indicators could therefore be used to provide an "aqua-alarm" function.

Another very important potential use for FCM concerns monitoring ballast waters from ships for invasive marine species. Invasive marine species have been identified as one of the greatest threats to the world's oceans. These species enter into their new environments via ships' ballast water, attached to ships' hulls and via other vectors. Ships use ballast water to maintain their balance, allowing for differences in cargo weight and bunker fuel oil. To minimize the risks of ballast water transport, the International Maritime Organization (IMO) has adopted a convention for ballast water management on board ships with discharge standards for treated ballast water that should be complied with (Anonymous 2004). This can be done by ballast water exchange in the open sea or by using ballast water treatment installations. The standards include limits for discharge of phytoplankton-sized organisms, as well as pathogens. FCM technology (Veldhuis et al. 2006) can be used for rapid, online monitoring of the performance of ballast water treatment installations, including sizing, enumerating, and assessing the viability of the individual organisms.

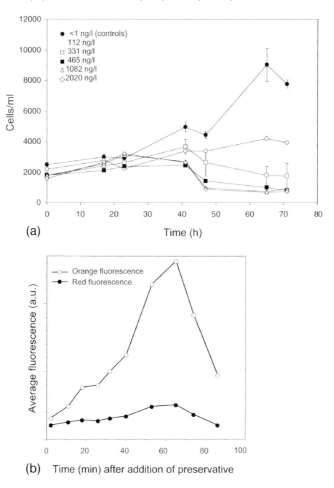

Fig. 13.9 (a) Cell numbers as measured by flow cytometry of eukaryotic phytoplankton exposed to different concentrations of Irgarol® 1051, a phytocide and photosynthesis blocker, during a 72-h period (from Readman et al. 2004). (b) Average autofluorescence of Irgarol®-treated *Chroomonas salina* cells in culture (Plymouth Marine Laboratory) measured by flow cytometry at 532 nm excitation.

Acknowledgments

We thank A. Cunningham (Strathclyde University, Glasgow, UK) and W. K. W. Li (Bedford Institute of Oceanography, Dartmouth, Nova Scotia, Canada) for contributing Sections 13.7 and 13.6.3 to this chapter, respectively. We thank the Alfred

Wegener Institute (Bremerhaven, Germany) for giving us the opportunity to measure cultures with the kind assistance of R. Groben and for hospitality on their research ship Heincke, and F. Colijn (Institute for Coastal Research, GKSS Geesthacht, Germany) for inviting us on their cruise. We also thank K. Salonen (University of Jyväskylä, Finland) for the opportunity to join the lake experiment at Lammi Biological Station. R. Dugdale (Tiburon Centre, San Francisco State University, California USA) and M. D. DuRand (Memorial University of Newfoundland, St. John's, Canada) kindly provided graphs for this chapter.

References

Anonymous **2004**, *The International Convention for the Control and Management of Ships' Ballast Water and Sediments (the Ballast Water Management Convention)*, International Maritime Organization.

Archer, S. D., Widdicombe, C. E., Tarran, G. A., Rees, A. P., Burkill, P. H. **2001**, *AME* 24, 225–241.

Babin, M., Cullen, J. J., Roesler, C. S., Donaghay, P. L., Doucette, G. J., Kahru, M., Lewis, M. R., Scholin, C. A., Sieracki, M. E., Sosik, H. M. **2005**, *Oceanography* 18, 210–227.

Balfoort, H. W., Berman, T., Maestrini, S. Y., Wenzel, A., Zohary, T. **1992**, *Hydrobiologia* 238, 89–97.

Baretta, J. W., Baretta-Bekker, J. G., Ruardij, P. **1998**, *ICES J. Marine Sci.* 55, 756–766.

Biegala, I. C., Not, F., Vaulot, D., Simon, N. **2003**, *Appl. Environ. Microbiol.* 69, 5519–5529.

Boddy, L., Morris, C. W., Wilkins, M. F., Tarran, G. A., Burkill, P. H. **1994**, *Cytometry* 15, 283–293.

Boddy, L., Morris, C. W., Wilkins, M. F., Al-Haddad, L., Tarran, G. A., Jonker, R. R., Burkill, P. H. **2000**, *Marine Ecol. – Prog. Series* 195, 47–59.

Boelen, P., De Beer, M. K., Kraay, G. W., Veldhuis, M. J. W., Buma, A. G. J. **2000**, *Marine Ecol. – Prog. Series* 193, 1–9.

Brunet, C., Casotti, R., Aronne, B., Vantrepotte, V. **2003**, *J. Plankton Res.* 25, 1413–1425.

Burkill, P. H., Archer, S. D., Robinson, C., Nightingale, P. N., Groom, S. B., Tarran, G. A., Zubkov, M. V. **2002**, *Deep-Sea Res. Part II – Top. Studies Oceanogr.* 49, 2863–2885.

Carr, M. R., Tarran, G. A., Burkill, P. H. **1994**. The application of multivariate statistical methods to the identification of phytoplankton from flow cytometric data: a EurOPA subproject in *OCEANS 94 – OSATES: Oceans Engineering for Today's Technology and Tomorrow's Preservation*, (vol. I), ed. G. B. Cannelli, E. D'Ottavi, IEEE, New York, pp. 570–562.

Casotti, R., Brunet, C., Aronne, B., Ribera d'Alcalà, M. **2000**, *Marine Ecol. – Prog. Series* 195, 15–27.

Chisolm, S. W., Olson, R. J., Zettler, E. R., Goericke, R., Waterbury, J. B., Welschmeyer, N. A. **1988**, *Nature* 334, 340–343.

Christaki, U., Jacquet, S., Dolan, J. R., Vaulot, D., Rassoulzadegan, F. **1999**, *Limnol. Oceanogr.* 44, 52–61.

Collier, J. L. **2000**, *J. Phycol.* 36, 628–644.

Cucci, T. L., Shumway, S. E., Brown, W. S., Newell, C. R. **1989**, *Cytometry* 10, 659–670.

Cunningham, A. **1990a**, *J. Plankton Res.* 12, 149–160.

Cunningham, A. **1990b**, *J. Microb. Methods* 11, 27–36.

Cunningham, A., Buonaccorsi, G. A. **1992**, *J. Phytoplankton Res.* 14, 223–234.

Cunningham, A., McKee, D., Craig, S., Tarran, G. A., Widdicombe, C. **2003**, *J. Marine Syst.* 43, 51–59.

Demers, S. (ed.) **1991**, *Particle Analysis in Oceanography*, NATO ASI Series, G: Ecological Sciences, (vol. 27), Springer-Verlag, Berlin, Heidelberg.

Dignum, M., Hoogveld, H. L., Matthijs, H. C. P., Laanbroek, H. J., Pel, R. **2004**, *FEMS Microbiol. Ecol.* 48, 29–38.

Donaghay, P. L. **2004**, Profiling systems for understanding the dynamics and impacts of thin layers of harmful algae in stratified coastal waters in *Proceedings of the 4th Irish Marine Biotoxin Science Workshop*, Marine Institute, Dublin, pp. 44–53.

Donaghay, P. L., Osborn, T. R. **1997**, *Limnol. Oceanogr.* 42, 1283–1296.

Dubelaar, G. B. J., Gerritzen, P. L. **2000**, *Scient. Marina* 64, 255–265.

Dubelaar, G. B. J., Visser, J. W. M., Donze, M. **1987**, *Cytometry* 8, 405–412.

Dubelaar, G. B. J., Groenewegen, A. C., Stokdijk, W., van den Engh, G. J., Visser, J. W. M. **1989**, *Cytometry* 10, 529–539.

Dubelaar, G. B. J., Cunningham, A., Groenewegen, A. C., Klijnstra, J., Boddy, L., Wilkins, M. F., Jonker, R. R., Ringelberg, J. **1995**, A European optical plankton analysis system: flow cytometer based technology for automated phytoplankton identification and quantification in *Marine Science and Technology, Second MAST days and EUROMAR Market. Project Reports*. (vol. 2), ed. M. Weydert, E. Lipiatou, R. Goñi, C. Fragakis, M. Bohle-Carbonell, K. G. Barthel, Commission of the European Community, Luxembourg, pp. 946–956.

Dubelaar, G. B. J., Gerritzen, P. L., Beeker, A. E. R., Jonker, R. R., Tangen, K. **1999**, *Cytometry* 37, 247–254.

Dubelaar, G. B. J., Paul, J. F., Jonker, G., Jonker, R. R. **2004**, *J. Environ. Monitor.* 6, 946–952.

DuRand, M. D., Green, R. E., Sosik, H. M., Olson, R. J. **2002**, *J. Phycol.* 38, 1132–1142.

Dusenberry, J. A., Olson, R. J., Chilsom, S. W. **2001**, *Deep-Sea Res. Part I – Oceanogr. Res. Papers* 48, 1443–1458.

Frankel, S. L., Binder, B. J., Chisholm, S. W., Shapiro, H. M. **1990**, *Limnol. Oceanogr.* 35, 1164–1169.

Gasol, J. M., Del Giorgio, P. A. **2000**, *Scient. Marina* 64, 197–224.

Giovannoni, S. J., Stingl, U. **2005**, *Nature* 437, 343–348.

Hofstraat, J. W., van Zeijl, W. J. M., Peeters, J. C. H., Peperzak, L., Dubelaar, G. B. J. **1990**, Flow cytometry and other optical methods for characterization and quantification of phytoplankton in seawater in *Environment and Pollution Measurements Sensors and Systems, S.P.I.E. Proceedings 1269*, ed. H. O. Nielsen, International Society for Optical Engineering, Bellingham, WA, pp. 116–133.

Hofstraat, J. W., de Vreeze, M. E. J., van Zeijl, W. J. M., Peperzak, L., Peeters, J. C. H., Balfoort, H. W. **1991**, *J. Fluor.* 1, 249–265.

Hüller, R., Schmidlechner, S., Gloßner, E., Schaub, S., Kachel, V. **1991**, *Cytometry Suppl.* 5, 53.

Jonker, R. R., Meulemans, J. T., Dubelaar, G. B. J., Wilkins, M. F., Ringelberg, J. **1995**, *Water Sci. Technol.* 32, 177–182.

Jonker, R. R., Groben, R., Tarran, G. A., Medlin, L., Wilkins, M., Garcia, L., Zabala, L., Boddy, L. **2000**, *Scient. Marina* 64, 225–234.

Li, W. K. W. **1994**, *Limnol. Oceanogr.* 39, 169–175.

Li, W. K. W. **1995**, *Marine Ecol. – Prog. Series* 122, 1–8.

Li, W. K. W. **2002**, *Nature* 419, 154–157.

Li, W. K. W., Dickie, P. M. **2001**, *Cytometry*, 44, 236–246.

Li, W. K. W., Harrison, W. G., Head, E. J. H. **2006a**, *Science* 311, 1157–1160.

Li, W. K. W., Harrison, W. G., Head, E. J. H. **2006b**, *Proc. Roy. Soc. B – Biol. Sci.* 273, 1953–1960.

Mackas, D. L., Denman, K. L., Abbot, M. R. **1985**, *Bull. Marine Sci.* 37, 652–674.

Mills, M. M., Ridame, C., Davey, M. S., La Roche, J., Geider, R. J. **2004**, *Nature* 429, 292–294.

Morel, A. **1991**, Optics of marine particles and marine optics in *Particle Analysis in Oceanography*, ed. S. Demers, NATO ASI Series, G: Ecological Sciences, (vol. 27), Springer-Verlag, Berlin, Heidelberg, pp. 141–188.

Not, F., Latasa, M., Marie, D., Cariou, T., Vaulot, D., Simon, N. **2004**, *Appl. Environ. Microbiol.* 70, 4064–4072.

Olson, R. J., Frankel, S. L., Chisholm, S. W. **1983**, *J. Exp. Marine Biol. Ecol.* 68, 129–144.

Olson, R. J., Zettler, E. R., Anderson, O. K. **1989**, *Cytometry* 10, 636–644.

Olson, R. J., Chekalyuk, A. M., Sosik, H. M. **1996**, *Limnol. Oceanogr.* 41, 1253–1263.

Olson, R. J., Shalapyonok, A., Sosik, H. M. **2003**, *Deep-Sea Res. Part I – Oceanogr. Res. Papers* 50, 301–315.

Owens, T. G. **1991**, Energy transformation and fluorescence in photosynthesis in *Particle Analysis in Oceanography*, ed. S. Demers, NATO ASI Series, G: Ecological Sciences, (vol. 27), Springer-Verlag, Berlin, Heidelberg, pp. 101–140.

Paasche, E., Throndsen, J. **1970**, *Nytt Mag. Bot.* 17, 209–212.

Peeters, J. C. H., Dubelaar, G. B. J., Ringelberg, J., Visser, J. W. M. **1989**, *Cytometry* 10, 522–528.

Perry, M. J., Porter, S. M. **1989**, *Limnol. Oceanogr.* 34, 727–1738.

Platt, T. **1989**, *Cytometry* 10, 500.

Premazzi, G., Bertona, F., Binda, S., Bowe, G., Rodari, E. **1992**, *Application of Innovative Methods for Phytoplankton Analysis*, EUR 14806, European Communities – JRC.

Readman, J. W., Devilla, R. A., Tarran, G. A., Llewellyn, C. A., Fileman, T. W., Easton, A., Burkill, P. H., Mantoura, R. F. C. **2004**, *Marine Environ. Res.* 58, 353–358.

Ribera d'Alcalà, M., Conversano, F., Corato, F., Licandro, P., Mangoni, O., Marino, D., Mazzocchi, M. G., Modigh, M., Montresor, M., Nardella, M., Saggiomo, V., Sarno, D., Zingone, A. **2004**, *Scient. Marina* 68 (Suppl. 1), 65–83.

Richardson, A. J., Schoeman, D. S. **2004**, *Science* 305, 1609–1612.

Romari, K., Vaulot, D. **2004**, *Limnol. Oceanogr.* 49, 784–798.

Roth, B. L., Poot, M., Yue, S. T., Millard, P. J. **1997**, *Appl. Environ. Microbiol.* 63, 2421–2431.

Rutten, T. P. A., Sandee, B., Hofman, A. R. T. **2005**, *Cytometry* 64A, 16–26.

Sieracki, C. K., Sieracki, M. E., Yentsch, C. S. **1998**, *Marine Ecol. – Prog. Series* 168, 285–296.

Sieracki, M. E., Haugen, E. M., Cucci, T. L. **1995**, *Deep-Sea Res. Part I – Oceanogr. Res. Papers* 42, 1399–1410.

Simon, N., Barlow, R. G., Marie, D., Partensky, F., Vaulot, D. **1994**, *J. Phycol.* 30, 922–935.

Simon, N., Brenner, J., Edvardsen, B., Medlin, L. K. **1997**, *Eur. J. Phycol.* 32, 393–401.

Smayda, T. J. **1998**, *ICES J. Marine Sci.* 55, 562–573.

Spinrad, R. W., Brown, F. J. **1986**, *Appl. Optics* 25, 1930–1934.

Stramski, D., Mobley, C. D. **1997**, *Limnol. Oceanogr.* 42, 538–549.

Tarran, G. A., Burkill, P. H. **1992**, Flow cytometry at sea in *Flow Cytometry in Microbiology*, ed. D. Lloyd, Springer-Verlag, London, pp. 143–158.

Tarran, G. A., Zubkov, M. V., Sleigh, M. A., Burkill, P. H., Yallop, M. **2001**, *Deep-Sea Res. Part II – Top. Studies Oceanogr.* 48, 963–985.

Tarran, G. A., Heywood, J. L., Zubkov, M. V. **2006**, *Deep-Sea Res. Part II – Top. Studies Oceanogr.* 53, 1516–1529.

Urbach, E., Robertson, D. L., Chisholm, S. W. **1992**, *Nature* 335, 267–270.

Utermohl, H. **1958**, *Mitt. Int. Vereinig. Limnol.* 9, 1–38.

Vaulot, D., Courties, C., Partensky, F. **1989**, *Cytometry* 10, 629–636.

Vaulot, D., Le Gall, F., Marie, D., Guillou, L., Partensky, F. **2004**, *Nova Hedwigia* 79, 49–70.

Veldhuis, M. J. W., Wassmann, P. **2005**, *Harmful Algae* 4, 805–809.

Veldhuis, M. J. W., Kraay, G. W., Timmermans, K. R. **2001**, *Eur. J. Phycol.* 36, 167–177.

Veldhuis, M. J. W., Fuhr, F., Boon, J. P., ten Hallers-Tjabbes, C. C. **2006**, *Environ. Technol.* 27, 909–921.

Vrieling, E. G., Anderson, D. M. **1996**, *J. Phycol.* 32, 1–16.

Vrieling, E. G., Gieskes, W. W. C., Rademaker, T. W. M., Vriezekolk, G., Peperzak, L., Veenhuis, M. **1995**, Flow cytometric identification of the ichthyotoxic dinoflagellate *Gyrodinium aureolum* in the central North Sea in *Harmful Marine Algal Blooms*, ed. P. Lassus, G. Arzul, E. Erard, P. Gentien, C. Marcaillou, Lavoisier Science Publishers, Paris, pp. 743–748.

Vrieling, E. G., Vriezekolk, G., Gieskes, W. W. C., Veenhuis, M., Harder, W. **1996**, *J. Plankton Res.* 18, 1503–1512.

Vrieling, E. G., Van de Poll, W. H., Vriezekolk, G., Gieskes, W. W. C. **1997**, *J. Sea Res.* 37, 91–100.

Waterbury, J. B., Watson, S. W., Guillard, R. R., Brand, L. E. **1979**, *Nature* 277, 293–294.

Wietzorrek, J., Stadler, M., Kachel, V. **1994**, *Proc. Oceans* 1, 688–693.

Wilkins, M. F., Boddy, L., Morris, C. W., Jonker, R. R. **1996**, *CABIOS* 12, 9–18.

Wilkins, M. F., Boddy, L., Morris, C. W., Jonker, R. R. **1999**, *Appl. Environ. Microbiol.* 65, 4404–4410.

Zubkov, M. V., Sleigh, M. A., Burkill, P. H., Leakey, R. J. G. **2000**, *J. Plankton Res.* 22, 685–711.

Zubkov, M. V., Fuchs, B. M., Tarran, G. A., Burkill, P. H., Amann, R. **2003**, *Appl. Environ. Microbiol.* 69, 1299–1304.

14
Cell Cycle Analysis in Plants

Martin Pfosser, Zoltan Magyar, and Laszlo Bögre

Overview

The cell division cycle is a fundamental process, and flow cytometry methods provide an important tool for rapid measurement of cellular DNA contents, and thus to determine the distribution of cell cycle phases and DNA ploidy levels. In this chapter we provide a historical account of how flow cytometry has played a role in plant cell cycle research. We describe synchronization methods for plant cell cycle and summarize approaches that have been devised to measure cell cycle parameters in developing organs. Expression of fluorescent proteins is revolutionizing the field allowing live cell imaging, and thus the *in vivo* monitoring of cell divisions within organs, as well as the use of flow cytometry and cell sorting methods to determine molecular changes and DNA ploidy levels in specific cell types.

14.1
Introduction

The cell cycle is the universal process by which cells duplicate their mass and their DNA content, segregate chromosomes and divide into daughter cells, and thus underlies the growth and development of all living organisms. Most of the basic molecular mechanisms that control the cell cycle are conserved among all eukaryotes and therefore apply to yeast as well as to animal and plant cells. Investigators of the plant cell cycle have profited from results obtained with animal and yeast cells and vice versa. Homologs of the yeast and animal cell cycle regulators have been found to exist in plants with similar functions as their yeast and animal counterparts (Criqui and Genschik 2002; De Veylder et al. 2003; Dewitte and Murray 2003; Gutierrez et al. 2002; Menges et al. 2005). Interest in cell cycle research during recent decades have brought overwhelming insights into the molecular mechanisms regulating cell division but have also raised intriguing new questions for future research with regard to how the cell cycle is co-ordinated with growth and development in multicellular organisms.

Flow Cytometry with Plant Cells. Edited by Jaroslav Doležel, Johann Greilhuber, and Jan Suda
Copyright © 2007 WILEY-VCH Verlag GmbH & Co. KGaA, Weinheim
ISBN: 978-3-527-31487-4

Most of the results obtained in cell cycle research rely on the ability to measure cell cycle progression and to determine the position of cells within the cell cycle with high precision. Without a flow cytometer, this task was met predominantly by a time-consuming and cumbersome procedure employing radioisotope-labeled cells and autoradiography (Howard and Pelc 1951; Rogers 1973). Despite the fact that autoradiography is largely outdated today, such experiments resulted in the finding that DNA replication is discontinuous during the cell cycle (Swift 1950) and thus laid the basis for subdivision of the cell cycle into four major phases: presynthetic interphase or gap 1 (G_1 phase), DNA synthesis phase (S phase), postsynthetic interphase or gap 2 preceding mitosis (G_2 phase), and mitosis (M phase). All these phases can be identified by a distinct nuclear DNA content of the cells. We take here diplophasic (somatic) higher plant tissue for example (Fig. 14.1). During S phase the chromosomal DNA is doubled and therefore cells are characterized by a gradual increase in DNA content from the basic value which is referred to as the 2C level, to the 4C level (Swift 1950). During M phase, when sister chromatids separate in anaphase, the DNA content drops to the 2C level in the two newly formed daughter cells. For haplophasic tissues the DNA quantities cycle between the 1C and 2C levels.

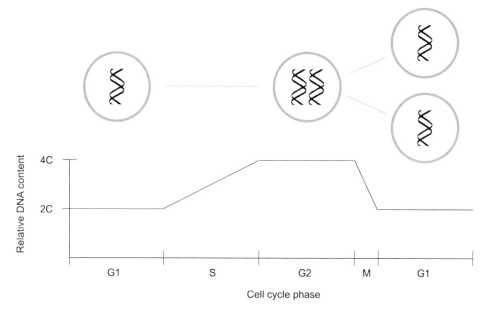

Fig. 14.1 Schematic representation of cell cycle phases and corresponding DNA content values in diplophasic higher plant cells. Proliferating cells undergo cell division following a series of cell cycle phases characterized by distinct DNA content values. During G_1 phase cells prepare for DNA replication and show 2C DNA content levels. S phase-cells are characterized by a gradual increase in DNA content from the basic level to the 4C level. During M phase the DNA content again drops to the 2C level in the two newly formed daughter cells.

This important finding provided the foundation for the development of complementary techniques to autoradiography such as microspectrophotometry and microfluorometry. These techniques were used for the quantitative measurement of DNA and other constituents in single cells and thus had the potential to determine their position within the cell cycle. However, the methods were also cumbersome and time consuming and cell cycle analysis was biased by prior visual selection of subsequently measured cells. The development of flow cytometers finally provided the investigator with the possibility of accurately and rapidly measuring cell constituents in a previously unknown manner. By using flow cytometry (FCM), cell selection is largely unbiased and it is possible to analyze thousands of cells instead of only a few cells as in static methods of DNA content measurements, making the analysis statistically much more significant. Protocols for cell cycle analysis by FCM have been developed for a variety of applications and the accuracy and speed of the analysis and available software packages to analyse the data (e.g. ModFit, Verify, Flomax) has made FCM the method of choice for monitoring cell cycle progression for almost all researchers working in this field (for recent examples see Citterio et al. 2006; Jang et al. 2005; Kadota et al. 2005a; Magyar et al. 2005; Menges and Murray 2002; Sano et al. 2006, and references therein).

14.2
Univariate Cell Cycle Analysis in Plant Cells

Unlike animal and human cells, plant cells do not lend themselves easily to FCM analysis because of the rigid cellulose cell wall, which prevents dissociation of plant tissues and cell cultures into single particle suspensions required for analysis. For cell cycle analysis, two methods are most often applied for transforming plant tissues and cell cultures into measurable particle suspensions: (i) isolation of nuclei by mechanical dissociation of tissues, and (ii) enzymatic digestion of cell walls and conversion of plant cells into protoplasts (Yanpaisan et al. 1999). A rapid and efficient method of preparing single particle suspensions is an even more critical issue for cell kinetic studies when the progression of cells through the cell cycle is to be analyzed. Consequently, and for a variety of plant cells, mechanical disruption and subsequent analysis of isolated nuclei has become the universal method of sample preparation for FCM cell cycle analysis (Galbraith et al. 1983; Gualberti et al. 1996).

A variety of fluorescent dyes was successfully employed to accurately measure the DNA content of cells by flow cytometers (for a review of DNA stains see Crissman and Hirons 1994; Darzynkiewicz et al. 2004; and Chapter 4). Although many fluorescent dyes are available for use in staining DNA in cells, the most common are the minor groove-binding, ultraviolet (UV)-excited and blue-fluorescing stain 4′,6-diamidino-2-phenylindole (DAPI), and the DNA intercalating, green-excited and red-fluorescing stain propidium iodide (PI). The suitability of a particular dye or staining procedure is evaluated in terms of the coefficient of variation (CV) of the G_1 peak of the DNA content histogram. Both dyes yield high resolution DNA content measurements and the choice is mainly determined by the type of excitation light source available. Lowest CV values for the G_1 peak have been reported by staining of DNA with DAPI (Doležel and Göhde 1995; Pfosser et al. 1995).

DNA content histograms of asynchronously cycling and synchronized cell populations have to be deconvoluted in order to determine phase durations or percentages of cells in specific phases of the cell cycle. Such algorithms usually employ the fitting of Gaussian distributions to model the G_1 and the G_2/M fluorescence peaks and either a polynomial or trapezoid function to model the S phase (Bagwell 1993; Dean and Jett 1974). Alternatively, a series of Gaussian distributions can be used to fit the S phase, which works particularly well with synchronized or disturbed cell populations (Fried et al. 1980).

14.3
BrdUrd Incorporation to Determine Cycling Populations

Univariate DNA content measurements as described above continue to be applied most frequently for rapid cell cycle analysis. If cell populations appear to be homogeneous with all cells cycling at equal rates, the proportion of cells in a particular cell cycle phase is directly related to the relative duration of that phase. However, this kind of direct relationship breaks down in heterogeneous cell populations (Naill and Roberts 2005), and therefore univariate DNA content measurements are not suited to detailed studies of cell cycle traverse rates and phase transition times (Lucretti et al. 1999; Sgorbati et al. 1991). Multiparametric analyses of DNA and RNA, or DNA, RNA and protein have been used to discriminate between cycling and quiescent cells both in animal and plant systems (Bergounioux et al. 1988; Darzynkiewicz et al. 1980a, 1980b). When analyzing nuclei of *Petunia hybrida* (Solanaceae) stained with acridine orange for simultaneous measurement of DNA and RNA, Bergounioux et al. (1988) showed that, as in animal cell systems, plant cells in G_1 with low RNA contents were unable to complete DNA replication. A limitation of staining cells with acridine orange is its cytotoxicity, which makes it impossible to monitor cell cycle progression after labeling. More widely applicable techniques based on the incorporation of the thymidine analog 5-bromo-2-deoxyuridine (BrdU) into DNA during replication and subsequent simultaneous detection of DNA and BrdU have been developed for animal cells (Dolbeare et al. 1983) and then transferred to plant systems (Lucretti et al. 1999; Moretti et al. 1992; Naill and Roberts 2005; Yanpaisan et al. 1999). Although different methodological approaches for discriminating between cycling and quiescent cells based on BrdU incorporation have been devised (see Lucretti et al. 1999), detection of incorporated BrdU is today almost exclusively achieved by employing monoclonal antibodies to BrdU (Fig. 14.2). By using the BrdU technique, Yanpaisan et al. (1998) were able to demonstrate that in a *Solanum aviculare* (Solanaceae) suspension culture with a dry weight doubling time of 2 days, the duration of the cell cycle of active cells was only 1 day and that the maximum

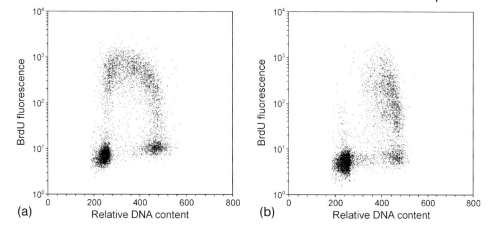

Fig. 14.2 Cell cycle analysis in *Vicia faba* (Fabaceae) root tips after BrdU incorporation. Roots were incubated with 30 µM BrdU solution at 25 °C for 1 h. After that roots were incubated in BrdU-free medium. Nuclei were isolated immediately (a) and 4 h (b) after BrdU removal. Incorporated BrdU was detected via indirect immunofluorescence. Nuclei were stained with the DNA-specific fluorescent dye DAPI to determine relative nuclear DNA contents. The movement of the BrdU-positive population from S into G_2/M phase of the cell cycle can be clearly seen on cytograms of BrdU/DNA content (J. Bartoš, unpublished data).

proportion of active cells never exceeded 52%. In addition to the detailed analysis of cell cycle kinetics, identification of non-cycling cells by bivariate DNA-BrdU analysis also has important applications in operation and control of bioreactors for large-scale culture of plant cells and production of secondary metabolites (Naill and Roberts 2005; Yanpaisan et al. 1999).

14.4 Cell Cycle Synchronization Methods: Analysis of Cell Cycle Transitions in Cultured Plant Cells

Plant cell suspension cultures provided the primary source of material for studying the molecular basis of cell cycle regulation in plants because of the ease of manipulation, and the relatively homogeneous population of cells in the absence of cell differentiation. Many of the tools used to study cell cycle progression, however, rely on the isolation and analysis of transcripts or proteins in sufficient amounts, which can only be accomplished by employing synchronized cells. Methods have been developed to synchronize plant cell cultures for specific cell cycle phases that firstly relied on the removal and subsequent re-supply of compounds required for growth, such as phosphate (Amino et al. 1983; Dahl et al. 1995), nitrate (King et al. 1973), hormones (Kodama et al. 1991; Nishida et al. 1992; Trehin et al. 1998) or sucrose (Riou-Khamlichi et al. 1999, 2000). Mimosine and anisomycin are drugs that inhibit protein synthesis, and both were used to reversibly arrest cell cycle progression in plant cells (Gonzales-Fernandes et al.

1974; Perennes et al. 1993). These drugs should arrest the cell cycle through the same pathway as nutrient limitation. There are multiple checkpoints in plants for cell growth and nutrient limitations, as was studied by following the synchronous cell divisions in onion root tips after pulse application of the transcriptional inhibitor alpha amanitin or the protein synthesis inhibitor anisomycin (De la Torre et al. 1974; Gonzales-Fernandes et al. 1974). The molecular mechanism of how nutrient or energy limitation is sensed by the cells and communicated to halt cell division is not well understood in plants, but based on studies in yeast it is thought that cells somehow measure the protein translation efficiency, and couple this to cell cycle progression (Ingram and Waites 2006; Jorgensen and Tyers 2004). An inherent problem with nutrient limitation is the reversibility, which greatly depends on the timing before cells enter autophagy. Alternatively, and most effectively, cell cycle progression can be synchronized by applying reversible blocks at different stages of the cell cycle using specific inhibitors (Binarová et al. 1998; Fukuda et al. 1994; Glab et al. 1994; Gould 1984; Ito et al. 1997; Magyar et al. 1993, Nagata et al. 1992, Perennes et al. 1993; Pfosser 1989; Planchais et al. 1997, 2000; Roudier et al. 2000).

The block of cell cycle in S phase relies on inhibition of DNA synthesis, which is most commonly achieved with hydroxyurea which inhibits the activity of ribonucleotide diphosphate reductase, thus depriving the cells of newly-synthesized deoxyribonucleotide triphosphates, or with aphidicolin which inhibits DNA polymerases α and δ (Sala et al. 1980). Many publications interpret the result of the DNA synthesis inhibitors as a block in G_1 to S phase transition. However, it is important to keep in mind that even if DNA synthesis is not fully blocked, cells are sensitive to the presence of unreplicated DNA, and activate checkpoints that halt cell cycle progression either at G_1 to S or at G_2 to M phase transitions to allow completion or repair of DNA synthesis. At which of the phases the checkpoint is activated might depend on the timing, duration and concentration of drug application. Some DNA replicates early in the cell cycle, and thus faulty replication at these forks triggers the G_1 to S phase block, while other segments of chromosomes replicate late during S phase, and thus halting these replication forks will trigger the G_2 to M checkpoint. The existence of different DNA synthesis checkpoints might also explain the very different synchronization results obtained, depending on whether the inhibitor is applied to logarithmically growing cultures, or to freshly diluted cultures from the stationary phase. Theoretically, for complete synchrony it would be necessary to halt the cell cycle for the duration of cell doubling time, but the considerations on checkpoint mechanisms highlight the need for optimizing every synchronization experiment for drug concentration and timing of application (Rouse and Jackson 2002). The frailty of checkpoints means that after a certain time they are lifted and cells might continue cell cycle progression with potentially unrepaired chromosomes (del Campo et al. 1997). Prolonged activation of DNA synthesis checkpoints can trigger apoptosis in animal cells, making the arrest irreversible and the cells non-viable (Rouse and Jackson 2002). Whether such mechanisms exist in plant cells is not known. The review by Planchais et al. (2000) collated the information on experimental conditions and drug concentrations.

Inhibitors of microtubule polymerization disrupt the mitotic spindle. Cells have a sensitive monitoring mechanism for the alignment of chromosomes at the metaphase plate by sensing the tension exerted through the pulling force of the bipolar spindle on the kinetochores. The lack of tension with faulty microtubules activates a mitotic checkpoint to block the splitting of chromosomes by blocking the anaphase-promoting complex, a proteolytic mechanism that dissolves the cohesion between sister chromatids at the meta- to anaphase transition (Nasmyth 2005). Colchicine is a plant alkaloid that functions as a microtubule drug, but plant microtubules are relatively insensitive to it. However, there are a number of herbicides that act through efficient and specific inhibition of microtubule polymerization, such as oryzalin, propyzamide and amiprosphos-methyl (APM) (Anthony et al. 1998). Among these drugs, propyzamide was found to be reversible, which allows its removal and thus the synchronous onset of meta- to anaphase transition, and progression through G_1 phase (Nagata and Kumagai 1999; Samuels et al. 1998). The metaphase checkpoint arrest is also frail, and therefore the timing and duration of drug application has a paramount influence on the successful accumulation of cells in mitosis, as evidenced by a high mitotic index. Because of the frailty caused by checkpoint adaptation, efficient synchronization methods combine the use of DNA synthesis inhibitors and microtubule drugs applied in succession at optimized timing (Doležel et al. 1999; Nagata and Kumagai 1999; Samuels et al. 1998). For experimental conditions and drug concentrations see Doležel et al. (1999) and Planchais et al. (2000).

As described above, the meta- to anaphase transition relies on the proteolytic cleavage of the cohesion glue between sister chromatids. Proteosome inhibitors, such as MG132, epoxomycin or lactacystein, block the onset of the proteolysis of cohesins and thus efficiently halt cells in metaphase without the disruption of microtubules (Genschik et al. 1998; Weingartner et al. 2004). Because there are many cellular processes that rely on the proteosome, the timing of drug application is critical for efficient metaphase arrest.

Eukaryotes have accumulated a vast amount of DNA and this imposes serious topological tasks on the cells to organize and segregate DNA to daughter cells. This is facilitated by different classes of topoisomerase enzymes that catalyze the decatenation of DNA. ICRF 193 is a drug that specifically blocks topoisomerase II. It was found that plant and animal cells have a checkpoint that senses DNA topology and blocks cell cycle progression in prophase (Gimenez-Abian et al. 2002a). Thus, ICRF 193 can be utilized to synchronize cells before they enter into mitosis.

Cell cycle transitions are regulated by the conserved cyclin-dependent protein kinase (CDK) complexes. Fairly specific ATP analog inhibitors have been developed for CDKs, such as olomoucine, roscovitine, and bohemine. In plants these inhibitors effectively block the A-type CDKs, that operate both at the G_1 to S and G_2 to M transitions (Binarová et al. 1998). Correspondingly, roscovitine was found to arrest the cell cycle at both transitions, which could be reversed to a certain degree (Binarová et al. 1998; Planchais et al. 1997). Prolonged roscovitine treatment was found to induce apoptosis in plant cells (P. Binarová and L. Bögre, unpublished results).

Table 14.1 Drugs and compounds known to affect plant cell cycle transitions.

Compound/agent	Target/mechanism	Cell cycle phase	Concentration	Reference
Phosphate starvation	Nutrient	G_1		Amino et al. 1983; Dahl et al. 1995
Nitrate starvation	Nutrient	G_1		King et al. 1973
Sucrose starvation	Nutrient	G_1		Menges and Murray 2002
Mimosine	Protein synthesis	G_1	200 µM	Perennes et al. 1993
Anisomycin	Protein synthesis	Multiple	4 µM	Gonzales-Fernandes et al. 1974
Alpha amanitin	Transcription	Multiple	11–54 µM	De la Torre et al. 1974
Auxin starvation	Growth hormone	G_1		Nishida et al. 1992
Lovastatin	Cytokinin synthesis	G_2/M	10 µM	Laureys et al. 1998
Jasmonic acid	Stress hormone	G_1, G_2	100 µM	Swiatek et al. 2002, 2004
Abscisic acid	Stress hormone	G_1	200 µM	Swiatek et al. 2002
Menadione	Oxidative stress	G_2/M	20–100 µM	Reichheld et al. 1999
Cryptogein	Elicitor	G_1, G_2, apoptosis	500 nM	Kadota et al. 2004, 2005b
Heat, H_2O_2, $KMnO_4$	Abiotic stress	G_1/S, G_2/M	30 °C, 0.5 mM, 100 µM	Jang et al. 2005; Sano et al. 2006
Indomethacin	cAMP production	G_1/S	28 µM	Ehsan et al. 1999
Hydroxyurea	Ribonucleotide diphosphate reductase	S	5 mM, 1.25–2.5 mM	Magyar et al. 1993; Doležel et al. 1999
Aphidicolin	DNA polymerases α and δ	S	15–30 µM	Nagata and Kumagai 1999; Menges 2002
Oryzalin	Microtubule	M	15 µM	Anthony et al. 1998
Propyzamide	Microtubule	M	3 µM	Samuels et al. 1998; Nagata and Kumagai 1999
Amiprosphos-methyl	Microtubule	M	2.5–10 µM	Doležel et al. 1999
MG132	Proteosome	M	100 µM	Genschik et al. 1998; Weingartner et al. 2004
Epoxomycin	Proteosome	M	1 µM	Weingartner et al. 2004

14.4 Cell Cycle Synchronization Methods: Analysis of Cell Cycle Transitions in Cultured Plant Cells

Table 14.1 (continued)

Compound/agent	Target/mechanism	Cell cycle phase	Concentration	Reference
Lactacystein	Proteosome	M	5 µM	Weingartner et al. 2004
ICRF 193	Topoisomerase II	G_2	18 µM	Gimenez-Abian et al. 2002a
Olomoucine	CDKA[a]	G_1, G_2	50–200 µM	Binarová et al. 1998
Roscovitine	CDKA[a]	G_1, G_2	50–100 µM	Planchais et al. 1997; Binarová et al. 1998
Bohemine	CDKA[a]	G_1, G_2	50–100 µM	Binarová et al. 1998
Staurosporine	Protein kinases	G_2	20 µM	Katsuta and Shibaoka 1992
K252-a	Protein kinases	Pre-prophase	2 µM	Katsuta and Shibaoka 1992; Weingartner et al. 2001
Okadaic acid	PP2A[b]	M	0.1 µM	Weingartner et al. 2003
Endothal	PP2A[b]	M	1 µM	Ayaydin et al. 2000
Caffeine	ATM checkpoint kinase	M	0.75 mM	Pelayo et al. 2001; Weingartner et al. 2003

[a] Cyclin-dependent kinase A.
[b] Protein phosphatase 2A.

Entry into mitosis relies on the timely activation of CDK, which is regulated by activating dephosphorylation and inhibitory phosphorylation events. General protein kinase inhibitors, such as staurosporine and K252-a, were found to arrest the cycle before entering into mitosis when applied to aphidicolin-synchronized cells in G_2 phase. K252-a arrests cells in G_2 phase and cells with pre-prophase band microtubules accumulate as a result (Katsuta and Shibaoka 1992; Weingartner et al. 2001).

Conversely, okadaic acid, a protein phosphatase 2A (PP2A) inhibitor, induces premature mitosis by interfering with the inactivating CDK phosphorylation (Weingartner et al. 2003). Endothal, another PP2A-specific inhibitor, was also found to prematurely activate a plant mitotic CDK, resulting in microtubule and chromosome condensation abnormalities (Ayaydin et al. 2000). Caffeine is another drug that acts on the checkpoint-induced regulation of CDK activity in G_2 phase by inhibiting the ATM kinase, which works upstream of the two checkpoint kinases, Chk1 and Chk2 (Rouse and Jackson 2002). The caffeine target, ATM kinase, is conserved in plants, and caffeine was shown to override the DNA synthesis checkpoint, causing cells to unduly enter into mitosis (Pelayo et al. 2001; Weingartner et al. 2003). For a list of compounds known to affect plant cell cycle transition see Table 14.1.

Table 14.2 Cell culture, protoplast and root tip experimental systems for cell cycle synchronization and cell cycle studies.

Species	Material	Synchronizing agent(s)	References
Tobacco (*Nicotiana tabacum*)	Cell line BY-2	Aphidicolin	Nagata et al. 1992
		Propyzamide	Samuels et al. 1998; Nagata and Kumagai 1999
		Roscovitine	Planchais et al. 1997
		Auxin starvation	Chen et al. 2001; Magyar et al. 2005; Laureys et al. 1998
Tobacco (*Nicotiana tabacum*)	Cell line cv. Virginia Bright Italia-0	Auxin starvation	Campanoni and Nick 2005
Arabidopsis (*Arabidopsis thaliana*)	Cell line M02	Aphidicolin	Menges and Murray 2002
		Sucrose starvation	Menges and Murray 2002
Alfalfa (*Medicago sativa*)	Cell suspension cultures	Hydroxyurea	Magyar et al. 1993
		Aphidicolin	Bögre et al. 1997, Magyar et al. 1997
		Phosphate starvation	Dahl et al. 1995
		Aphidicolin, propyzamide	Bögre et al. 1999
Madagascar periwinkle (*Catharanthus roseus*)	Cell suspension cultures	Aphidicolin	Ito et al. 1997
		Phosphate starvation	Amino et al. 1983
		Auxin	Nishida et al. 1992
Acer (*Acer* sp.)	Cell suspension cultures	Nitrate starvation	King et al. 1974
Soybean (*Glycine max*)	Cell suspension cultures	Cytokinin starvation	Mader and Hanke 1996
Carrot (*Daucus carota*)	Cell suspension cultures	Auxin	Lloyd 1999
Parsley (*Petroselinum crispum*)	Cell suspension cultures	Elicitors	Logemann et al. 1995
Petunia (*Petunia hybrida*)	Mesophyll protoplasts	Auxin, cytokinin	Trehin et al. 1998
Alfalfa (*Medicago sativa*)	Cell suspension protoplasts	Auxin	Pasternak et al. 2002
Tobacco (*Nicotiana tabacum*)	Mesophyll protoplasts	Cell wall inhibitor	Galbraith et al. 1981
Zinnia (*Zinnia elegans*)	Mesophyll cells	Auxin, cytokinin	McCann et al. 2001

Table 14.2 *(continued)*

Species	Material	Synchronizing agent(s)	References
Field bean (*Vicia faba*)	Root tips	Hydroxyurea, APM	Doležel et al. 1992
		Roscovitine	Binarová et al. 1998
Pea (*Pisum sativum*)	Root tips	Hydroxyurea, APM	Doležel et al. 1999
Alfalfa (*Medicago sativa*)			
Barley (*Hordeum vulgare*)			
Rye (*Secale cereale*)			
Wheat (*Triticum aestivum*)			
Corn (*Zea mays*)			
Onion (*Allium cepa*)	Root tips	Hydroxyurea	Pelayo et al. 2001

Only a few plant cell lines can be synchronized to a high degree (Table 14.2). The tobacco BY-2 cell line is widely used in cell cycle studies and is considered to be the plant equivalent of HeLa cells (Nagata et al. 1992; Samuels et al. 1998). Using the drug aphidicolin, BY-2 cells can be arrested in early S phase, resulting in an S-phase synchrony of 90% after release from the aphidicolin block and in a mitotic index of 45–50% (Nagata and Kumagai 1999; Sorrell et al. 2001). A double cell cycle block was also realized by subsequent applications and release from the S-phase blocker, aphidicolin, and the mitotic blocker, propyzamide. This method allowed up to 90% metaphase-arrested cells to be obtained, which after the removal of propyzamide, synchronously went through cytokinesis (Nagata and Kumagai 1999; Samuels et al. 1998). The BY-2 experimental system was used to study a wide variety of cell cycle processes, including phase-specific gene expression (Combettes et al. 1999; Ito et al. 2001; Reichheld et al. 1996; Sorrell et al. 1999), CDK activities (Porceddu et al. 2001; Sorrell et al. 2001), microtubule rearrangements (Hasezawa and Nagata 1991), and cell plate formation (Samuels et al. 1995).

The rapid accumulation of experimental tools and molecular resources with the completed *Arabidopsis thaliana* (Brassicaceae) genome sequence offers vast opportunities, such as genome-wide studies of cell cycle gene regulation (Menges et al. 2002, 2003, 2005), and proteomics studies (Peck 2005). For a long time, cell cycle research in this model plant was hampered by the lack of established cell culture and synchronization methods. This had been superbly overcome with the establishment of a fast growing *Arabidopsis* culture that can be readily synchronized by aphidicolin arrest or sucrose starvation (Menges and Murray 2002). The syn-

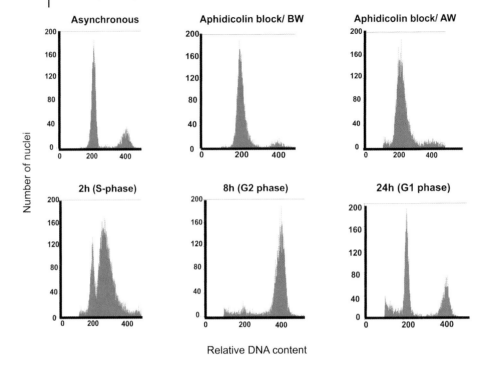

Fig. 14.3 Cell cycle synchronization of *Arabidopsis* cell culture using aphidicolin. The *Arabidopsis* cell line Ler (May and Leaver 1993) was synchronized by aphidicolin arrest and release essentially as described by Menges et al. (2002). Histograms of nuclear DNA content were obtained after flow cytometric analysis of Ler cultures growing asynchronously (Asynchronous), after 18 h arrest with 0.5 mg l^{-1} aphidicolin (Aphidicolin block/BW), immediately after the removal of aphidicolin (Aphidicolin block/AW), and 2, 8 and 24 h after the release from aphidicolin that resulted in synchronous transit of the majority of cells though the S, G_2 and G_1 phases of the cell cycle (Z. Magyar and L. Bögre, unpublished data).

chrony after release from aphidicolin arrest is around 95% in S phase, but the transition through mitosis becomes much less synchronous (Fig. 14.3). We have found that light-exposed *Arabidopsis* cultures lose the synchrony because of the arrest before the entry into mitosis more than cultures grown in the dark (Z. Magyar and L. Bögre, unpublished results).

Moreover, methods have been developed for stable *Agrobacterium*-mediated transformation of this *Arabidopsis* culture, and for a simple inexpensive cryopreservation to maintain transformed cell lines (Menges and Murray 2004). This *Arabidopsis* cell culture requires both auxin (naphthyl acetic acid) and cytokinin (kinetin) for its growth, and is thus ideal for hormone studies. The culture also retained chloroplasts that become fully green in light, and thus studies of chloroplast biology in a homogeneous cell suspension system are possible. Methods have also been developed for the transient transformation and expression of genes using *Agrobacterium* transfection (Koroleva et al. 2005) and polyethylene glycol-mediated transformation of protoplasts (De Sutter et al. 2005).

Further synchronizable systems worth mentioning include cultures of *Acer* (Aceraceae; King et al. 1974), *Catharanthus roseus* (Apocynaceae; Amino et al. 1983; Nishida et al. 1992), alfalfa (*Medicago sativa*, Fabaceae; Magyar et al. 1993), and soybean (*Glycine max*, Fabaceae; Mader and Hanke 1996).

14.5
Plant Protoplasts to Study the Cell Cycle

Tobacco leaf protoplasts were one of the very first cell cycle models that were pursued by FCM (cf. Chapter 10). Placed in culture, tobacco mesophyll protoplasts initiated cell wall formation, entered into the cell division cycle, began DNA synthesis within 30 h and over a period of 2 days produced clusters of undifferentiated cells (Galbraith et al. 1981). Interestingly, blocking cell wall synthesis with the cellulose synthase inhibitor 2,6-dichlorobenzonitrile had little or no effect on the re-initiation of the cell cycle of leaf protoplasts in culture (Galbraith et al. 1981). Mesophyll protoplasts of *Petunia hybrida* display almost exclusively 2C DNA content and enter the mitotic cell cycle after a few hours when incubated in a medium supplemented with 2,4-dichlorophenoxyacetic acid (2,4-D) and N6-benzyl-adenine (BA). Both hormones together are required to pass the cell cycle control point of CDK activation, as characterized by inhibition of CDK activities using the drug roscovitine (Trehin et al. 1998).

One of the characteristics of plant development is that somatic cell differentiation is reversible. This can be best demonstrated in isolated protoplasts where somatic plant cells are stimulated to express their totipotency and form embryos through the developmental pathway of somatic embryogenesis, or through organogenesis. In alfalfa, protoplast-derived cells cultured at high 2,4-D concentrations were shown to develop into embryogenic structures (Dudits et al. 1991; Pasternak et al. 2002). The synergistic and antagonistic effect of auxin and stress factors on alfalfa protoplast division and somatic embryogenesis was characterized (Otvos et al. 2005; Pasternak et al. 2002).

14.6
Root Meristems for Cell Cycle Synchronization

The linear growth pattern of roots makes them an ideal model with which to study organ growth and cell divisions. The cell cycle duration was found to be highly constant in root meristems, while the proportion of cells actively dividing is highly influenced by environmental conditions (Baskin 2000). A protocol for cell cycle synchronization in root tips of *Vicia faba* (Fabaceae) was developed based on the use of the DNA synthesis inhibitor hydroxyurea (Doležel et al. 1992). FCM data indicated that about 90% of root tip cells were synchronized.

On average, mitotic indices exceeding 50% were obtained with this method. Synchronized cells may be accumulated at metaphase using a mitotic spindle inhibitor to achieve metaphase indices exceeding 50%. Modifications of the protocol for *Medicago sativa*, *Pisum sativum* (both Fabaceae), *Hordeum vulgare*, *Secale cereale*, *Triticum aestivum*, and *Zea mays* (all Poaceae) were shown to result in reproducible high cell cycle synchrony (Doležel et al. 1999). Methods have also been developed for cell cycle synchronization of onion root tip cells with extreme efficiency and reproducibility, providing a model to study plant cell division cycles (De la Torre et al. 1989). The extreme high synchrony allowed binucleate and tetranucleate cells to be produced by timed inhibition of cytokinesis through pulsed treatment with caffeine during mitosis (Gimenez-Abian et al. 2002a).

14.7
Study of Cell Cycle Regulation by using Synchronized Cell Cultures and Flow Cytometry

As mentioned above, progression through the cell cycle is driven by conserved heterodimeric kinases comprising regulatory subunits, designated as cyclins, and catalytic subunits known as cyclin-dependent kinases (CDKs). Plants possess different classes of CDKs and cyclins. A-type CDKs contain a conserved PSTAIRE cyclin binding motif, and function throughout the cell cycle, similar to their yeast and mammalian counterparts. Correspondingly, A-type CDKs were found to be active throughout the cell cycle (Bögre et al. 1997; Magyar et al. 1997), and overexpression of a mutant kinase-inactive form of CDKA blocked cell cycle progression both at G_1 to S and at G_2 to M phases (Hemerly et al. 1995). B-type CDKs are plant specific and it was shown by cell synchronization experiments in cultured alfalfa cells that these were expressed and active in the G_2 phase of the cell cycle (Magyar et al. 1997). This is supported by the finding that overexpression of a kinase-inactive CDKB form blocks only the entry into mitosis (Porceddu et al. 2001). The expression of various D-type cyclins often depends on plant hormones, growth conditions, and development. Hormonal stimulation of cell cycle re-entry in leaves was followed by FCM, and was correlated with the expression of an A-type alfalfa cyclin, CycMs3, and a D-type cyclin, CycMs4 (Dahl et al. 1995; Meskiene et al. 1995). Both were found to precede the G_1 to S transition. Most A- and B-type cyclins have a cell cycle-dependent expression pattern, being restricted to the G_2 to M phases. The cell cycle phase-specific expression was initially demonstrated for an alfalfa B-type cyclin, CycMs2, using synchronized cell culture and FCM (Hirt et al. 1992). By making CycMs2 expression inducible, it was shown that the CycMs2 amounts are rate limiting for the cells to enter into mitosis. Forced expression of CycMs2 resulted in an advanced entry into mitosis in a synchronized tobacco cell culture (Weingartner et al. 2003). CycMs2 or its associated CDK also appears to be the target of checkpoint controls for DNA damage and DNA catenation, as shown by FCM measurements and observing cells in mitosis (Gimenez-Abian et al. 2002b). The activity of CDKs are tightly regulated by phosphorylation through the CDK-activating kinases (CAK), responsible for the phosphorylation of a conserved threonine in the T-loop around position 160, and by homologs of the fission yeast kinase wee1 and the cdc25 which carry out inhibitory phosphorylation and dephosphorylation, respectively at the Thr14Tyr15 positions. Furthermore, the activities of CDKs are influenced by the binding of small protein inhibitors, called CDK inhibitors (CKIs) or based on their homology to the animal KIP^{p27} protein, the KIP-related proteins (KRPs). It was found that overexpression of KRP1 and KRP2 blocks cell proliferation as well as endoreduplication, and severely retards plant growth (De Veylder et al. 2001). The activities of CDKA and CDKB complexes throughout the cell cycle were followed in synchronized cell cultures of alfalfa (Bögre et al. 1997; Magyar et al. 1997), *Arabidopsis* (Stals et al. 2000), and tobacco (Sorrell et al. 2001). Specific chemical inhibitors for CDKs, such as olomoucine, roscovitine and bohemine are most effective against the A-type CDKs in plants, and were shown by FCM to block cell cycle progression at both G_1 to S and G_2 to M transition points (Binarová et al. 1998; Planchais et al. 1997).

One of the major downstream targets for CDKs is the sequential phosphorylation of the retinoblastoma protein (RB) at multiple sites that results in the inactivation of RB and the release of active E2F-DP transcription factors, inducing a wave of transcriptional activity essential for the passage through S and M phases. Cell cycle-dependent phosphorylation of plant RB-related protein (RBR) by CDKs was also demonstrated (Nakagami et al. 2002). Mutation in *Arabidopsis thaliana* RBR1 is gametophytically lethal, producing megagametophytes with excessive nuclear proliferation, showing its function as a suppressor of proliferation by preventing the expression of genes necessary for DNA replication and mitosis (Ebel et al. 2004). Correspondingly, the virus-induced silencing of the tobacco (*Nicotiana tabacum*) RB homolog RBR1 led to prolonged cell proliferation and surprisingly also induced DNA endoreduplication in tobacco leaf cells (Park et al. 2005). However, it should be noted that this result relies on FCM measurements of DNA content, which cannot distinguish whether the DNA amount increased due to endoreduplication or endomitosis (mitotic restitution). Chromosome counts in mitotic cells or fluorescence *in situ* hybridization with chromosome-specific probes are required to distinguish between these two possibilities. The *Arabidopsis* genome encodes three E2F proteins, E2FA, E2FB, and E2FC, that form heterodimers with one of the two dimerization partner (DP) proteins, DPA or DPB (Vandepoele et al. 2002). The individual *Arabidopsis* E2Fs differ in their function. E2FA in conjunction with DPA promotes cell proliferation. E2FC is likely to be a repressor because it has a shortened C-terminal transactivation domain, its overexpression results in decreased expression of the S phase genes, and it inhibits cell division leading to enlarged cells (Inze 2005). It was found that in synchronized tobacco cell cultures followed by FCM measurements, the overexpression of E2FB promoted both the G_1 to S and G_2 to M phase transitions, leading to shortened cell cycle duration, and extremely small cell sizes (Magyar et al. 2005). E2FB protein accumulation is controlled by auxin, and elevated E2FB levels can render plant cell proliferation auxin-independent. Thus, E2FB might be one of the targets through which auxin could influence cell proliferation. Other mechanisms for cell cycle-dependent regulation of gene expression are known for mitotic cyclins (Ito et al. 1998) and histone genes (Shen and Gigot 1997), and were studied by using the highly synchronizable tobacco BY-2 cell culture (Nagata and Kumagai 1999).

The sequencing of the entire *Arabidopsis* genome has led to the cataloguing of conserved plant cell cycle regulators, revealing around 80 core cell cycle components (Menges et al. 2005; Vandepoele et al. 2002). The establishment of synchronization methods in *Arabidopsis* cell cultures and the availability of genome-wide gene expression profiling techniques opened the way for the study of global cell cycle regulation of genes in *Arabidopsis* (Hennig et al. 2003; Menges et al. 2002, 2003, 2005). This study not only allowed the cell cycle regulators to be grouped into functional categories, based on their expression, but also made possible the identification of genes that are co-regulated with cell cycle genes, or genes that are specifically expressed in certain time-windows during the cell cycle. For instance, it was established that 82 *Arabidopsis* genes share the G_2/M regulatory pattern, about half being new candidate mitotic genes with previously unknown function.

14.8
Cell Cycle and Plant Development

Plant growth is an area of considerable research interest because it has implications for crop production, for understanding evolutionary divergence in plant size and shape, and the adaptation of plants to changing environments. Growth of plant organs is generally due to the interplay of two processes: cell division and cell expansion. Plants cannot move and their growth and development reflects their need to adapt to the local environment. Thus, organogenesis in plants differs from that in animals in that it is mainly postembryonic. Many plant organs have the capacity for indeterminate growth and possess zones of proliferating cells, called meristems. Cells within meristems provide a continuous supply of cells which go through a series of divisions with strict size control, resulting in an increasing population of proliferating cells. We can now begin to understand the intricate genetic network that keeps cells within meristems in an undifferentiated proliferating state and allows them to differentiate when leaving the meristematic zone (Weigel and Jurgens 2002). When leaving the zone of meristematic division these cells exit the cell cycle and start to differentiate, a process that is typically accompanied by a massive increase in cell size, in some species often also by an increase in their DNA content through endoreduplication (see Chapter 15), the expansion of vacuoles and the loosening of cross-links between cell wall polymers (Sugimoto-Shirasu and Roberts 2003). The timing of the transition from proliferative growth to differentiation largely determines the output of cell numbers and thereby governs the growth rate of the organ as a whole (Beemster et al. 2003; Horiguchi et al. 2005). This is underpinned by the fact that differences in organ size among species tend to reflect cell number variations rather than variations in cell sizes (Mizukami 2001).

Furthermore, organ growth and morphogenesis in plants show a remarkable plasticity which allows adaptation to changing environmental conditions, including light, temperature, and nutrient status. For example, Bonzai cultivation practices lead to the most extreme forms of environmentally-induced plant dwarfism, with leaves becoming up to 50 times smaller, but remarkably the size of the constituent cells remains largely unaltered (Körner et al. 1989). All these examples underline the need to understand how cell proliferation is regulated in plants in order to understand how plants grow and develop. What mechanisms ensure the remarkable cell size homeostasis of plants? How is the production of cells matched to organ growth demands? How is cell proliferation regulated in meristems and what are the mechanisms that regulate the exit from cell cycle during differentiation, or allow cells to switch to an altered cell cycle, that is, endoreduplication? Which control points are used in plants for cell cycle arrest and how are these coupled with developmental and environmental signals?

To find the answer to these questions, we must develop methods that allow us to study cell cycle parameters in developing plant tissues. Combination of non-destructive imaging technologies, isolation techniques of cell and tissue types from organs and their molecular study (e.g. gene expression), and cell type-specific measurements of DNA content that might be combined with measurements of other parameters, such as BrdU incorporation or protein abundance, are the methods of the future.

14.9
Flow Cytometry of Dissected Tissues in Developmental Time Series

As outlined above, the most interesting and relevant information concerning cell cycle parameters is that gained during plant development. The difficulty in these experiments is to obtain tissues and cells in specific developmental stages, and then to release nuclei from these cells for DNA content measurements. Precisely dissecting plant organs and tissues in a developmental time series and measuring their nuclear DNA content has produced useful information concerning how the cell cycle is regulated during development, and data on the developmental regulation of the switch from proliferative growth to endoreduplication. Examples, where FCM measurements were applied to follow cell cycle parameters, are leaf development (Beemster et al. 2005), tomato fruit development (Joubes et al. 1999), maize seed development (Leblanc et al. 2002; Leiva-Neto et al. 2004; Schweizer et al. 1995), seed development and size in pea (Lemontey et al. 2000), and seed germination (Barroco et al. 2005; Fujikura et al. 1999).

14.10
Cell Type-specific Characterization of Nuclear DNA Content by Flow Cytometry

Univariate cell cycle analysis has its limitation, because it relies on measuring dispersed populations of cells. Using fluorescence marker proteins, such as the green fluorescent protein (GFP), it has become possible to mark specific cells within tissues by targeting the expression of GFP with cell- and tissue-specific, developmentally regulated, promoters (Laplaze et al. 2005). This technology has led to the development of biparametric FCM methods to measure nuclear DNA content in specific cell types (Zhang et al. 2005; Chapter 17). The method relies on the nuclear targeting of GFP by its fusion to the coding region of a histone 2A gene (HTA6), and expression of this HTA6-GFP under the control of developmentally-regulated promoters. The fusion of GFP to a chromatin-associated protein circumvents the diffusion of GFP out of the nucleus during the experimental procedure. Using tissue-specific promoters it was shown in *Arabidopsis* that cells in the meristem, phloem companion cells and style exclusively contained 2C and 4C nuclei while endodermal cells that had undergone endoreduplication, predominantly contained 4C and 8C nuclei. These data proved the existence of cell type-specific patterns of C-values, and suggest that increasing nuclear DNA content represents one strategy evolved in plants to specify cell types. In future it should be possible to develop further multiparametric FCM methods combining the identification of C-values of nuclei of specific cell types with the determination of the occurrence of S phase, relying on antibody-based detection of BrdU (Lucretti et al. 1999), occurrence of G_2/M cells, relying on labeling with phospho-histone 3 antibody or any other protein markers indicative of developmental processes or cell cycle stages. A systematic method for cloning GFP-open reading frame (ORF) fusions and assessing their subcellular localization in *Arabidopsis thaliana* cells was also described based on *Agrobacterium*-mediated transformation to *Arabidopsis* cultured cells (Koroleva et al. 2005). Using a chromatin-bound protein–GFP fusion the authors assessed, by biparametric FCM analysis, the DNA content of cells versus GFP fluorescence and found that the 2C and 4C cells equally expressed the transiently transformed GFP fusion. This approach will be useful in the search for proteins that exhibit cell cycle phase-specific abundance, for example due to altered stability during the various phases of the cell cycle. It is worth noting, however, that the method is restricted to nuclear- and chromatin-associated proteins.

14.11
Other Methods and Imaging Technologies to Monitor Cell Cycle Parameters and Cell Division Kinetics in Developing Organs

FCM measurements only inform us about the percentage of cells with defined DNA content, but cannot alone give the rate of cell division (measures how fast a cell progresses through the cell cycle) or the rate of cell production (measures the rate of increase of cell number within the population and is proportional to the number of dividing cells multiplied by their rate of division; Baskin 2000).

Methods that have been used to quantify cell division rate can be grouped as being either cytological, in which the rate of accumulation of cells in a particular phase of the cell cycle is determined based on some form of cytological labeling, or kinematic, in which the cell division rate is determined from the movement of cells through the cell cycle. A commonly used parameter to quantify cell division is the mitotic index, which is the percentage of the total number of cells in a sample that are in mitosis. This cannot measure the division rate directly, but dividing the cell production rate by the number of dividing cells can provide an estimate (Baskin 2000). The cytological approach labels cells that are in a particular stage of the cell cycle and follows their fate subsequently. Cells may be pulse labeled (e.g. with tritiated thymidine or BrdU) and the extent of labeling quantified in each cell cycle phase over the time span covering the duration of a number of cell cycles. Alternatively, cells may be labeled continuously to determine how fast they accumulate in a specific cell cycle phase; for example, metaphases accumulate during exposure to colchicine or labeled mitoses during exposure to tritiated thymidine. These methods suffer from a number of drawbacks, for example that not all cells in tissues label equally, the disruptive effect of drugs like colchicine, and the movement of cells from one cell cycle phase to the other during the labeling period.

In contrast to the cytological means, kinematic methods can measure rates of cell division non-invasively. Moreover, this approach can measure not only local rates of cell division but also local rates of expansion. The kinematic approach is ideal for roots because of their linear organization. For calculating the rate of cell division, the data needed for solving the equation of continuity are the spatial profiles of velocity and of cell length. The most straightforward approach is to measure each, velocity by recording the displacement of marks on the root, and cell length by microscopical measurements, preferably on living material to avoid shrinkage due to fixation, embedding, or sectioning. The spatial profile of cell length, velocity, relative elongation rate, and cell division rate fully characterize the growth parameters of the root (Beemster and Baskin 1998; Beemster et al. 2003).

Although changes in cell division rate undoubtedly occur and presumably play important roles in physiology and development, the widespread constancy of the cell division rate suggests that this is a robust parameter, programmed into meristematic cells at a deep level. This general constancy highlights the importance of the number of dividing cells within meristems. To regulate the number of dividing cells, the plant organ must control the exit from the cell cycle at the base of the meristem. This control could be exerted spatially, by maintaining proliferation up to a certain position, or temporally, by maintaining proliferation for a certain number of cell cycles (Baskin 2000). Recently, the retinoblastoma-related protein 1 was discovered as an important regulator which controlled the exit from the proliferation zone and entry into differentiation programme. It could provide an important control mechanism for developmental and environmental inputs that determine meristem size (Wildwater et al. 2005).

As has been described above for the mitotic index as a measure of cell division rate, it is also possible to identify the frequency of cells in a particular phase of the cell cycle by *in situ* hybridization with cell cycle-regulated genes as a probe. In this method the quantity of mRNA is measured within cells in a tissue. Thus, the labeling for a mRNA that shows cell cycle-specific gene expression will identify cells in that particular cell cycle phase. This technique was first applied to flower meristems of *Antirrhinum majus* (Scrophulariaceae; Fobert et al. 1994). Alternatively, reporter constructs were created that are expressed phase-specifically in plant cells. The most commonly used reporter system is to express the β-glucuronidase (GUS), with which the mitosis-specific cyclin B1 would be expressed. This is achieved by the translational fusion of the cyclin B1 promoter and the N-terminal fragment of cyclin B1 protein containing the so-called destruction box, with the GUS reporter gene. This construct in plants confines the expression of GUS to a time-window from late G_2 phase (where the cyclin B1 promoter becomes activated) to the meta-/anaphase transition (where cyclin B1 is degraded; Colon-Carmona et al. 1999; Donnelly et al. 1999). Detailed understanding of the dynamic patterns of cell division, both of number of divisions and their orientation in a multicellular organism, is central to the understanding of morphogenesis. One of the major limitations in understanding growth in both plants and animals has been the inability to monitor cell behavior in real time. Methods have now been devised to monitor cell division in developing organs in real time (Kurup et al. 2005; Reddy et al. 2004; Wildwater et al. 2005). These rely on the expression of fluorescent markers that allow, for instance, the visualization of nuclei and mitosis, through the expression of histone fused to the yellow fluorescent protein tag and thus, when incorporated into chromatin, labels nuclei and chromosomes in live cells. Expression of plasma-membrane located proteins tagged with GFP can provide the precise outline of cells. Cell cycle phase-specific markers such as the cyclinB1;1-GFP label cells in specific phases, in the case of Cyclin B1 in late G_2 to M phase. Time-lapse images of live tissues through confocal microscopy are used to collect cell division data in developing organs. Once cell positions can be extracted by cell-finding algorithms, it should be possible to integrate cell co-ordinates in time-lapse observations and calculate cell division rates in developmental space. The live imaging technique has led to the development of a spatial and temporal map of cell division patterns and integration of cell behavior over time to visualize growth in flower meristems (Reddy et al. 2004), roots (Kurup et al. 2005; Wildwater et al. 2005), and developing endosperm (Boisnard-Lorig et al. 2001).

14.12
Concluding Remarks

Flow cytometry and cell sorting have played an important role in the study of cell cycle progression, but this is just the beginning. These methods in future will allow the monitoring of cell cycle progression in specific cell types, by following the expression of fluorescent proteins using multiparametric analysis (Zhang et al. 2005). Cell sorting will allow molecular changes in specific cell types to be monitored during development (Birnbaum et al. 2005). Live cell imaging makes possible the reconstruction of cell division patterns in space and time, and the collection of data for mathematical modeling of organ growth and thus the reconstruction of virtual plants (Jonsson et al. 2006).

Acknowledgments

We are grateful to Dr Jan Bartoš (Olomouc) for sharing unpublished data and providing Fig. 14.2.

References

Amino, S., Fujimura, T., Komamine, A. **1983**, *Physiol. Plant.* 59, 393–396.

Anthony, R. G., Waldin, T. R., Ray, J. A., Bright, S. W., Hussey, P. J. **1998**, *Nature* 393, 260–263.

Ayaydin, F., Vissi, E., Meszaros, T., Miskolczi, P., Kovacs, I., Feher, A., Dombradi, V., Erdodi, F., Gergely, P., Dudits, D. **2000**, *Plant J.* 23, 85–96.

Bagwell, B. C. **1993**, Theoretical aspects of flow cytometry data analysis in *Clinical Flow Cytometry*, ed. K. E. Bauer, R. E. Duque, T. V. Shankey, Williams & Wilkins, Baltimore, pp. 41–61.

Barroco, R. M., Van Poucke, K., Bergervoet, J. H., De Veylder, L., Groot, S. P., Inze, D., Engler, G. **2005**, *Plant Physiol.* 137, 127–140.

Baskin, T. I. **2000**, *Plant Mol. Biol.* 43, 545–554.

Beemster, G. T., Baskin, T. I. **1998**, *Plant Physiol.* 116, 1515–1526.

Beemster, G. T., Fiorani, F., Inze, D. **2003**, *Trends Plant Sci.* 8, 154–158.

Beemster, G. T., De Veylder, L., Vercruysse, S., West, G., Rombaut, D., Van Hummelen, P., Galichet, A., Gruissem, W., Inze, D., Vuylsteke, M. **2005**, *Plant Physiol.* 138, 734–743.

Bergounioux, C., Perennes, C., Brown, S. C., Gadal, P. **1988**, *Planta* 175, 500–505.

Binarová, P., Doležel, J., Dráber, P., Heberle-Bors, E., Strnad, M., Bögre, L. **1998**, *Plant J.* 16, 697–707.

Birnbaum, K., Jung, J. W., Wang, J. Y., Lambert, G. M., Hirst, J. A., Galbraith, D. W., Benfey, P. N. **2005**, *Nature Methods* 2, 615–619.

Bögre, L., Zwerger, K., Meskiene, I., Binarová, P., Czizmadia, V., Planck, C., Wagner, E., Hirt, H., Heberle-Bors, E. **1997**, *Plant Physiol.* 113, 841–852.

Bögre, L., Calderini, O., Binarová, P., Mattauch, M., Till, S., Kiegerl, S., Jonak, C., Pollaschek, C., Barker, P., Huskisson, N. S., Hirt, H., Heberle-Bors, E. **1999**, *Plant Cell* 11, 101–114.

Boisnard-Lorig, C., Colon-Carmona, A., Bauch, M., Hodge, S., Doerner, P., Bancharel, E., Dumas, C., Haseloff, J., Berger, F. **2001**, *Plant Cell* 13, 495–509.

Campanoni, P., Nick, P. **2005**, *Plant Physiol.* 137, 939–948.

Chen, J. G., Shimomura, S., Sitbon, F., Sandberg, G., Jones, A. M. **2001**, *Plant J.* 28, 607–617.

Citterio, S., Piatti, S., Albertini, E., Aina, R., Varotto, S., Barcaccia, G. **2006**, *Exp. Cell Res.* 312, 1050–1064.

Colon-Carmona, A., You, R., Haimovitch-Gal, T., Doerner, P. **1999**, *Plant J.* 20, 503–508.

Combettes, B., Reichheld, J. P., Chaboute, M. E., Philipps, G., Shen, W. H., Chaubet-Gigot, N. **1999**, *Methods Cell Sci.* 21, 109–121.

Criqui, M. C., Genschik, P. **2002**, *Curr. Opin. Plant Biol.* 5, 487–493.

Crissmann, H. A., Hirons, G. T. **1994**, *Methods Cell Biol.* 41, 195–209.

Dahl, M., Meskiene, I., Bögre, L., Ha, D. T., Swoboda, I., Hubmann, R., Hirt, H., Heberle-Bors, E. **1995**, *Plant Cell* 7, 1847–1857.

Darzynkiewicz, Z., Sharpless, T., Staiano-Coico, L., Melamed, M. R. **1980a**, *Proc. Natl Acad. Sci. USA* 77, 6696–6699.

Darzynkiewicz, Z., Traganos, F., Melamed, M. R. **1980b**, *Cytometry* 1, 98–109.

Darzynkiewicz, Z., Crissmann, H., Jacobberger, J. W. **2004**, *Cytometry* 58A, 21–32.

De la Torre, C., Fernandez Gomez, M. E., Gimenez-Martin, C., Gonzales-Fernandes, A. **1974**, *J. Cell Sci.* 14, 461–473.

De la Torre, C., Gonzales-Fernandes, A., Gimenez-Martin, G. **1989**, *J. Cell Sci.* 94, 259–265.

De Sutter, V., Vanderhaeghen, R., Tilleman, S., Lammertyn, F., Vanhoutte, I., Karimi, M., Inze, D., Goossens, A., Hilson, P. **2005**, *Plant J.* 44, 1065–1076.

De Veylder, L., Beeckman, T., Beemster, G. T., Krols, L., Terras, F., Landrieu, I., van der Schueren, E., Maes, S., Naudts, M., Inze, D. **2001**, *Plant Cell* 13, 1653–1668.

De Veylder, L., Joubes, J., Inze, D. **2003**, *Curr. Opin. Plant Biol.* 6, 536–543.

Dean, P. N., Jett, J. H. **1974**, *J. Cell Biol.* 60, 523–527.

Del Campo, A., Gimenez-Martin, G., Lopez-Saez, J. F., de la Torre, C. **1997**, *Eur. J. Cell Biol.* 74, 289–293.

Dewitte, W., Murray, J. A. H. **2003**, *Annu. Rev. Plant Biol.* 54, 235–264.

Dolbeare, F., Gratzner, H., Pallavicini, M. G., Gray, J. W. **1983**, *Proc. Natl Acad. Sci. USA* 80, 5573–5577.

Doležel, J., Göhde, W. **1995**, *Cytometry* 19, 103–106.

Doležel, J., Čihalíková, J., Lucretti, S. **1992**, *Planta* 188, 93–98.

Doležel, J., Čihalíková, J., Weiserová, J., Lucretti, S. **1999**, *Methods Cell Sci.* 21, 95–107.

Donnelly, P. M., Bonetta, D., Tsukaya, H., Dengler, R. E., Dengler, N. G. **1999**, *Dev. Biol.* 215, 407–419.

Dudits, D., Bögre, L., Gyorgyey, J. **1991**, *J. Cell Sci.* 99, 475–484.

Ebel, C., Mariconti, L., Gruissem, W. **2004**, *Nature* 429, 776–780.

Ehsan, H., Roef, L., Witters, E., Reichheld, J. P., Van Bockstaele, D., Inze, D., Van Onckelen, H. **1999**, *FEBS Lett.* 458, 349–353.

Fobert, P. R., Coen, E. S., Murphy, G. J., Doonan, J. H. **1994**, *EMBO J.* 13, 616–624.

Fried, J., Perez, A. G., Clarkson, B. **1980**, *Exp. Cell Res.* 126, 63–74.

Fujikura, Y., Doležel, J., Čihalíková, J., Bögre, L., Heberle-Bors, E., Hirt, H., Binarová, P. **1999**, *Seed Sci. Res.* 9, 297–304.

Fukuda, H., Ito, M., Sugiyama, M., Komamine, A. **1994**, *Int. J. Dev. Biol.* 38, 287–299.

Galbraith, D. W., Mauch, T. J., Shields, B. A. **1981**, *Physiol. Plant.* 51, 380–386.

Galbraith, D. W., Harkins, K. R., Maddox, J. R., Ayres, N. M., Sharma, D. P., Firoozabady, E. **1983**, *Science* 220, 1049–1051.

Genschik, P., Criqui, M. C., Parmentier, Y., Derevier, A., Fleck, J. **1998**, *Plant Cell* 10, 2063–2076.

Gimenez-Abian, J. F., Clarke, D. J., Gimenez-Martin, G., Weingartner, M., Gimenez-Abian, M. I., Carballo, J. A., Diaz, D., Bögre, L., De, L. **2002a**, *Eur. J. Cell Biol.* 81, 9–16.

Gimenez-Abian, J. F., Weingartner, M., Binarová, P., Clarke, D. J., Anthony, R. G., Calderini, O., Heberle-Bors, E., Moreno Diaz de la Espina, S., Bögre, L., De la Torre, C. **2002b**, *Cell Cycle* 1, 187–192.

Glab, N., Labidi, B., Qin, L. X., Trehin, C., Bergounioux, C., Meijer, L. **1994**, *FEBS Lett.* 353, 207–211.

Gonzalez-Fernandez, A., Gimenez-Martin, G., Fernandez-Gomez, M. E., de la Torre, C. **1974**, *Exp. Cell Res.* 88, 163–170.

Gould, A. R. **1984**, *CRC Crit. Rev. Plant Sci.* 1, 315–344.

Gualberti, G., Doležel, J., Macas, J., Lucretti, S. **1996**, *Theor. Appl. Genet.* 92, 744–751.

Gutierrez, C., Ramirez-Parra, E., Castellano, M. M., Del Pozo, J. C. **2002**, *Curr. Opin. Plant Biol.* 5, 480–486.

Hasezawa, S., Nagata, T. **1991**, *Bot. Acta* 104, 206–211.

Hemerly, A., Engler, J. D., Bergounioux, C., Van Montagu, M., Engler, G., Inze, D., Ferreira, P. **1995**, *EMBO J.* 14, 3925–3936.

Hennig, L., Menges, M., Murray, J. A., Gruissem, W. **2003**, *Plant Mol. Biol.* 53, 457–465.

Hirt, H., Mink, M., Pfosser, M., Bögre, L., Gyorgyey, J., Jonak, C., Gartner, A., Dudits, D., Heberle-Bors, E. **1992**, *Plant Cell* 4, 1531–1538.

Horiguchi, G., Ferjani, A., Fujikura, U., Tsukaya, H. **2005**, *J. Plant Res.* 118, 223–227.

Howard, A., Pelc, S. R. **1951**, *Exp. Cell Res.* 2, 178–187.

Ingram, G. C., Waites, R. **2006**, *Curr. Opin. Plant Biol.* 9, 12–20.

Inze, D. **2005**, *EMBO J.* 24, 657–662.

Ito, M., Marieclaire, C., Sakabe, M., Ohno, T., Hata, S., Kouchi, H., Hashimoto, J.,

Fukuda, H., Komamine, A., Watanabe, A. **1997**, *Plant J.* 11, 983–992.

Ito, M., Iwase, M., Kodama, H., Lavisse, P., Komamine, A., Nishihama, R., Machida, Y., Watanabe, A. **1998**, *Plant Cell* 10, 331–341.

Ito, M., Araki, S., Matsunaga, S., Itoh, T., Nishihama, R., Machida, Y., Doonan, J. H., Watanabe, A. **2001**, *Plant Cell* 13, 1891–1905.

Jang, S. J., Shin, S. H., Yee, S. T., Hwang, B., Im, K. H., Park, K. Y. **2005**, *Mol. Cell* 20, 136–141.

Jonsson, H., Heisler, M. G., Shapiro, B. E., Meyerowitz, E. M., Mjolsness, E. **2006**, *Proc. Natl Acad. Sci. USA* 103, 1633–1638.

Jorgensen, P., Tyers, M. **2004**, *Curr. Biol.* 14, R1014–R1027.

Joubes, J., Phan, T. H., Just, D., Rothan, C., Bergounioux, C., Raymond, P., Chevalier, C. **1999**, *Plant Physiol.* 121, 857–869.

Kadota, Y., Watanabe, T., Fujii, S., Higashi, K., Sano, T., Nagata, T., Hasezawa, S., Kuchitsu, K. **2004**, *Plant J.* 40, 131–142.

Kadota, Y., Furuichi, T., Sano, T., Kaya, H., Gunji, W., Muramaki, Y., Muto, S., Hasezawa, S., Kuchitsu, K. **2005a**, *Biochem. Biophys. Res. Comm.* 336, 1259–1267.

Kadota, Y., Watanabe, T., Fujii, S., Maeda, Y., Ohno, R., Higashi, K., Sano, T., Muto, S., Hasezawa, S., Kuchitsu, K. **2005b**, *Plant Cell Physiol.* 46, 156–165.

Katsuta, J., Shibaoka, H. **1992**, *J. Cell Sci.* 103, 397–405.

King, P. J., Mansfield, K. J., Street, H. E. **1973**, *Can. J. Bot.* 51, 1807–1823.

King, P. J., Cox, B. J., Fowler, M. W., Street, H. E. **1974**, *Planta* 117, 109–122.

Kodama, H., Ito, M., Hattori, T., Nakamura, K., Komamine, A. **1991**, *Plant Physiol.* 95, 406–411.

Körner, C., Menendez-Riedl, S. P., John, P. C. L. **1989**, *Austral. J. Plant Physiol.* 16, 443–448.

Koroleva, O. A., Tomlinson, M. L., Leader, D., Shaw, P., Doonan, J. H. **2005**, *Plant J.* 41, 162–174.

Kurup, S., Runions, J., Kohler, U., Laplaze, L., Hodge, S., Haseloff, J. **2005**, *Plant J.* 42, 444–453.

Laplaze, L., Parizot, B., Baker, A., Ricaud, L., Martiniere, A., Auguy, F., Franche, C., Nussaume, L., Bogusz, D., Haseloff, J. **2005**, *J. Exp. Bot.* 56, 2433–2442.

Laureys, F., Dewitte, W., Witters, E., Van Montagu, M., Inze, D., Van Onckelen, H. **1998**, *FEBS Lett.* 426, 29–32.

Leblanc, O., Pointe, C., Hernandez, M. **2002**, *Plant J.* 32, 1057–1066.

Leiva-Neto, J. T., Grafi, G., Sabelli, P. A., Dante, R. A., Woo, Y. M., Maddock, S., Gordon-Kamm, W. J., Larkins, B. A. **2004**, *Plant Cell* 16, 1854–1869.

Lemontey, C., Mousset-Declas, C., Munier-Jolain, N., Boutin, J. P. **2000**, *J. Exp. Bot.* 51, 167–175.

Lloyd, C. **1999**, *Bioessays* 21, 1061–1068.

Logemann, E., Wu, S. C., Schroder, J., Schmelzer, E., Somssich, I. E., Hahlbrock, K. **1995**, *Plant J.* 8, 865–876.

Lucretti, S., Nardi, L., Nisini, P. T., Moretti, F., Gualberti, G., Doležel, J. **1999**, *Methods Cell Sci.* 21, 155–166.

Mader, J. C., Hanke, D. E. **1996**, *J. Plant Growth Reg.* 15, 95–102.

Magyar, Z., Bako, L., Bögre, L., Dedeoglu, D., Kapros, T., Dudits, D. **1993**, *Plant J.* 4, 151–161.

Magyar, Z., Meszaros, T., Miskolczi, P., Deak, M., Feher, A., Brown, S., Kondorosi, E., Athanasiadis, A., Pongor, S., Bilgin, M., Bako, L., Koncz, C., Dudits, D. **1997**, *Plant Cell* 9, 223–235.

Magyar, Z., De Veylder, L., Atanassova, A., Bako, L., Inze, D., Bögre, L. **2005**, *Plant Cell* 17, 2527–2541.

May, M. J., Leaver, C. J. **1993**, *Plant Physiol.* 103, 621–627.

McCann, M. C., Stacey, N. J., Dahiya, P., Milioni, D., Sado, P. E., Roberts, K. **2001**, *Plant Physiol.* 127, 1380–1382.

Menges, M., Murray, J. A. **2002**, *Plant J.* 30, 203–212.

Menges, M., Murray, J. A. **2004**, *Plant J.* 37, 635–644.

Menges, M., Hennig, L., Gruissem, W., Murray, J. A. **2002**, *J. Biol. Chem.* 277, 41987–42002.

Menges, M., Hennig, L., Gruissem, W., Murray, J. A. **2003**, *Plant Mol. Biol.* 53, 423–442.

Menges, M., de Jager, S. M., Gruissem, W., Murray, J. A. **2005**, *Plant J.* 41, 546–566.

Meskiene, I., Bögre, L., Dahl, M., Pirck, M., Ha, D. T., Swoboda, I., Heberle-Bors, E., Ammerer, G., Hirt, H. **1995**, *Plant Cell* 7, 759–771.

Mizukami, Y. **2001**, *Curr. Opin. Plant Biol.* 4, 533–539.

Moretti, F., Lucretti, S., Doležel, J. **1992**, *Eur. J. Histochem.* 36, 367.

Nagata, T., Kumagai, F. **1999**, *Methods Cell Sci.* 21, 123–127.

Nagata, T., Nemoto, Y., Hasezawa, S. **1992**, *Int. Rev. Cytol.* 132, 1–30.

Naill, M., C., Roberts, S. C. **2005**, *Biotechnol. Bioengin.* 90, 491–500.

Nakagami, H., Kawamura, K., Sugisaka, K., Sekine, M., Shinmyo, A. **2002**, *Plant Cell* 14, 1847–1857.

Nasmyth, K. **2005**, *Phil. Trans. Roy. Soc. Lond. Series B – Biol. Sci.* 360, 483–496.

Nishida, T., Ohnishi, N., Kodama, H., Komamine, A. **1992**, *Plant Cell Tiss. Organ Cult.* 28, 37–43.

Otvos, K., Pasternak, T. P., Miskolczi, P., Domoki, M., Dorjgotov, D., Szucs, A., Bottka, S., Dudits, D., Feher, A. **2005**, *Plant J.* 43, 849–860.

Park, J. A., Ahn, J. W., Kim, Y. K., Kim, S. J., Kim, J. K., Kim, W. T., Pai, H. S. **2005**, *Plant J.* 42, 153–163.

Pasternak, T. P., Prinsen, E., Ayaydin, F., Miskolczi, P., Potters, G., Asard, H., Van Onckelen, H. A., Dudits, D., Feher, A. **2002**, *Plant Physiol.* 129, 1807–1819.

Peck, S. C. **2005**, *Plant Physiol.* 138, 591–599.

Pelayo, H. R., Lastres, P., De la Torre, C. **2001**, *Planta* 212, 444–453.

Perennes, C., Qin, L. X., Glab, N., Bergounioux, C. **1993**, *FEBS Lett.* 333, 141–145.

Pfosser, M. **1989**, *J. Plant Physiol.* 134, 741–745.

Pfosser, M., Amon, A., Lelley, T., Heberle-Bors, E. **1995**, *Cytometry* 21, 387–393.

Planchais, S., Glab, N., Trehin, C., Perennes, C., Bureau, J. M., Meijer, L., Bergounioux, C. **1997**, *Plant J.* 12, 191–202.

Planchais, S., Glab, N., Inze, D., Bergounioux, C. **2000**, *FEBS Lett.* 476, 78–83.

Porceddu, A., Stals, H., Reichheld, J. P., Segers, G., De Veylder, L., Barroco, R. P., Casteels, P., Van Montagu, M., Inze, D., Mironov, V. **2001**, *J. Biol. Chem.* 276, 36354–36360.

Reddy, G. V., Heisler, M. G., Ehrhardt, D. W., Meyerowitz, E. M. **2004**, *Development* 131, 4225–4237.

Reichheld, J. P., Chaubet, N., Shen, W. H., Renaudin, J. P., Gigot, C. **1996**, *Proc. Natl Acad. Sci. USA* 93, 13819–13824.

Reichheld, J. P., Vernoux, T., Lardon, F., Van Montagu, M., Inze, D. **1999**, *Plant J.* 17, 647–656.

Riou-Khamlichi, C., Huntley, R., Jacqmard, A., Murray, J. A. **1999**, *Science* 283, 1541–1544.

Riou-Khamlichi, C., Menges, M., Healy, J. M., Murray, J. A. **2000**, *Mol. Cell Biol.* 20, 4513–4521.

Rogers, A. W. **1973**, *Techniques of Autoradiography*, Elsevier, Amsterdam.

Roudier, F., Fedorova, E., Gyorgyey, J., Feher, A., Brown, S., Kondorosi, A., Kondorosi, E. **2000**, *Plant J.* 23, 73–83.

Rouse, J., Jackson, S. P. **2002**, *Science* 297, 547–551.

Sala, F., Parisi, B., Burroni, D., Amileni, A. R., Pedrali-Noy, G., Spadari, S. **1980**, *FEBS Lett.* 117, 93–98.

Samuels, A. L., Giddings, T. H., Jr., Staehelin, L. A. **1995**, *J. Cell Biol.* 130, 1345–1357.

Samuels, A. L., Meehl, J., Lipe, M., Staehelin, L. A. **1998**, *Protoplasma* 202, 232–236.

Sano, T., Higaki, T., Handa, K., Kadota, Y., Kuchitsu, K., Hasezawa, S., Hoffmann, A., Endter, J., Zimmermann, U., Hedrich, R., Roitsch, T. **2006**, *FEBS Lett.* 580, 597–602.

Schweizer, L., Yerk-Davis, G. L., Phillips, R. L., Srienc, F., Jones, R. J. **1995**, *Proc. Natl Acad. Sci. USA* 92, 7070–7074.

Sgorbati, S., Sparvoli, E., Levi, M., Galli, M., Citterio, S., Chiatante, D. **1991**, *Physiol. Plant.* 81, 507–512.

Shen, W. H., Gigot, C. **1997**, *Plant Mol. Biol.* 33, 367–379.

Sorrell, D. A., Combettes, B., Chaubet-Gigot, N., Gigot, C., Murray, J. A. **1999**, *Plant Physiol.* 119, 343–352.

Sorrell, D. A., Menges, M., Healy, J. M., Deveaux, Y., Amano, C., Su, Y., Nakagami, H., Shinmyo, A., Doonan, J. H., Sekine, M., Murray, J. A. **2001**, *Plant Physiol.* 126, 1214–1223.

Stals, H., Casteels, P., Van Montagu, M., Inze, D. **2000**, *Plant Mol. Biol.* 43, 583–593.

Sugimoto-Shirasu, K., Roberts, K. **2003**, *Curr. Opin. Plant Biol.* 6, 544–553.

Swiatek, A., Lenjou, M., Van Bockstaele, D., Inze, D., Van Onckelen, H. **2002**, *Plant Physiol.* 128, 201–211.

Swiatek, A., Azmi, A., Stals, H., Inze, D., Van Onckelen, H. **2004**, *FEBS Lett.* 572, 118–122.

Swift, H. **1950**, *Proc. Natl Acad. Sci. USA* 36, 643–654.

Trehin, C., Planchais, S., Glab, N., Perennes, C., Tregear, J., Bergounioux, C. **1998**, *Planta* 206, 215–224.

Vandepoele, K., Raes, J., De Veylder, L., Rouze, P., Rombauts, S., Inze, D. **2002**, *Plant Cell* 14, 903–916.

Weigel, D., Jurgens, G. **2002**, *Nature* 415, 751–754.

Weingartner, M., Binarová, P., Dryková, D., Schweighofer, A., David, J. P., Heberle-Bors, E., Doonan, J., Bögre, L. **2001**, *Plant Cell* 13, 1929–1943.

Weingartner, M., Pelayo, H. R., Binarová, P., Zwerger, K., Melikant, B., De La Torre, C., Heberle-Bors, E., Bögre, L. **2003**, *J. Cell Sci.* 116, 487–498.

Weingartner, M., Criqui, M. C., Meszaros, T., Binarová, P., Schmit, A. C., Helfer, A., Derevier, A., Erhardt, M., Bögre, L., Genschik, P. **2004**, *Plant Cell* 16, 643–657.

Wildwater, M., Campilho, A., Perez-Perez, J. M., Heidstra, R., Blilou, I., Korthout, H., Chatterjee, J., Mariconti, L., Gruissem, W., Scheres, B. **2005**, *Cell* 123, 1337–1349.

Yanpaisan, W., King, N. J. C., Doran, P. M. **1998**, *Biotechnol. Bioengin.* 58, 515–528.

Yanpaisan, W., King, N. J. C., Doran, P. M. **1999**, *Biotechnol. Adv.* 17, 3–27.

Zhang, C., Gong, F. C., Lambert, G. M., Galbraith, D. W. **2005**, *Plant Methods* 1, 7–12.

15
Endopolyploidy in Plants and its Analysis by Flow Cytometry

Martin Barow and Gabriele Jovtchev

Overview

Endpolyploidization is a common process in plants that enables the multiplication of the entire genome in somatic cells. Endopolyploidy can easily be investigated by flow cytometry, and this chapter gives suggestions for its analysis and evaluation. In most seed plants that exhibit endopolyploidy, it is systemic and expressed to different degrees in a given organ and tissue characteristic for a species. Differences between families, species, varieties, ecotypes, life strategies, organs, and tissues are described. Furthermore, the impacts of genome size, environmental factors and phytohormones on endopolyploidy are discussed. Awareness of all these factors is important when endopolyploidy is to be studied to guarantee the reliability of the results. Finally, possible applications of endopolyploidy analysis and its significance in plant breeding and biotechnology are outlined.

15.1
Introduction

Endopolyploidy is defined as the occurrence of elevated ploidy levels of cells within an organism generated either by endoreduplication or by endomitosis (definition adapted from Rieger et al. 1991). Geitler (1939) introduced the term "endomitosis" to describe the regular and controlled event of chromatin reduplication during cell differentiation without nuclear division where changes in chromatin structure occur which resemble those taking place during mitosis. This concept was initially based on observations on polyploidization and nuclear structural changes in Heteroptera, in which *a-priori* a prochromosomal interphase structure prevails. The terms endomitosis and endomitotic polyploidization were later also extended by him to the situation in somatic tissues of plants, and the meaning of the term endomitosis was broadened, to include endomitosis with mitotic contraction and nuclear growth without any mitotic contraction. Geitler (1953, p. 23) even describes polytene chromosomes of Diptera as formed by identical chromonemata of *endomitotic* origin, but meaning multiplication without any mitotic contraction. Actually, polyploidization in somatic tissues of plants mostly occurs without structural changes resembling mitosis (D'Amato 1989; Nagl 1978) and is then better referred to as *endoreduplication*, a term introduced by Levan and Hauschka (1953) but not used by Geitler at those times. Thus the terminology was and still is somewhat confusing.

The exit from endomitosis (i.e. with chromosome condensation) can occur at any stage of the endocycle (endopro- to endotelophase) depending on the species, cell type and developmental stage (Anisimov 2005; Edgar and Orr-Weaver 2001; Winkelmann et al. 1987). According to D'Amato (1989), only endomitosis generates endopolyploidy with doubled chromosome numbers, whereas polyteny is a result of endoreduplication without separation of sister chromatids. However, endomitosis may also result in chromosomes of which the sister chromatids do not separate completely after reduplication (Edgar and Orr-Weaver 2001). Furthermore, in plants, sister chromatids are generally not neatly aligned in parallel arrays, but only at certain sites, particularly at the chromocenters (Doležal and Tschermak-Woess 1955; Schubert et al. 2006; Tschermak-Woess 1956) and are thus endopolyploid rather than polytene. Different modes of cohesion of sister chromatids can be found after endoreduplication. The cohesion may be retained after the first round of reduplication, at least at a few sites, but further chromatid doublings may double the number of chromocenters, or cohesion may be loosened after some time in mature cells, or different levels of cohesion may coexist within one tissue or cell (Doležal and Tschermak-Woess 1955; Schubert et al. 2006; Šesek et al. 2005; Tschermak-Woess 1956). In plants, chromosomes structurally resembling polytene chromosomes of some animals (i.e. cable-like threads composed of aligned sister chromatids) are almost exclusively found in cells of the ovule or developing seed (Nagl 1981; Tschermak-Woess 1963). As the latter author points out, this phenomenon does not depend on the high endopolyploidy levels, which are also very often attained in these cells, since it also occurs when ploidy levels as low as those in vegetative. Cohesion in such chromosomes is basically restricted to heterochromatic segments (Nagl 1978), in contradistinction to the complete alignment in animal polytene chromosomes. It is questionable, whether the term polyteny is justified for plants at all, even when chromatin bundles are formed in endopolyploid nuclei, because the *continuously* banded morphology of the typical polytene or giant chromosomes of Diptera (Beermann 1962), Collembola (Cassagnau 1968) and some Ciliata (Ammermann 1971) is never attained (Nagl 1978; Tschermak-Woess 1963).

In seed plants, endopolyploidy is a common feature (D'Amato 1964; Tschermak-Woess and Hasitschka 1953). For most species, it occurs in almost all organs and tissues (D'Amato 1964; Barow and Meister 2003; Tschermak-Woess 1956; Tschermak-Woess and Hasitschka 1953). The degree of endopolyploidy is indicated by the C-levels, where 1C is the DNA content of reduced gametes (see Chapter 4). Generally, in higher plants endopolyploidy levels range from 4C to 64C, whereby the 4C level can be ambiguous regarding its status as just *replicated* or already *endopolyploid*. Exceptionally high DNA contents up to 24 576C (D'Amato 1998; Nagl 1978) are documented for certain cells of the ovule or developing seed, that is, for antipodal cells of the embryo sac, suspensor cells of the developing embryo, endosperm, endosperm haustoria, and fruit elaiosomes, as well as for the tapetum of anthers.

15.2
Methods to Analyze Endopolyploidy

Microscopic methods as well as flow cytometry (FCM) can be used to assess endopolyploidy in plants. Microscopic techniques allow the analysis of single cells or defined tissues, whereas FCM does not discriminate between cells of different origin but results in higher accuracy since more cells can be evaluated within a reasonably short time.

15.2.1
Microscopy

15.2.1.1 Chromosome Counts

The first investigators of endopolyploidy had to rely on chromosome counts since techniques to quantify DNA content were not yet available. As endopolyploidization in plants does not comprise chromosome condensation, two different approaches were applied in the past to induce mitosis in endopolyploid cells. The first was a wounding technique invented and applied by Lothar Geitler and his students in Vienna, the first description of which was published by Grafl (1939). Plants were wounded by a cut and the tissue surrounding the wound analyzed microscopically after 2 to 10 days. The second approach exploited the fact that mitotic cell division could be induced in endopolyploid cells (e.g. root cells of *Rhoeo discolor* (Commelinaceae; Huskins 1947) and *Allium cepa* (Alliaceae; D'Amato and Avanzi 1948)) by auxin treatment as first documented (although misinterpreted as induction of endomitosis) by Levan (1939).

The number of chromocentres after DAPI staining (Fras and Maluszynska 2004) and fluorescence signals after fluorescence *in situ* hybridization (FISH) with probes that bind specifically to particular chromosome regions (centromeres of chromosomes 2 and 4, nucleolar organizing region, and arbitrarily chosen single copy regions of the DNA; Baroux et al. 2004) were recently used to assess endopolyploidy in callus and initial explants, and endosperm of *Arabidopsis thaliana* (Brassicaceae), respectively. However, as sister chromatids may partly remain aligned in endopolyploid cells of *A. thaliana in vivo*, particularly at the centromeres, but also at other regions of the chromosomes, the probes must be chosen carefully to reveal endopolyploidy (Schubert et al. 2006).

15.2.1.2 Feulgen Microdensitometry, Fluorescence Microscopy, Image Analysis

Microspectrofluorometry and Feulgen microdensitometry represent methods to measure DNA content and therewith endopolyploidy levels in defined cells and tissues, respectively. Endopolyploidy in *Helianthus annuus* (Asteraceae) was measured by Feulgen densitometry by Nagl and Capesius (1976), and endopolyploidy in caryopses of *Zea mays* (Poaceae) and seedlings of *Glycine max* (Fabaceae) were studied by image densitometry by Kladnik et al. (2005) and Stepinski (2003/4), respectively.

Melaragno et al. (1993), Gilissen et al. (1996) and Berta et al. (2000) measured the fluorescence intensity of nuclei after staining with DAPI in epidermal and root cells.

Jovtchev et al. (2006) observed epidermis cells either directly using Differential Interference Contrast (DIC) or after DAPI staining. Images were acquired using an epifluorescence microscope equipped with a color CCD camera. The microscope was integrated into a Digital Optical 3D Microscope system to generate 3D extended focus images and stacks of optical sections through the tissue as the basis for measuring cell and nucleus size.

15.2.2
Flow Cytometry

Generally, FCM is the method of choice to investigate nuclear DNA content in plants since it is fast and accurate, as first shown by Galbraith et al. (1983). The experimental design of endopolyploidy studies is essentially the same as that described for genome size analysis, involving tissue homogenization (i.e. usually chopping with a razor blade) in a staining buffer containing a DNA fluorochrome (see Chapter 4).

If the potential of a plant or a plant organ to exhibit endopolyploidy is to be investigated, fully expanded organs need to be studied to insure that the endopolyploidization program has been completed. However, mature plants tend to have higher levels of compounds which interfere with DNA staining. Therefore, the nuclear isolation buffer should be supplemented with substances such as mercaptoethanol and polyvinylpyrrolidone (PVP) to counteract self-tanning (see Chapter 4).

DNA intercalating fluorochromes as well as base-specific fluorochromes are suitable since comparisons are only made between different endopolyploidy levels within one plant. For the same reason no internal standard is needed. However, for some species or samples an (internal) standard (preferably a non-endopolyploid species for the sake of clarity) may help to identify the 2C peak of the species under investigation. This approach is particularly advisable for species with small genomes (like *Arabidopsis thaliana*) where a 2C peak often falls in the range of autofluorescing debris or plastids in the histogram of DNA content, and may thus remain obscured if it is not high enough. Sometimes, for example in *A. thaliana*, debris may form a peak on its own below the real 2C peak, whose position fits into the exponential sequence of endopolyploid nuclei and might be

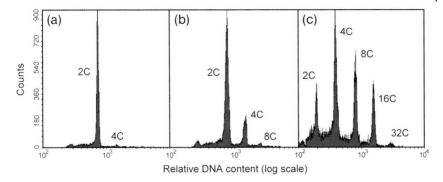

Fig. 15.1 Typical FCM histograms of relative fluorescence of isolated nuclei released from (a) organs without endopolyploidy (petal of *Lactuca sativa*, mean C-level = 2.04 and cycle value = 0.02); (b) medium endopolyploidy (leaf of *Allium ledebourianum*, mean C-level = 3.34 and cycle value = 0.50); and (c) high endopolyploidy (cotyledon of *Cucumis sativus*, mean C-level = 7.32 and cycle value = 1.50). The instrument gain was adjusted differently for the different samples in order to display all the peaks.

mistaken for the 2C peak, leading to overestimation of the extent of endopolyploidization.

15.2.2.1 Evaluation of Histograms

The easiest way to evaluate FCM data with respect to endopolyploidy is to measure the peak height in a histogram on logarithmic scale. In a logarithmic histogram, all peaks theoretically have the same width (given the same CV for all peaks) and therefore the peak height should correspond to the number of nuclei forming a given peak (Barow and Meister 2003; Givan 2001; Mishiba and Mii 2000).

Typical histograms for species with low and high levels of endopolyploidy, and those lacking endopolyploidy are shown in Fig. 15.1. For most species and organs, the scale is sufficient to display all peaks, but this may be difficult in highly endopolyploid organs, for instance in cotyledons of *Arabidopsis thaliana* with up to eight peaks (= up to 256C).

In plants lacking endopolyploidy, a very small peak at the position of a 4C peak comprising approximately 3% of the nuclei analyzed is often discernible even in mature, fully differentiated and expanded organs. Fifty to seventy percent of recorded events of this peak are 2C nuclei which stick together, the rest consisting of replicated nuclei with the 4C DNA content (Barow and Meister 2003). Because of the very low percentage of these nuclei, they should not be considered as the result of endopolyploidization. In immature organs of plants with endopolyploidy, this peak may also comprise nuclei in the G_2 or mitotic stage of the mitotic cell cycle.

In principle, nuclei sticking together may also contribute to the peaks of higher endopolyploidy levels. However, this number is generally so small that it can be neglected during histogram evaluation. Only in some cases are small peaks generated in intermediate positions, corresponding to 6C, 12C and so forth, by nuclei sticking together. This has to be considered when identifying the highest endopolyploidy peak. If intermediate peaks occur, the highest endopolyploidy peak can only be real if it is at least as high as the next lower intermediate peak. Because of diverse shapes (round, spindle or disc) of the nuclei of some species independent of the endopolyploidy level, different sizes of nuclei of different endopolyploidy levels with a large overlap between adjacent categories, and the low number of nuclei sticking together, a doublet discrimination to exclude such nuclei from analysis is probably hardly feasible. Doublet discrimination is mainly used for cell cycle analysis to distinguish between cells in G_2 and doublets of cells in G_1, both of which give the same pulse height (representing the DNA content) but different pulse widths (representing the time a particle needs to pass the laser interrogation point). Different shapes of nuclei with the same DNA content would result in a wider range of pulse width signals even within one ploidy level, thereby making doublet discrimination difficult.

15.2.2.2 Quantification of the Degree of Endopolyploidy

Apart from displaying histograms of DNA content and calculating the percentage of nuclei of particular endopolyploidy levels (Gendreau et al. 1999), two parameters may be used to indicate the mean degree of endopolyploidization of an organ or tissue: (i) the mean C-level (Engelen-Eigles et al. 2000; Mishiba and Mii 2000), and (ii) the so-called cycle value (Barow and Meister 2003).

The mean C-level is calculated from the number of nuclei in each represented ploidy level multiplied by the corresponding ploidy levels. The sum is then divided by the total number of nuclei measured.

Mean C-level (for diplophasic organisms)
$$= (2 \times n_{2C} + 4 \times n_{4C} + 8 \times n_{8C} + 16 \times n_{16C} \ldots)/$$
$$(n_{2C} + n_{4C} + n_{8C} + n_{16C} \ldots),$$

where $n_{2C}, n_{4C}, n_{8C}, \ldots$ are numbers of nuclei with the corresponding C-levels $(2C, 4C, 8C, \ldots)$.

The cycle value indicates the mean number of endoreduplication cycles per nucleus of the cells investigated. It is calculated from the number of nuclei of particular ploidy levels multiplied by the number of endoreduplication cycles necessary to reach the corresponding ploidy level. The sum is then divided by the total number of nuclei measured. In diplophasic organisms, the 4C level is classified as the first endopolyploid level (cycle value 1), although this is only correct if mitotically-active cells are not present or if their number is negligible (that is, in mature organs). Correspondingly, in haplophasic organisms (e.g. bryophytes), the 2C level is the first endopolyploid level (also with cycle value 1).

Cycle value (for diplophasic organisms)
$$= (0 \times n_{2C} + 1 \times n_{4C} + 2 \times n_{8C} + 3 \times n_{16C} \ldots)/$$
$$(n_{2C} + n_{4C} + n_{8C} + n_{16C} \ldots),$$

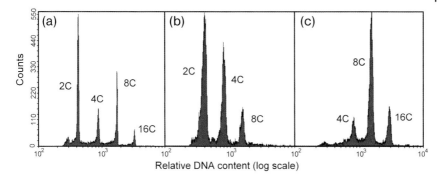

Fig. 15.2 Histograms showing different levels of endopolyploidy in the *Raphanus sativus* petal. (a) Entire petal with mean C-level = 4.82 and cycle value = 0.89; (b) distal part of the petal (limb) only with mean C-level = 3.61 and cycle value = 0.66; and (c) proximal part of the petal (claw) only with mean C-level = 8.99 and cycle value = 2.05.

where $n_{2C}, n_{4C}, n_{8C}, \ldots$ are numbers of nuclei with the corresponding C-levels $(2C, 4C, 8C, \ldots)$.

The mean C-level may be used to describe the possible impact of endopolyploidy on a plant as a consequence of increased nuclear DNA content. This increase is accompanied by a larger volume of endopolyploid cells, and might be important for growth, development and phenotype of cells and tissues (reviewed in Barow 2006). However, the mean C-level over-emphasizes high ploidy levels due to the exponential character of ploidy incremental steps. This problem is avoided by the cycle value, which represents an appropriate parameter to compare the degree of endopolyploidization in different species (Figs. 15.1, 15.2 and 15.6).

A binomial-like distribution is generally the typical pattern observed on FCM histograms (our unpublished data). Obvious deviations may indicate that very different levels of endopolyploidization occur in different parts of the organ under study. Figure 15.2 shows histograms of the entire petal of *Raphanus sativus* (Brassicaceae) as well as of the base (petal claw) and of the morphologically different apical area (petal limb) analyzed separately. As documented by Kudo and Kimura (2002) for *Brassica oleracea* (Brassicaceae), the base exhibits a higher endopolyploidy than the upper part. The authors state that "the elongate morphology of the proximal epidermal cells appears to relate to their rapid expansion which pushes the petal out of the bud as it opens".

In most organs of endopolyploid species, differences in endopolyploidy levels (yet less obvious) can be found (see Section 15.4.4).

15.3
Occurrence of Endopolyploidy

Endopolyploidy is not uniformly distributed in the plant kingdom but seems to be a characteristic feature of some plant groups (e.g. families), while being rare in others. Furthermore, species, and even varieties and ecotypes, organs and tissues may differ in their degree of endopolyploidy. Therefore, it is important to take into account the nature of the plant material that is being investigated in endopolyploidy studies.

15.3.1
Endopolyploidy in Species

While very abundant in angiosperms, endopolyploidy seems to be absent in gymnosperms (Barow and Meister 2003; D'Amato 1989; Nagl 1978). Very few data are available on endopolyploidy in algae. Nevertheless, it was found in Characeae (Kwiatkowska et al. 1998; Maszewski 1991) and Phaeophyceae (Garbary and Clarke 2002).

Angiosperm families seem to be very consistent in regard to whether or not their species exhibit systemic endopolyploidy (Barow and Meister 2003; Czeika 1956; Fenzl and Tschermak-Woess 1954; Olszewska and Osiecka 1982; Tschermak-Woess and Hasitschka 1953). Also the degree of endopolyploidization seems to be characteristic for a given family (Barow and Meister 2003). For instance, Cucurbitaceae, Chenopodiaceae/APG: Amaranthaceae and many species of Brassicaceae show very high, Solanaceae, Alliaceae and Fabaceae intermediate, and Poaceae low levels of endopolyploidy. An analysis of variance (ANOVA) based on data from over 50 species and including genome size, life strategy, organ type and family as explanatory variables showed the tightest correlation between endopolyploidy and family affiliation (Barow and Meister 2003). However, since there is no clear relationship between endopolyploidy and phylogeny above the family level (Barow 2006), the authors speculated that related species embark on similar life strategies and occupy habitats with similar characteristics, which probably also requires adaptation with regard to endopolyploidy. Angiosperm families with predominantly endopolyploid and non-endopolyploid species, respectively, are listed in Tables 15.1 and 15.2. Exceptions may exist, for instance in Ranunculaceae (Barow and Meister 2003; Fenzl and Tschermak-Woess 1954; Tschermak-Woess 1963; Tschermak-Woess and Doležal 1953) and Rosaceae (Bradley and Crane 1955; D'Amato 1998), supposedly in species that deviate in their ecological characteristics from the predisposition of their family (Barow and Meister 2003; Jovtchev et al. 2006b).

15.3.2
Endopolyploidy in Ecotypes and Varieties

Endopolyploidization may differ between ecotypes and varieties of the same species. For example, frequencies of 8C and 16C nuclei in root cortex cells varied between 18 different *Arabidopsis thaliana* accessions (Beemster et al. 2002). Endopolyploidy levels also differed significantly in epidermal cells (Cavallini et al. 1997) and endosperm (Larkins et al. 2001) of different *Zea mays* lines and cotyledons of five different pea (*Pisum sativum*, Fabaceae) varieties (Lemontey et al. 2000).

Table 15.1 Plant families with predominantly endopolyploid species.

Plant family	Reference
Aizoaceae	De Rocher et al. 1990; Tschermak-Woess 1956; Wulf 1940
Alliaceae	Barow and Meister 2003; Olszewska and Osiecka 1982
Asphodelaceae	Fenzl and Tschermak-Woess 1954; Jähnl 1947; Lauber 1947
Brassicaceae	Barow and Meister 2003; Reitberger 1949; Siwinska and Maluszynska 2003; Tschermak-Woess and Hasitschka 1954
Chenopodiaceae/APG: Amaranthaceae	Barow and Meister 2003; Tschermak-Woess and Doležal 1953; Wulf 1936
Caryophyllaceae	Tschermak-Woess and Hasitschka 1954
Commelinaceae	Fenzl and Tschermak-Woess 1954
Crassulaceae	Czeika 1956; De Rocher et al. 1990; Jähnl 1947
Cucurbitaceae	Barow and Meister 2003; Tschermak-Woess and Hasitschka 1954
Euphorbiaceae	Tschermak-Woess and Hasitschka 1954
Fabaceae	Barow and Meister 2003; Scott and Smith 1998
Hyacinthaceae	Fenzl and Tschermak-Woess 1954
Hydrocharitaceae	Tschermak-Woess and Hasitschka 1954
Iridaceae	Olszewska and Osiecka 1982
Orchidaceae	Jones and Kuehnle 1998; Lee et al. 2004; Lim and Loh 2003; Lin et al. 2001
Poaceae, subfamily Pooidae	Barow and Meister 2003; Griffiths et al. 1994; Olszewska and Osiecka 1982
Potamogetonaceae	Tschermak-Woess and Hasitschka 1954
Solanaceae	Barow and Meister 2003
Urticaceae	Barow and Meister 2003

15.3.3
Endopolyploidy in Different Life Strategies

The level of endopolyploidization may reflect the life strategy and ecological adaptation of a plant species (Barow and Meister 2003). Endopolyploidy is more frequent in annual and biennial than in perennial herbs and seems to be absent in woody species (Fig. 15.3). Among short-lived plants, those adapted to infertile soils seem less likely to exhibit endopolyploidy (Barow and Meister 2003), al-

Table 15.2 Plant families with predominantly non-endopolyploid species.

Plant family	Reference
Amaryllidaceae	Olszewska and Osiecka 1982
Apiaceae	D'Amato 1998; Nagl 1978; Tschermak-Woess and Doležal 1953
Araceae	Olszewska and Osiecka 1982
Asteraceae	Barow and Meister 2003; Czeika 1956; Fenzl and Tschermak-Woess 1954
Fagaceae	Barow and Meister 2003
Ginkgoaceae	Barow and Meister 2003; Nagl 1978
Lamiaceae	Barow and Meister 2003
Liliaceae	Barow and Meister 2003; Fenzl and Tschermak-Woess 1954
Papaveraceae	Tschermak-Woess and Doležal 1953
Pinaceae	Barow and Meister 2003; Nagl 1978
Ranunculaceae	Barow and Meister 2003; Fenzl and Tschermak-Woess 1954
Rosaceae	Barow and Meister 2003

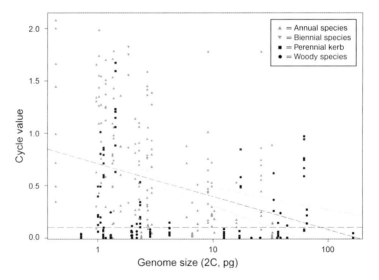

Fig. 15.3 Relation between endopolyploidy (cycle value), genome size and life strategy for 54 species out of 16 seed plant families and up to 12 organs per species (symbols in the same column). The correlation coefficient between genome size and cycle value is $r = -0.259$ ($P < 0.001$, short dashed line). Dotted line, confidence interval of 95%. Long dashed line, endopolyploidy cut-off value at the cycle value of 0.1. Data taken from Barow and Meister (2003).

though these findings are largely based on Central European species and might differ in species from other climatic regions. Furthermore, endopolyploidy seems to be a common characteristic of succulent plants (Czeika 1956; De Rocher et al. 1990; Emshwiller 2002).

15.3.4
Endopolyploidy in Organs

The level of endopolyploidy is generally organ-specific (Barow and Meister 2003; Galbraith et al. 1991; Gilissen et al. 1993,) and, interestingly, endopolyploid species share a similar pattern of between-organ differences (Fig. 15.4). The highest levels are found in cotyledons, stamina and lower leaf stalks, while upper leaves and roots generally exhibit the lowest levels of endoploidy. Gynoecia are probably not endopolyploid until the pollination and onset of ripening (Barow and Meister 2003; Geitler 1953; Lagunes-Espinoza et al. 2000). On average, lower organs exhibit higher levels of endopolyploidy than upper organs, particularly when corresponding organs are compared, for instance leaves or leaf stalks (Barow and

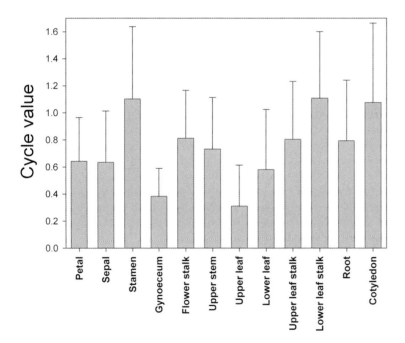

Fig. 15.4 Mean endopolyploidy (cycle value) and standard deviation per organ for endopolyploid species out of eight angiosperm families. Values of 11–31 species per organ are included. Plants were grown in the greenhouse under standard conditions. Samples from 5–10 individuals per organ and species were separately prepared and analyzed. Data taken from Barow and Meister (2003).

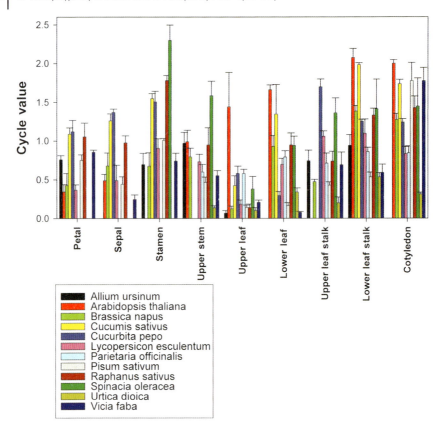

Fig. 15.5 Mean endopolyploidy (cycle value) and standard deviation shown separately for different organs of 12 species. Plants were grown in the greenhouse under standard conditions. Samples from 5–10 individuals per organ and species were separately prepared and analyzed. The figure shows that endopolyplodization is generally a tightly-regulated process resulting in a distinct endopolyploidy level of each organ of a species. Despite considerable differences between species, a general tendency for a given organ becomes obvious (e.g. high levels in cotyledons and lower leaf stalks, and low levels in upper leaves).

Meister 2003; Galbraith et al. 1991; Smulders et al. 1994). Generally, endopolyploidization of organs is a relatively tightly-controlled process (Fig. 15.5).

Within a certain tissue of a given organ, a gradient of endopolyploidy levels may occur. Endopolyploidy decreases from inner to outer cell layers in fruits (Lauber 1947; Sinnott 1939), succulent leaves of *Othonna crassifolia* (Asteraceae; Czeika 1956), stems, where the endopolyploidy level may be higher for pith cells than for cortex cells (Czeika 1956; Fenzl and Tschermak-Woess 1954), cotyledons of *Pisum sativum* (Scharpé and van Parijs 1973), and endosperm of *Zea mays* and *Ecballium elaterium* (Cucurbitaceae; D'Amato 1998; Kowles and Phillips 1988; Kowles et al. 1990; see also Lur and Setter 1993). In *Bryonia dioica* (Cucurbitaceae), endopolyploidy levels are higher (up to 16C) in the periphery of the pith of the stem than in the center with predominantly 2C (Tschermak-Woess 1956). In *Vicia faba* (Fabaceae; Coleman 1950) and *Zebrina purpusii* (Commelinaceae; Tschermak-Woess 1956), higher endopolyploidy levels are found in nodes than in internodes. The degree of endopolyploidy decreases towards the vascular bundles of many fruits (Lauber 1947) and of *Cereus spachianus* (Cactaceae; Fenzl and Tschermak-Woess 1954), of which the companion cells remain in 2C or 4C. In leaves of *Vanda* (Orchidaceae; Lim and Loh 2003) and *Arabidopsis thaliana* (De Veylder et al. 2001; Pyke et al. 1991), the endopolyploidy decreases from tip to base, whereas in cotyledons of *Cucumis sativus* (Cucurbitaceae; Gilissen et al. 1993) and leaves of *Spathoglottis plicata* (Orchidaceae; Yang and Loh 2004), it decreases from base to the tip. In *Triticum durum* (Poaceae), the degree of endopolyploidy of epidermal cells of the first foliage leaf is higher in the middle than at the tip and the base (Cionini et al. 1983).

Epidermal tissues except trichomes often exhibit low or no endopolyploidy. Among 15 succulents, Czeika (1956) found only four species in which the epidermal cells exceeded 4C. Epidermal cells of cotyledons of *Pisum sativum* remain in 2C and exceptionally 4C (Scharpé and van Parijs 1973). Also in fruits, epidermal tissue generally consists of only 2C or 2C and 4C cells (Lauber 1947). In stems, epidermal pavement cells and often the first two or three subepidermal cell layers remain in 2C (Fenzl and Tschermak-Woess 1954).

However, in *Arabidopsis thaliana*, the ploidy levels of epidermal pavement cells range from 2C to 8C in stems and from 2C to 16C in leaves (Melaragno et al. 1993).

Guard cells of stomata seem to remain exclusively in 2C (Butterfass 1963; Czeika 1956; Melaragno et al. 1993; Tschermak-Woess 1956). Interestingly, a decrease in endopolyploidy has been found towards stomata in the epidermal pavement cells of *Portulaca grandiflora*, *Anacampseros filamentosa* (both Portulacaceae; Czeika 1956) and *Arabidopsis thaliana* (Melaragno et al. 1993).

In contrast to pavement cells, trichomes often exhibit high ploidy levels, ranging from 4C up to 256C (Melaragno et al. 1993; Tschermak-Woess and Hasitschka 1954).

For the root cortex, Tschermak-Woess (1956) assumed three types of layering of ploidy within the cortical cylinder. In the first type (e.g. in *Aloe*, Asphodelaceae, with further species mentioned by Tschermak-Woess and Doležal 1953), inner and outer cell layers consist exclusively or predominantly of 2C, and endopolyploid cells are found in the middle. In the second type (e.g. in *Rhoeo discolor*, Commelinaceae), the inner cell layers are endopolyploid and the outer layers are predominantly in 2C, while no clear layering is discernible in the third type (e.g. in Chenopodiaceae/APG: Amaranthaceae), but cells with higher ploidy predominate in the middle of the cortical cylinder. In the central cylinder of roots, 2C and polyploid cells are intermingled. The root pith retains the diploid state in most species investigated (Tschermak-Woess and Doležal 1953). However, the authors assumed high endopolyploidy levels in some monocots because of their large pith cells.

Whereas in most tissues the upper limit of endopolyploidy is 32C or 64C, a few tissues and cell types (suspensor cells, endosperm haustoria, endosperm, antipodal cells, fruit elaiosome) may exhibit extraordinarily high levels of endopolyploidy ranging from about 512C to 24 576C (Nagl 1976a, 1978; D'Amato 1998).

15.4
Factors Modifying the Degree of Endopolyploidization

The levels of endopolyploidy are modified during ontogenesis in response to environmental conditions. Therefore, growth conditions need to be controlled and reported when endopolyploidy is being studied. In plant breeding and *in vitro* cultures, genome size as well as phytohormone applications and concentrations, respectively, need to be considered.

15.4.1
Genome Size and Endopolyploidy

A negative correlation exists between endopolyploidy levels and genome size in angiosperms (Fig. 15.3; Barow and Meister 2003; De Rocher et al. 1990; Nagl 1976b). However, this correlation is weak, especially when woody species are included (often possessing small genomes and generally lacking endopolyploidy). Moreover, other factors such as phylogeny and life strategy have a stronger impact than genome size on whether or not a species exhibits endopolyploidy (Barow and Meister 2003; Jovtchev et al. 2006b).

However, artificially-generated tetraploid accessions of *Zea mays* (Biradar et al. 1993), *Lycopersicon esculentum* (Solanaceae; Smulders et al. 1994) and *Portulaca grandiflora* (Mishiba and Mii 2000) revealed an immediate reduction in endopolyploidy when compared to the corresponding, supposedly genetically identical, diploid accessions, although the pattern of endopolyploidy was similar in both di- and tetraploids. In *Beta vulgaris* (Chenopodiaceae/APG: Amaranthaceae), even the endopolyploidy pattern was different between diploid plants on the one hand and tri- and tetraploid plants on the other (all belonging to different cultivars or breeding lines) for roots and hypocotyls but not for cotyledons, petioles and leaves (Sliwinska and Lukaszewska 2005). Contrarily, diploid and artificial tetraploid accessions of *Solanum tuberosum* (Solanaceae; Pijnacker et al. 1989) and diploid and tetraploid ecotype of *Arabidopsis thaliana* (Gendreau et al. 1998) exhibited the same extent of endopolyploidization.

Investigating 10 diploid and two tetraploid accessions of *Arabidopsis thaliana* (Brassicaceae), a diploid and a tetraploid accession of *Koeleria macrantha*, *Dactylis glomerata* and *Phleum pratense* as well as a decaploid *Koeleria pyramidata* (all Poaceae), Jovtchev et al. (2006b) found higher degrees of endopolyploidy in the diploid than in the corresponding polyploid accession or species, respectively. They suppose that in polyploids the degree of endopolyploidy is regulated down over many generations by selection.

15.4.2
Environmental Factors

As early as 1951, the variable leaf shape and stature of *Kalanchoë blossfeldiana* (Crassulaceae) in response to different *light* conditions were related to different degrees of endopolyploidization (Witsch and Flügel 1951). Thicker leaves and larger cells under short day conditions in this species coincided with higher endopolyploidy. Furthermore, etiolated seedlings of several endopolyploid species revealed higher degrees of endopolyploidy than those found in seedlings grown in light (Galli 1988; Gendreau et al. 1998; Giles and Myers 1964; Van Oostveldt and Van Parijs 1975). This process is mediated by phytochromes A and B (Gendreau et al. 1998; Van Oostveldt et al. 1978).

Endopolyploidy of epidermal cells of the first foliage leaf of *Triticum durum* in the variety "Creso", but not the variety "Capelli", was higher for plants grown in the dark than for those grown in white light (Cavallini et al. 1995). Imposing shade on potato plants at the stage of tuber initiation leads to lower endopolyploidy in tubers (Chen and Setter 2003).

Temperature has different effects on endopolyploidization depending on the adaptation of a given species and the findings seem to be rather equivocal.

Tschermak-Woess and Hasitschka (1954) found that endopolyploidy in trichomes of *Urtica pilulifera* and *U. caudata* (Urticaceae) grown at very low temperatures decreased by three levels whereas cell size was the same as in trichomes of plants cultivated under higher temperatures. Likewise, in *Lolium multiflorum* (Poaceae) grown at 7, 15, 20 and 25 °C, lower temperatures resulted in (not significantly) lower endopolyploidy in bulliform epidermal cells (Griffiths et al. 1994). Bulliform cells are thin-walled epidermal cells with a large vacuole in the trough between two ridges of the leaf blade that can rapidly expand or contract due to changes in water content, thereby controlling the folding or rolling of a leaf.

However, investigating *Arabidopsis thaliana*, *Brassica napus* and *Sinapis arvensis* grown at 14, 18, and 22 °C, Jovtchev et al. (2006b) found opposite tendencies in all these species, but with a significant negative correlation only in *Arabidopsis thaliana*.

Exposure of developing maize endosperm to high temperature (35 instead of 25 °C) for 4 or 6 days (starting at the fourth day after pollination) represses endoreduplication (Engelen-Eigles et al. 2000).

Low *water* supply decreases endopolyploidy in maize endosperm (Artlip et al. 1995), whereas a certain combination of *nutrition* and irrigation may enhance endopolyploidy in leaves of *Beta vulgaris* (Butterfass 1966). In a *Sorghum bicolor* (Poaceae) variety with inducible tolerance to salinity, exposure to sublethal salt concentrations in early development results in significantly higher endopoly-

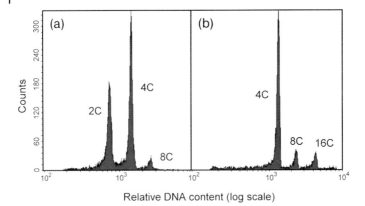

Fig. 15.6 Endopolyploidy of (a) only mature roots of *Pisum sativum* with mean C-level = 3.55 and cycle value = 0.72, and (b) only root nodules with mean C-level = 5.53 and cycle value = 1.29. The same instrument setting was used for both analyses. Endopolyploidy is considerably enhanced in root nodules as a result of the bacterial infection.

ploidy in the root cortex but not in the root vascular cylinder or leaves (Ceccarelli et al. 2006). A higher phosphate supply leads to higher endopolyploidy in roots and leaves of *Allium porrum* (Fusconi et al. 2005).

15.4.3
Symbionts and Parasites

Symbionts and parasites may influence the degree of endopolyploidy in roots. Endopolyploidization triggered by Rhizobiaceae soil bacteria is an integral part of symbiotic cell differentiation in the root nodules of legumes (Fig. 15.6; Cebolla 1999; Kondorosi et al. 2000).

Tomato root cells colonized by mycorrhizal fungi from the genus *Glomus* show increased endopolyploidy (Berta et al. 2000), whereas endopolyploidy in roots cells of *Allium porrum* and *Pisum sativum* is not affected by the infection itself (Lingua et al. 2001a, 2001b).

Nematodes of the genus *Meloidogyne* induce endoreduplication in roots of host plants (Williamson and Hussey 1996).

In endopolyploid species, cells of crown galls often show a higher level of endopolyploidy than the tissue of origin (D'Amato 1964). In sunflower, however, gall cells remain diploid. Hesse (1971) in his review on nodule, gall and tumor building species of bacteria, fungi (Plasmodiophoromycetes, Oomycetes, Ascomycetes, Basidiomycetes), Nematodes, Acarina, Coleoptera, Homoptera, Hymenoptera, Lepidoptera and Diptera, describes three different mechanisms leading to endopolyploidy caused by symbionts and parasites: (i) endoreduplication, (ii) disturbance of mitosis resembling endomitosis although less well regulated since different stages of mitosis can be affected in adjacent cells, and (iii) cell fusion leading to multinucleate cells, which may also contain endopolyploid restitution nuclei. From the compilation of data for 163 species, he concludes that in a particular group one mechanism is (almost) exclusively exhibited, such as endoreduplication by Cynipidae (Hymenoptera), disturbance of mitosis or multinucleate cells by Tylenchidae (Nematodes) or no change of endopolyploidy by greenflies (in a broader sense) and Coccidae (both Homoptera), Tenthredinidae (Hymenoptera), Taphrinales (Ascomycetes), and Uredinales (Basidiomycetes). However, different nodule and crown gall-evoking bacteria may or may not cause endopolyploidy as well as different species of Tetrapodili (Acarina), Psyllidae (Homoptera) and Cecidomyidae (Diptera) which may or may not induce endopolyploidy or multinucleate cells.

15.4.4
Phytohormones

The impact of the above-mentioned environmental factors on endopolyploidy is mediated by phytohormones.

Auxins are known to increase endopolyploidy. In apricot fruits, 2,4,5-trichlorophenoxyacetic acid enhances endopolyploidy by approximately 30–50% (Bradley and Crane 1955). In maize endosperm, endoreduplication coincides with a rapid increase in the concentration of indolyl-acetic acid, IAA (Lur and Setter 1993). Naphtaleneacetic acid (NAA) induces endoreduplication in embryos of the orchid *Vanda* growing in Murashige Skoog medium (Lim and Loh 2003), single cell cultures of *Nicotiana tabacum* (Solanaceae; Valente et al. 1998) and in *in vitro* cultured explants of the orchid *Cymbidium* (Fuji et al. 1999).

In cell suspension cultures of *Doritaenopsis* (Orchidaceae), endopolyploidy is increased by 2,4-dichlorophenoxyacetic acid (2,4-D), Picloram, and high (10 mg l^{-1}) doses of NAA but not by IAA and indole-3-butyric acid (IBA) (Mishiba et al. 2001).

However, in *Mammillaria san-angelis* (Cactaceae), NAA, IAA, IBA, 2,4-D, and Picloram had no clear effect on endopolyploidy in plants regenerated from cell cultures (Palomino et al. 1999).

Gibberellins may enhance endopolyploidy in plants as well, although less clearly than auxins.

Gibberellic acid (GA) increased endopolyploidization in etiolating seedlings of the variety "Finale" but not the variety "Rondo" of *Pisum sativum* (Callebaut et al. 1982). In contrast, seedlings of the dwarf *Pisum sativum* variety "Kleine Rheinländerin" grown in light show increased endopolyploidy after treatment with GA$_3$ (Mohamed and Bopp 1980). Correspondingly, hypocotyls of GA-deficient mutants of *Arabidopsis thaliana* exhibit lower endopolyploidy than the wild type (Gendreau et al. 1999). In cultured embryos of the orchid *Vanda*, only high concentrations of gibberellin cause a slight increase of endopolyploidy (Lim and Loh 2003). Also in *Chara vulgaris* (Characeae), GA seems to enhance endopolyploidy (Kwiatkowska et al. 1998).

Ethylene induces an extra round of endoreduplication in hypocotyls of *Arabidopsis thaliana* seedlings (Gendreau et al. 1999). It also triggers DNA synthesis and impedes mitosis in epidermal cells of *Cucumis sativus* seedlings grown in light (Dan et al. 2003).

Brassinosteroids, growth-promoting plant steroid hormones, were shown to support endopolyploidization in *Arabidopsis thaliana* (Hase et al. 2005).

Cytokinins are, unlike auxins, gibberellins and ethylene, clear antagonists of endopolyploidization. In cell suspension cultures of the orchid *Doritaenopsis*, benzyladenine and 1-phenyl-3-(1,2,3-thiadiazol-5-yl)urea slightly decreased endopolyploidy (Mishiba et al. 2001). 6-Benzylaminopurine induces divisions of endopolyploid cells in single cell cultures of *Nicotiana tabacum*, in which endoreduplication had been driven by NAA. These cells divide (without DNA replication) until they reach the 2C/4C state and then continue with a normal mitotic cycle (Valente et al. 1998). Correspondingly, *in vivo*, the ratio of auxins to cytokinins seems to be crucial for the switch from cell division to endoreduplication (Bryant and Francis 2001; Lur and Setter 1993). Furthermore, in many cases the impact of phytohormones on endopolyploidy might be a secondary effect associated with growth stimulation or inhibition rather than a specific effect on endopolyploidization itself (Dewitte and Murray 2003; Gendreau et al. 1998). However, expression of cell cycle regulators such as cyclins and inhibitors of cyclin-dependent kinases, which mainly drive the mitosis and the endocycle, was shown to be modulated in response to phytohormones (Arias et al. 2006; Dewitte and Murray 2003; Larkins et al. 2001; Mironov et al. 1999; Stals and Inzé 2001).

15.5
Dynamics of Endopolyploidization

Several authors studied the dynamics of endopolyploidization, for instance in maize endosperm (Schweizer et al. 1995), orchid flowers (Lee et al. 2004), lupin pods (Lagunes-Espinoza et al. 2000), cucumber plants (Gilissen et al. 1993), cabbage petals (Kudo and Kimura 2002), and the first leaves of *Arabidopsis thaliana* (Beemster et al. 2005). Starting with all cells going through the mitotic cycle, cells with the higher ploidy levels occur in a stepwise manner one after the other following the onset of endoreduplication. Since cell division is terminated after some time while endoreduplication still continues, 2C nuclei are no longer generated but become nuclei of a higher ploidy level. Starting with the lowest ploidy, the number of nuclei of particular ploidy levels decreases correspondingly, because they are converted into the next highest endopolyploidy category. This again happens stepwise, except for the highest ploidy, and is less pronounced for higher than for lower ploidy levels.

To describe the progression (transition) of cells from one ploidy level to the next higher level, a set of differential equations (one equation per each possible transition) was employed. Lee et al. (2004) modified the model obtained by Schweizer et al. (1995) by changing the value of the transition rate by a Fermi function, also known as a logistic or sigmoid function, to optimize the equation for their object of study.

15.6
Endopolyploidy and Plant Breeding

Endopolyploidy offers new potentials in plant breeding for both horti- and agriculture.

In general, proportionality exists between cell volume and endopolyploidy level for a particular tissue of a given species. Therefore, it might be speculated that enhanced endopolyploidy could result in larger organs in crop plants (e.g. larger fruits), or in ornamental plants (e.g. larger flowers or leaves). Furthermore, there are speculations that endopolyploidy might enhance the metabolic and synthetic potential in endopolyploid cells, again resulting in higher yields in crop plants.

In addition, assessment of endopolyploidy is particularly important in plant cell and tissue cultures to check for undesirable somaclonal variation.

15.6.1
Endopolyploidy in Crop Plants

Endopolyploidy is exhibited by many crop plants (e.g. species of the Brassicaceae, Chenopodiaceae/APG: Amarantaceae, Cucurbitaceae, Fabaceae, Solanaceae, and some Poaceae; Barow and Meister 2003; see also Section 15.3.2) and is present in agriculturally important plant organs, such as potatoes (Chen and Setter 2003), tomatoes (Bergervoet et al. 1996), maize (Schweizer et al. 1995), spinach (Barow and Meister 2003), peas (Lemontey et al. 2000; Scharpé and van Parijs 1973), and apricots (Bradley and Crane 1955).

Because of the correlation between cell size and DNA content in both meristematic (Price et al. 1973) and endopolyploid cells (Jovtchev et al. 2006; Melaragno et al. 1993; reviewed in Barow 2006), plant organ size depends on cell number, genome size and endopolyploidy. Correspondingly, Lemontey et al. (2000) found a strong correlation between mature seed weight and mean C-level for nine pea (*Pisum sativum*) varieties and reciprocal hybrids. Therefore, it is to be expected that a higher yield of crops may be achieved by increasing endopolyploidy levels (Butterfass 1966; Nagl 1978), either by traditional breeding or by manipulation of corresponding genes. However, the control of plant organ growth and size is more complex, and a certain organ size is often reached even when the mitotic cell cycle is inhibited or cell growth is changed (reviewed in Mizukami 2001). Furthermore, Butterfass (1966) reports that in sugar beet (*Beta vulgaris*) grown in Central Europe, diploid plants with predominating 4C and 8C cells in leaves give the highest yields in wet years, whereas plants with predominantly 2C cells give the highest yields in dry years; in average conditions, plants with 4C leaf cells give the highest yields. These results suggest an optimal degree of endopolyploidy in a given crop plant for a given climate. Therefore, Butterfass (1966) argues that if tetraploid sugar beets are produced to combine different genomes, breeders should aim to suppress endopolyploidization in order to achieve approximately the same absolute amount of DNA per cell in tetraploid plants as in their diploid counterparts.

Results on effects of environmental stress on the mitotic cell cycle and endopolyploidization imply a lower sensitivity of the latter, with respect to water-deficit (e.g. in developing maize endosperm; Artlip et al. 1995; Setter and Flannigan 2001) or shade (e.g. during tuber induction in potato; Chen and Setter 2003). However, endopolyploidization is a part of the cell differentiation process, which generally coincides with the onset of vacuolization. Therefore, it may be possible that the corresponding stress induces earlier differentiation and maturation of an organ, thereby leading to earlier cessation of the mitotic cell cycle and onset of endoreduplication rather than initiating endopolyploidization more or less specifically.

Several authors speculated that endopolyploidy might be a means to enhance the metabolic and synthetic capacity of cells (Nagl 1978). However, cell volumes of endopolyploid cells are generally proportional to their DNA contents within a given tissue (Barow 2006; Jovtchev et al. 2006). Therefore, the amount of DNA templates *per cell volume*, which is supposed to allow a higher metabolic activity in endopolyploid cells, is not increased. Consequently, the metabolic potential *per cell volume* is not enhanced in such cells, at least not by this simple mechanism.

15.6.2
In vitro Culture and Plant Regeneration

Mixoploidy, polyploidy, and aneuploidy can occur in *in vitro* plant cultures and are particularly undesirable in regenerants. One cause for somaclonal variation might be endopolyploidy in initial explants (Fras and Maluszynska 2004; Iantcheva et al. 2001; Yang and Loh 2004) or endoreduplication during *in vitro* culture (Nontaswatsri and Fukai 2005; Valente et al. 1998). The degree of endopolyploidy may increase with the age of the culture (Fras and Maluszynska 2004; Nontaswatsri and Fukai 2005; Valente et al. 1998) and therewith the likelihood of polyploid regenerants. However, generally only a low percentage of regenerants is polyploid, indicating that mainly cells in 2C or 4C of the mitotic cell cycle are able to regenerate while the majority of the differentiated cells with 4C and higher endopolyploidy levels are mostly excluded from regeneration. Teixeira da Silva (2005) found in tobacco that juvenile *in vitro* plant material results in a greater amount of physiologically normal, harvestable shoots and a higher shoot regeneration capacity. Nontaswatsri and Fukai (2005) report that 94.4% of *Dianthus* (Caryophyllaceae) shoots regenerated from callus were diploid plants with corresponding endopolyploidy levels, whereas only a few plants were tetraploid and octoploid and even fewer plants showed 6C as the lowest endopolyploidy level. To further reduce the risk of poly- or mixoploid regenerants, Iantcheva et al. (2001) analyzed endopolyploidy levels in petals and leaves of five *Medicago* (Fabaceae) species to identify the organ type with lower endopolyploidy (i.e. leaves in this particular case). In addition, Fras and Maluszynska (2004) found slightly different endopolyploidy levels and patterns during primary and secondary callus cultures. Interestingly, Smulders et al. (1995) reported that seedlings of direct regenerants from different organs of tomato plants show higher standard deviations of endopolyploidy to different extents depending on the organ. Variances were particularly high for cotyledons but reverted back to a normal level after two generations, demonstrating a long-term effect of *in vitro* culture.

Like endopolyploidy, aneuploidy *in vitro* increases with the increased culture time (Fras and Maluszynska 2004). Some aneuploid cells may originate *in vitro* from irregular mitosis of endopolyploid cells (Valente et al. 1998) since chromosome orientation and segregation is complicated in such cells. However, endoreduplication can probably be controlled and directed by choosing appropriate concentrations and combinations of phytohormones (Valente et al. 1998; see also Section 15.4.4).

15.7 Conclusions

Endopolyploidy is a common feature in the plant kingdom but not universally exhibited in all plant families and species. Generally, endopolyploidy is systemic and relatively tightly controlled, with characteristic levels in different organs and tissues of a given species. Nevertheless, the degree of endopolyploidy may be modified as a response to different environmental conditions. These adaptations are mediated by phytohormones. Knowledge and control of the factors influencing the endopolyploidy pattern (e.g. phylogeny, organ and tissue type, genome size, environmental factors, etc.) is essential to achieve reliable results. Endopolyploidy is also of practical importance in improving the yield and quality in agricultural crops, breeding ornamental plants, and during *in vitro* culture.

Since large amounts of data can be generated in a relatively short time, flow cytometry has revolutionized endopolyploidy research and greatly enhanced our knowledge about the distribution and frequency of endopolyploidy across the plant kingdom. With representative samplings at both the species and tissue levels, statistically well-founded assessment of relationships with ecological, physiological and phylogenetic variables has become possible. Furthermore, reliable insights into the dynamics and development of endopolyploidy during ontogeny have been gained. Recently, FCM has also contributed to the elucidation of endoreduplication at the molecular level and flow sorting greatly facilitated investigations on chromatin organization and structure in nuclei of particular endopolyploidy categories.

Difficulties with identification of tissue and/or cell type-specific nuclei are often considered a significant drawback, which may hinder further use of FCM in endopolyploidy studies. However, it has been shown recently (Zhang et al. 2005) that a biparametric analysis of the cell type-specific expression of a nuclear-targeted Green Fluorescent Protein (GFP) and nuclear DNA content allowed specific cell types to be distinguished and facilitated the examination of endopolyploidy in *Arabidopsis thaliana* with a previously unattainable level of resolution. We anticipate that similar analyses will open up a new era in endopolyploidy research using flow cytometry.

References

Ammermann, D. **1971**, *Chromosoma* 33, 209–238.

Anisimov, A. P. **2005**, *Cell Biol. Int.* 29, 993–1004.

Arias, R. S., Filichkin, S. A., Strauss, S. H. **2006**, *Trends Biotechnol.* 24, 267–273.

Artlip, T. S., Madison, J. T., Setter, T. L. **1995**, *Plant, Cell Environ.* 18, 1034–1040.

Baroux, C., Fransz, P., Grossniklaus, U. **2004**, *Planta* 220, 38–46.

Barow, M. **2006**, *BioEssays* 28, 271–281.

Barow, M., Meister, A. **2003**, *Plant, Cell Environ.* 26, 571–584.

Beemster, G. T. S., De Vusser, K., De Tavernier, E., De Bock, K., Inzé, D. **2002**, *Plant Physiol.* 129, 854–864.

Beemster, G. T. S., De Veylder, L., Vercruysse, S., West, G., Rombaut, D., Van Hummelen, P., Galichet, A., Gruissem, W., Inzé, D. **2005**, *Plant Physiol.* 138, 734–743.

Beermann, W. **1962**, *Riesenchromosomen*. Protoplasmatologia VI/D, Springer-Verlag, Wien.

Bergervoet, J. H. W., Verhoeven, H. A., Gilissen, L. J. W., Bino, R. J. **1996**, *Plant Sci.* 116, 141–145.

Berta, G., Fusconi, A., Sampo, S., Lingua, G., Perticone, S., Repetto, O. **2000**, *Plant Soil* 226, 37–44.

Biradar, D. P., Rayburn, A. L., Bullock, D. G. **1993**, *Ann. Bot.* 71, 417–421.

Bradley, M. V., Crane, J. C. **1955**, *Am. J. Bot.* 42, 273–281.

Bryant, J., Francis, D. **2001**, Plant growth regulators and the control of S-phase in *The Plant Cell Cycle and its Interfaces*, ed. D. Francis, Sheffield Academic Press.

Butterfass, T. **1963**, *Ber. Deutsch. Bot. Gesellsch.* 76, 123–134.

Butterfass, T. **1966**, *Mitt. Max-Planck-Gesellsch. Förder. Wissen.* 1952, 47–58.

Callebaut, A., Van Oostveldt, P., Van Parijs, R. **1982**, *Planta* 156, 553–559.

Cassagnau, P. **1968**, *Chromosoma* 24, 42–58.

Cavallini, A., Baroncelli, S., Lercari, B., Cionini, G., Rocca, M., D'Amato, F. **1995**, *Protoplasma* 186, 57–62.

Cavallini, A., Natali, L., Cionini, G., Balconi, C., D'Amato, F. **1997**, *Theor. Appl. Genet.* 94, 782–787.

Cebolla, A., Vinardell, J. M., Kiss, E., Boglarka, O., Roudier, F., Kondorosi, A., Kondorosi, E. **1999**, *EMBO J.* 18, 4476–4484.

Ceccarelli, M., Santantonio, E., Marmottini, F., Amzallag, G. N., Cionini, P. G. **2006**, *Protoplasma* 227, 113–118.

Chen, C.-T., Setter, T. L. **2003**, *Ann. Bot.* 91, 373–381.

Cionini, P. G., Cavallini, A., Baroncelli, S., Lercai, B., D'Amato, F. **1983**, *Protoplasma* 118, 6–43.

Coleman, L. C. **1950**, *Can. J. Res.* 28, 382–391.

Czeika, G. **1956**, *Österr. Bot. Zeit.* 103, 536–366.

D'Amato, F. **1964**, *Caryologia* 17, 41–52.

D'Amato, F. **1989**, *Caryologia* 42, 183–211.

D'Amato, F. **1998**, Chromosome endoreduplication in plant tissue development and function in *Plant Cell Proliferation and its Regulation in Growth and Development*, ed. J. A. Bryant and D. Chiatante, John Wiley and Sons, pp. 153–166.

D'Amato, F., Avanzi, M. G. **1948**, *Nuovo Giorn. Bot. Ital.* 55, 161–213.

Dan, H., Imaseki, H., Wasteneys, G. O., Kazama, H. **2003**, *Plant Physiol.* 133, 1726–1731.

De Rocher, E. J., Harkins, K. R., Galbraith, D. W., Bohnert, H. J. **1990**, *Science* 250, 99–101.

De Veylder, L., Beeckman, T., Beemster, G. T. S., Krols, L., Terras, F., Landrieu, I., Van der Schueren, E., Maes, S., Naudts, M., Inzé, D. **2001**, *Plant Cell* 13, 1653–1667.

Dewitte, W., Murray, J. A. H. **2003**, *Annu. Rev. Plant Biol.* 54, 235–264.

Doležal, R., Tschermak-Woess, E. **1955**, *Österr. Bot. Zeit.* 102, 158–185.

Edgar, B. A., Orr-Weaver, T. L. **2001**, *Cell* 105, 297–306.

Emshwiller, E. **2002**, *Ann. Bot.* 89, 741–753.

Engelen-Eigles, G., Jones, R. J., Phillips, R. L. **2000**, *Plant, Cell Environ.* 23, 657–663.

Fenzl, E., Tschermak-Woess, E. **1954**, *Österr. Bot. Zeit.* 101, 140–164.

Fras, A., Maluszynska, J. **2004**, *Genetica* 121, 145–154.

Fuji, K., Kawamo, M., Kako, S. **1999**, *J. Jpn. Soc. Hort. Sci.* 68, 41–48.

Fusconi, A., Lingua, G., Trotta, A., Berta, G. **2005**, *Mycorrhiza* 15, 313–321.

Galbraith, D. W., Harkins, K. R., Maddox, J. M., Ayres, N. M., Sharma, D. P., Firoozabady, E. **1983**, *Science* 220, 1049–1051.

Galbraith, D. W., Harkins, K. R., Knapp, S. **1991**, *Plant Physiol.* 96, 985–989.

Galli, M. G. **1988**, *Ann. Bot.* 62, 287–293.

Garbary, D. J., Clarke, B. **2002**, *Bot. Mar.* 45, 211–216.

Geitler, L. **1939**, *Chromosoma* 1, 1–22.

Geitler, L. **1953**, Endomitose und endomitotische Polyploidisierung in *Protoplasmatologia – Handbuch der Protoplasmaforschung*, (vol. VI/C), ed. L. V. Heilbrunn, and F. Weber, Springer-Verlag, Wien.

Gendreau, E., Höfte, H., Grandjean, O., Brown, S., Traas, J. **1998**, *Plant J.* 13, 221–230.

Gendreau, E., Orbovic, V., Höfte, H., Traas, J. **1999**, *Planta* 209, 513–516.

Giles, K. W., Myers, A. **1964**, *Biochim. Biophys. Acta* 87, 460–477.

Gilissen, L. J. W., Van Staveren, M. J., Creemersmolenaar, J., Verhoeven, H. A. **1993**, *Plant Sci.* 91, 171–179.

Gilissen, L. J. W., Van Staveren, M. J., Hakkert, J. C., Smulders, M. J. M. **1996**, *Plant Physiol.* 111, 1243–1250.

Givan, A. L. **2001**, *Methods Cell Biol.* 63, 19–50.

Grafl, I. **1939**, *Chromosoma* 1, 265–275.

Griffiths, P. D., Ougham, H. J., Jones, R. N. **1994**, *New Phytologist* 128, 339–345.

Hase, Y., Fujioka, S., Yoshida, S., Sun, G., Umeda, M., Tanaka, A. **2005**, *J. Exp. Bot.* 56, 1263–1268.

Hesse, M. **1971**, *Österr. Bot. Zeit.* 119, 454–463.

Huskins, L. **1947**, *Am. Naturalist* 81, 401–434.

Iantcheva, A., Vlahova, M., Trinh, T. H., Brown, S. C., Slater, A., Elliott, M. C., Atanassov, A. **2001**, *Plant Sci.* 160, 621–627.

Jähnl, G. **1947**, *Chromosoma* 3, 48–51.

Jones, W. E., Kuehnle, A. R. **1998**, *Lindleyana* 13, 11–18.

Jovtchev, G., Schubert, V., Meister, A., Barow, M., Schubert, I. **2006**, *Cytogenet. Genome Res.* 114, 77–82.

Jovtchev, G., Barow, M., Meister, A., Schubert, J. **2006b**, *Environmental and Experimental Botany*, accepted.

Kladnik, A., Chamusco, K., Chourey, P. S., Dermastia, M. **2005**, *Period. Biol.* 107, 11–16.

Kondorosi, E., Roudier, F., Gendreau, E. **2000**, *Curr. Opin. Plant Biol.* 3, 488–492.

Kowles, R. V., Phillips, R. L. **1988**, *Int. Rev. Cytol.* 112, 97–135.

Kowles, R. V., Srienc, F., Phillips, R. L. **1990**, *Dev. Genet.* 11, 125–132.

Kudo, N., Kimura, Y. **2002**, *J. Exp. Bot.* 53, 1017–1023.

Kwiatkowska, M., Wojtczak, A., Poplonska, K. **1998**, *Plant Cell Physiol.* 39, 1388–1390.

Lagunes-Espinoza, L. C., Huyghe, C., Bousseau, D., Barre, P., Papineau, J. **2000**, *Ann. Bot.* 86, 185–190.

Larkins, B. A., Dilkes, B. P., Dante, R. A., Coelho, C. M., Woo, Y.-M., Liu, Y. **2001**, *J. Exp. Bot.* 52, 183–192.

Lauber, H., **1947**, *Österr. Bot. Zeitschr.* 94, 30–60.

Lee, H.-C., Chiou, D.-W., Chen, W.-H., Markhart, A. H., Chen, Y.-H., Lin, T.-Y. **2004**, *Plant Sci.* 166, 659–667.

Lemontey, C., Mousset-Déclas, C., Munier-Jolin, N., Boutin, J.-P. **2000**, *J. Exp. Bot.* 51, 167–175.

Levan, A. **1939**, *Hereditas* 25, 87–96.

Levan, A., Hauschka, T. **1953**, *J. Natl Cancer Inst.* 14, 1–43.

Lim, W. L., Loh, C. S. **2003**, *New Phytologist* 159, 279–287.

Lin, S., Lee, H.-C., Chen, W.-H., Chen, C.-C., Kao, Y.-Y., Fu, Y.-M., Chen, Y.-H., Lin, T.-Y. **2001**, *J. Am. Soc. Hort. Sci.* 126, 195–199.

Lingua, G., Fusconi, A., Berta, G. **2001a**, *Eur. J. Histochem.* 45, 9–20.

Lingua, G., Fusconi, A., Trotta, A., Berta, G. **2001b**, Effect of AM symbiosis and phosphate on root ploidy in *Third International Conference on Mycorrhizas (ICOM3) "Diversity and Intergration in Mycorrhizas"*, Adelaide, July 8–13, http://www.waite.adelaide.edu.au/Soil_Water/3ICOM_ABSTs/Abstracts/L/G.Lingua.htm.

Lur, H.-S., Setter, T. S. **1993**, *Plant Physiol.* 103, 273–280.

Maszewski, J. **1991**, *Plant Systemat. Evol.* 177, 39–52.

Melaragno, J. E., Mehrotra, B., Coleman, A. W. **1993**, *Plant Cell* 5, 1661–1668.

Mironov, V., De Veylder, L., Van Montagu, M., Inzé, D. **1999**, *Plant Cell* 11, 509–521.

Mishiba, K., Mii, M. **2000**, *Plant Sci.* 156, 213–219.

Mishiba, K., Okamoto, T., Mii, M. **2001**, *Physiol. Plant.* 112, 142–148.

Mizukami, Y. **2001**, *Curr. Opin. Plant Biol.* 4, 533–539.

Mohamed, Y., Bopp, M. **1980**, *Zeit. Pflanzenphysiol.* 98, 25–33.

Nagl, W. **1976a**, *Annu. Rev. Plant Physiol.* 27, 39–69.

Nagl, W. **1976b**, *Nature* 261, 614–615.

Nagl, W. **1978**, *Endopolyploidy and Polyteny in Differentiation and Evolution*, North-Holland Publishing Company, Amsterdam, New York, Oxford.

Nagl, W., Capesius, I. **1976**, *Plant Systemat. Evol.* 125, 261–268.

Nagl, W. **1981**, *Int. Rev. Cytol.* 73, 21–53.

Nontaswatsri, C., Fukai, S. **2005**, *Plant Cell, Tiss. Organ Cult.* 83, 351–355.

Olszewska, M. J., Osiecka, R. **1982**, *Biochem. Physiol. Pflanz.* 177, 319–336.

Palomino, G., Doležel, J., Cid, R., Brunner, I., Méndez, I., Rubluo, A. **1999**, *Plant Sci.* 141, 191–200.

Pijnacker, L. P., Sree Ramulu, K., Dijkhuis, P., Ferwerda, M. A. **1989**, *Theor. Appl. Genet.* 77, 102–110.

Price, H. J., Sparrow, A. H., Nauman, A. F. **1973**, *Experientia* 29, 1028–1029.

Pyke, K. A., Marrison, J. L., Leech, R. M. **1991**, *J. Exp. Bot.* 42, 1407–1416.

Reitberger, A. **1949**, *Die Naturwissensch.* 36, 380.

Rieger, R., Michaelis, A., Green, M. M. **1991**, *Glossary of Genetics: Classical and Molecular*, Springer Verlag, Berlin.

Scharpé, A., van Parijs, R. **1973**, *J. Exp. Bot.* 24, 216–222.

Schubert, V., Klatte, M., Pečinka, A., Meister, A., Jasenčakowa, Z., Schubert, I. **2006**, *Genetics* 172, 467–475.

Schweizer, L., Yerk-Davis, G. L., Phillips, R. L., Srienc, F., Jones, R. J. **1995**, *Proc. Natl Acad. Sci. USA* 92, 7070–7073.

Scott, R. C., Smith, D. L. **1998**, *Bot. J. Linn. Soc.* 128, 15–44.

Šesek, P., Kump, B., Bohanec, B. **2005**, *Acta Biol. Cracov. Series Bot.* 47, 93–99.

Setter, T. L., Flannigan, B. A. **2001**, *J. Exp. Bot.* 52, 1401–1408.

Sinnott, E. W. **1939**, *Am. J. Bot.* 26, 179–189.

Sliwinska, E., Lukaszewska, E. **2005**, *Plant Sci.* 168, 1067–1074.

Siwinska, D., Maluszynska, J. **2003**, Genome size change during plant development – does it depend on original ploidy level? In *C-Value – Plant Genome Size Discussion Meeting, 11th and 12th September 2003, Poster Abstracts*, Royal Botanic Gardens, Kew, UK, p. 27.

Smulders, M. J. M., Rus-Kortekaas, W., Gilissen, L. J. W. **1994**, *Plant Sci.* 97, 53–60.

Smulders, M. J. M., Rus-Kortekaas, W., Gilissen, L. J. W. **1995**, *Plant Sci.* 106, 129–139.

Stals, H., Inzé, D. **2001**, *Trends Plant Sci.* 6, 359–364.

Stepinski, D. **2003/4**, *Biol. Plant.* 47, 333–339.

Teixeira da Silva, J. A. **2005**, *Plant Sci.* 169, 1046–1058.

Tschermak-Woess, E., Doležel, R. **1953**, *Oesterr. Bot. Ecit.* 100, 358–402.

Tschermak-Woess, E. **1956**, *Protoplasma* 46, 798–834.

Tschermak-Woess, E. **1963**, Strukturtypen der Ruhekerne von Pflanzen und Tieren in *Protoplasmatologia – Handbuch der Protoplasmaforschung*, (vol. V/1), ed. L. V. Heilbrunn, and L. Victor, Springer-Verlag, Wien/New York.

Tschermak-Woess, E., Hasitschka, G. **1953**, *Chromosoma* 5, 574–614.

Tschermak-Woess, E., Hasitschka, G. **1954**, *Österr. Bot. Zeit.* 101, 79–117.

Valente, P., Tao, W., Verbelen, J.-P. **1998**, *Plant Sci.* 134, 203–215.

Van Oostveldt, P., Van Parijs, R. B. **1975**, *Planta* 124, 287–295.

Van Oostveldt, P., Boeken, G., Van Parijs, R. **1978**, *Photochem. Photobiol.* 27, 217–221.

Williamson, V. M., Hussey, R. S. **1996**, *Plant Cell* 8, 1735–1745.

Winkelmann, M., Pfitzer, P., Schneider, W. **1987**, *Klin. Wochen.* 65, 1115–1131.

Witsch, H. v., Flügel, A. **1951**, *Die Naturwissen.* 38, 138.

Wulf, H. D. **1936**, *Planta* 26, 275–290.

Wulf, H. D. **1940**, *Ber. Deutsch. Bot. Gesellsch.* 58, 400–410.

Yang, M., Loh, C. S. **2004**, *BMC Cell Biol.* 5, 33–40.

Zhang, C., Gong, F. C., Lambert, G. M., Galbraith, D. W. **2005**, *Plant Methods*, 1, 1–12.

16
Chromosome Analysis and Sorting

Jaroslav Doležel, Marie Kubaláková, Pavla Suchánková,
Pavlína Kovářová, Jan Bartoš, and Hana Šimková

Overview

Flow karyotyping is the classification of mitotic metaphase chromosomes by flow cytometry according to DNA content and AT:GC ratio. The analysis is performed at high speed and chromosomes that are resolved as single peaks on the resulting frequency distributions (flow karyotypes) can be isolated in large quantities by flow sorting. Methods for chromosome analysis and sorting (flow cytogenetics) were originally developed for humans and some animal species. Their adaptation for plants was delayed by difficulties in preparing suspensions of intact chromosomes and in discriminating individual chromosome types. Mechanical homogenization of synchronized root tips provided a high-yielding method for sample preparation and is now widely used. The use of cytogenetic stocks, such as chromosome deletion, translocation and alien addition lines, provides a means of discriminating and sorting particular chromosomes and chromosome arms. So far, chromosome analysis and sorting has been reported in 18 plant species, including important crops such as barley, maize, and wheat. Flow karyotyping has also permitted quantitative detection of structural and numerical chromosome changes and chromosome polymorphisms. The applications of chromosome sorting have included physical mapping using PCR, high-resolution cytogenetic mapping, production of recombinant DNA libraries, and targeted isolation of DNA markers. The availability of subgenomic DNA libraries greatly simplifies the analysis of complex genomes and facilitates positional gene cloning and genome sequencing. In this chapter we describe the origins of flow cytogenetics and explain its principles. We review in detail the development of the methodology for plants, describe various applications, and assess their potential in genomics and other areas of research.

Flow Cytometry with Plant Cells. Edited by Jaroslav Doležel, Johann Greilhuber, and Jan Suda
Copyright © 2007 WILEY-VCH Verlag GmbH & Co. KGaA, Weinheim
ISBN: 978-3-527-31487-4

16.1
Introduction

The nuclear genomes of plants are organized into linear chromosomes. The number and morphology of chromosomes collectively define a karyotype, one of the characteristics of a species. While some karyotype alterations can be lethal, others may lead to reproductive isolation and accompany the origin of new species. Karyotype alterations are studied by cytogenetics, a scientific discipline that analyzes chromosome structure, behavior and function. Cytogenetics has a long history dating back to the end of 19th century, and ever since then it has advanced the understanding of plant genome structure and function. Continuous development of new methods and instrumentation has been crucial in achieving this progress.

Chromosomes can be analyzed in a number of ways and at different degrees of resolution. Visible light and fluorescence microscopy are the most common techniques. Development of molecular cytogenetics and computer technology has facilitated detailed analysis of chromosome structure at the DNA level. Coupled with computer-controlled microscopy and automatic image analysis, the instrumentation makes it possible to analyze a large number of metaphase plates in a short time. It should be noted, however, that human intervention is required in decision making about the presence of karyotype rearrangements, and that the commercially available software packages are usually limited to human karyotyping. The molecular structure of chromosomes can be analyzed at higher resolution on chromatin fibers released from nuclei, while ultrastructural analysis can be done using transmission and scanning electron microscopy.

Microscopical methods differ in principles and technical details, but a common feature is the analysis of chromosomes fixed on a flat surface. This may pose limitations on the throughput of analysis and chromosome manipulation. For the latter, the only possibilities are manual microdissection and laser capture microdissection, with a throughput of only a few chromosomes per working day. Methods for chromosome analysis and sorting using flow cytometry (FCM) or flow cytogenetics were developed to overcome these restrictions. As FCM analyzes chromosomes in aqueous suspension, they can be measured at high speed, typically 10^3 per second. An obvious drawback is that information on the chromosome complements of individual cells is lost. This disadvantage is counterbalanced by the possibility of analyzing large populations of chromosomes and isolating particular chromosomes for subsequent analyses of DNA and protein composition, and for gene transfer. Currently, there is no other method available that allows isolation of intact plant chromosomes in large quantities and high purity. This makes flow cytogenetics a unique method, which nicely interfaces with genetics, genomics and proteomics.

The obvious potential for plant genome analysis and manipulation led to the development of flow cytogenetics for a variety of plant species. In this chapter we explain the principles of flow cytogenetics and look back at the origins of this exciting methodology. In the main part of the chapter we review the development of methods for chromosome analysis and sorting in plants, describe their applications, and assess their potential in various areas of research.

16.2 How Does it Work?

The basic principles of flow cytometry and sorting are described in Chapters 1 and 2. Like other types of samples for FCM, chromosome samples must also be in the form of a liquid suspension of single particles. No attempts have been made to analyze meiotic chromosomes and, as a rule, only mitotic metaphase chromosomes are examined. The first step in sample preparation is cell cycle synchronization and accumulation of cells in metaphase. The use of tissues enriched for metaphases is a prerequisite to achieve a sufficient concentration of chromosomes in the sample. The second step, which usually follows immediately after metaphase accumulation, is the release of metaphase chromosomes from synchronized cells.

Prior to analysis, the sample is stained with one or two DNA fluorescent dyes. As the differences in relative DNA or base content between chromosomes may be small, it is critically important that the analysis of fluorescence intensity is performed at a high resolution. The coefficients of variation (CVs) of fluorescence peaks should be lower than 2%. This requires careful instrument alignment and adjustment of excitation light intensity to saturate fluorescence emission. As explained in Chapter 2, sample core diameter should be kept small to achieve high resolution. This can be done with orifices of 50–80 µm diameter and low sample flow rates. It is thus important that the chromosome concentration in the sample is high so that they can be analyzed at a reasonable speed (10^3 per second and higher). In some species, chromosomes may be longer than the height of the spot of the excitation light beam (typically 15 µm, see Chapter 2), and the fluorescence pulse area is used to measure their total fluorescence.

The analysis of tens of thousands of chromosomes takes only a few minutes, and the results are displayed as histograms of relative fluorescence intensity, or flow karyotypes. In the ideal situation, each chromosome is represented by a distinct peak on the flow karyotype (Fig. 16.1). The use of two different fluorescent dyes differing in AT/GC preference is critical in animal and human flow cytogenetics to resolve chromosomes that differ in AT/GC ratio but not in DNA content. In this case the results are displayed as cytograms or bivariate flow karyotypes. The data are acquired after linear amplification of fluorescence signals. However, logarithmic amplification can be useful to display peaks of chromosomes and nuclei in the same histogram. Any change in relative chromosome frequency in the analyzed population is reflected in the flow karyotype by a change in chromosome peak area. Alteration of relative DNA content results in an altered peak position. This provides the basis for the application of flow karyotyping to quantitative analysis of chromosome numerical changes, structural rearrangements, and polymorphisms.

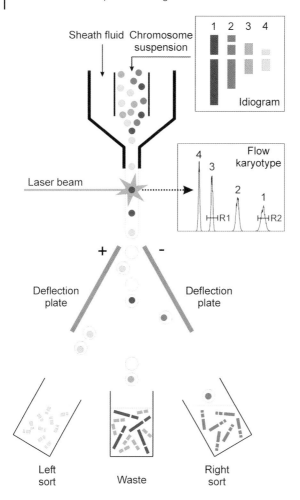

Fig. 16.1 A simplified scheme of a flow cytometer and sorter. An instrument of this type can sort two different chromosomes simultaneously. The example considers chromosome analysis and sorting in a hypothetical species with $2n = 8$, in which all chromosomes differ in size and hence DNA content. Each of the four chromosomes is represented by a single peak on the flow karyotype (a flow karyotype is a distribution of relative chromosome DNA content, where the x axis = relative DNA content, and the y axis = number of events). The example shows sorting of chromosomes 1 and 3 with their respective sort windows R2 and R1 indicated on the flow karyotype.

Conventional flow cytometry works with a so-called zero resolution, and the analysis of fluorescence pulses only yields information on peak height, width, and area (see Chapter 2). In a so-called slit-scan FCM (see Chapter 2) chromosomes are measured by flowing through a thin laser beam (~1.5 μm thick) and the digital profile of the fluorescence intensity along the chromosome is recorded. Although it was shown to be useful for identifying chromosomes by their banding patterns, and to detect centromere position (Bartholdi et al. 1989; Lucas et al. 1983), the method never became widely used. Another approach to identifying particular chromosomes is to label specific antigens on chromosomes using fluorescently-labeled antibodies (Levy et al. 1991). Again, this strategy has only rarely been used. A chromosome or a group of chromosomes that can be discriminated on a flow karyotype as a distinct peak can be isolated by sorting (Fig. 16.1). In order to avoid overlap between two peaks, difference in relative chromosome fluorescence intensity should be at least five-fold the CV of the peaks. This underlines the need for high resolution analysis.

As discussed in Chapter 2, there are different designs of flow sorters. However, only electrostatic droplet sorters are recommended for chromosome work. Their advantage is low dilution of the sorted fraction and no need for further steps to recover the sorted population, which is critical for the preservation of chromosomal DNA and proteins and avoids loss of sorted particles. The volume of the sorted drops depends on the droplet generation frequency, sheath fluid pressure, and nozzle diameter, and is best determined empirically. For example, with a fluid pressure of 20 psi (pound per square inch), frequency of 35 kHz, and 70 µm nozzle diameter, the volume of one drop is ca. 1.7 nl. Chromosomes in droplets can be sorted directly into PCR tubes and large collecting vessels, or onto nylon filters and microscope slides. The latter option facilitates rapid examination of the identity and purity in sorted fractions.

The speed at which chromosomes can be sorted depends on several factors, such as the relative concentration of a chromosome in the sample, the resolution of its peak, and the type of flow sorter. In the first generation of commercially-available sorters, which worked with a low pressure of about 10 psi and generated about 20 000 drops per second, it was possible to purify human chromosomes of a single type at rates of up to 20 per second. However, some applications require large numbers of sorted chromosomes, exceeding 10^6 copies. Sorting with these machines would be impractical and time consuming. This was one of the reasons why the so-called high-speed flow sorters were developed. They work with high pressure, generate up to 100 000 drops per second (see Chapter 2), and their sort rates with human chromosomes reach 100 chromosomes per second. Some droplet sorters can sort up to four different chromosome types simultaneously. However, this option is rarely used and usually only one or two chromosome types are sorted from one sample.

16.3
How it All Began

The development of plant flow cytogenetics was stimulated by the success witnessed in animal and human genetics and genomics. It is therefore appropriate to look back at the origins of this exciting field of research. As described in Chapter 1, flow cytometry and sorting became well established during the mid-1970s. Procedures for quantitative analysis of nuclear DNA content after staining with fluorescent dyes were developed, as well as electrostatic droplet sorting. FCM was used routinely with cells and nuclei, and the time was ripe for chromosomes. The first successful attempt was published independently by two groups who analyzed chromosomes from Chinese hamster cell lines (Gray et al. 1975; Stubblefield et al. 1975). Suspensions of chromosomes were stained with ethidium bromide and distributions of fluorescence intensity were obtained, containing peaks of individual chromosomes or groups of chromosomes with a similar DNA content. This breakthrough indicated the potential for rapid classification of chromosomes according to DNA content and for separation of purified populations of individual chromosomes.

The success with Chinese hamster cells stimulated development of flow cytogenetics in some other animals (Carrano et al. 1976; Stohr et al. 1980) and in humans (Carrano et al. 1979). Initial experiments involved chromosome staining with a single DNA fluorochrome, and not all chromosomes could have been resolved on flow karyotypes, as the peaks of similarly-sized chromosomes overlapped. Introduction of bivariate flow karyotyping after staining with two dyes differing in base-pair preference, such as Hoechst 33258 and chromomycin A3, marked a dramatic improvement in resolution (Gray et al. 1979). Although other approaches were tested, bivariate flow karyotyping using Hoechst and chromomycin became a universal method for use with human samples, where it facilitated discrimination of all chromosomes except chromosomes 9–12 (Langlois et al. 1982), and in a large number of animal species (Ferguson-Smith 1997). The use of chromosomes from hybrid cell lines (Cremer et al. 1982) and individuals heterozygous for a deletion involving centromeric heterochromatin (Harris et al. 1985) permitted the sorting of chromosomes that could not be discriminated in normal karyotypes.

Flow karyotyping was shown to be suitable for the detection of chromosome rearrangements (Cooke et al. 1988) and polymorphisms (Green et al. 1984). Although these observations generated hopes of automatic detection of aberrant chromosomes (Boschman et al. 1992), they were not fulfilled as cheaper and more sensitive methods of molecular biology and cytogenetics became available. On the other hand, chromosome sorting was found to be of paramount importance, with applications including physical gene mapping (Lebo et al. 1979), creation of chromosome-specific DNA libraries (Van Dilla and Deaven 1990), development of DNA markers (Shepel et al. 1998), generation of chromosome painting probes, which allow fluorescent labeling of particular chromosomes (Cremer et al. 1988; Pinkel et al. 1988), and development of artificial chromosomes for gene transfer (de Jong et al. 1999). Today, chromosome sorting is most frequently used for preparation of painting probes for chromosome identification and examination of chromosome rearrangements in humans (Blennow 2004; Langer et al. 2004), and for comparative analysis of chromosome evolution in animals (Svartman et al. 2004; Tian et al. 2004). A new and powerful application to study the composition and breakpoints of aberrant chromosomes combines chromosome sorting and DNA microarray technology (Gribble et al. 2004).

16.4
Development of Flow Cytogenetics in Plants

The impact that flow cytogenetics made in animal and human genetics and genomics generated much interest among plant geneticists. However, due to the intrinsic differences between animal and plant cells, it was not possible to apply the existing protocols for sample preparation. As a result, it took 9 years after the first reports on animal chromosome sorting (Gray et al. 1975; Stubblefield et al. 1975) before chromosome analysis and sorting of plant chromosomes were reported (de Laat and Blaas 1984).

16.4.1
Preparation of Suspensions of Intact Chromosomes

Preparation of samples for flow cytogenetics in plants is hampered by two difficulties that are not encountered in animals and humans. The first is the enrichment of a mitotic metaphase cell population, and the second is the release of intact chromosomes from cells with rigid walls. Both have a direct impact on the quality of a chromosome suspension sample, and deserve a detailed discussion. With the exception of meristematic tissues, a plant body does not contain tissues rich in cells that can be easily synchronized. Various biological systems were tested to overcome this difficulty (Table 16.1).

16.4.1.1 Biological Systems for Chromosome Isolation

Plant cell suspension cultures *in vitro* were popular during the early days of plant flow cytogenetics, perhaps due to their apparent similarity to animal cell cultures. Large numbers of cells that could be manipulated and a single-cell character of cultures were the key advantages considered. In their pioneering work, de Laat and Blaas (1984) used a cell line of *Haplopappus gracilis* (Asteraceae). Subsequently, suspension cultures were employed for chromosome preparations in two species of tomato (Arumuganathan et al. 1991, 1994) and in common wheat (Schwarzacher et al. 1997; Wang et al. 1992). These studies revealed inherent limitations. The cultures were found to be heterogeneous (Arumuganathan et al. 1991), karyologically unstable (Schwarzacher et al. 1997), and rapidly growing cultures could not be established in all species.

In an alternative approach, Conia et al. (1987, 1989) prepared chromosome suspensions from protoplasts derived from leaf mesophyll cells of petunia and tobacco. Mesophyll cells are arrested in the G_1 phase of the cell cycle, and protoplasts isolated after the removal of cell walls can be stimulated to enter the cell cycle synchronously by culturing them in a nutrient medium. The resulting population of synchronized cells can be accumulated in metaphase. As the metaphase indices (percentage of cells in metaphase) thus obtained are low (Conia et al. 1987), and because viable protoplasts cannot be obtained in many species, this method was not used by others.

Table 16.1 Overview of flow cytometric analysis and sorting of mitotic chromosomes in plants.

Species (Family)	Common name	Material	$n^{[a]}$	Method[b]	Number of chromosomes discriminated		References
					Standard karyotype[c]	Cytogenetic stock[d]	
Avena sativa (Poaceae)	Oat	Root meristems	21	M	0	1	Li et al. 2001, 2004
Cicer arietinum (Fabaceae)	Chickpea	Root meristems	8	M	5	–	Vláčilová et al. 2002
Haplopappus gracilis (Asteraceae)		Suspension cells	2	P	2	–	De Laat and Blaas 1984; De Laat and Schel 1986
Hordeum vulgare (Poaceae)	Barley	Root meristems	7	M	1 (2)	3	Lee et al. 2000; Lysák et al. 1999
Lycopersicon esculentum (Solanaceae)	Tomato	Suspension cells	12	P	0	–	Arumuganathan et al. 1991
Lycopersicon pennellii (Solanaceae)		Suspension cells	12	P	2	–	Arumuganathan et al. 1991, 1994
Melandrium album (Caryophyllaceae)	White campion	Hairy root meristems	12	P	2	–	Kejnovský et al. 2001; Veuskens et al. 1995
Nicotiana plumbaginifolia (Solanaceae)	Tobacco	Mesophyll protoplasts	10	P	0	–	Conia et al. 1989

16.4 Development of Flow Cytogenetics in Plants

Species (Family)	Common name	Tissue					References
Oryza sativa (Poaceae)	Rice	Root meristems	12	M	0	–	Lee and Arumuganathan 1999
Petunia hybrida (Solanaceae)	Garden petunia	Mesophyll protoplasts	7	P	1	–	Conia et al. 1987
Picea abies (Pinaceae)	White spruce	Root meristems	12	M	3	–	Überall et al. 2004
Pisum sativum (Fabaceae)	Garden pea	Root meristems	7	M	2	4	Gualberti et al. 1996; Neumann et al. 2002
		Hairy root meristems	7	M	2	4	Neumann et al. 1998
Secale cereale (Poaceae)	Rye	Root meristems	7	M	1	2[e]	Kubaláková et al. 2003
Triticum aestivum (Poaceae)	Common wheat	Suspension cells	21	P	0	–	Schwarzacher et al. 1997; Wang et al. 1992
		Root meristems	21	M	1 (2)	3[f]	Gill et al. 1999; Janda et al. 2004; Kubaláková et al. 2002; Lee et al. 1997; Šafář et al. 2004; Vrána et al. 2000
Triticum durum (Poaceae)	Durum wheat	Root meristems	14	M	0	2[g]	Kubaláková et al. 2005

Table 16.1 (continued)

Species (Family)	Common name	Material	n[a]	Method[b]	Number of chromosomes discriminated		References
					Standard karyotype[c]	Cytogenetic stock[d]	
Vicia faba (Fabaceae)	Field bean	Root meristems	6	M	1	6	Doležel and Lucretti 1995; Lucretti et al. 1993; Macas et al. 1993
Vicia sativa (Fabaceae)	Common vetch	Root meristems	6	M	2	–	Kovářová et al. 2006
Zea mays (Poaceae)	Maize	Root meristems	10	M	2 (3)	0[h]	Lee et al. 1996, 2002

[a] Number of chromosomes in a haploid set.
[b] Preparation of chromosome suspensions: M = mechanical homogenization; P = protoplast lysis.
[c] Number of chromosomes that could be discriminated unambiguously. In some genotypes, a higher number of chromosomes could be discriminated due to polymorphism (given in parentheses).
[d] Maximum number of chromosomes discriminated in one line. Different chromosomes may be discriminated from specific chromosome translocation, deletion and addition lines.
[e] Rye chromosomes 2R–7R could be discriminated from wheat–rye chromosome addition lines.
[f] Sorting of all chromosome arms except 3BL and 5BL is possible in hexaploid wheat from individual (di)telosomic lines.
[g] All chromosome arms may be sorted from individual (di)telosomic lines.
[h] Two oat–maize chromosome addition lines were used to sort maize chromosomes 1 and 9.

The third, and so far the most successful, approach involves the preparation of chromosomes from synchronized meristem root tips of seedlings. It was originally developed for field bean (Doležel et al. 1992) and its main advantages are easy handling of seedlings during the synchronization procedure, a high degree of synchrony, and karyological stability. Since the first report, the protocol has been used for a number of species (Table 16.1). Working with white campion, Veuskens et al. (1995) adapted the root-tip method for genetically transformed "hairy" roots cultured *in vitro*. This option may be useful when seeds of special genotypes are not available and/or difficult to maintain.

16.4.1.2 Cell Cycle Synchronization and Metaphase Accumulation

With the exception of some cell suspension cultures that can be synchronized by nutrient starvation (Arumuganathan et al. 1991), and leaf mesophyll protoplasts (Conia et al. 1987, 1989), mitotic synchrony is induced by the action of DNA synthesis inhibitors. So far, aphidicolin and hydroxyurea, which accumulate cycling cells at the G_1/S interface, have been the only inhibitors used. Upon release from the block, the cells transit the S and G_2 phases and enter mitosis synchronously. The concentrations of inhibitors, the length of the treatment, and the recovery time need to be optimized (Doležel et al. 1999a, 1999b). Mitotic indices exceeding 50% were reported in synchronized root tips of seedlings (Kubaláková et al. 2003, 2005; Lysák et al. 1999). Lower mitotic synchrony was achieved in hairy root cultures (Neumann et al. 1998), cell suspensions (de Laat and Blaas 1984), and synchronized mesophyll protoplasts (Conia et al. 1987).

A synchronous wave of cells passing through mitosis can be arrested at metaphase by inhibiting the mitotic spindle. De Laat and Blaas (1984) and subsequently others (Arumuganathan et al. 1991; Wang et al. 1992) used colchicine, a traditional spindle poison used in human, animal and plant cytogenetics. However, this alkaloid has a lower affinity for plant tubulins, and synthetic herbicides such as amiprophos methyl (Doležel et al. 1992), oryzalin (Veuskens et al. 1995) and trifluralin (Lee et al. 1997), which are effective at micromolar concentrations, are preferred. Doležel et al. (1999b) recommended only a 2-h treatment, to avoid the splitting of chromosomes into chromatids that occurs after longer treatments. Subsequent incubation in ice water reduced the number of clumps in chromosome suspensions of some species (Lysák et al. 1999; Vláčilová et al. 2002).

16.4.1.3 Preparation of Chromosome Suspensions

While it is relatively easy to release chromosomes from animal cells after their lysis in a hypotonic buffer, it is very difficult to isolate intact chromosomes from plant cells encapsulated in rigid cell walls and embedded in three-dimensional tissues. When working with cell cultures, a logical step was to remove cell walls using hydrolytic enzymes such as pectinases and cellulases, and to release chromosomes by lysing protoplasts in a hypotonic buffer (de Laat and Blaas 1984).

The use of the protoplast lysis method was not limited to cell suspensions and Veuskens et al. (1995) adapted it to isolate chromosomes from synchronized meristem tips of hairy roots. The problem with this method is that during the enzymatic treatment metaphase chromosomes can split into chromatids and eventually decondense (Veuskens et al. 1995). Moreover, its application is limited to species from which protoplasts can be prepared.

After unsuccessful attempts to isolate protoplasts from root tips of field bean, Doležel et al. (1992) developed an alternative protocol for chromosome isolation. By analogy with the procedure for preparation of suspensions of nuclei (Galbraith et al. 1983), they released chromosomes by chopping root tips with a sharp scalpel. Separation of chromosome clumps was achieved by syringing. To stabilize chromosome structure, the roots were fixed mildly with formaldehyde prior to chopping. Interestingly, the fixation increased chromosome yield dramatically (Doležel et al. 1992). However, the extent of the fixation is critical to obtaining good chromosome suspensions (Doležel et al. 1999b).

A few years after the original report, chopping was replaced by mechanical homogenization, which made the method less laborious and suitable for small root tips (Gualberti et al. 1996). This variant became the most frequently used method for the preparation of chromosome suspensions in plants (Doležel et al. 2004). In the same year that Gualberti et al. (1996) published their variation of Doležel's protocol, Lee et al. (1996) came up with another modification, in which formaldehyde fixation was omitted. However, an advantage of fixation is the possibility of arresting the cell cycle at the right stage of metaphase accumulation. Fixed chromosomes are more resistant to mechanical shearing forces during isolation, cytometric analysis and sorting. This has important consequences for the use of sorted chromosomes and the quality of their DNA.

Chromosomes and their DNA need to be protected from degradation after their release from the protective cellular environment. This can be achieved by using a proper isolation buffer. Although a variety of buffers was used in the early days (Doležel et al. 1994), only two buffers are used currently. The polyamine-based LB01 buffer (Doležel et al. 1989) and its variants, involving higher pH and composition aimed at preserving high molecular weight DNA (Šimková et al. 2003), are the most popular. The second most frequently used buffer is based on magnesium sulfate (Lee et al. 1996).

A chromosome suspension suitable for FCM should be free of cellular and chromosomal debris, chromatids and chromosome clumps. Chromosome concentration is equally important to achieve high-resolution flow karyotypes and high sort rates (see Section 16.2), and depends on metaphase synchrony and efficacy of the chromosome isolation protocol. In addition to the optimized protocol for releasing intact chromosomes from synchronized cells, it is important to use highly synchronized tissues with metaphase indices around 50% or higher (Doležel et al. 1999b). The range of reported concentrations in chromosome samples varies from 1.6×10^5 to 1×10^6 chromosomes per millilitre (Doležel et al. 1992; Gill et al. 1999).

16.4.2
Chromosome Analysis

FCM analysis of plant chromosomes has been reported in 18 plant species, including major crops (Table 16.1). In the first experiment, de Laat and Blaas (1984) chose a good model to work with, *Haplopappus gracilis*, with only two pairs of chromosomes, which differ significantly in size. Despite the low resolution, the analysis of ethidium-stained chromosomes resulted in flow karyotypes with two peaks representing both chromosome types. Only subsequent experiments with other plant species revealed one of the major obstacles to plant flow cytogenetics. Typically, only one or few chromosomes from the chromosome complement form distinct peaks on the flow karyotype (Table 16.1). Peaks of other chromosomes overlap and form composite peaks (Fig. 16.2).

16.4.2.1 Bivariate Analysis of AT and GC Content

Bivariate analysis of chromosome suspensions stained with two dyes differing in base-pair preference, a standard method in animal and human flow cytogenetics, was found to be of little help in plants. With a few exceptions (Arumuganathan

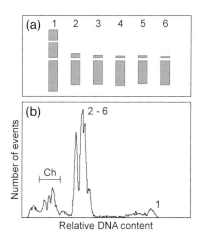
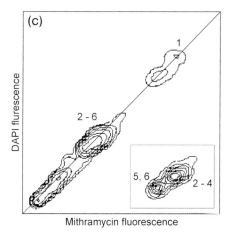

Fig. 16.2 Monovariate and bivariate flow karyotyping in field bean (*Vicia faba*, $2n = 12$). (a) The idiogram shows that the karyotype comprises a large metacentric chromosome (1), and five acrocentric chromosomes (2–6) of similar size. (b) Monovariate flow karyotype obtained after analysis of the DAPI-stained chromosomes. The karyotype consists of a composite peak of chromosomes 2–6 that cannot be resolved, and a peak representing chromosome 1 that can be sorted. Note the presence of peaks representing chromatids (Ch). (c) Bivariate flow karyotype obtained after analysis of chromosomes stained simultaneously with DAPI (AT preference) and mithramycin (GC preference). The chromosome peaks lie on a straight diagonal line, indicating negligible differences in the AT/GC ratio among the chromosomes. The insert shows the region of the flow karyotype containing acrocentric chromosomes 2–6.

et al. 1991, 1994), it showed no improvement over monovariate analysis in resolving additional chromosomes (Kovářová et al. 2007; Lee et al. 2000; Lucretti and Doležel 1997; Schwarzacher et al. 1997; Fig. 16.2). This failure could be explained by the similarity in overall AT/GC ratio between chromosomes, most probably due to the prevalence of dispersed repetitive DNA sequences in plant genomes.

16.4.2.2 Fluorescent Labeling of Repetitive DNA

A different approach to identifying specific chromosomes by FCM would be to label fluorescently repetitive DNA sequences that exhibit chromosome-specific quantitative distribution. Early attempts to use fluorescence *in situ* hybridization failed due to excessive clumping induced by the procedure (J. Doležel et al., unpublished data). This stimulated Macas et al. (1995) to develop a method called primed *in situ* DNA labeling *en* suspension (PRINSES), which allowed fluorescent labeling of repetitive DNA sequences on chromosomes in suspension. Pich et al. (1995) used PRINSES to label field bean chromosomes in suspension and showed that bivariate analysis of chromosome DNA content and signals from labeled repetitive DNA enabled discrimination of similar-sized chromosomes. Unfortunately, PRINSES is difficult to perform quantitatively (J. Doležel et al., unpublished data) and no other successful use of PRINSES has been reported so far.

16.4.2.3 The Use of Cytogenetic Stocks

Facing these difficulties, Lucretti et al. (1993) and Doležel and Lucretti (1995) pioneered an alternative and powerful strategy to discriminate higher numbers of chromosomes. When working with field bean, they showed that the use of translocation lines with altered chromosome lengths and hence relative DNA content facilitated discrimination of many, and sometimes all, of the chromosomes in the complement. After other strategies aiming at resolving individual chromosomes had been shown to be unfeasible, this approach became the main strategy in plant flow cytogenetics. Various cytogenetic stocks, including chromosome translocation (Kubaláková et al. 2002; Lysák et al. 1999; Neumann et al. 1998), deletion (Gill et al. 1999; Kubaláková et al. 2002, 2005), and alien addition lines (Kubaláková et al. 2003; Li et al. 2001), have been used extensively (Figs 16.3 and 16.4). The only limitation of this approach is a dependence on special cytogenetic stocks.

16.4.2.4 Assignment of Chromosomes to Peaks on Flow Karyotypes

The identity of peaks on flow karyotypes has been determined in a number of ways. An indirect approach involved theoretical models to predict peak positions on a flow karyotype (Conia et al. 1989; Doležel 1991). The models were based on relative chromosome length or DNA content, and were useful in planning the experiments and assessing the possibility of discriminating specific chromosomes at a given resolution (Lee et al. 2000; Neumann et al. 1998; Vláčilová et al. 2002). A direct approach was to sort particles from individual peaks and identify them. This was done on particles sorted onto nylon membranes by dot-blotting (Arumu-

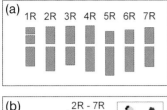
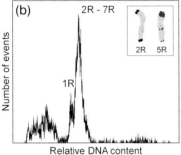

Fig. 16.3 Flow karyotyping in rye (*Secale cereale*, 2n = 14) and the effect of chromosome translocation. (a) The idiogram demonstrates that the seven rye chromosomes do not differ significantly in size. (b) Flow karyotype of cv. "Selgo" obtained after analysis of DAPI-stained chromosomes. The karyotype consists of a composite peak of chromosomes 2R–7R that cannot be resolved, and a shoulder representing the partially resolved chromosome 1. The insert shows wild-type chromosomes 2R and 5R after FISH with a probe for pSc119.2 repeats (dark bands). (c) Idiogram of a line carrying the translocation chromosomes T2RS·2RL–5RL (T1) and T5RS·5RL–2RL (T2). The reciprocal translocation resulted in one short (T1) and one long (T2) chromosome. Arrowheads indicate the positions of the breakpoints. (d) Flow karyotype of a line carrying chromosomes T1 and T2. Their peaks are clearly resolved and the chromosomes can be sorted. The insert shows T1 and T2 chromosomes after FISH with a probe for pSc119.2 repeats (dark bands).

ganathan et al. 1994), by PCR with primers for chromosome-specific markers (Lysák et al. 1999; Vrána et al. 2000), and by microscopical observation (de Laat and Blaas 1984; Lucretti et al. 1993).

16.4.3
Chromosome Sorting

In principle, a peak representing a chromosome or a group of chromosomes to be sorted is selected and the cytometer instructed to sort particles from that peak (Fig. 16.1). In reality, the chromosome suspensions are contaminated to various extents with chromosome fragments, whose sizes may range from very small particles to complete arms, chromatids and their fragments, and various doublets and clumps. These particles usually form a background continuum of debris on which chromosome peaks are superimposed (Fig. 16.2). Moreover, specific fragments and aggregates may be present at high frequency and form peaks of their

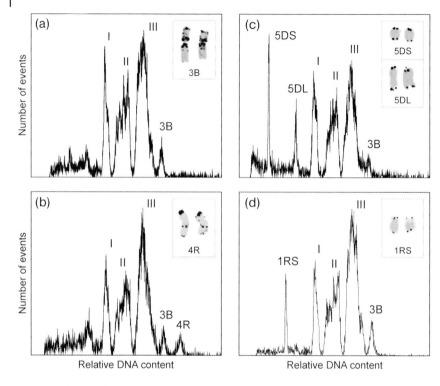

Fig. 16.4 The use of cytogenetic stocks of hexaploid wheat (*Triticum aestivum*, 2n = 6x = 42) to sort wheat chromosome arms, and alien chromosomes and chromosome arms. (a) Flow karyotype of the cultivar "Chinese Spring" with a standard karyotype obtained after analysis of DAPI-stained chromosomes. Three composite peaks (I–III) representing groups of chromosomes (Kubaláková et al. 2002) and a peak of chromosome 3B can be resolved. The insert shows flow-sorted chromosome 3B after FISH with a probe for the GAA microsatellite (dark bands). (b) Flow karyotype of a wheat chromosome addition line carrying rye chromosome 4 (4R). The peak of 4R is resolved and the chromosome can be sorted. The insert shows flow-sorted chromosome 4R after FISH with a probe for pSc119.2 (dark bands). (c) Double ditelosomic line of cv. "Chinese Spring" carrying short and long arms of wheat chromosome 5D in the form of telocentric chromosomes 5DS and 5DL, respectively. Peaks of both telosomes are resolved. The insert shows flow-sorted telosomes 5DS and 5DL after FISH with a probe for telomeric repeats (dark bands). (d) Ditelosomic line of cv. "Chinese Spring" carrying telocentric chromosome 1RS (short arm of rye chromosome 1). The peak of 1RS is clearly resolved. The insert shows flow-sorted telosome 1RS after FISH with a probe for telomeric repeats (dark bands).

own. This holds true for arms of large chromosomes, which are prone to breakage (Lucretti et al. 1993), and especially chromatids, which form peaks at half the fluorescence intensities of whole chromosomes (Conia et al. 1989). If the relative fluorescence intensity of fragments and clumps is close to that of intact chromosomes, they are hard to discriminate and may contaminate the sorted fraction.

16.4.3.1 Estimating the Purity in Sorted Fractions

Determination of purity in sorted fractions is more demanding than the assignment of peaks to chromosomes. Dot blotting (Arumuganathan et al. 1994) and PCR (Vrána et al. 2000) are not suitable as they do not permit identification of different chromosomes, chromosome fragments and clumps, and do not quantify their frequency in one assay. This is best done by microscopical observation. However, a simple examination of chromosome morphology may not be sufficient because the degree of chromosome condensation varies among the chromosomes in the sample, and chromosome length and arm ratio may change during sorting due to mechanical shearing forces and uneven drying on a slide. As the traditional banding techniques generally do not provide a sufficient number of bands in plants, fluorescence *in situ* hybridization (FISH) and primed *in situ* DNA labeling (PRINS) appear to be the methods of choice (Lucretti et al. 1993; Kubaláková et al. 2000), and have been used extensively. Unlike PRINS, FISH can be performed easily with two probes, and this may be advantageous in some cases. Kubaláková et al. (2005) noted that in durum wheat, two-color FISH with a probe for telomeric repeats facilitated unambiguous discrimination of sorted telocentric chromosomes from broken chromosome arms without telomere sequences at the centromeric end. The same approach is valuable in other species (Fig. 16.7a).

16.4.3.2 Improving the Sort Purity

With the aim of improving the sort purity, other parameters in addition to fluorescence intensity can be used to help in chromosome discrimination. The most common is analysis of forward-angle light scatter. Conia et al. (1987) were the first to report on improved discrimination of chromosomes, nuclei and debris by simultaneous analysis of forward scatter and fluorescence intensity. Lucretti et al. (1993), who analyzed chromosomes of field bean, faced the problem of distinguishing doublets of short acrocentric chromosomes from a large metacentric chromosome. The doublets had similar DNA content but approximately half the length of a metacentric chromosome. This provided an opportunity to discriminate metacentric chromosomes based on their length measured as fluorescence pulse width. Simultaneous analysis of fluorescence pulse area and pulse width has subsequently been used extensively (Kejnovský et al. 2001; Kubaláková et al. 2003; Li et al. 2001; Neumann et al. 2002; Vrána et al. 2000).

16.4.3.3 Two-step Sorting

As demonstrated by Lucretti et al. (1993), the purity of sorted fractions could be improved by two-step sorting. During the first sort run, recovery is preferred over purity, and the sample is enriched for the chromosomes of interest. One of the major effects is that due to the mechanical shearing forces most of the aggregates present in the sample dissociate. During the second run, sort stringency is preferred over yield, and chromosomes can be sorted at high purity. Two-step sorting is laborious, and not suitable if large numbers of sorted chromosomes are required. On the other hand, it is useful to sort small numbers of chromosomes at very high purities (Kejnovský et al. 2001), and if the frequency of the chromosome of interest in the sample is low.

16.4.3.4 Purities and Sort Rates Achieved

The above discussion demonstrates that it is hard, if not impossible, to eliminate contamination by particles with the same or very similar fluorescence, length, and light scatter parameters as the sorted chromosome. This underlines the importance of sample quality and analysis at the highest possible resolution. In various experiments, plant chromosomes have been sorted at different purities, and it is not meaningful to compare individual experiments as the purity depends on many factors, including the rate of analysis and the resolution of the chromosome peak. However, purities exceeding 90%, and sometimes approaching 100%, were frequently reported (Kubaláková et al. 2005; Li et al. 2001; Vrána et al. 2000). Similarly, the chromosome sort rates varied depending on whether the purity or yield was the preferred parameter. In the case of large-scale sorting, our results indicate that it is possible to sort several hundred thousand chromosomes in a working day with purities around 90% (Janda et al. 2004, 2006).

16.5
Applications of Flow Cytogenetics

The previous sections of this chapter have outlined the possibilities and methodological limitations of flow cytogenetics in plants. It is now time to consider the applications. A flow cytometer is both an analytical and a preparative tool. It offers possibilities for chromosome analysis and flow karyotyping, as well as for isolation of particular chromosomes by flow sorting.

16.5.1
Flow Karyotyping

Following the early experiments (Arumuganathan et al. 1991; Conia et al. 1987; de Laat and Blaas 1984), the work of Lucretti et al. (1993) indicated that it was possible to detect numerical and structural chromosome changes by flow karyotyping. Although this potential was clearly articulated by Doležel et al. (1994), it took some time before it was verified in other species. Identification of numerical chromosome changes by flow karyotyping was reported by Lee et al. (2000), who were able to detect trisomy of chromosome 6 in barley. Several reports described the identification of alien chromosome and/or chromosome arm additions. For example, Li et al. (2001) resolved maize chromosome 6 in an oat-maize chromosome addition line. Flow karyotyping of wheat-rye chromosome addition lines showed the ability not only to detect alien rye chromosomes but also to monitor their frequency in the population (Kubaláková et al. 2003). Recently, we were able to discriminate peaks of barley chromosome arms on flow karyotypes of wheat-barley telosome addition lines (Suchánková et al. 2006).

The suitability of flow karyotyping for identification of chromosome deletions and translocations was demonstrated not only in field bean (Doležel and Lucretti 1995; Lucretti et al. 1993), but also in garden pea (Neumann et al. 1998), barley (Lysák et al. 1999), rye (Kubaláková et al. 2003), durum wheat (Kubaláková et al. 2005), and common wheat (Gill et al. 1999; Kubaláková et al. 2002; Vrána et al. 2000). A survey of 58 varieties and landraces of common wheat led to the discovery of a translocation chromosome 5BL·7BL in seven of them, in which its presence was not previously known (Kubaláková et al. 2002). The 5BL·7BL chromosome is the longest in the karyotype and can easily be detected. In the wheat varieties "Panthus" and "Sida", an altered chromosome 4B with significantly increased length was detected. The nature of the structural change is unknown at present (Kubaláková et al. 2002). In addition to the normal or A chromosomes, some plants carry accessory or B chromosomes. The size of the B chromosomes that are present in some genotypes of rye is about half that of the A chromosomes, so that their presence can easily be detected. When flow karyotyping 22 varieties of rye, Kubaláková et al. (2003) found B chromosomes in the variety "Adams", in which such chromosomes had not been expected.

The possibility of detecting structural rearrangements, such as translocations and deletions, depends on the effect they have on chromosome length or fluorescence intensity, and on the resolution of the analysis. Some results show that in specific cases the alteration in chromosome structure can be detected even if the chromosome peak is not fully resolved. For example, a translocation chromosome 1RS·1BL in wheat is not resolvable as a distinct peak. However, its presence results in a diagnostic change in the flow karyotype (a shoulder on the composite peak III), and lines carrying 1RS·1BL can be identified (Kubaláková et al. 2002). In common wheat, chromosome 4D cannot be identified as a distinct peak, as it overlaps with peaks of chromosomes 1D and 6D, forming a composite peak. In varieties "Mona" and "Rexia" the composite peak is split and chromosome 4D forms a separate peak, probably due to a minor increase in DNA content (Kubaláková et al. 2002).

As in humans, flow karyotyping can also detect chromosome polymorphism in plants. This was demonstrated in barley (Lee et al. 2000), maize (Lee et al. 2002), rye (Kubaláková et al. 2003), and wheat (Kubaláková et al. 2002). It is interesting to note that the "fingerprint" patterns of flow karyotypes were characteristic for the varieties and were heritable.

Conventional cytogenetics characterizes karyotypes in individuals. This is not common in flow cytogenetics, where a large number of chromosomes is required for the analysis, and a sample is typically prepared from several individuals. However, as shown by Gualberti et al. (1996) and Lee et al. (1996), it is possible to prepare a measurable sample from only one root tip. We were able to analyze chromosomes prepared from single root tips in barley, wheat and rye (J. Doležel et al., unpublished data). On the other hand, the analysis of chromosomes from a large number of individuals may be advantageous in some cases. One example is the determination of the frequency of a particular chromosome in a population of plants (Kubaláková et al. 2003).

Despite the obvious potential for identifying numerical and structural chromosome changes and chromosome polymorphisms, as with animal and human flow cytogenetics, plant flow karyotyping will probably never become a widely used method. The main reasons include the cost of the equipment, lack of morphological information about the chromosomes, limited sensitivity, and the inability to characterize the chromosome complements of single cells. On the other hand, the possibility of isolating large numbers of chromosomes by sorting makes flow cytogenetics a unique and extremely powerful research tool.

16.5.2
Chromosome Sorting

The applications of sorted chromosomes can be classified according to the number of chromosomes that are required and the environment into or onto which the chromosomes are sorted (Fig. 16.5). Historically, the first applications of flow-sorted chromosomes were based on a small number of chromosomes (10^2–10^4), which could be sorted in a few seconds or minutes. Applications requiring higher numbers of chromosomes (10^5–10^6) have been developed more recently.

16.5.2.1 Physical Mapping and Integration of Genetic and Physical Maps

One way of mapping DNA sequences to specific chromosomes is to use flow-sorted chromosomes as a template for PCR with sequence-specific primers (Fig. 16.6). The fact that only a few hundred chromosomes are required makes flow sorting an attractive tool. Macas et al. (1993) were the first to demonstrate the feasibility of this approach. They localized seed storage protein genes to specific field bean chromosomes, and integrated genetic and physical maps of the crop. To overcome problems with discrimination of each of the six chromosomes of field bean, they sorted chromosomes from a line with a standard karyotype and from three translocation lines. The advantage of this strategy was that it permitted localization of DNA sequences at a subchromosomal level. Lysák et al. (1999) employed the same strategy to localize RFLP markers to regions of barley chromosomes. Kejnovský et al. (2001) used PCR on chromosomes sorted from a dioecious plant (white campion) to localize male-specifically expressed genes on sex chromosomes and autosomes, and Vláčilová et al. (2002) assigned genetic linkage groups of chickpea to specific chromosomes. Integration of genetic and physical maps was also reported by Neumann et al. (2002) in garden pea. This work completed efforts extending over many years to assign all seven chromosomes of pea to genetic linkage groups.

16.5.2.2 Cytogenetic Mapping

A common way to determine the genomic distribution of DNA sequences on chromosomes is FISH on mitotic metaphase spreads which, however, has some negative consequences for sensitivity and throughput. The use of flow-sorted chromosomes for FISH offers a powerful alternative. Thousands of copies of a particular chromosome can be placed on a slide in a few minutes. This pro-

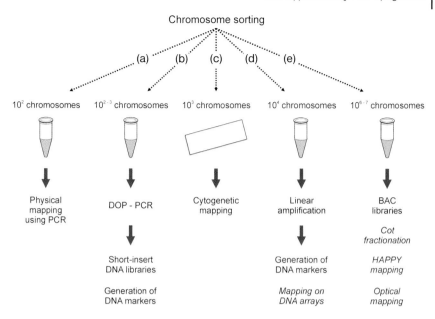

Fig. 16.5 Examples of applications of flow-sorted chromosomes in plant genome analysis. (a) A small number of chromosomes (10^2) is sufficient as a template for physical mapping using PCR. (b) 10^2–10^3 chromosomes are used as a template for DOP-PCR. A short-insert DNA library produced from amplified DNA is suitable for generation of molecular markers. (c) About 10^3 chromosomes sorted onto a microscope slide allow high throughput mapping of DNA sequences to chromosomes by FISH and immunofluorescent localization of proteins. (d) DNA from 10^4 chromosomes can be amplified to microgram quantities for hybridization on DNA arrays and generation of chromosome-specific DNA markers. (e) 10^6–10^7 chromosomes are needed to prepare micrograms of high-molecular weight DNA for direct cloning and other applications that may include Cot-based cloning and sequencing (Peterson et al. 2002), HAPPY mapping (Thangavelu et al. 2003), and optical mapping (Aston et al. 1999). The applications currently under development are listed in italics.

vides an opportunity to analyze more chromosomes than in metaphase spreads. It also greatly increases throughput by eliminating the time needed to locate metaphases suitable for evaluation. The sorted chromosomes are free of cell wall and cytoplasmic remnants, and sequences as short as 1.9 kb can be localized (Janda et al. 2006).

Analysis of the genomic distribution of DNA sequences using FISH on sorted chromosomes was reported by Kubaláková et al. (2002), who analyzed inter- and intra-varietal polymorphism in the distribution of GAA clusters in wheat chromosomes 3B and 5BL·7BL. Because of the high number of sorted chromosomes that could be analyzed by FISH, Kubaláková et al. (2003) were able to detect rarely-occurring translocations between A and B chromosomes of rye. The ease of FISH mapping on sorted chromosomes facilitated development of a molecular karyotype of durum wheat (Kubaláková et al. 2005). Construction of physical

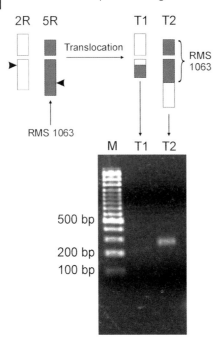

Fig. 16.6 Physical mapping in rye (*Secale cereale*) using PCR on sorted chromosomes. While chromosomes 2R and 5R cannot be resolved by flow karyotyping, products of a reciprocal translocation T2RS·2RL–5RL (T1) and T5RS·5RL–2RL (T2) can be sorted (see Fig. 16.3). This facilitates physical mapping of DNA markers by PCR to subchromosomal regions. Agarose gel electrophoresis was performed with products of PCR with primers for chromosome 5R-specific marker RMS 1063 (V. Korzun, personal communication) on 100 copies of chromosomes T1 and T2. Arrowheads point to translocation breakpoints. M = 100 bp DNA ladder.

contig maps is facilitated by cytogenetic mapping, which may confirm contig position on a chromosome, orient contigs with respect to the centromere and telomere, and estimate contig gaps (Harper and Cande 2000). Several studies showed the suitability of sorted chromosomes for FISH mapping of BAC (Bacterial Artificial Chromosome) clones, which are used to develop physical maps (Janda et al. 2004, 2006; Šafář et al. 2004). To avoid hybridization of dispersed repeats, which are present in most wheat BAC clones, Janda et al. (2006) used low-copy subclones instead of BACs for FISH. The success of this approach was made possible by the ability to localize short DNA sequences on sorted chromosomes (Figs 16.7a, b).

The limit of spatial resolution of FISH signals on mitotic metaphase chromosomes varies between 1 and 3 Mbp (Heng and Tsui 1998), which greatly limits the suitability of FISH for resolving the order of closely spaced contigs and estimating the size of small contig gaps. To overcome these limitations, Valárik et al. (2004) introduced a protocol for longitudinal stretching of sorted chromosomes.

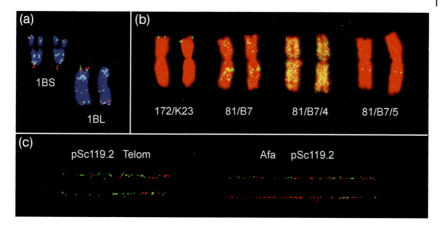

Fig. 16.7 Fluorescence *in situ* hybridization (FISH) on flow-sorted chromosomes (panels a, b) and stretched flow-sorted chromosomes (panel c) of common wheat (*Triticum aestivum*). (a) Simultaneous localization of GAA microsatellite (turquoise) and telomeric (red) repeats on telosomic chromosomes 1BS and 1BL facilitates unambiguous discrimination of sorted telocentric chromosomes from broken chromosome arms lacking telomere sequences at the centromeric end. (b) Localization of wheat BAC clones 172/K23 and 81/B7 on chromosome 3B (yellow signals). While BAC 172/K23 localizes to a discrete locus, repetitive DNA sequences present in BAC 81/B7 result in dispersed signals and prevent localization of the BAC. This could only be achieved by selecting a subclone free of repeats. While FISH with subclone 81/B7/4 carries dispersed repeat(s), subclone 81/B7/5 appears free of such repeats and could be localized to a discrete locus on 3B. (c) Simultaneous localization of the pSc119.2 repeat (yellow-green) and telomeric repeats (red), and of Afa-family repeats (yellow-green) and pSc119.2 repeats (red) reveals their genomic arrangement in the distal part of the short arm of chromosome 3B. Prior to FISH, the chromosomes were stretched after flow sorting to achieve higher spatial resolution. In all the images, the probes were labeled with digoxigenin or biotin, and the sites of hybridization were detected by anti-digoxigenin-FITC (yellow) and streptavidin-Cy3 (red), respectively. The chromosomes were counterstained with propidium iodide (red) or DAPI (blue).

The use of stretched chromosomes improved spatial resolution down to approximately 70 kb. Thus, FISH on stretched flow-sorted chromosomes fills the gap in spatial resolution between mitotic chromosomes and DNA fibers (Fig. 16.7c).

Preparation of probes for chromosome painting is the most frequent application of flow cytogenetics in animals and humans (Langer et al. 2004). The obvious potential for genome analysis stimulated several research groups to test the use of sorted chromosomes to prepare painting probes in plants. Unfortunately, the paints developed so far have not been chromosome-specific, and hybridized to all chromosomes in the complement (J. Doležel et al., unpublished data). Fuchs et al. (1996) concluded that this was due to the large proportion of dispersed repetitive DNA sequences in plant genomes, and an insufficient signal intensity of unique chromosome-specific sequences. It is worth noting that dispersed DNA sequences were also blamed for the failure of bivariate flow karyotyping in plants (see Section 16.4.2.1).

16.5.2.3 Analysis of Chromosome Structure

The fact that isolated chromosomes are free of cell wall and cytoplasmic remnants makes them suitable not only for mapping DNA sequences by FISH and PRINS, but also for the analysis of ultrastructure and protein composition. The suitability of isolated chromosomes for analysis of their structure using scanning electron microscopy was first demonstrated by Schubert et al. (1993). The same work also documented efficient localization of chromosomal proteins using immunostaining. Binarová et al. (1998) took advantage of the specificity and sensitivity of immunofluorescent staining on isolated chromosomes, and demonstrated, for the first time, the presence of γ-tubulin in the plant kinetochore/centromeric region. Ten Hoopen et al. (2000) used isolated chromosomes of field bean and barley to localize a set of kinetochore proteins by immunofluorescent staining, and demonstrated phylogenetic conservation of the SKP1 and CBF5 protein domains in plants.

16.5.2.4 Targeted Isolation of Molecular Markers

Molecular markers are useful in marker-assisted breeding, positional gene cloning, and comparative genome analysis. The markers are usually developed from genomic resources, making it difficult to saturate particular genome regions. The possibility of obtaining DNA from flow-sorted chromosomes offers an attractive opportunity to develop markers from specific regions. Targeted preparation of molecular markers from flow-sorted chromosomes had already been reported by Arumuganathan et al. (1994), who developed 11 new RFLP markers for chromosome 2 of tomato. As microsatellite markers are more versatile, Koblížková et al. (1998) introduced a procedure for constructing chromosome-specific DNA libraries enriched up to 100-fold for a particular microsatellite. In a continuation of this work, Požárková et al. (2002) established an efficient protocol for targeted retrieval of microsatellite markers, and demonstrated its usefulness for the development of markers from a particular region of field bean chromosome 1. The markers were useful in assembling a composite genetic map of field bean (Román et al. 2004). The use of flow-sorted chromosomes for development of microsatellite markers is attractive, as only 250 chromosomes are necessary to establish enriched DNA libraries. Moreover, the chromosome specificity of candidate markers can be verified rapidly prior to their genetic mapping by PCR, using sorted chromosomes as a template (Požárková et al. 2002).

16.5.2.5 Recombinant DNA Libraries

As discussed above, chromosome-specific DNA libraries are useful for targeted retrieval of molecular markers. Short-insert DNA libraries, which can be generated from DNA amplified using DOP-PCR (Degenerate Oligonucleotide Primed PCR; Telenius et al. 1992), are suitable for this work. The amplification step considerably reduces the number of chromosomes that need to be sorted. Thus, Arumuganathan et al. (1994) created a tomato chromosome 2-specific library from only 1000 sorted chromosomes, while Požárková et al. (2002) constructed a field bean chromosome 1-specific DNA library from only 250 chromosomes. In the same species, Macas et al. (1996) established DNA libraries from seven chromosomes that were sorted from two different translocation lines. As a result of this achievement, field bean remains the only plant species whose genome is available in chromosome-specific libraries. Collectively, the libraries cover the field bean genome more than once.

PCR amplification results in short DNA fragments, a fact that may limit the application of the resulting DNA libraries. If DNA libraries with larger inserts are required, the amplification step should be omitted. However, this requires that large numbers of chromosomes are sorted and enough DNA is collected to allow for direct cloning. The first library of this type was created by Wang et al. (1992) from chromosome 4A of common wheat. They sorted 10^5 chromosomes, digested their DNA using the methylation-sensitive restrictase *Hpa*II, and cloned them in a pUBs1 vector. Only about 300 recombinant clones were obtained in this particular experiment. Working with maize, Li et al. (2004) sorted about 6×10^5 copies of chromosome 1 and created a DNA library cloned in a lambda vector. The library comprises 1.2×10^5 clones, and with an average insert size of 12.6 kb, it represents a 0.9-fold coverage of the chromosome.

Development of physical contig maps, clone-by-clone sequencing, and positional gene cloning requires DNA libraries with large insert sizes, such as those cloned in a BAC vector (Shizuya et al. 1992). Although their construction requires microgram quantities of high molecular weight DNA, this is not a serious problem if genomic DNA is used. Until recently, it was not clear if such libraries could be produced from particular chromosomes. Two problems were foreseen: the quality of DNA recovered from sorted chromosomes and the ability to sort enough chromosomes. In 2000, we reported that DNA of sorted wheat chromosomes was of high molecular weight (Vrána et al. 2000). Further improvements of the protocol by Šimková et al. (2003) set the stage for construction of chromosome-specific BAC libraries (Fig. 16.8). To preserve their structure, chromosomes are embedded in agarose plugs after flow sorting.

A breakthrough experiment was reported by Šafář et al. (2004), who created the first BAC library from flow-sorted plant chromosomes. The library, specific for wheat chromosome 3B, was created from 1.8×10^6 sorted chromosomes, has an average insert size of 103 kb, and represents 6.2-fold coverage of the chromosome. This library of 67 968 clones is a unique resource for wheat genomics. Its availability permitted rapid progress in creating a physical contig map of the chromosome (Paux et al. 2006). The small size and the specificity of the library greatly facilitated map-based cloning of major QTL (Quantitative Trait Locus) for *Fusarium* head blight resistance and a durable rust-resistance gene in polyploid wheat (Kota et al. 2006; Liu et al. 2005). The possibility of creating BAC libraries from flow-sorted chromosomes was confirmed by Janda et al. (2004), who created a composite BAC library specific for wheat chromosomes 1D, 4D and 6D.

By representing small and defined parts of nuclear genomes, chromosome-specific BAC libraries greatly facilitate genome sequencing and gene cloning. Dissecting a genome into chromosome-specific DNA libraries facilitates division of labor in a genome sequencing project (Gill et al. 2004). If available, BAC li-

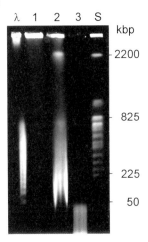

Fig. 16.8 Pulsed field gel electrophoresis of DNA prepared from flow-sorted wheat chromosomes. Chromosomal DNA in agarose plugs was incubated in a buffer containing 0 (lane 1), 5 (lane 2) or 25 (lane 3) units of HindIII restriction enzyme for 5 min (lanes 1 and 2) or 2 h (lane 3) at 37 °C. The results indicate that DNA of flow-sorted chromosomes is intact (megabase-sized) and accessible to restriction enzymes. This makes it suitable for construction of large-insert DNA libraries. The electrophoresis was run on a 1% agarose gel in 0.5 × TBE buffer, at 13.5 °C, 6 V cm^{-1} with a 90-s switch time, for 20 h. Lane λ, lambda ladder; lane S, chromosomes of *Saccharomyces cerevisiae* used as size markers.

braries specific for chromosome arms would be even more useful. By developing a BAC library from the short arm of wheat chromosome 1B, which represents only about 1.9% of the hexaploid wheat genome, Janda et al. (2006) demonstrated the possibility of constructing such libraries. This work also showed the utility of chromosome deletion stocks for sorting chromosome arms. Similar work by Šimková et al. (unpublished data) relied on the use of an alien telocentric chromosome addition line. The authors sorted 9.6 million copies of the short arm of rye chromosome 1R (1RS) from a wheat–rye telosome addition line. DNA from flow-sorted chromosome arms was used to create a BAC library that covers the 1RS arm 19-fold.

16.6
Conclusions and Future Prospects

From humble beginnings more than 20 years ago, plant flow cytogenetics has developed into a productive research area with many important applications. By analogy with animal and human flow cytogenetics, the use of FCM to analyze structural and numerical chromosome changes will remain limited to special cases, the most frequent being the discrimination and sorting of particular chromosomes and chromosome arms. On the other hand, chromosome sorting by FCM is currently the only method that can be used to isolate particular chromosomes in large numbers and at high purity. As such it finds numerous applications.

Introduction of a protocol for preparation of liquid chromosome suspensions from root meristems facilitated development of flow cytogenetics in a number of species. Despite the relative simplicity of the protocol, its optimization is time consuming and laborious. Consequently, establishment of flow cytogenetics for a new species needs to be well justified. Moreover, chromosomes in many plant species are of similar size and flow karyotyping may fail to discriminate them. While this may not be a problem in some applications, such as the purification of chromosomes for the analysis of chromatin structure and protein analysis using immunocytochemistry and proteomics, the prevailing use of flow cytogenetics is the isolation of particular chromosomes. For these reasons, the development of flow cytogenetics is warranted mainly for economically important crops and important model species, provided their karyotypes comprise chromosomes differing in size, or cytogenetic stocks (such as chromosome deletion, translocation and addition lines) are available. Good examples are legumes (field bean, garden pea, and chickpea) and cereals (barley, rye, wheat, and maize). As these crops are characterized by large genomes, flow cytogenetics can play an important role in their analysis.

Construction of chromosome- and chromosome arm-specific BAC libraries is and probably will remain the most attractive use of flow-sorted chromosomes. Libraries of this type simplify the assembly of physical contig maps for clone-by-clone sequencing and positional gene cloning. An ambitious plan is to create BAC libraries for each of the 42 chromosome arms of hexaploid wheat and sequence the complex wheat genome in a targeted and stepwise manner, working with a particular chromosome arm at a time. Such libraries would also facilitate international collaboration and division of labor (Gill et al. 2004). As the DNA of sorted chromosomes is of high molecular weight, they should be suitable for other uses. HAPPY mapping is an *in vitro* technique that defines the order and spacing of DNA markers directly on native DNA (Thangavelu et al. 2003). As HAPPY mapping is suitable for development of maps from genomes of up 1000 Mbp, chromosome sorting would be required to provide fractions from complex genomes. High-molecular weight DNA is also required for optical mapping (Aston et al. 1999), and DNA from sorted chromosomes would simplify optical mapping in complex genomes.

The potential of flow cytogenetics for genome analysis was expanded recently by the advent of whole genome amplification (WGA) protocols that employ DNA polymerase phi29 (Dean et al. 2001; Lizardi et al. 1998). WGA is superior to DOP-PCR in that it is representative and produces microgram quantities of DNA with longer DNA fragments (ca. 10–30 kb) from only 1–10 ng of template DNA (ca. 10^3–10^4 chromosomes). DNA produced after WGA should be suitable for construction of short insert DNA libraries to be used in targeted retrieval of DNA markers, including the use of DArT technology (Wenzl et al. 2004). It is worth mentioning that WGA produces enough DNA for hybridization of DNA arrays

(Fiegler et al. 2003), another application of flow-sorted chromosomes that deserves further exploration.

We envisage extensive use of sorted chromosomes in cytogenetic mapping. Obvious advantages are the high sensitivity, the possibility of analyzing hundreds of copies of the same chromosome on one slide, and high spatial resolution on stretched chromosomes. In fact, various applications of flow cytogenetics can be integrated to improve further the efficacy of genome mapping. For example, a particular chromosome is sorted, its DNA used to create a BAC library, which is then used to retrieve marker-tagged clones, which are mapped by FISH on the same chromosome after sorting hundreds of copies onto a microscope slide.

To conclude, more than 20 years of evolution of flow cytogenetics has resulted in a versatile toolbox for targeted and effective analysis of plant genomes at various levels and degrees of resolution. A disadvantage of flow cytogenetics is the high cost of the equipment. It is therefore not realistic to expect that many laboratories will adopt this technology. However, this may not be necessary, as with only few exceptions, the applications of flow cytogenetics do not rely on high numbers of sorted chromosomes. It can therefore be envisaged that a few centers of flow cytogenetics would be established that would distribute sorted plant chromosomes worldwide.

Acknowledgments

This work was supported by the Czech Science Foundation (grant awards 521/04/0607, 521/05/P257 and 521/05/H013) and the Ministry of Education, Youth and Sports of the Czech Republic (grant awards LC06004 and ME844).

References

Arumuganathan, K., Slattery, J. P., Tanksley, S. D., Earle, E. D. **1991**, *Theor. Appl. Genet.* 82, 101–111.

Arumuganathan, K., Martin, G. B., Telenius, H., Tanksley, S. D., Earle, E. D. **1994**, *Mol. Gen. Genet.* 242, 551–558.

Aston, C., Mishra, B., Schwartz, D. C. **1999**, *Trends Biotechnol.* 17, 297–302.

Bartholdi, M. F., Meyne, J., Johnston, R. G., Cram, L. S. **1989**, *Cytometry* 10, 124–133.

Binarová, P., Hause, B., Doležel, J., Dráber, P. **1998**, *Plant J.* 14, 751–757.

Blennow, E. **2004**, *Chromosome Res.* 12, 25–33.

Boschman, G. A., Manders, E. M. M., Rens, W., Slater, R., Aten, J. A. **1992**, *Cytometry* 13, 469–477.

Carrano, A. V., Gray, J. W., Moore, D. H., Minkler, J. L., Mayall, B. H., Van Dilla, M. A., Mendelsohn, M. L. **1976**, *J. Histochem. Cytochem.* 24, 348–354.

Carrano, A. V., Gray, J. W., Langlois, R. G., Burkhart-Schultz, K. J., Van Dilla, M. A. **1979**, *Proc. Natl Acad. Sci. USA* 76, 1382–1384.

Conia, J., Bergounioux, C., Perennes, C., Müller, P., Brown, S., Gadal, P. **1987**, *Cytometry* 8, 500–508.

Conia, J., Müller, P., Brown, S., Bergounioux, C., Gadal, P. **1989**, *Theor. Appl. Genet.* 77, 295–303.

Cooke, A., Gillard, E. F., Yates, J. R. W., Mitchell, M. J., Aitken, D. A., Weir, D. M., Affara, N. A., Ferguson-Smith, M. A. **1988**, *Hum. Genet.* 79, 49–52.

Cremer, C., Gray, J. W., Ropers, H. H. **1982**, *Hum. Genet.* 60, 262–266.

Cremer, T., Lichter, P., Borden, J., Ward, D. C., Manuelidis, L. **1988**, *Hum. Genet.* 80, 235–246.

Dean, F. B., Nelson, J. R., Giesler, T. L., Lasken, R. S. **2001**, *Genome Res.* 11, 1095–1099.

De Jong, G., Telenius, A. H., Telenius, H., Perez, C. F., Drayer, J. I., Hadlaczky, G. **1999**, *Cytometry* 35, 129–133.

De Laat, A. M. M., Blaas, J. **1984**, *Theor. Appl. Genet.* 67, 463–467.

De Laat, A. M. M., Schel, J. H. N. **1986**, *Plant Sci.* 47, 145–151.

Doležel, J. **1991**, *Biológia* 46, 1059–1064.

Doležel, J., Lucretti, S. **1995**, *Theor. Appl. Genet.* 90, 797–802.

Doležel, J., Binarová, P., Lucretti, S. **1989**, *Biol. Plant.* 31, 113–120.

Doležel, J., Číhalíková, J., Lucretti, S. **1992**, *Planta* 188, 93–98.

Doležel, J., Lucretti, S., Schubert, I. **1994**, *Crit. Rev. Plant Sci.* 13, 275–309.

Doležel, J., Číhalíková, J., Weiserová, J., Lucretti, S. **1999a**, *Methods Cell Sci.* 21, 95–107.

Doležel, J., Macas, J., Lucretti, S. **1999b**, Flow analysis and sorting of plant chromosomes in *Current Protocols in Cytometry*, ed. J. P. Robinson, Z. Darzynkiewicz, P. N. Dean, L. G. Dressler, A. Orfao, P. S. Rabinovitch, C. C. Stewart, H. J. Tanke, L. L. Wheeless, John Wiley & Sons, Inc., New York, pp. 5.3.1–5.3.33.

Doležel, J., Kubaláková, M., Bartoš, J., Macas, J. **2004**, *Chromosome Res.* 12, 77–91.

Ferguson-Smith, M. A. **1997**, *Eur. J. Hum. Genet.* 5, 253–265.

Fiegler, H., Gribble, S. M., Burford, D. C., Carr, P., Prigmore, E., Porter, K. M., Clegg, S., Crolla, J. A., Dennis, N. R., Jacobs, P., Carter, N. P. 2003, *J. Med. Genet.* 40, 664–670.

Fuchs, J., Houben, A., Brandes, A., Schubert, I. **1996**, *Chromosoma* 104, 315–320.

Galbraith, D. W., Harkins, K. R., Maddox, J. M., Ayres, N. M., Sharma, D. P., Firoozabady, E. **1983**, *Science* 220, 1049–1051.

Gill, K. S., Arumuganathan, K., Lee, J. H. **1999**, *Theor. Appl. Genet.* 98, 1248–1252.

Gill, B. S., Appels, R., Botha-Oberholster, A. M., Buell, C. R., Bennetzen, J. L., Chalhoub, B., Chumley, F., Dvorak, J., Iwanaga, M., Keller, B., Li, W., McCombie, W. R., Ogihara, Y., Quetier, F., Sasaki, T. **2004**, *Genetics* 168, 1087–1096.

Gray, J. W., Carrano, A. V., Steinmetz, L. L., Van Dilla, M. A., Moore, D. H., Mayall, B. H., Mendelsohn, M. L. **1975**, *Proc. Natl Acad. Sci. USA* 72, 1231–1234.

Gray, J. W., Langlois, R. G., Carrano, A. V., Burkhart-Schultz, K., Van Dilla, M. A. **1979**, *Chromosoma* 73, 9–27.

Green, D. K., Fantes, J. A., Buckton, K. E., Elder, J. K., Malloy, P., Carothers, A., Evans, H. J. **1984**, *Hum. Genet.* 66, 143–146.

Gribble, S. M., Fiegler, H., Burford, D. C., Prigmore, E., Yang, F., Carr, P., Ng, B. L., Sun, T., Kamberov, E. S., Makarov, V. L., Langmore, J. P., Carter, N. P. **2004**, *Chromosome Res.* 12, 35–43.

Gualberti, G., Doležel, J., Macas, J., Lucretti, S. **1996**, *Theor. Appl. Genet.* 92, 744–751.

Harper, L. C., Cande, W. Z. **2000**, *Funct. Integrat. Genomics* 1, 89–98.

Harris, P., Boyd, E., Ferguson-Smith, M. A. **1985**, *Hum. Genet.* 70, 59–65.

Heng, H. H. Q., Tsui, L. C. **1998**, *J. Chromat. A* 806, 219–229.

Janda, J., Bartoš, J., Šafář, J., Kubaláková, M., Valárik, M., Číhalíková, J., Šimková, H., Caboche, M., Sourdille, P., Bernard, M., Chalhoub, B., Doležel, J. **2004**, *Theor. Appl. Genet.* 109, 1337–1345.

Janda, J., Šafář, J., Kubaláková, M., Bartoš, J., Kovářová, P., Suchánková, P., Pateyron, S., Číhalíková, J., Sourdille, P., Šimková, H., Faivre-Rampant, P., Hřibová, E., Bernard, M., Lukaszewski, A., Doležel, J., Chalhoub, B. **2006**, *Plant J.* 47, 977–986.

Kejnovský, E., Vrána, J., Matsunaga, S., Souček, P., Široký, J., Doležel, J., Vyskot, B. **2001**, *Genetics* 158, 1269–1277.

Koblížková, A., Doležel, J., Macas, J. **1998**, *BioTechniques* 25, 32–38.

Kota, R., Spielmeyer, W., Doležel, J., Šafář, J., Paux, E., McIntosh, R. A., Lagudah, E. S. **2006**, Towards isolating the durable stem rust resistance gene Sr2 from hexaploid wheat (*Triticum aestivum* L.) in *Abstracts of the International Conference "Plant and Animal Genome XIV"*, Sherago International, Inc., San Diego, p. 181.

Kovářová, P., Navrátilová, A., Macas, J., Doležel, J. **2007**, *Biol. Plant.* 51, 43–48.

Kubaláková, M., Lysák, M. A., Vrána, J., Šimková, H., Číhalíková, J., Doležel, J. **2000**, *Cytometry* 41, 102–108.

Kubaláková, M., Vrána, J., Číhalíková, J., Šimková, H., Doležel, J. **2002**, *Theor. Appl. Genet.* 104, 1362–1372.

Kubaláková, M., Valárik, M., Bartoš, J., Vrána, J., Číhalíková, J., Molnár-Láng, M., Doležel, J. **2003**, *Genome* 46, 893–905.

Kubaláková, M., Kovářová, P., Suchánková, P., Číhalíková, J., Bartoš, J., Lucretti, S., Watanabe, N., Kianian, S. F., Doležel, J. **2005**, *Genetics* 170, 823–829.

Langer, S., Kraus, J., Jentsch, I., Speicher, M. R. **2004**, *Chromosome Res.* 12, 15–23.

Langlois, R. G., Yu, L. C., Gray, J. W., Carrano, A. V. **1982**, *Proc. Natl Acad. Sci. USA* 79, 7876–7880.

Lebo, R. V., Carrano, A. V., Burkhart-Schultz, K., Dozy, A. M., Yu, L. C., Kan, Y. W. **1979**, *Proc. Natl Acad. Sci. USA* 76, 5804–5808.

Lee, J. H., Arumuganathan, K. **1999**, *Molecules Cells* 9, 436–439.

Lee, J. H., Arumuganathan, K., Kaeppler, S. M., Kaeppler, H. F., Papa, C. M. **1996**, *Genome* 39, 697–703.

Lee, J. H., Arumuganathan, K., Yen, Y., Kaeppler, S. M., Kaeppler, H. F., Baenziger, P. S. **1997**, *Genome* 40, 633–638.

Lee, J. H., Arumuganathan, K., Chung, Y. S., Kim, K. Y., Chung, W. B., Bae, K. S., Kim, D. H., Chung, D. S., Kwon, O. C. **2000**, *Molecules Cells* 10, 619–625.

Lee, J. H., Arumuganathan, K., Kaeppler, S. M., Park, S. W., Kim, K. Y., Chung, Y. S., Kim, D. H., Fukui, K. **2002**, *Planta* 215, 666–671.

Levy, H. P., Schultz, R. A., Ordonez, J. V., Cohen, M. M. **1991**, *Cytometry*, 12, 695–700.

Li, L. J., Arumuganathan, K., Rines, H. W., Phillips, R. L., Riera-Lizarazu, O., Sandhu, D., Zhou, Y., Gill, K. S. **2001**, *Theor. Appl. Genet.* 102, 658–663.

Li, L. J., Arumuganathan, K., Gill, K. S., Song, Y. C. **2004**, *Hereditas* 141, 55–60.

Liu, S., Pumphrey, M. O., Zhang, X., Gill, B. S., Stack, R. W., Gill, J. S., Doležel, J., Chalhoub, B., Anderson, J. **2005**, Toward positional cloning of Qfhs.ndsu-3BS, a major QTL for *Fusarium* head blight resistance in wheat in Abstracts of the International Conference "Plant and Animal Genome XIII", Sherago International, Inc., San Diego, p. 71.

Lizardi, P. M., Huang, X. H., Zhu, Z. R., Bray-Ward, P., Thomas, D. C., Ward, D. C. **1998**, *Nature Genet.* 19, 225–232.

Lucas, J. N., Gray, J. W., Peters, D. C., Van Dilla, M. A. **1983**, *Cytometry* 4, 109–116.

Lucretti, S., Doležel, J. **1997**, *Cytometry* 28, 236–242.

Lucretti, S., Doležel, J., Schubert, I., Fuchs, J. **1993**, *Theor. Appl. Genet.* 85, 665–672.

Lysák, M. A., Číhalíková, J., Kubaláková, M., Šimková, H., Künzel, G., Doležel, J. **1999**, *Chromosome Res.* 7, 431–444.

Macas, J., Doležel, J., Lucretti, S., Pich, U., Houben, A., Wobus, U., Schubert, I. **1993**, *Chromosome Res.* 1, 107–115.

Macas, J., Doležel, J., Gualberti, G., Pich, U., Schubert, I., Lucretti, S. **1995**, *BioTechniques* 19, 402–408.

Macas, J., Gualberti, G., Nouzová, M., Samec, P., Lucretti, S., Doležel, J. **1996**, *Chromosome Res.* 4, 531–539.

Neumann, P., Lysák, M. A., Doležel, J., Macas, J. **1998**, *Plant Sci.* 137, 205–215.

Neumann, P., Požárková, D., Vrána, J., Doležel, J., Macas, J. **2002**, *Chromosome Res.* 10, 63–71.

Paux, E., Roger, D., Badaeva, E., Gay, G., Bernard, M., Sourdille, P., Feuillet, C. **2006**, *Plant Journal* 48, 463–474.

Peterson, D. G., Schulze, S. R., Sciara, E. B., Lee, S. A., Bowers, J. E., Nagel, A., Jiang, N., Tibbitts, D. C., Wessler, S. R., Paterson, A. H. **2002**, *Genome Res.* 12, 795–807.

Pich, U., Meister, A., Macas, J., Doležel, J., Lucretti, S., Schubert, I. **1995**, *Plant J.* 7, 1039–1044.

Pinkel, D., Landegent, J., Collins, C., Fuscoe, J., Seagraves, R., Lucas, J., Gray, J. **1988**, *Proc. Natl Acad. Sci. USA* 85, 9138–9142.

Požárková, D., Koblížková, A., Román, B., Torres, A. M., Lucretti, S., Lysák, M. A., Doležel, J., Macas, J. **2002**, *Biol. Plant.* 45, 337–345.

Román, B., Satovic, Z., Požárková, D., Macas, J., Doležel, J., Cubero, J. I., Torres, A. M. **2004**, *Theor. Appl. Genet.* 108, 1079–1088.

Šafář, J., Bartoš, J., Janda, J., Bellec, A., Kubaláková, M., Valárik, M., Pateyron, S., Weiserová, J., Tušková, R., Číhalíková, J., Vrána, J., Šimková, H., Faivre-Rampant, P., Sourdille, P., Caboche, M., Bernard, M.,

Doležel, J., Chalhoub, B. **2004**, *Plant J.* 39, 960–968.

Schubert, I., Doležel, J., Houben, A., Scherthan, H., Wanner, G. **1993**, *Chromosoma* 102, 96–101.

Schwarzacher, T., Wang, M. L., Leitch, A. R., Miller, N., Moore, G., Heslop-Harrison, J. S. **1997**, *Theor. Appl. Genet.* 94, 91–97.

Shepel, L. A., Lan, H., Brasic, G. M., Gheen, M. E., Hsu, L. C., Haag, J. D., Gould, M. N. **1998**, *Mamm. Genome* 9, 622–628.

Shizuya, H., Birren, B., Kim, U. J., Mancino, V., Slepak, T., Tachiiri, Y., Simon, M. **1992**, *Proc. Natl Acad. Sci. USA* 89, 8794–8797.

Šimková, H., Číhalíková, J., Vrána, J., Lysák, M. A., Doležel, J. **2003**, *Biol. Plant.* 46, 369–373.

Stohr, M., Hutter, K., Frank, M., Futterman, G. **1980**, *Histochemistry* 67, 179–190.

Stubblefield, E., Cram, S., Deaven, L. **1975**, *Exp. Cell Res.* 94, 464–468.

Suchánková, P., Kubaláková, M., Kovářová, P., Bartoš, J., Číhalíková, J., Molnár-Láng, M., Endo, R. T., Doležel, J. **2006**, *Theor. Appl. Genet.* 113, 651–659.

Svartman, M., Stone, G., Page, J. E., Stanyon, R. **2004**, *Chromosome Res.* 12, 45–53.

Telenius, H., Carter, N. P., Bebb, C. E., Nordenskjöld, M., Ponder, B. A. J., Tunnacliffe, A. **1992**, *Genomics* 13, 718–725.

Ten Hoopen, R., Manteuffel, R., Doležel, J., Malysheva, L., Schubert, I. **2000**, *Chromosoma* 109, 482–489.

Thangavelu, M., James, A. B., Bankier, A., Bryan, G. J., Dear, P. H., Waugh, R. **2003**, *Plant Biotechnol. J.* 1, 23–31.

Tian, Y., Nie, W., Wang, J., Ferguson-Smith, M. A., Yang, F. **2004**, *Chromosome Res.* 12, 55–63.

Űberall, I., Vrána, J., Bartoš, J., Šmerda, J., Doležel, J., Havel, L. **2004**, *Biol. Plant.* 48, 199–203.

Valárik, M., Bartoš, J., Kovářová, P., Kubaláková, M., de Jong, H., Doležel, J. **2004**, *Plant J.* 37, 940–950.

Van Dilla, M. A., Deaven, L. L. **1990**, *Cytometry* 11, 208–218.

Veuskens, J., Marie, D., Brown, S. C., Jacobs, M., Negrutiu, I. **1995**, *Cytometry* 21, 363–373.

Vláčilová, K., Ohri, D., Vrána, J., Číhalíková, J., Kubaláková, M., Kahl, G., Doležel, J. **2002**, *Chromosome Res.* 10, 695–706.

Vrána, J., Kubaláková, M., Šimková, H., Číhalíková, J., Lysák, M. A., Doležel, J. **2000**, *Genetics* 156, 2033–2041.

Wang, M. L., Leitch, A. R., Schwarzacher, T., Heslop-Harrison, J. S., Moore, G. **1992**, *Nucleic Acids Res.* 20, 1897–1901.

Wenzl, P., Carling, J., Kudrna, D., Jaccoud, D., Huttner, E., Kleinhofs, A., Kilian, A. **2004**, *Proc. Natl Acad. Sci. USA* 101, 9915–99.

17
Analysis of Plant Gene Expression Using Flow Cytometry and Sorting

David W. Galbraith

Overview

Flow cytometry (FCM) and cell sorting is emerging as a pre-eminent technology for the analysis of global plant gene expression. It combines unparalleled accuracy with unique capabilities for dissecting the contributions of different cell types, and the underlying regulatory mechanisms. Further, it is well suited to integration with other methods, such as microarrays, that are central to the analysis of global gene expression. This chapter provides a general overview of the methods that are available for the global analysis of plant gene expression, shows how FCM can be used to monitor patterns of gene expression in plants, discusses the role of stochasticity of gene expression, and describes how flow sorting can be used for selective isolation of specific cell types for gene expression analysis. Finally, a discussion of future directions suggests new research avenues in which flow cytometry and sorting will play pivotal roles.

17.1
Introduction

"Gene expression" as a compound noun in its most general sense links genotype to phenotype. Paradoxically, as we have gained more understanding of the structure and function of living organisms, our grasp of concepts previously considered as largely understood has loosened; in the case of the word *genotype*, we now are beginning to recognize the potential contributions of epigenetic factors, as well as of previously unrecognized transcripts, and not simply the DNA sequences, to the store of genomic information that is inherited between generations, and employed for elaboration of the organism during development and in response to interactions with the environment. In the case of the word *phenotype*, we correspondingly are becoming sensitive to the notion that phenotypes can be described only in terms of the methods and technologies available to recognize them. Taken together, these raise questions as how best to define a gene, how

Flow Cytometry with Plant Cells. Edited by Jaroslav Doležel, Johann Greilhuber, and Jan Suda
Copyright © 2007 WILEY-VCH Verlag GmbH & Co. KGaA, Weinheim
ISBN: 978-3-527-31487-4

adequately our available technology can discern different phenotypes, and how these might be linked together by the action of regulatory networks.

Implicit in the concept of regulatory networks is the existence of combinatorial interactions between genes that results in different cellular phenotypes. Stated in its simplest form, the expression of different combinations of genes, measured across the entire genome, reflects this cellular phenotype and represents a molecular phenotype that precisely defines that cell state (Hughes et al. 2000). From this, it follows that the ability to measure this molecular phenotype is of paramount importance, as is equally the availability of multiple examples of different cellular states over which these measurements can be made. If sufficient information is extracted, it follows that information concerning the structure of the regulatory networks can be determined (Hughes et al. 2000). The ultimate goal, of course, is a sampling of all possible different states, from which a complete and comprehensive description of genome-wide gene regulation can be derived.

In terms of the relationship of flow cytometry (FCM) and cell sorting to the analysis of global gene regulation, three aspects are apparent within the context discussed above, and they represent the three parts of this review. The first starts with a description of the recent developments in methods for analysis of global gene expression, and discusses how these technologies can best be married to those of FCM and cell sorting. The second part of this review concerns the ability of FCM to provide a unique means with which to monitor patterns of gene expression in plants, thereby providing novel phenotypes that would not otherwise be recognized or used for measurements. Many of these phenotypes are revealed using transgenic technologies, which are briefly detailed, followed by a discussion of results obtained in the analysis of wild-type and transgenic plants. FCM also has a valuable role to play in the measurement of the stochasticity of gene expression and this aspect is discussed. The third, and final part, concerns the use of FCM and cell sorting for the manipulation of specific cell types (defined in the broadest sense) such that their individual contributions to gene expression can be determined. Again, transgenic technologies are central to this type of activity. In a separate chapter in this volume (Chapter 10), I have described some special cases of analysis of gene expression through construction of cDNA libraries, involving natural (i.e. non-protoplast) single cell suspensions.

17.2
Methods, Technologies, and Results

17.2.1
Current Methods for Global Analysis of Gene Expression

Over the last decade, the thrust in methods development has been towards analyses that are global in scope and provide data in a high throughput manner. The most widely adopted methods can be divided into two groups according to the underlying physical principles of analysis, which obtain their specificity for individual transcripts based on hybridization and sequencing, respectively.

17.2.1.1 Methods Based on Hybridization

DNA microarrays represent the first and most widely adopted of these methods. DNA microarrays are produced through immobilization of DNA sequences at defined locations on the surfaces of solid supports, typically glass slides. Immobilization is done either through mechanical deposition ("spotted" microarrays; Schena et al. 1995) or by step-wise synthesis using photolithographic or droplet deposition methods. Affymetrix (Fodor et al. 1991), NimbleGen (Singh-Gasson et al. 1999), and Agilent (http://www.agilent.com) are the major suppliers of microarrays produced using *in situ* synthesis. For spotted arrays, features are typically 100 µm in diameter, and are spaced at around 180 µm, center-to-center. This allows printing of around 30 000 elements per standard microscope slide. Smaller feature sizes can be produced using smaller pins, and it is anticipated that as many as 50 000 features will be routinely printed. This generally exceeds the number of genes for most plant genomes. Affymetrix and NimbleGen arrays have square features which are considerably smaller. For example, the commercially available Affymetrix *Arabidopsis thaliana* (Brassicaceae) genome (ATH-1) microarray comprises 18-µm features, with nominal DNA lengths of 25 bases. NimbleGen arrays are more flexible in terms of array element lengths, and their feature size is 16 µm. The large total capacity of these microarrays (the Nimble-Gen platform is advertised as providing as many as 786 000 different features on a single microarray) is offset by the use of probe sets, rather than single elements, to query specific gene transcripts. For Affymetrix, the ATH-1 microarray employs 11 probe pairs per transcript sequence, which means only 22 000 genes are represented per microarray, or about 70% of the genome. The restriction in array capacity imposed by the use of probe pairs relates to the method of analysis recommended by Affymetrix. This method, termed the Microarray Analysis Suite version 5.0 (MAS5.0), involves comparisons of signals emerging from DNA elements that are perfect matches (PM) to the cognate transcript with corresponding mismatch (MM) elements in which the central base is replaced. There is considerable discussion as to the optimal method for data analysis, and a clear consensus has not yet emerged (Qin et al. 2006; Shedden et al. 2005; Zakharkin et al. 2005). Evidently, methods that do not require MM elements (e.g. the Robust Multichip Average (RMA); (Irizarry et al. 2003) and gcRMA, a variant of RMA in which the background is modeled based on the GC content of the probe (Wu et al. 2004); see Schmid et al. (2005) for extensive use of gcRMA) provide the potential to double the gene information content of a single microarray. Whatever analysis method is employed, Affymetrix has more recently moved to an 11-µm feature size, and is planning a further decrease to a 5-µm feature. This provides sufficient space for producing tiling arrays, defined (in the case of Affymetrix) as indexed 25-mer sequences, scanning across the genome in as low as 1-bp increments, at reasonable costs. Further GeneChip designs are also likely to focus on the measurement of the activities of open reading frames (ORFs), rather than annotated gene transcripts, thereby allowing flexible identification of changes in transcript topology associated with alternative splicing. Of the two companies providing microarrays employing photolithographic methods, the principal advantage offered by NimbleGen is that it uses a Digital Micromirror Device (DMD), a solid-state integrated circuit containing an array of miniature steerable mirrors, to reflect the deprotecting ultraviolet light onto the microarray surface. The unblocking sequence required for *in situ* DNA synthesis then simply involves use of software to program the movements of the DMD, which can be readily modified for each new microarray. In contrast, Affymetrix achieves spatial unblocking through the design and construction of a series of photolithographic masks, new masks being required for each array layout. Agilent arrays achieve flexibility similar to those of NimbleGen, through sequential deposition of activated nucleotides at the array element locations using a printer employing the ink-jet principle. Expression array elements are 60-mers, and have diameters of 110 μm. The next generation of array elements will have diameters of 60 μm, and ultimately it is planned to synthesize 30-μm elements, with corresponding increases in element densities.

For interfacing with FCM and cell sorting, the primary consideration is the amount of RNA that can be extracted from the sorted protoplasts, cells or subcellular organelles. We have found that *Arabidopsis* root protoplasts contain in the range of 2–20 pg total RNA per cell (Birnbaum et al. 2003), with nuclei containing about 0.2 pg total RNA (C. Zhang and D. W. Galbraith, unpublished data). Since a total of about 250 000 objects is a reasonable practical upper limit for sorting (based largely on preparation needs and the proportions of the desired cell type within the tissues of interest), this represents about 0.5–5 μg total RNA for protoplasts and 10 to 100-fold less for nuclei. Amplification is therefore essential for microarray target production. A number of different techniques are available for this purpose (for reviews see Brandt 2005; Nygaard and Hovig 2006), and these provide amplification factors as large as 10^{11}-fold (Nygaard and Hovig 2006). Employing amplification raises concerns regarding maintenance of fidelity of representation of the transcripts, overall reproducibility, avoidance of 5′-truncation, cost, and so on (Brandt 2005; Nygaard and Hovig 2006). Methods based on the use of *in vitro* transcription (IVT; Van Gelder et al. 1990) have gained particular acceptance, being generally considered to introduce fewer artifacts than methods based on the polymerase chain reaction (PCR). One round of IVT is in fact routinely employed in the Affymetrix GeneChip expression platform.

Other high throughput methods based on hybridization have included Northerns (Brown et al. 2001), and RT-PCR (Czechowski et al. 2004), but these have not been widely employed due to issues of costs and/or convenience.

17.2.1.2 Methods Based on Sequencing

Simply counting the frequencies of Expressed Sequence Tags (ESTs) within cDNA libraries provides a direct means to measure transcript abundance (see for example Fernandes et al. 2002), and this approach, colloquially termed "electronic

Northerns", can be used for analysis of gene expression. However, the accuracy with which EST frequency reports transcript level, and hence gene expression, depends both on clonability of the individual sequences and on the sampling depth. Given the historical expense of conventional sequencing, two methods were developed to reduce these costs, and these have become widely adopted. Termed Serial Analysis of Gene Expression (SAGE) and Massively Parallel Signature Sequencing (MPSS), these methods achieve economies of sequencing primarily based on a reduction in the lengths of individual sequence reads. This is based on the recognition that individual transcripts can be unambiguously identified by very short sequence reads (this is calculated as $x = 4^N$ such that for $N > 9$, $x > 262\,144$ which exceeds the predicted number of transcripts for most eukaryotic organisms; Saha et al. 2002; Velculescu et al. 1995). SAGE achieves economies of scale by identifying short (9–11 bp; Matsumura et al. 1999) or longer (20–26 bp; Matsumura et al. 2005; Saha et al. 2002) tags, located at the first *Nla*III site 5' to the polyA$^+$ tail of each transcript, using the activity of Type IIs restriction enzymes, such as *Bms*FI or *Fok*I in the former case and *Mme*I or *Eco*P15I in the latter, to define these tags. The tags are combined in pairs, amplified through PCR, excised and concatenated, and cloned. Conventional capillary sequencing then identifies multiple ditags within single runs, these being precisely indexed within the overall sequence by the presence of residual restriction enzyme recognition sites. For a recent review of the method, see Harbers and Carninci (2005).

MPSS employs a unique bead-based method for capture of individual transcript-specific tags (the 3'-most *Dpn*II site 5' to the polyA$^+$ tail of a cDNA molecule; Meyers et al. 2004), each bead carrying about 100 000 copies of a single sequence. The beads are then immobilized within a specialized flow cell, within which they are packed as a two-dimensional monolayer through which the sequencing reagents can flow. Although only short reads are possible (up to 20 bases), this is done in parallel for all the beads that are immobilized, resulting in extremely high total rates of sequence production. The accuracy of SAGE and MPSS in measuring transcript abundances and thereby calculating differential gene expression depends on the depth of tag sampling. Much work is also underway concerning the most appropriate statistical methods for extrapolating differential expression from SAGE data (see for example Lu et al. 2005; Man et al. 2000).

Interfacing flow sorting to SAGE and MPSS requires only that sufficient RNA be available for production of libraries of adequate diversity. In the case of MPSS, 20 µg total RNA is required for each sample (B. Meyers, personal communication). SAGE libraries can be made from lower amounts of RNA (about 50 ng of polyA$^+$; Gowda et al. 2004).

17.2.1.3 Emerging Sequencing Technologies

Given the fact that sampling-depth scales directly and inversely with sequencing costs, it is appropriate to explore the specifications of the next generation of sequencing technologies that have recently been described (Margulies et al. 2005; Shendure et al. 2005). The technology developed by 454 Life Sciences is based on parallel PCR amplification of 300–500-bp DNA fragments which are ligated to specific linkers and individually immobilized on 28-µm beads within aqueous droplets in an oil emulsion (Margulies et al. 2005). Each droplet is arranged to contain one bead carrying a single DNA fragment, and this enables clonal amplification of the DNA within the droplet. The amplified DNA is then employed for parallel pyrosequencing reactions using picoliter reactor wells carried on a fibre optic slide, each well accommodating a single bead and its associated amplified DNA. The presence or absence of signal during each cycle of pyrosequencing generates base-called sequences. A run of 4.5 h produces approximately 200 000 to 300 000 high-quality reads with an average of 100 nucleotides per read, resulting in 20 to 30 million bases per sequencing reaction. The methods developed by the Church laboratory (Shendure et al. 2005) employ a conceptually similar approach involving parallel, low-volume, bead-based PCR for clonal amplification of individual DNA molecules. This is followed by a novel method of ligation-based sequencing involving immobilization on polyacrylamide gels. The methods outlined on the Solexa website (www.solexa.com) describe a method of sequencing that relies on immobilization of tagged DNA fragments on a solid surface, which are then subjected to PCR *in situ*, and sequenced at the many locations using 2-D fluorescence imaging to provide highly parallel data output.

Although there is some debate about the true costs of the new sequencing technologies (Church et al. 2006), it is clear that highly parallel sequencing methods have considerable potential to impact the area of expression profiling. This is because the ability to determine short sequences, a general feature of the emerging methods, is quite sufficient to unambiguously define specific transcripts, as detailed for the SAGE and MPSS methods described above. Therefore the only limitation on their general use in expression profiling, through counting of individual sequence tags, becomes the cost of the sequencing methods as required to provide sufficient depth of tag counting. Our best guess at the moment is that costs need to drop by at least a factor of 10–100 for the methods to be competitive with microarrays. Nonetheless, given that Moore's law (the 1965 prediction by Gordon Moore, co-founder of Intel, that the density of transistors on semiconductor chips would double every 18 months, hence that the costs of computation would inversely scale at the same rate) appears to apply to sequencing capabilities and costs, a drop in this magnitude is inevitable, and it is reasonable to plan for this event. For interfacing to plant flow sorting, the amounts of RNA required for Solexa sequencing is about 5 µg per sample.

17.2.1.4 Other -omics Disciplines and Technologies

The number of different -omics disciplines has expanded greatly over the last decade, to the extent that there is even a journal incorporating that name as its title. For the purposes of this review, I will briefly mention only one other discipline, proteomics, since the available technologies, instruments, and associated methods for this discipline are relatively mature. Proteomics, defined as the global analysis of cellular proteins, has much the same scope and all-encompassing sweep as does genomics. From the technical end, it possesses the intrinsic advantage of mass-spectrometric instrumentation that is inherently and extraordinarily sensitive (Shen and Smith 2005), offset to a certain degree by a lack of amplification methods for proteins comparable to those available for nucleic acids (PCR, rolling circle amplification, *in vitro* transcription, etc.). Considerable interest is emerging concerning measurement of alterations in the abundances of proteins and of modifications to these proteins (such as phosphorylation status). FCM provides an excellent means for multiparametric analysis of proteins (see for example Irish et al. 2004; Sachs et al. 2005; reviewed in Irish et al. 2006), and it seems reasonable that this type of approach could be applicable to plant systems. In terms of the applicability of flow sorting for providing cell samples for protein characterization based on mass spectrometry, advances in proteomics technologies include moving liquid chromatography to the nanoscale level, and combining it with high-sensitivity, high-resolution Fourier transform ion cyclotron resonance mass spectrometry. This currently allows the analysis of protein mixtures at the low nanogram level, with individual protein identification sensitivity being at the low zeptomole level (Shen and Smith 2005). This sets the stage for proteomics based on single cells or small populations of cells, and is therefore clearly compatible with flow sorting.

17.2.2
Using Flow Cytometry to Monitor Gene Expression and Cellular States

17.2.2.1 Transgenic Markers Suitable for Flow Cytometry and Sorting

Pre-eminent as transgenic markers suited for FCM and cell sorting are the Fluorescent Proteins (FPs) of which the Green Fluorescent Protein (GFP) is the prototype (for recent reviews, see Galbraith 2004; Shaner et al. 2005). FPs share the felicitous property of autocatalytic fluorophore formation, employing the sidechains of the primary sequences for this purpose. This, in essence, means that transgenic expression of the coding sequence is the only mandated prerequisite for production of fluorescence. Needless to say, various factors have been found to increase the amounts of fluorescence produced per cell, notably alterations to the primary sequence to alter the spectral characteristics of the FP, to improve its folding capability, translatability, stability and solubility, and to decrease the impact of its expression on the viability of the recipient organism (Galbraith 2004; Tsien 1998). GFP is particularly suited to FCM since the absorption maximum of the popular S65T mutant (Tsien 1998) is almost exactly that of the 488-nm laser line found in most flow cytometers. A number of different mutants of GFP have been produced, and a wide variety of novel FPs and corresponding mutant forms is commercially available (Galbraith 2004; Shaner et al. 2005).

Introduction of FP coding sequences into plant cells can be either transient or transgenic. In the former case, expression is achieved either through transfection of protoplasts, or through bombardment of intact cells using DNA-coated particles (Newell 2000). In the latter case, the same methods can be used to introduce DNA into the cells, but with subsequent imposition of selection ensuring that progeny cells and/or plants have the transgenics stably integrated into the genome. However, by far the most widely employed method for the production of transgenic plants involves use of *Agrobacterium tumefaciens* (Gelvin 2000; Newell 2000). This microorganism has the capability to promiscuously transfer plasmid DNA sequences into recipient organisms (Lacroix et al. 2006), and if selectable or screenable markers are included, transgenics can be recovered. With various caveats, FP expression in plants has not been found to affect viability, similar to the situation in other organisms (Galbraith 2004). FPs can also be targeted to various subcellular compartments with reasonable impunity (Cutler et al. 2000; Koroleva et al. 2005; Tian et al. 2004). Tzfira et al. (2005) have recently described a series of vectors for facile production of N- and C-terminal FP-fusions, and my laboratory has detailed (Zhang et al. 2005) similar tagging vectors and their sequences have been deposited in GenBank. A valuable resource has been the collection of *Arabidopsis thaliana* enhancer trap lines displaying cell type specific GFP expression (Haseloff 1999); for examples of how these can be used in FCM and sorting, see below. Similar collections are now being produced for rice (Johnson et al. 2005; Kumar et al. 2005), and should be equally useful.

FCM analysis of GFP expression was first reported using transfected maize and tobacco leaf protoplasts (Galbraith et al. 1995; Sheen et al. 1995). Since that time, although a number of different reports have appeared concerning FP expression in plants (see Galbraith (2004) for a recent review), the number of reports of FCM analysis has been much more limited. Desikan et al. (1999) first outlined the use of FCM for the detection of GFP expressed in transfected rice callus protoplasts under the control of the abscisic acid (ABA)-inducible *Em* promoter. Further work provided validation of the FCM assay (Hagenbeek and Rock 2001), and, through its use, uncovered details concerning ABA signal transduction in protoplasts (Gampala et al. 2001; Hagenbeek et al. 2000). Koroleva et al. (2005) provided some uniparametric flow histograms illustrating GFP fluorescence in transformed cell lines, but a lack of experimental details make this work hard to evaluate.

Halweg et al. (2005) recently employed FCM to examine the heterogeneity of gene expression within protoplasts prepared from transgenic tobacco cell cultures. This work employed counterstaining by propidium iodide to identify viable protoplasts which were then examined for GFP fluorescence as a function of the sequences present in the different constructs used for producing the transgenic cells (Fig. 17.1). The flow methods provide a sensitive and accurate means to identify variation in GFP expression both within and between transgenic lines. Inclusion of matrix attachment regions (MARs) in the GFP constructs increased the probability that GFP would be expressed, and also increased the magnitude of its expression.

17.2.2.2 Subcellular Targeting as a Means for Transgenic Analysis

As discussed previously, the nucleus is the site of storage of much of the information underlying gene expression. Adopting a focus on the nucleus therefore represents a reasonable way to attempt dissection of the flow of information from its source in the nuclear genome (Galbraith 2003). Targeting FPs to the nucleus can

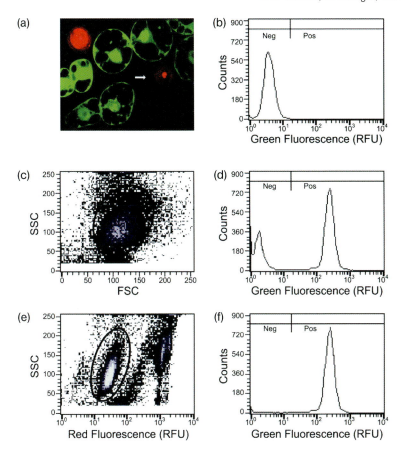

Fig. 17.1 Propidium iodide (PI) staining of protoplast preparations for improved measurement of GFP using flow cytometry. (a) A 1-μm thick confocal section of a PI-stained protoplast preparation of a tobacco cell line uniformly and stably expressing GFP (green color). Intact protoplasts exclude propidium iodide, whereas free nuclei and nuclei in protoplasts with compromised plasma membrane integrity stain brightly with red. GFP is absent from protoplasts in which nuclei stain brightly with PI. The arrow indicates a missing portion of the plasma membrane. (b) Uniparametric analysis of green fluorescence emission by wild-type protoplasts. Fluorescence was measured in relative fluorescence units (RFU). In the wild-type cells, the green fluorescence intensity is below 17 RFU. Therefore transformed cells having green fluorescence below 17 RFU are considered GFP negative (Neg), and cells with green fluorescence above 17 RFU are considered GFP positive (Pos). (c) GFP-transformed protoplasts stained with PI subjected to biparametric analysis of side scatter (SSC) versus forward scatter (FSC). Gated events inside the oval are plotted in panel d. (d) Uniparametric analysis of green fluorescence emission of transformed protoplasts, gated as described in panel c. (e) The same data from the stably transformed protoplasts stained with PI were subjected to biparametric analysis of side light scatter versus red fluorescence. Gated events inside the oval are plotted in panel f. These gated events have background levels of PI fluorescence similar to that of wild-type cells. (f) Green fluorescence histogram of gated, GFP-expressing protoplasts from panel e. Note that after exclusion of protoplasts and debris that have high levels of PI fluorescence, the lesser peak corresponding to wild-type fluorescence was lost. Modified from Halweg et al. (2005, Fig. 2), with permission.

be achieved in one of two ways; fusion of nuclear localization signals (NLSs) to FPs will only lead to their accumulation within the nucleoplasm if arrangements are made to increase the size of the FP beyond the exclusion limit of the nuclear pore (Grebenok et al. 1997a, 1997b).

Alternatively, FPs can be fused to structural components of the nucleus which provide both the NLSs and immobilization within the nucleus. In recent work from my laboratory, we have described (Zhang et al. 2005) the subcellular targeting of GFP, in the form of a histone 2A fusion, to highlight the nuclei of different cell types found within the *Arabidopsis* root. The chimeric histone–GFP fusion is retained within the nuclei following tissue homogenization, and the individual nuclei can be characterized by two-dimensional flow analysis after DAPI staining (Fig. 17.2). The *Arabidopsis* root is characterized by endoreduplicated cells, having DNA contents of 2C–16C and possibly higher. Interestingly, different cell types exhibit different patterns of endoreduplication and/or cell cycle arrest. Thus, whereas those cells marked by activity of the *sul2* promoter (phloem companion cells) are 2C and 4C, those marked by activity of the *SCARECROW* promoter (endodermal cells) are 4C and 8C (Zhang et al. 2005). This suggests an interaction between the cell cycle, differentiation of specific cell types, and the maintenance of stem cell identity.

17.2.3
Using Flow Sorting to Measure Gene Expression and Define Cellular States

17.2.3.1 Protoplast and Cell Sorting Based on Endogenous Properties

Plant protoplasts and natural single cell populations possess intrinsic light scatter and fluorescence properties that can be used, in combination with the choice of specific source tissues, to define cellular states. The ability of flow sorting to selectively purify subpopulations from within heterogeneous mixtures then allows characterization of gene expression associated with these cellular states. For example, the natural fluorescence of chlorophyll, combined with staining with fluorescein diacetate to identify living cells, can be employed to differentiate between epidermal and mesophyll leaf protoplasts (Harkins et al. 1990). We further employed this approach to selectively purify mesophyll and epidermal protoplasts produced from transgenic plants expressing β-glucuronidase (GUS) under the control of promoters that are constitutive (cf. the cauliflower mosaic virus (CaMV) 35S promoter) or light-regulated (the promoters from genes encoding the small subunit of ribulose-bisphosphate carboxylase (*rbcS*) and the light-harvesting photosystem II chlorophyll a/b-binding protein (*Lhcb*)). Expression of GUS regulated by the *rbcS* and *Lhcb* promoters was only observed in the mesophyll protoplasts, whereas the CaMV promoter appeared active in both mesophyll and epidermal protoplasts. Similar cell type-specific patterns of expression were observed when wild-type protoplasts were transfected with the three different GUS constructs. Norflurazon treatment, which eliminates chlorophyll from all cell types, severely abrogated GUS expression regulated by the *rbcS* and *Lhcb* promoters, but had little effect on expression regulated by the CaMV promoter. This

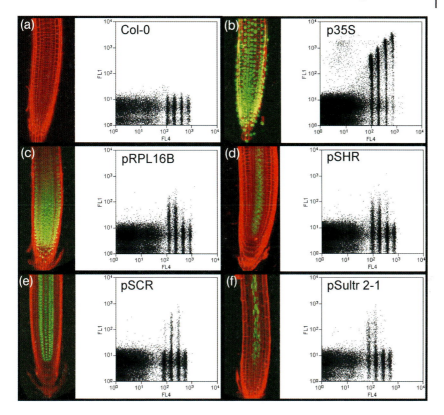

Fig. 17.2 Confocal and biparametric flow cytometric analyses of wild-type and transgenic plants of *Arabidopsis thaliana* expressing nuclear GFP. (a) Wild-type plant. (b–f) Transgenic plants expressing nuclear GFP regulated by (b) p35S, (c) pRPL16B, (d) pSHR, (e) pSCR, and (f) pSultr2-1. Flow cytometry was done using a Cytomation MoFlo flow cytometer with laser excitation at 365/488 nm, and biparametric detection of DAPI fluorescence (418–482 nm; FL4; log units), and GFP fluorescence (505–530 nm; FL1; log units), with triggering based on 90°-light scatter. Populations of endoreduplicated nuclei, typical of the *Arabidopsis* root, appear as discrete clusters equally spaced along the logarithmic abscissa according to DNA content. Those labeled with GFP appear as additional clusters placed at higher ordinate values. For confocal microscopy, the cell walls were counterstained by dipping the roots into a solution of propidium iodide (1 μg ml^{-1} in water) for 2 min. Abbreviations: p35S, CaMV 35S promoter; pSCR, *SCARECROW* promoter; pSHR, *SHORTROOT* promoter; pRPL16B, ribosomal protein large subunit 16B promoter; pSultr2-1, sulfate transporter 2-1 promoter. From Zhang et al. (2005, Fig. 4), with permission.

work was extended by Meehan et al. (1996) who found that GUS activities under the transcriptional control of the *Lhcb* promoter were correlated with cell size and chlorophyll content in green leaf protoplasts prepared either from transgenic or wild-type *Arabidopsis* plants. They were further able to sort white and green leaf protoplasts from variegated leaves of *immutans* plants, establishing that white protoplasts were considerably smaller than their green counterparts, and that green leaf cells of *immutans* have higher levels of chlorophyll than their wild-type counterparts.

In the companion Chapter 10, I have discussed the use of flow sorting for the isolation of specific single cell types (microspores, pollen, sperm) for production of cDNA libraries. In principle, flow sorting could be employed for the isolation of protoplasts, transgenic or otherwise, for similar work, although this does not yet seem to have occurred. Likewise, it should be possible to employ flow sorted protoplasts as inputs for alternative means for expression analysis such as SAGE and MPSS. The following section describes progress in the expression profiling of flow sorted protoplasts using microarrays.

17.2.3.2 Protoplast Sorting Based on Transgenic Markers

Birnbaum et al. (2003) employed five different transgenic *Arabidopsis* lines, in which expression of GFP highlighted within the root the endodermis, the endodermis *plus* the cortex, the epidermis, the lateral root cap, and the stele. Protoplasts, prepared from the roots, were flow sorted to provide purified populations comprising the different cell types. RNA was then extracted, and used for hybridization to Affymetrix ATH-1 GeneChips. The roots were separately dissected into regions roughly corresponding to the root tip, the elongation zone, and the maturation zone; together this provided 15 subzones (Fig. 17.3), within which differential gene expression patterns could be clustered using standard statistical methods. Most of the variation inherent in the root could be explained by eight to 10

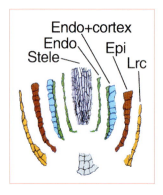
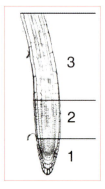

Fig. 17.3 Dividing the *Arabidopsis* root into specific subzones for expression measurements. Fifteen expression subzones were generated through the sorting and analysis of protoplasts prepared from five different transgenic plant lines, in which GFP expression was restricted to the five indicated groupings of cell types, and combining this with mechanical partitioning of the roots into three developmental stages, according to physical distance from the root-tip. Abbreviations: Endo, endodermis; Epi, epidermis; Lrc, lateral root cap (from Birnbaum et al. 2003).

major patterns, termed localized expression domains (LEDs), which could then be mapped to physical locations in the root. Examining the distributions of the gene ontologies within the clusters allowed assignment of functions, the recognition of the involvement of phytohormones in root development and responses to the environment, the characterization of potential functional redundancies for pyramiding knock-outs, and the identification of co-regulation as a function of chromosomal location. Above all, the dataset provides an atlas of expression values for the *Arabidopsis* root, which is available on the web for searching purposes.

This work was recently extended to a study of the quiescent center (QC) using GFP expression directed by the *AGAMOUS-LIKE 42* promoter and first intron to highlight this cell type (Nawy et al. 2005). Sorting of protoplasts was more challenging in this case, due to the low number of positive events. RNA from sorted protoplasts, hybridized to Affymetrix GeneChips, identified a set of genes whose transcription was specifically upregulated in the QC cells. Within this set, confirmation of QC-specific expression patterns was obtained for seven transcription factor genes by creating additional transgenic plants expressing transcriptional GFP fusions from these promoters. Details for the methods underlying these two reports have been separately published (Birnbaum et al. 2005).

The particular value of flow sorting in this type of work (Birnbaum et al. 2003; Nawy et al. 2005) is that it enables the recognition and separation of cells in different states (Hughes et al. 2000), based on expression of a specific marker gene, and identification of the molecular phenotype corresponding to this state. The degree to which complex transcriptional regulatory networks can be dissected depends fundamentally on the number of different states that can be measured. Given that techniques of enhancer trapping and gene trapping have the potential to provide a very large, if not unlimited, number of examples of FP-highlighting of different cell types, this technology has much future potential. Clearly it requires that the process of protoplast production does not affect gene expression; this was addressed at least in part through identification and exclusion from analysis of the limited number of genes that were consistently upregulated by the protoplasting process (Birnbaum et al. 2003). Another essential requirement is that the tissue of interest be susceptible to production of protoplasts, which is not true of all tissues, cell types, or plant species. For this reason, sorting of FP-highlighted nuclei may represent an alternative strategy (see below).

17.2.3.3 Sorting of Nuclei Based on Transgenic Markers

The concept of measuring gene expression through analysis of the complement of RNA transcripts found within isolated nuclei (Galbraith 2003) relies on the supposition that transcript polyadenylation, and, to a lesser extent, processing, occurs co-transcriptionally. It also supposes that the transcript composition of the nucleus reflects, or at the very least be relevant, to the transcript composition of the cytoplasm. Extending this concept to examine cell type-specific gene expression requires the FP targeting to the nucleus (described above and in Grebenok et al. 1997a, 1997b; Zhang et al. 2005), followed by production of cellular homogenates by chopping (Galbraith et al. 1983). Nuclei are then stained using DAPI, and the minor subset of GFP-positive nuclei identified and sorted. Care must be taken to eliminate cytoplasmic contamination from the sorted nuclei prior to or after sorting, since the dilution inherent to flow sorting is most likely insufficient to offset the greater levels of RNA within the cytoplasm than within the nucleus (C. Zhang, G. M. Lambert, and D. W. Galbraith, unpublished data). Initial experiments have indicated that the amplification steps required for target preparation for microarrays result in highly reproducible hybridization signals, and this suggests that the overall strategy will provide a useful alternative to protoplast sorting.

17.3
Prospects

17.3.1
Combining Flow and Image Cytometry

Instruments have recently become available from at least two companies (e.g. Amnis, and Beckman-Coulter) that combine the capabilities of flow and image cytometry. They provide multiparametric quantification of fluorescent signals which can also be assigned to particular cellular locations, on a cell-by-cell basis. The application of these instruments to protoplasts should allow comprehensive analyses of treatments that result in alterations to the locations of subcellular markers. These could include, for example, transfection experiments to define transcripts that induce such changes, as well as the use of inducible gene expression systems for time-course analyses, and could be combined with advanced optical techniques such as Fluorescence Resonance Energy Transfer (FRET) to monitor protein–protein interactions. FRET measurements rely on the non-radiative transfer of energy from "donor" to "acceptor" fluorochromes, resulting in an increased proportion of fluorescent light being produced at the necessarily longer wavelength of the emission spectrum of the acceptor fluorochrome, at the expense of that of the shorter wavelength of the donor fluorochrome. FRET requires overlap of the donor emission and acceptor excitation spectra, alignment of the fluorescence dipoles, and close proximity of the donor and acceptor molecules. FRET employing various combinations of FPs has become particularly popular (for a recent review, see Giepmans et al. 2006).

17.3.2
Use of Protoplasts for Confirmatory Studies

The ease with which plant protoplasts can be transfected means that they can be used for confirmation of various characteristics of genes identified through other experiments (see for example Choi et al. 2005). At this point, coupling confirmatory studies to FCM analysis is not widespread, but this area should expand as appreciation increases of the power of FCM to provide quantitative information concerning gene expression. The further development of inducible and transactivatable gene expression systems for plants provides obvious additional potential lines of enquiry involving FCM (Moore et al. 2006; Rutherford et al. 2005). Further, I anticipate applications of cell sorting should become more widespread for the isolation of protoplasts exhibiting unusual patterns of expression, perhaps following mutagenesis or transfection of effector nucleic acids or more specialized reagents, such as Molecular Beacons (Stewart 2005).

17.3.3
Analysing Noise in Gene Expression

Noise is defined as variation in the measured output of gene expression across cells that otherwise appear genetically and/or phenotypically identical (Raser and O'Shea 2004, 2005). Noise results as a consequence of differences in inherent gene expression capacity, for example cell size, metabolic status, or local microenvironment, or may be related to temporally-predictable events such as passage through the cell division cycle. In addition, the small sizes of cells mean that mRNA molecules and the regulatory proteins that govern their transcription can be present and active at very low numbers within each cell (Bengtsson et al. 2006), to the extent that stochastic fluctuations come into play (Colman-Lerner et al. 2005; Raser and O'Shea 2004, 2005). Stochastic fluctuations also affect the partitioning of organelles during cell division. The operation of stochastic processes inherently therefore gives rise to additional noise within regulatory pathways. There is increasing interest in determining the different sources of noise within gene expression, in estimating their proportional contributions, and in discovering the mechanisms whereby cells and tissues can suppress the potentially randomizing effects of noise on development.

Experiments to measure noise and noise sources typically employ combinations of FP reporters, the coding regions of which are placed downstream of defined regulatory sequences and transferred into transgenic organisms. Comparison of the magnitudes of the FP signals on a cell-by-cell basis allows separation of noise into an "intrinsic" component (one that creates differences between reporters), and an "extrinsic" component (one that affects the two reporters similarly within single cells, but that creates differences between cells). The precise type of noise (whether intrinsic or extrinsic) depends on the transcriptional context of the reporters (Fig. 17.4); clearly, additional contexts could be devised that would report noise components associated with translation, and protein degradation, amongst other things. FCM is particularly suitable for studying stochastic and noise effects in transgenic organisms expressing FPs, since multiparametric measurements are simple to implement and perform (Raser and O'Shea 2004). Its application to the study of higher plants should provide interesting insights.

(a)

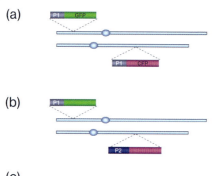

(b)

(c)

(d)

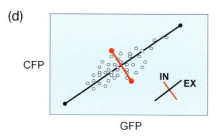

Fig. 17.4 Strategies for measurement of noise in gene expression. (a) Two reporters (GFP, the Green Fluorescent Protein, and CFP, the Cyan Fluorescent Protein) are placed under the transcriptional control of identical promoter sequences (P1), and are transgenically integrated into separate chromosomal locations. Noise measurements in this case reflect differences associated with the sites of integration. (b) The reporters are integrated in different chromosomal locations, but also comprise different transcriptional regulatory sequences (P1, P2). Noise measurements reflect a combination of locus-specific and promoter-specific effects. (c) The reporters are placed within the same regulatory context at homologous loci on the two diploid alleles (this requires homologous recombination, which may be difficult to achieve). Noise measurements reflect allele-specific effects. (d) Idealized biparametric flow cytometric measurements of GFP versus CFP expression. Extrinsic (EN) and intrinsic (IN) noise are measured orthogonally as indicated, and can be expressed as coefficients of variation.

Acknowledgments

Work in the Galbraith laboratory has been supported by grants from the National Science Foundation, and the US Departments of Agriculture and of Energy, and by the University of Arizona Agriculture Experiment Station. I thank Georgina Lambert for editorial suggestions.

References

Bengtsson, M., Ståhlberg, A., Rorsman, P., Kubista, M. **2006**, *Genome Res.* 15, 1388–1392.

Birnbaum, K., Shasha, D. E., Wang, J. Y., Jung, J. W., Lambert, G. M., Galbraith, D. W., Benfey, P. N. **2003**, *Science* 302, 1956–1960.

Birnbaum, K., Jung, J. W., Wang, J. Y., Lambert, G. M., Hirst, J. A., Galbraith, D. W., Benfey, P. N. **2005**, *Nature Methods* 2, 1–5.

Brandt, S. P. **2005**, *J. Exp. Bot.* 56, 495–505.

Brown, A. J. P., Planta, R. J., Restuhadi, F., Bailey, D. A., Butler, P. R., Cadahia, J. L., et al. **2001**, *EMBO J.* 20, 3177–3186.

Choi, M. S., Kim, M. C., Yoo, J. H., Moon, B. C., Koo, S. C., Park, B. O., Lee, J. H., Koo, Y. D., Han, H. J., Lee, S. Y., Chung, W. S., Lim, C. O., Cho, M. J. **2005**, *J. Biol. Chem.* 280, 40820–40831.

Church, G., Shendure, J., Porreca, G. **2006**, *Nature Biotechnol.* 24, 139.

Colman-Lerner, A., Gordon, A., Serra, E., Chin, T., Resnekov, O., Endy, D., Pesce, C. G., Brent, R. **2005**, *Nature* 437, 699–706.

Cutler, S. R., Ehrhardt, D. W., Griffitts, J. S., Somerville, C. R. **2000**, *Proc. Natl Acad. Sci. USA* 97, 3718–3723.

Czechowski, T., Bari, R. P., Stitt, M., Scheible, W. R., Udvardi, M. K. **2004**, *Plant J.* 38, 366–379.

Desikan, R., Hagenbeek, D., Neill, S. J., Rock, C. D. **1999**, *FEBS Lett.* 456, 257–262.

Fernandes, J., Brendel, V., Gai, X., Lal, S., Chandler, V. L., Elumalai, R., Galbraith, D. W., Pierson, E., Walbot, V. **2002**, *Plant Physiol.* 128, 896–910.

Fodor, S. P. A., Read, J. L., Pirrung, M. C., Stryer, L., Lu, A. T., Solas, D. **1991**, *Science* 251, 767–773.

Galbraith, D. W. **2003**, *Comp. Funct. Genomics* 4, 208–215.

Galbraith, D. W. **2004**, *Methods Cell Biol.* 75, 153–169.

Galbraith, D. W., Harkins, K. R., Maddox, J. R., Ayres, N. M., Sharma, D. P., Firoozabady, E. **1983**, *Science* 220, 1049–1051.

Galbraith, D. W., Grebenok, R. J., Lambert, G. M., Sheen, J. **1995**, *Methods Cell Biol.* 50, 3–12.

Gampala, S. S. L., Hagenbeek, D., Rock, C. D. **2001**, *J. Biol. Chem.* 276, 9855–9860.

Gelvin, S. B. **2000**, *Annu. Rev. Plant Physiol. Plant Mol. Biol.* 51, 223–256.

Giepmans, B. N. G., Adams, S. R., Ellisman, M. H., Tsien, R. Y. **2006**, *Science* 312, 217–224.

Gowda, M., Jantasuriyarat, C., Dean, R. A., Wang, G. L. **2004**, *Plant Physiol.* 134, 890–897.

Grebenok, R. J., Pierson, E. A., Lambert, G. M., Gong, F.-C., Afonso, C. L., Haldeman-Cahill, R., Carrington, J. C., Galbraith, D. W. **1997a**, *Plant J.* 11, 573–586.

Grebenok, R. J., Lambert, G. M., Galbraith, D. W. **1997b**, *Plant J.* 12, 685–696.

Hagenbeek, D., Rock, C. D. **2001**, *Cytometry* 45, 170–179.

Hagenbeek, D., Quatrano, R. S., Rock, C. D. **2000**, *Plant Physiol.* 123, 1553–1560.

Halweg, C., Thompson, W. F., Spiker, S. **2005**, *Plant Cell* 17, 418–429.

Harbers, M., Carninci, P. **2005**, *Nature Methods* 2, 495–502.

Harkins, K. R., Jefferson, R. A., Kavanagh, T. A., Bevan, M. W., Galbraith, D. W. **1990**, *Proc. Natl Acad. Sci. USA* 87, 816–820.

Haseloff, J. **1999**, *Methods Cell Biol.* 58, 139–151.

Hughes, T. R., Marton, M. J., Jones, A. R., Roberts, C. J., Stoughton, R., Armour, C. D., Bennett, H. A., Coffey, E., Dai, H., He, Y. D., Kidd, M. J., King, A. M., Meyer, M. R., Slade, D., Lum, P. Y., Stepaniants, S. B., Shoemaker, D. D., Gachotte, D., Chakraburtty, K., Simon, J., Bard, M., Friend, S. H. **2000**, *Cell* 102, 109–126.

Irish, J. M., Hovland, R., Krutzik, P. O., Perez, O. D., Bruserud, O., Gjertsen, B. T., Nolan, G. P. **2004**, *Cell* 118, 217–228.

Irish, J. M., Kotecha, N., Nolan, G. P. **2006**, *Nature Rev. Cancer* 6, 146–155.

Irizarry, R. A., Bolstad, B. M., Collin, F., Cope, L. M., Hobbs, B., Speed, T. P. **2003**, *Nucleic Acids Res.* 31, Art. No. e15.

Johnson, A. A. T., Hibberd, J. M., Gay, C., Essah, P. A., Haseloff, J., Tester, M., Guiderdoni, E. **2005**, *Plant J.* 41, 779–789.

Koroleva, O. A., Tomlinson, M. L., Leader, D., Shaw, P., Doonan, J. H. **2005**, *Plant J.* 41, 162–174.

Kumar, C. S., Wing, R. A., Sundaresan, V. **2005**, *Plant J.* 44, 879–892.

Lacroix, B., Tzfira, T., Vainstein, A., Citovsky, V. **2006**, *Trends Genet.* 22, 29–37.

Lu, J., Tomfohr, J. K., Kepler, T. B. **2005**, *BMC Bioinformatics* 6, Art. No. 165.

Man, M. Z., Wang, X. N., Wang, Y. X. **2000**, *Bioinformatics* 16, 953–959.

Margulies, M., Egholm, M., Altman, W. E., Attiya, S., Bader, J. S., Bemben, L. A., Berka, J., Braverman, M. S., et al. **2005**, *Nature* 437, 376–380.

Matsumura, H., Nirasawa, S., Terauchi, R. **1999**, *Plant J.* 20, 719–726.

Matsumura, H., Ito, A., Saitoh, H., Winter, P., Kahl, G., Reuter, M., Kruger, D. H., Terauchi, R. **2005**, *Cell. Microbiol.* 7, 11–18.

Meehan, L., Harkins, K., Chory, J., Rodermel, S. **1996**, *Plant Physiol.* 112, 953–963.

Meyers, B. C., Tej, S. S., Vu, T. H., Haudenschild, C. D., Agrawal, V., Edberg, S. B., Ghazal, H., Decola, S. **2004**, *Genome Res.* 14, 1641–1653.

Moore, I., Samalova, M., Kurup, S. **2006**, *Plant J.* 45, 651–683.

Nawy, T., Lee, J. Y., Colinas, J., Wang, J. Y., Thongrod, S. C., Malamy, J. E., Birnbaum, K., Benfey, P. N. **2005**, *Plant Cell* 17, 1908–1925.

Newell, C. A. **2000**, *Mol. Biotechnol.* 16, 53–65.

Nygaard, V., Hovig, E. **2006**, *Nucleic Acids Res.* 34, 996–1014.

Qin, L. X., Beyer, R. P., Hudson, F. N., Linford, N. J., Morris, D. E., Kerr, K. F. **2006**, *BMC Bioinformatics* 7, Art. No. 23.

Raser, J. M., O'Shea, E. K. **2004**, *Science* 304, 1811–1814.

Raser, J. M., O'Shea, E. K. **2005**, *Science* 309, 2010–2013.

Rutherford, S., Brandizzi, F., Townley, H., Craft, J., Wang, Y. B., Jepson, I., Martinez, A., Moore, I. **2005**, *Plant J.* 43, 769–788.

Sachs, K., Perez, O., Pe'er, D., Lauffenburger, D. A., Nolan, G. P. **2005**, *Science* 308, 523–529.

Saha, S., Sparks, A. B., Rago, C., Akmaev, V., Wang, C. J., Vogelstein, B., Kinzler, K. W., Velculescu, V. E. **2002**, *Nature Biotechnol.* 19, 508–512.

Schena, M., Shalon, D., Brown, P. O., Davis, R. W. **1995**, *Science* 270, 467–470.

Schmid, M., Davison, T. S., Henz, S. R., Pape, U. J., Demar, M., Vingron, M., Scholkopf, B., Weigel, D., Lohmann, J. U. **2005**, *Nature Genet.* 37, 501–506.

Shaner, N. C., Steinbach, P. A., Tsien, R. Y. **2005**, *Nature Methods* 2, 905–909.

Shedden, K., Chen, W., Kuick, R., Ghosh, D., Macdonald, J., Cho, K. R., Giordano, T. J., Gruber, S. B., Fearon, E. R., Taylor, J. M. G., Hanash, S. **2005**, *BMC Bioinformatics* 6, Art. No. 26.

Sheen, J., Hwang, S., Niwa, Y., Kobayashi, H., Galbraith, D. W. **1995**, *Plant J.* 8, 777–784.

Shen, Y. F., Smith, R. D. **2005**, *Expert Rev. Proteomics* 2, 431–447.

Shendure, J., Porreca, G. J., Reppas, N. B., Lin, X. X., McCutcheon, J. P., Rosenbaum, A. M., Wang, M. D., Zhang, K., Mitra, R. D., Church, G. M. **2005**, *Science* 309, 1728–1732.

Singh-Gasson, S., Green, R. D., Yue, Y., Nelson, C., Blattner, F., Sussman, M. R., Cerrina, F. **1999**, *Nature Biotechnol.* 17, 974–978.

Stewart, C. N. **2005**, *Trends Plant Sci.* 10, 390–396.

Tian, G. W., Mohanty, A., Chary, S. N., Li, S. J., Paap, B., Drakakaki, G., Kopec, C. D., Li, J. X., Ehrhardt, D., Jackson, D., Rhee, S. Y., Raikhel, N. V., Citovsky, V. **2004**, *Plant Physiol.* 135, 25–38.

Tsien, R. Y. **1998**, *Annu. Rev. Biochem.* 11, 328–329.

Tzfira, T., Tian, G. W., Lacroix, B., Vyas, S., Li, J. X., Leitner-Dagan, Y., Krichevsky, A., Taylor, T., Vainstein, A., Citovsky, V. **2005**, *Plant Mol. Biol.* 57, 503–516.

Van Gelder, R. N., von Zastrow, M. E., Yool, A., Dement, W. C., Barchas, J. D., Eberwine, J. H. **1990**, *Proc. Natl Acad. Sci. USA* 87, 1663–1667.

Velculescu, V. E., Zhang, L., Vogelstein, B., Kinzler, K. W. **1995**, *Science* 270, 484–487.

Wu, Z., Irizarry, R. A., Gentleman, R., Martinez-Murillo, F., Spencer, F. **2004**, *J. Am. Stat. Assoc.* 99, 909–917.

Zakharkin, S. O., Kim, K., Mehta, T., Chen, L., Barnes, S., Scheirer, K. E., Parrish, R. S., Allison, D. B., Page, G. P. **2005**, *BMC Bioinformatics* 6, Art. No. 214.

Zhang, C. Q., Gong, F. C., Lambert, G. M., Galbraith, D. W. **2005**, *Plant Methods* 1, 1–12.

18
FLOWer: A Plant DNA Flow Cytometry Database

João Loureiro, Jan Suda, Jaroslav Doležel, and Conceição Santos

Overview

The ever-increasing number of articles on flow cytometric analysis of plant genomes highlights the need to collect the available information and make it accessible in one comprehensive database. This goal was materialized in the Plant DNA Flow Cytometry Database (FLOWer), a project aimed at gathering an exhaustive list of articles on flow cytometry of nuclear DNA content and providing a comprehensive overview of published data. DNA-based studies clearly dominate applications of flow cytometry in plant sciences, which often give a false impression of a well-established method devoid of pitfalls. However, many particulars of the methodology are still under discussion and quality standards have not yet been universally accepted. This chapter demonstrates the usefulness of the FLOWer database as a tool for providing unbiased and quantitative data on taxonomic representation, nuclear isolation buffers, standardization, including reference DNA standards, DNA fluorochromes and measures of result quality. In addition, issues related to the objective(s) of the studies, type of instrument(s) used, scientific journals, and countries of origin of the authors may also be assessed and quantified. The database is freely accessible for public use on the Internet (http://flower.web.ua.pt/) and users can undertake their own searches and analyses. The database is regularly updated by the authors who appreciate receiving newly published papers relevant to plant DNA flow cytometry.

18.1
Introduction

Flow cytometry (FCM) is a powerful approach for measuring optical properties (light scatter and fluorescence) of single particles (cells, protoplasts, nuclei, and chromosomes) in suspension. It has been increasingly applied in plant sciences since the late 1980s, with the estimation of DNA ploidy level and genome size be-

Flow Cytometry with Plant Cells. Edited by Jaroslav Doležel, Johann Greilhuber, and Jan Suda
Copyright © 2007 WILEY-VCH Verlag GmbH & Co. KGaA, Weinheim
ISBN: 978-3-527-31487-4

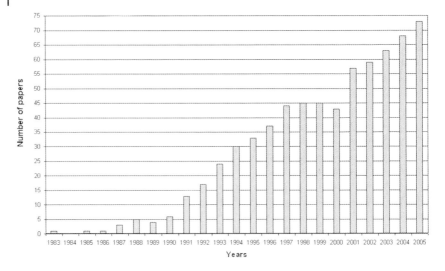

Fig. 18.1 Year distribution of the articles included in the Plant DNA Flow Cytometry (FLOWER) database.

ing the most frequent applications (Doležel and Bartoš 2005). Both uses rely on the determination of DNA amounts in cell nuclei.

Plant scientists are attracted by the numerous advantages of this technique (e.g. ease of sample preparation, rapidity of analysis, and no requirement for dividing cells) and, as expected, the number of articles has been continuously increasing over the years (Fig. 18.1). Nevertheless, there are also some weak points, which may hinder further, and perhaps more extensive use of FCM in plant sciences, such as the high cost of the instruments, difficulties in the analysis of some plant tissues/species and, in the case of estimation of genome size, occasionally contradicting results obtained in different laboratories.

The scientific community engaged in genome size analysis is conscious of such problems and has made several efforts to overcome them. A set of recommendations to achieve precise and highly comparable FCM results was thoroughly debated and finally approved at the Plant Genome Size Workshops held in the Royal Botanic Gardens, Kew in 1998 and 2003, and at the International Botanical Congress in Vienna in 2005. Among the "best practice" recommendations, a proper choice of calibration standard(s), fluorochrome(s) and buffer(s), and the awareness of potential negative effect of secondary compounds were discussed (see http://www.rbgkew.org.uk/cval/pgsm/; and Chapters 4, 7 and 12). Despite the experience of authorities participating at the Plant Genome Size Workshops, no quantitative data supporting the decisions were available and no large-scale survey of FCM literature, which could help to elucidate controversial topics and identify additional methodological issues, had been carried out. In addition, knowledge on how and to what extent the recommendations were followed is essential for the assessment of result credibility (crucial particularly for newcomers to the

FCM arena). Finally, an acquaintance with the ever-increasing number of plant FCM articles published in an ever-expanding list of journals is beyond the grasp of most FCM users and thus makes any exhaustive comparative study difficult.

To cope with the above-mentioned issues, we built and released a comprehensive Plant DNA Flow Cytometry Database (with the acronym FLOWER). The database serves as a basic source of information for plant FCM users, providing bibliographic citations together with relevant data concerning methodology, material and instrumentation. The database aims to cover a full range of DNA FCM applications in plant sciences. Currently (July 2006), it harbours more than 700 entries and is regularly updated. To make data easily accessible to the public, the FLOWER database is available in a dynamic webpage format on the Internet (http://flower.web.ua.pt/). The basic structure (searchable and output fields) is presented in Table 18.1. The database allows researchers to undertake quantitative analyses of various parameters, to access insights into the use of FCM in plant sciences over the years and to assess the reliability of individual articles based on methodological details and observance of best practice recommendations. The aim of this chapter is to describe the FLOWER database and demonstrate its usefulness. For that purpose, database outputs for the most relevant FCM parameters are briefly presented and discussed below.

18.2
Taxonomic Representation in DNA Content Studies

As might be expected, angiosperms are the most frequently analyzed group of plants in FCM studies (92.4% of all publications). Gymnosperms account for 4.2% of the entries while the proportion of other major taxonomic groups is much smaller and does not exceed 2.0%. Indeed, there are only three papers dealing with lycophytes, nine with monilophytes (i.e. horsetails and ferns) and four with bryophytes. To some extent, the number of DNA FCM articles reflects the diversity of a particular taxonomic group and the number of recognized species. Nevertheless, the relative proportion of angiosperms investigated (with ∼250 000 recognized species) is much lower than that of gymnosperms (with ∼730 recognized species). Similarly, lycophytes (∼900 recognized species), monilophytes (∼9000 recognized species) and bryophytes (∼18 000 recognized species) are also rather poorly represented.

Economically important (namely in agriculture and forestry) plant families dominate in FCM articles, with Poaceae, Fabaceae, Solanaceae and Brassicaceae together representing around 41.5% of the angiosperm database entries. In gymnosperms, the largest family, Pinaceae, prevails in FCM studies (77.8%) and the genus *Pinus* itself accounts for 42.4% of the entries.

More than three-quarters of the articles (76.8%) deal with herbaceous taxa. This is not surprising as they represent the highest number of recognized species, and most herbaceous species do not pose serious problems in FCM analyses. Woody species, which are generally considered more recalcitrant due to the presence

Table 18.1 Summary of searchable and output fields of the internet platform of the FLOWER database.

Searchable fields	Output fields	
Author	Author	
Year	Title	
Country	Year	
Nuclear isolation buffer	Country	
Fluorochrome	Nuclear isolation buffer	
Taxonomic fields	Buffer modification	
Main objective	Fluorochrome	
Standardization	Taxonomy	Plant group (bryophyte/lycophyte/monilophyte/gymnosperm/angiosperm)
Standard		
Flow cytometer		Family
Scientific journal		Species
		Growth type (herbaceous/woody/other)
	Main objective	
	Other objective	
	Standardization	Type (external/internal/pseudo-internal/no standardization/not applicable)
	Standard	Type (animal/plant)
		Species and cultivar
		2C nuclear DNA content
	Flow cytometer	Brand and model
	Scientific journal	
	Coefficient of variation	Given, range or not given
	DNA histograms	Shown or not shown
	Herbarium voucher	Available or not available

of secondary metabolites that may interfere with DNA staining (Loureiro et al. 2006a), were investigated in 20.1% of studies. Other recognized growth types (succulents, spore-bearing vascular plants and bryophytes) account for only 3.1% of the references.

18.3
Nuclear Isolation and Staining Buffers

Current methods to prepare nuclear suspensions for FCM analyses are mostly based on the breakthrough development of Galbraith et al. (1983). In their procedure, intact cell nuclei are released into the isolation or (isolation and staining) buffer simply by chopping a small amount of plant tissue with a razor blade. As reviewed in Chapter 4, the composition of the lysis buffer is crucial for obtaining precise, reliable, and high-resolution results. Given the diversity in tissue structure and chemical composition in the plant kingdom, it comes as no surprise that no single buffer works well with all species (as discussed by Doležel and Bartoš (2005) and experimentally confirmed by Loureiro et al. (2006b)). Nevertheless, the latter authors concluded that certain lysis buffers may consistently yield better results than others, at least when model species are analyzed.

Buffers undoubtedly represent one of the most important areas of the FLOWer database, offering both frequency analyses and assessment of various relationships and trends, such as which buffers have been used most frequently over the years, which buffers have been used by particular researchers and countries, and which buffers have been selected according to the type of plant material (herbaceous versus woody) under investigation.

Twenty-six different nuclear isolation and staining buffers were found in the literature excerpted. The chemical composition of the top 10 non-commercial buffers is presented in Chapter 4. The relative use of individual buffers is shown in Fig. 18.2. The six most popular buffers (Galbraith's, commercial buffers, $MgSO_4$, LB01, Otto's and Tris.$MgCl_2$ – arranged in descending order) collectively account for 72.6% of the references while the next group of five buffers and the remaining 15 buffers account for only 17.4 and 10.0% of references, respectively.

Analysis of temporal variation (over 5-year periods) in the use of the six most popular buffers (Fig. 18.3) shows that the relative contribution of the pioneering Galbraith's buffer has been decreasing over time. The same applies to LB01 and $MgSO_4$ buffers, which, after a period of frequent use in the 1990s, experienced a decline over the last 6 years. In contrast, the number of articles using commercial buffers is escalating and since 2001, these buffers represent the most popular choice. Such success is plausibly related to the fact that they are provided as ready-to-use kits. As the commercial buffers do not yield better results, we hypothesize that novices in plant DNA flow cytometry, who are unaware of the ease of preparation of other nuclear isolation buffers, are the main users of commercial products. Tris.$MgCl_2$ and Otto's buffers are also increasingly being used. While the former was the worst performing buffer in a comparative experimental

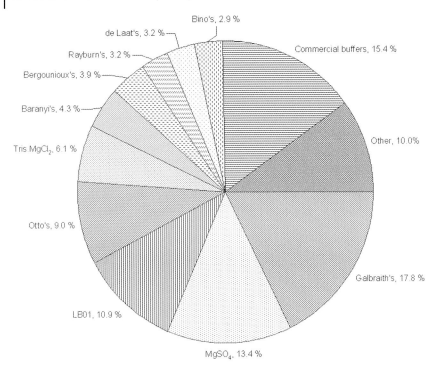

Fig. 18.2 The use of nuclear isolation buffers in plant DNA flow cytometry. The composition of each buffer is given in Chapter 4. Original references: Galbraith's buffer (Galbraith et al. 1983); $MgSO_4$ (Arumuganathan and Earle 1991); LB01 (Doležel et al. 1989); Otto's (Doležel and Göhde 1995; Otto 1981); $Tris.MgCl_2$ (Pfosser et al. 1995); Baranyi's (Baranyi and Greilhuber 1995); Bergounioux's (Bergounioux et al. 1986); Rayburn's (Rayburn et al. 1989); de Laat's (de Laat and Blaas 1984); Bino's (Bino et al. 1993).

study (Loureiro et al. 2006b), the latter is known for yielding DNA histograms with unsurpassed resolution in many plant species (Doležel and Bartoš 2005; Loureiro et al. 2006b). Oddly enough, it took about two decades for Otto's methodology to become widely adopted in plant sciences, considering that the buffer composition was first published in 1981 (Otto). In contemporary plant FCM, Otto's buffer became the third favorite just behind Galbraith's buffer, although still lagging behind commercial solutions.

Geographical survey of the use of a particular isolation buffer suggests that the choice is primarily correlated with a researcher's personal history and/or the laboratory's practice rather than with the buffer quality and/or species and tissue adequacy. The two prevailing buffers (Galbraith's and commercial buffers) also have the largest geographical coverage, being used in no less than 23 different countries. Nevertheless, there is a marked disproportion among the relative contribution of individual countries; nearly two-thirds of Galbraith's buffer hits come from the USA (28.5%), France (19.2%) and New Zealand (15.4%); while Japan

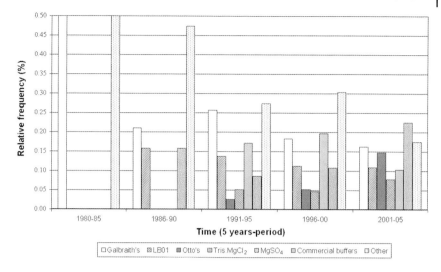

Fig. 18.3 The use of nuclear isolation buffers in plant DNA flow cytometry through the years. Data were grouped into 5-year periods.

(27.7%), Germany (15.2%) and USA (9.9%) account for more than half of the commercial buffers hits. This tendency is even more obvious for other buffers. The majority of LB01 users resides in the Czech Republic (34.2%), the country where the buffer originated, and in France (21.5%); $MgSO_4$ is most widely used in the USA (64.0%), where it was also developed; Otto's buffer is preferred in the Czech Republic, Slovenia and Belgium, with 46.1, 13.8 and 10.8% of database entries, respectively; and Tris.$MgCl_2$ buffers are mostly used in Japan (27.9%), Poland (27.9%) and the USA (16.3%). The most conspicuous example of a locally restricted use is Rayburn's buffer with 23 occurrences, but all from the same country (USA) and 91.3% of them even from the same research group. Once again, these data indicate that researchers generally use only one or two buffers throughout their publication history. When multiple buffers are employed, there is usually a favorite, which accounts for a substantial percentage of the references. However, such strict adherence to a particular methodology may have important consequences for the quality of the data obtained as no ideal buffer exists and testing several different alternatives prior to routine FCM investigation is always advisable.

Assessment of buffer selection according to the plant growth type did not show any clear preferences. Galbraith's and LB01 buffers were more often used for investigation of woody plants, while commercial buffers, $MgSO_4$, Otto's and Tris.$MgCl_2$ buffers predominated in the research on herbaceous species. However, no explanation for a particular choice was provided in the publications and it seems that it was merely standard laboratory practice that guided the selection of a buffer. As expected, minor modifications to buffer composition (e.g. addition of antioxidants) were often made when recalcitrant woody plants were analyzed by flow cytometry.

18.4
Standardization and Standards

The estimation of nuclear DNA content requires the use of a reference standard with known nuclear DNA content. C/Cx-value or DNA ploidy level of the plant to be analyzed is then inferred by comparing sample and standard peak positions. There are two basic types of standardization: external and internal. While in the former procedure, nuclei of the sample and standard are processed separately, the latter involves simultaneous isolation, staining and analysis. Although no extensive comparative study has been performed, internal standardization is generally recommended as the most reliable option (see also Chapter 4). Nevertheless, demands on standardization are usually less strict in DNA ploidy studies, at least when the aim of the study is to detect differences in the scale of whole chromosome sets (see Chapter 5).

Quantitative analysis of the type of standardization in ploidy-based studies revealed the following figures: internal 46.5%, external 7.8%, and no standardization 44.9%. In genome size studies, the proportion of internal standardization was much higher (91.8%) while external standardization was adopted in only 6.1% of articles; 2.1% of investigations were carried out with both approaches. Merging ploidy and genome size datasets indicates that 7.1% of publications use external standardization, which implies the successful adoption of the preferred internal standardization practice by most researchers.

Several requirements imposed on proper DNA reference standards, such as a close but non-overlapping genome size in relation to that of a target species (Bagwell et al. 1989; see also Chapter 4), led to the employment of many different standards and have fueled a discussion about the selection of a universal set of reference materials. As a comprehensive survey of reference standards has not yet been carried out, the FLOWer database can provide the first insights into the type of standards and the frequency of their use, and contribute to the identification of potential sources of variation.

Plant and animal reference standards were employed in 73.0 and 27.0% of articles, respectively. However, the use of animal standards such as chicken red blood cells (CRBCs), which is the main type of animal standard used with a 68.2% incidence, was not recommended by the 1997 Genome Size Workshop and further warnings were issued 6 years later. The plant FCM community responded positively to this recommendation and the contribution of CRBCs clearly decreased over time (Fig. 18.4). The use of CRBCs as a reference standard has been questioned mainly because there has been no general agreement regarding the size and stability of the chicken genome (see Chapter 4). The FLOWer database supports this contention and shows that published 2C-values vary from 1.88 pg (Chen et al. 2002) to 2.50 pg (Iannelli et al. 1998), with the most common 2C DNA value being 2.33 pg (87.3% of references).

Nevertheless, the problem of a non-identical genome size may persist even when plant reference standards are employed. Table 18.2 lists the most common plant standards with a range of 2C-values assigned by different authors. *Pisum*

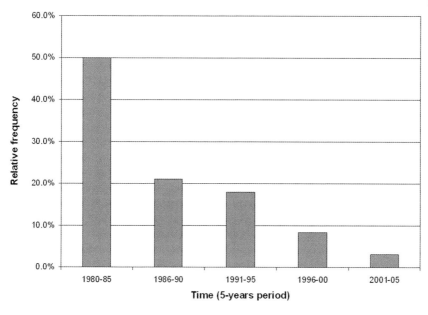

Fig. 18.4 The use of chicken red blood cells (CRBCs) as a reference standard in plant genome size estimations using flow cytometry over the years. Data were grouped into 5-year periods.

sativum (Fabaceae; 15.0%), *Hordeum vulgare* (Poaceae; 12.7%), *Petunia hybrida* (Solanaceae; 11.1%) and *Zea mays* (Poaceae; 8.6%) were the most popular standards, being used in 47.5% of the genome size estimation studies. DNA amounts in these plants vary from 2.85 pg/2C in *P. hybrida* to 11.26 pg/2C in *H. vulgare*. It may therefore be expected that a great many of the available nuclear DNA content estimates will lie within a corresponding range. However, a careful analysis of the plant DNA C-values database (Bennett and Leitch 2005) revealed that this was not the case and that many estimates actually fall into the lower end of such a DNA range. Figure 18.5 illustrates this, and reveals that the frequency of reference standards used for genome size estimations in the lower end of DNA range is appropriate. Moreover it is clear that reference standards covering the 5.0–15.0 pg/2C DNA range are overused. This may suggest that in some cases, the best standard for a given species was not chosen. Our data also highlights the necessity of reducing the number of reference species currently used for genome size estimations. For example, the two most frequently used standards, *Pisum sativum* and *Hordeum vulgare*, cover nearly identical DNA ranges, although the former species is a preferred primary reference standard (Doležel and Bartoš 2005; Loureiro et al. 2006b).

Also, the lack of agreement on which cultivars should be used in several popular reference species (e.g. *H. vulgare* and *P. sativum*) may potentially contribute to the heterogeneity of FCM estimates. On the other hand, different genome sizes

Table 18.2 The most popular plant DNA reference standards (without cultivar distinction) used for FCM estimation of genome size.

Plant DNA reference standards	Range of assigned 2C DNA contents		No. of papers	Frequency of use	
	Min–Max (pg)	Variation (%)		%	| = 1%
Arabidopsis thaliana (L.) Heynh.	0.14–0.32	128.6	4	1.3%	|
Oryza sativa L.	0.89–1.20	34.8	12	3.8%	||||
Vigna radiata (L.) R. Wilczek	1.06	–	7	2.2%	||
Raphanus sativus L.	1.11	–	5	1.6%	||
Lycopersicon esculentum Mill.	1.96–2.01	2.6	20	6.4%	||||||
Trifolium repens L.	2.07	–	6	1.9%	||
Glycine max Merr.	2.27–2.70	18.9	19	6.1%	||||||
Petunia hybrida Vilm.	2.85–3.35	17.5	35	11.1%	|||||||||||
Zea mays L.	5.00–5.47	9.4	27	8.6%	|||||||||
Pisum sativum L.	8.11–9.73	20.0	47	15.0%	|||||||||||||||
Hordeum vulgare L.	9.81–11.26	14.8	40	12.7%	|||||||||||||
Secale cereale L.	15.58–16.80	7.8	5	1.6%	||
Agave americana L.	15.90	–	7	2.2%	||
Vicia faba L.	25.95–26.90	3.7	8	2.5%	|||
Triticum aestivum L.	30.90–34.85	12.8	19	6.1%	||||||
Allium cepa L.	33.50–34.89	4.1	13	4.1%	||||
Other species	–	–	40	12.7%	|||||||||||||

may be assigned to the same reference cultivar. An illustrative example is *Pisum sativum* cv. Minerva Maple, with the following DNA values: 2C = 8.22 pg (three references; first cited in Joyner et al. 2001), 2C = 9.56 pg (six references; Price et al. 1998), 2C = 9.64 pg (one reference; Johnston et al. 1999), and 2C = 9.73 pg (seven references; Leitch et al. 2001). The difference in input values, amounting to 18.4%, may well be an underlying cause of the artifactual variation in genome size data among different research groups.

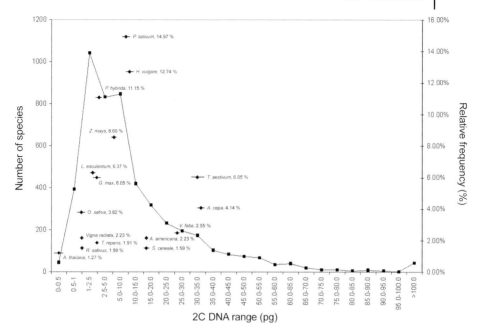

Fig. 18.5 Distribution of 2C-values for 5015 plant species (primary y axis) and the use of the most popular DNA reference standards in plant genome size estimations using flow cytometry (secondary y axis). For each reference standard, the frequency of use, its mean 2C DNA amount and a range of assigned 2C values is shown. Data on DNA amounts were taken from the Plant DNA C-values database (Bennett and Leitch 2005).

18.5
Fluorochromes

A range of DNA-specific fluorochromes has been used to study plant genomes. They are mostly grouped according to their binding properties: intercalation into double-stranded DNA (ethidium bromide, EB; propidium iodide, PI), preference for AT-rich regions (DAPI, Hoechst dyes) or preference for GC-rich regions of DNA (chromomycin, mithramycin and olivomycin). While the binding mode is of little importance in DNA ploidy studies, precise genome size estimates require intercalating dyes (as shown for the first time by Doležel et al. (1992)).

The analysis of fluorochrome data from the FLOWer database revealed that PI is the most frequently used fluorescent dye, with a 45.3% incidence. DAPI was employed in 38.2% of FCM studies while the frequency of any other fluorochrome did not exceed a 6% threshold. The obvious preference of DAPI among other base-specific fluorochromes such as chromomycin A_3, results from its lower toxicity, the likelihood of obtaining high-resolution histograms of DNA content and the fact that it can be used in cheaper, lamp-based instruments. Most

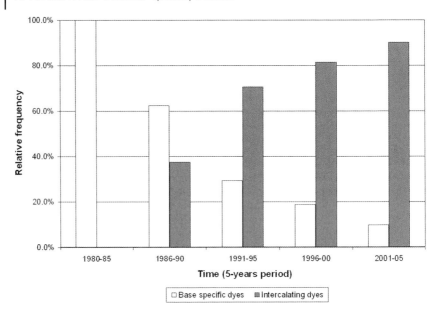

Fig. 18.6 The use of the two basic types of DNA fluorochromes (base-specific and intercalating dyes) in genome size estimations using flow cytometry over the years. Data were grouped into 5-year periods.

DAPI measurements refer to DNA ploidy estimations or base composition studies. In essays focused on absolute genome size estimations, PI reported in 71.1% of articles clearly surpasses the other intercalating dye, EB, which is mentioned in only 11.1% of reports. This disproportion may be related to the belief that PI produces histograms with lower coefficients of variation. Despite the known base preference, DAPI and other base-specific dyes were used in the remaining 17.8% of studies.

An assessment of temporal variation in the use of base-specific versus intercalating fluorochromes (Fig. 18.6) reveals that the former were preferred until the 1990s. Actually, early researchers paid little attention to the mode of binding and used any fluorochrome for a wide range of applications. Since the 1990s, a shift toward intercalating dyes is evident, plausibly triggered by the results of a comparative study of three fluorochromes performed by Doležel et al. (1992).

18.6
Quality Measures of Nuclear DNA Content Analyses

Coefficient of variation (CV) and the distribution of relative nuclear DNA content (DNA histogram) are the main tools for assessing the quality of FCM analyses and should therefore be presented in every publication. A literature survey, however, shows that the situation is not satisfactory, and CV values and DNA histograms were included in only 31.2 and 66.3% of articles, respectively. The corresponding figures change to 45.6% (CV) and 58.8% (histogram) when only genome size studies are evaluated, and to 21.9% (CV) and 69.8% (histogram) in ploidy-based studies alone. This difference may be driven by distinct requirements in the quality and design of both types of studies. While low CV is a crucial prerequisite for high-standard genome size work, FCM estimation of DNA ploidy level is generally less demanding. On the other hand, an FCM histogram represents the most straightforward proof of ploidy differentiation.

The FLOWER database also provides information on the range of CV values for DNA peaks. In 33.2% of the articles, the CV values were below 3.0%, in 39.6% they ranged from 3.0 to 5.0%, in 22.6% they ranged from 5.0 to 10.0%, and CV values above 10.0% were obtained in only 4.6% of the references. This analysis reveals that in published works, CV values mostly fall within the recommended range (i.e. below 5.0%; see Chapter 4 for further information on quality control and data presentation).

18.7
The Uses of DNA Flow Cytometry in Plants

The major applications of DNA flow cytometry are ploidy level and genome size estimations, and cell cycle analysis. Indeed, a survey of the literature stored in the FLOWER database revealed that a substantial proportion of plant FCM work dealt with DNA ploidy level (50.2%) and genome size (36.9%). The remaining uses cover cell cycle analysis (6.1%) and estimations of DNA base composition (4.1%). Other applications, which include sex determination in dioecious plants and technical and standardization experiments, account for only 2.8% of the studies. The low number of cell cycle studies is quite surprising considering the extensive use of FCM in human and animal cell cycle research.

18.8
Instrumentation

FCM users may also seek information regarding the contribution of particular brands and models of flow cytometers. Based on the number of articles, the leading brand used in plant sciences is Partec® which was mentioned in 44.1% of publications (the most successful model appears to be the Cell Analyser II), followed by Beckman-Coulter® (30.8% of the studies; most successful model, the EPICS V), and Becton-Dickinson® (19.2% of the reports; most successful model, the FACScan). Instruments from other manufacturers, which include the discontinued models from Leitz, Ortho Instruments and Phywe, and more recent offerings from Bio-Rad (now acquired by Apogee flow systems) and Dako, were used in only 5.7% of the studies. The prominent position of Partec may be related either to suitability for analysis of plant materials and/or to the relatively low price of their products. As project budgets in plant sciences are generally smaller than those in other fields where FCM is routinely employed (e.g. clinical studies), price is undoubtedly a significant criterion in instrument purchase.

18.9
Where Are the Results Published?

The FLOWER database also offers quantitative analyses of scientific journals in which plant DNA FCM studies were published. The top 10 journals are listed in Table 18.3. This synopsis may help authors to select the most appropriate periodical for their work. The year-trend overview of the top six scientific journals reveals that, with the exception of *Theoretical and Applied Genetics* (TAG), the number of published articles concerning plant DNA flow cytometry has been increasing over time (Fig. 18.7). *Plant Cell Reports* (PCR) and *Plant Cell, Tissue and Organ Culture* (PCTOC) experienced the highest increase in recent years. The former, together with *Annals of Botany* (AoB), has been the preferred journal for publication of plant FCM studies over the last 6 years, and the latter is placed third, after more than a 300% increase in the number of articles being published.

The spectrum of FCM applications covered by particular journals also deserves attention. While most papers in AoB concern genome size estimations, DNA ploidy level studies, particularly those related to *in vitro* cultures and transformation experiments, prevail in TAG, PCR and PCTOC. *Euphytica* is devoted to plant breeding and most of the FCM papers also fall into this category while *Plant*

Table 18.3 The 10 most popular scientific journals in plant DNA flow cytometry.

Scientific journal	No. of papers
Annals of Botany	57 (8.2%)
Theoretical and Applied Genetics	52 (7.5%)
Plant Cell Reports	49 (7.0%)
Plant Science	40 (5.7%)
Plant Cell, Tissue and Organ Culture	37 (5.5%)
Euphytica	36 (5.2%)
Plant Systematics and Evolution	25 (3.6%)
Crop Science	19 (2.7%)
Genome	18 (2.6%)
American Journal of Botany	16 (2.3%)
Other	342 (497%)

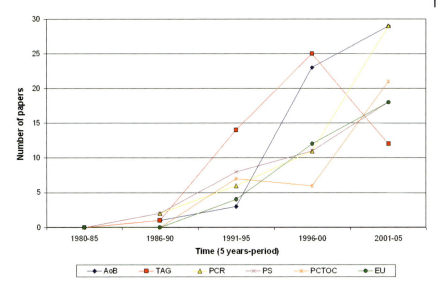

Fig. 18.7 Number of papers related to plant DNA flow cytometry published in the six most popular scientific journals, over the years. Data were grouped into 5-year periods. AoB, *Annals of Botany*; TAG, *Theoretical and Applied Genetics*; PCR, *Plant Cell Reports*; PS, *Plant Science*; PCTOC, *Plant Cell, Tissue and Organ Culture*; EU, *Euphytica*.

Science has a more general scope and publishes results both on genome size and DNA ploidy.

18.10 Conclusion

This chapter introduces a database of scientific publications which use DNA flow cytometry to study plant materials. The database with the acronym FLOWer is intended as a comprehensive, easily accessible and user-friendly source of information on plant FCM articles (search tool) as well as a platform for carrying out quantitative analyses of selected aspects important in FCM practice (survey tool). Excerpted methodology- and instrumentation-related data (such as types of nuclear isolation buffer, standards, and fluorochromes) form a basis for unbiased and statistically well-founded assessments of historical applications and approaches, methodological trends, developments, and the current state of affairs in plant FCM. Keyword filters offer rapid tracking of relevant information useful for both newcomers and experts. Evaluation of the reliability of results and close inspection of how the best practice recommendations were met can also be easily carried out. We believe that this ready-to-hand set of FCM articles will stimulate further use of DNA flow cytometry in plant sciences, contribute to the discussion on the best methodology, support the formulation of recommendations, and help to identify other hot topics. Availability of the interactive FLOWer database on the internet (http://flower.web.ua.pt/) guarantees accessibility to plant FCM-users worldwide and regular data updating.

Acknowledgments

J. L. was supported by fellowship FCT/BD/9003/2002. The Portuguese Foundation for Science and Technology is thanked for the grant project ref. POCTI/AGR/60672/2004.

References

Arumuganathan, K., Earle, E. **1991**, *Plant Mol. Biol. Rep.* 9, 229–233.

Bagwell, C. B., Baker, D., Whetstone, S., Munson, M., Hitchcox, S., Ault, K. A., Lovett, E. J. **1989**, *Cytometry* 10, 689–694.

Baranyi, M., Greilhuber, J. **1995**, *Plant Syst. Evol.* 194, 231–239.

Bennett, M. D., Leitch, I. J. **2005**, *Plant DNA C-values database (release 4.0, Oct. 2005)*, http://www.kew.org/genomesize/ homepage.

Bergounioux, C., Perennes, C., Miege, C., Gadal, P. **1986**, *Protoplasma* 130, 138–144.

Bino, R. J., Lanteri, S., Verhoeven, H. A., Kraak, H. L. **1993**, *Ann. Bot.* 72, 181–187.

Chen, J. F., Staub, J., Adelberg, J., Lewis, S., Kunkle, B. **2002**, *Euphytica* 123, 315–322.

de Laat, A. M. M., Blaas, J. **1984**, *Theor. Appl. Genet.* 67, 463–467.

Doležel, J., Bartoš, J. **2005**, *Ann. Bot.* 95, 99–110.

Doležel, J., Göhde, W. **1995**, *Cytometry* 19, 103–106.

Doležel, J., Binarová, P., Lucretti, S. **1989**, *Biol. Plant.* 31, 113–120.

Doležel, J., Sgorbati, S., Lucretti, S. **1992**, *Physiol. Plant.* 85, 625–631.

Galbraith, D. W., Harkins, K. R., Maddox, J. M., Ayres, N. M., Sharma, D. P., Firoozabady, E. **1983**, *Science* 220, 1049–1051.

Iannelli, D., Cottone, C., Viscardi, M., D'Apice, L., Capparelli, R., Bosell, M. **1998**, *Int. J. Plant Sci.* 159, 864–869.

Johnston, J., Bennett, M., Rayburn, A., Galbraith, D., Price, H. **1999**, *Am. J. Bot.* 86, 609–613.

Joyner, K. L., Wang, X. R., Johnston, J. S., Price, H. J., Williams, C. G. **2001**, *Can. J. Bot.* 79, 192–196.

Leitch, I. J., Hanson, L., Winfield, M., Parker, J., Bennett, M. D. **2001**, *Ann. Bot.* 88, 843–849.

Loureiro, J., Rodriguez, E., Doležel, J., Santos, C. **2006a**, *Ann. Bot.* 98, 515–527.

Loureiro, J., Rodriguez, E., Doležel, J., Santos, C. **2006b**, *Ann. Bot.* 98, 679–689.

Otto, F. **1981**, *Cytometry* 2, 189–191.

Pfosser, M., Amon, A., Lelley, T., Heberle-Bors, E. **1995**, *Cytometry* 21, 387–393.

Price, H. J., Morgan, P. W., Johnston, J. S. **1998**, *Ann. Bot.* 82, 95–98.

Rayburn, A. L., Auger, J. A., Benzinger, E. A., Hepburn, A. G. **1989**, *J. Exp. Bot.* 40, 1179–1183.

Index

a

A chromosomes 391
abscisic acid 243, 330, 412
absorption 2–5, 7, 21, 25, 252 f., 263, 295, 299, 312, 315 f., 411
Acer 332, 335
Acetabularia 276
acridine orange 6, 326
acriflavine 4
Actinidia 109, 139, 186
ADC, *see* analog-to-digital converter
adenosine triphosphate (ATP) 253, 261, 329
Affymetrix 407 f., 416 f.
AFLP, *see* amplified fragment length polymorphism
Agapanthus 118
Agilent 407 f.
agmatoploidy 113
Agrobacterium 49, 334 f., 340, 412
aleuron 54, 140, 143, 243
Alexandrium 308
alfalfa 132, 140, 332 f., 335, 336 f.
algae 59, 61, 156, 158 ff., 159, 251, 257 f., 263, 267 f., 271, 273 f., 276, 277 ff., 283 f., 289 f., 307 ff., 313, 356
  – brown 159, 267 ff., 271 ff., 276 f., 279 f., 289
  – green 251, 257, 267 f., 272, 276 ff., 280, 283 f., 289
  – harmful 287, 308 ff., 317
  – red 267, 271 f., 277–279, 289
alkaloid 48, 237, 329, 383
Alliaceae 56, 85, 178, 186–188, 203, 206, 276, 351, 356 f.
Allium 56, 85–88, 161, 178, 186–188, 202 f., 205 f., 276, 351, 353, 364
allopolyploid 55, 115, 118, 124
Aloe 361
Alstroemeria 124, 188
alveolates 267

amanitin 328
Amaranthaceae 74, 117, 124, 149, 188, 199, 203, 207, 209, 257, 356 f., 361 f.
Amaranthus 124
Amaryllidaceae 118, 358
amiprosphos-methyl (APM) 329, 333
amplified fragment length polymorphism (AFLP) 140 f., 154
amplifier 13, 32
Anabaena 293, 308
Anacampseros 361
analog-to-digital converter (ADC) 13, 30
Andropogon 51
Anemone 178, 188
aneuploid 41, 51 ff., 105 ff., 118, 149, 170, 172, 368 f.
angiosperms 53, 70 f., 74, 88 f., 117, 131 ff., 142 f., 146, 153, 155, 156 ff., 163 f., 169 ff., 171, 173, 202, 269, 272, 282 f., 356, 359, 362, 425
anisomycin 327 f., 330
ANN, *see* artificial neural network
annexin V 48, 240
annual 161 f., 290, 357
Anthoceros 159
anthocyan 74, 85, 89, 95
anthocyanin 48, 171, 237, 243
anthracene 4
antibody 1 ff., 10 ff., 15, 41 f., 53, 58, 217 f., 222 ff., 228, 243, 247, 262, 326, 340
  – fluorescent 3 ff., 15
  – monoclonal 1, 11 f., 243, 326
  – secondary 222 ff.
antioxidant 80, 90, 429
antipodal 134 ff., 351, 362
Antirrhinum 342
Aphanizomenon 308
aphidicolin 328, 331 ff.
Apiaceae 117, 358
APM, *see* amiprosphos-methyl

Flow Cytometry with Plant Cells. Edited by Jaroslav Dolezel, Johann Greilhuber, and Jan Suda
Copyright © 2007 WILEY-VCH Verlag GmbH & Co. KGaA, Weinheim
ISBN: 978-3-527-31487-4

Apocynaceae 48, 335
apomeiosis 132, 145 ff.
apomictic 52, 54, 74, 108, 132 ff., 150
apomixis 54, 75, 124, 132, 140, 147 ff., 148, 150
apoptosis 41, 48, 143, 238 ff., 242, 328 ff.
apospory 132, 135, 145 ff.
Arabidopsis 49, 71, 74, 83 f., 86 ff., 92, 99, 132, 143, 145 f., 149, 155, 160 f., 178 f., 185, 188, 204 ff., 235, 238, 240 f., 272, 276, 281 f., 332 ff., 337 f., 340, 351 ff., 356, 361 ff., 365 f., 369, 407 f., 412, 414 ff., 432
Arachis 97
Aralia 48, 243
arc lamp 3, 9 f., 12 f., 15, 25 f., 105, 110, 125, 255, 258
Archaea 288
Arecaceae 52
artificial neural network (ANN) 293, 297
Ascophyllum 279
asexual 122, 124, 132, 146
Asparagaceae 51, 188, 203
Asparagus 51, 188
Asphodelaceae 119, 357, 361
Aster 105
Asteraceae 57, 74, 95, 105, 115 ff., 123, 125, 140, 162, 170, 180, 189 ff., 196 f., 199 f., 203, 214, 221, 247, 263, 352, 358, 360, 379 f.
AT/GC ratio 41, 55 f., 203, 375, 385 f.
ATP, *see* adenosine triphosphate
Atriplex 117, 188, 207
Aureococcus 308
Aureoumbra 308
autofluorescence 4, 42, 45 ff., 111, 237 f., 240 f., 243 ff., 251, 255 ff., 258 ff., 271, 301, 311, 318
autonomous underwater vehicle (AUV) 287, 303, 315 f.
autopolyploid 115
autoradiography 324 f.
AUV, *see* autonomous underwater vehicle
auxin 135, 330, 332, 334 f., 337, 351, 365 f.
axial flow 22 ff.

b

B chromosomes 52, 391, 393
BAC, *see* bacterial artificial chromosome
Bacillariophyta 290, 293, 299
bacteria 4, 12, 25, 58, 61, 185, 205, 217 ff., 225 ff., 231, 288, 291, 301, 306, 310, 313, 317, 364 f.
bacterial artificial chromosome (BAC) 56 f., 394 f., 397 ff.
banana 50, 56, 118

barley 88, 132, 243, 333, 373, 380, 390 ff., 396, 399
base composition 55, 107, 177 ff., 183 ff., 202 ff., 211 f., 213 f., 272, 434 ff.
base frequency 177 f., 181, 183 f., 211
beam
 – attenuation 293, 295, 313
 – width 26, 112, 236
Beckman-Coulter 29, 418, 435
Becton-Dickinson 10, 29, 435
p-benzoquinone 258
berberine 48, 243
Beta 149, 188, 207, 362 f., 367
Betulaceae 162
biennial 357
biliproteins 289
binding
 – length 177 f., 180 ff., 205
 – mechanism 91, 182
biosystematics 51, 53
biparametric 50, 235, 238, 245, 340, 369, 413, 415, 420
Bituminaria 91
bivariate
 – analysis 327, 385 f.
 – flow karyotype 375
 – flow karyotyping 378
Boechera 52, 54, 142, 147, 149
bohemine 329, 331, 337
Botrytis 219 ff.
Brassica 47, 74, 123, 189, 240, 244, 247, 355, 363
Brassicaceae 47, 49, 52, 54, 71, 74, 116 f., 123, 132, 142, 160, 162, 178 f., 186, 188 f., 198 f., 203, 206, 238, 240, 272, 276, 333, 351, 355 ff., 362, 367, 407, 425
brassinosteroids 366
BrdU, *see* 5-bromo-2′-deoxyuridine
break-off point 8, 233
5-bromo-2′-deoxyuridine (BrdU) 53, 326 ff., 327, 339 ff.
Bryonia 360
bryophytes 44, 153, 156, 158 ff., 172 ff., 267, 269 ff., 277, 280 ff., 354, 425 ff.
buffer 46, 49, 68, 74 ff., 87 f., 90, 92 ff., 111, 139, 141, 148 f., 172, 272 f., 278, 281 f., 352, 383 f., 389, 423 f., 426 ff., 427, 429, 437
bundle sheath 58, 253 ff., 259, 261 ff.
buoyant density centrifugation 177, 204, 206 ff.

c

C_3 plants 260 ff.
C_4 plants 252, 254, 261 ff.
Cactaceae 361, 365

Caenorhabditis 85 f., 99, 155, 239
caffeine 90, 331, 336
Cajanus 97
Calcidiscus 278
calcofluor 47 f., 247
calibration 37 ff., 67, 85, 88, 169, 173, 258, 295, 312, 424
callus 75, 351, 368, 412
camptothecin 240
Cannabaceae 51 f.
carboxy fluorescein diacetate (cFDA) 219, 239
carotenoids 253, 255, 257
Caryophyllaceae 52, 78, 116, 195, 357, 368, 380
caryops 135, 143, 352
Catharanthus 48, 332, 335
Caulerpa 369
CCD camera 14 f., 352
CDK, *see* cyclin-dependent protein kinase
cDNA libraries 246, 406, 408, 416
cell cycle
– analysis 31, 52 f., 244 ff., 267, 273, 283, 311, 323 ff., 339, 354, 435
– kinetics 8, 48, 52, 327
– parameters 281, 283, 323, 339 ff.
– phase 141, 145, 323 f., 326 f., 330 f., 336, 340 ff.
– progression 53, 324 ff., 336 f., 342
– regulation 53, 327, 336 ff.
– studies 41, 59, 271, 280, 332 f., 335 ff., 435
– synchronization 53, 323, 327 ff., 331 ff., 338, 375, 383 ff.
cell division 232, 247, 280, 312, 323 f., 328, 335 ff., 340 ff., 351, 366, 419
cell nuclei 9, 46, 49 ff., 56 f., 67, 98, 141, 143 f., 178, 424, 427
cell sorter 8 ff., 13, 21, 30 f., 34
cell sorting 1, 7 ff., 10, 20 f., 34 ff., 36 f., 39, 231, 248, 323, 342, 405 f., 408, 414 ff., 419
cell suspension 44, 231, 246, 292, 327, 332, 334, 365 f., 379, 383 f., 406
cell wall 39, 44 f., 47, 48, 219, 231 ff., 247, 254, 268, 271 ff., 277, 307, 325, 332, 335, 338, 379, 383, 393, 396, 515
cellular state 232, 406, 411, 414
cellulase 232, 383
cellulose 47 f., 247, 325, 335
Celtis 79
Centaurea 115, 189
central cell 132, 134 f., 142, 144, 147
centromere 160, 351, 377, 394
ceramide 240
Ceratium 290

Ceratodon 281
cereal 85, 399
Cereus 361
cFDA, *see* carboxy fluorescein diacetate
CFP, *see* cyan fluorescent protein
Chaetoceros 290, 308
Chamerion 107, 110 ff., 117, 121, 123 f.
Chara 365
Characeae 356, 365
chelator 79
Chenopodiaceae 74, 117, 149, 188, 199, 203, 207, 209, 257, 356 f., 361 f., 367
chicken 68, 83, 86, 88 f., 272, 274, 278, 282, 430 f.
– red blood cells (CRBCs) 68, 83 f., 88, 272, 274 f., 277 f., 280, 430 f.
chickpea 380, 392, 399
Chinese hamster 378
Chionographis 113
Chlamydomonas 275
Chlorella 257, 268, 280, 290
chlorogenic acid 90, 171
Chlorophyceae 283
chlorophyll
– autofluorescence 46, 238, 243 ff., 255 ff., 301
– content 48, 243, 415
– fluorescence 57, 58, 254, 255, 257 ff., 262, 273, 297, 414
Chlorophyta 59, 153, 156, 158 f., 276 f., 289 f., 295, 297, 306, 315
Chondrus 189, 272, 277 ff.
chopping 47, 49, 56, 68, 75, 79, 95, 106 f., 109, 111, 139, 141, 148, 171, 272, 277 f., 282 f., 353, 384, 418, 427
chromatids 71, 278 f., 324, 329, 350 f., 383 ff., 387, 388
chromatin 48, 56 ff., 68, 74, 76, 78 ff., 90, 92, 111, 211, 240, 247, 340, 342, 349, 350, 369, 374, 399
chromatography 177, 204, 206 ff., 411
chromocentres 351
chromomycin A_3 11, 75, 169, 178, 181, 201, 277, 378, 433
chromosomal proteins 57, 395
chromosome
– analysis 5, 57, 373 ff., 379, 385 ff., 390
– count 50, 52, 103, 105, 113, 131, 139 ff., 174, 278, 337, 351 ff.
– isolation 379 ff., 384
– mitotic 18, 57 ff., 374, 380, 395
– number 51, 69 ff., 104, 106, 113, 118 f., 124, 132 f., 138 ff., 149, 160, 170, 270, 350
– polymorphism 57, 373, 391 f.

chromosome (cont.)
- rearrangement 51, 375, 378
- sorting 11, 57, 154, 373, 378 f., 387, 392 ff., 398 f.
- suspension 379, 382 ff., 387, 399
- translocation 57, 373, 382, 386, 387, 391 ff.
Chroomonas 258, 318
Chrysochromulina 273
Chrysophytes 267, 290
Cichorium 74, 189
Cirsium 117, 125, 189 f.
citrus 244
clearing technique 131, 135
C-level 70 f., 74, 89, 280, 350, 353 ff., 364, 367
CMOS camera 15
coccolithophore 278, 297, 303, 305, 315 f.
Coccolithus 278
coefficient of variation (CV) 14, 15, 52, 76, 78 f., 82, 84, 87, 96 f., 106 ff., 172, 326, 353, 375, 377, 426, 434 f.
Coffea 90 f., 170
cohesins 329
colchicine 341, 383
Collinsia 54
Commelinaceae 351, 357, 361
comparator circuit 30
compensation 11 ff., 33 ff., 219, 222
concavalin A 58
conversion factor 67, 72 ff.
Convolvulaceae 113
Coprosma 54, 146
core stream 23 f.
Coronosphaera 278
Coscinodiscus 290
Coulter
- counter 6 f.
- sizing 42, 293
- volume 8, 21, 31
coumarin 91, 93
Crassulaceae 79, 357, 363
CRBCs, see chicken red blood cells
Crepis 119, 180, 190 f.
critical scales 287, 302 ff., 315 f.
Cryptomonadaceae 291
Cryptophyta 258, 295, 297
Cucumis 353
Cucurbitaceae 52, 191, 195, 203, 207 f., 356 f., 360 f., 367
Cupressus 123, 141
Cuscuta 113
CV, see coefficient of variation
C-value 52, 54, 67, 69 f., 84 ff., 98 f., 132 ff., 140 f., 143, 145, 155 ff., 163 f., 166 ff., 171 f., 274, 278, 280 f., 283, 340, 430 f., 433

Cx-value 67, 69 f., 134, 136 ff., 274, 278, 430
cyan fluorescent protein (CFP) 420
cyanidin 91, 95
cyanobacteria 59, 288 ff., 293, 295, 308, 315
cycle value 71, 353 ff., 358 ff., 364
cyclin-dependent protein kinase (CDK) 329, 331, 333, 335 ff., 366
cyclins 329, 331, 336, 342, 366
Cyperaceae 113, 117, 254
Cyperus 254
cytogenetic
- mapping 373, 392 ff., 400
- stock 373, 380, 382, 386 ff., 399
cytogeography 119
cytogram 31 f., 42, 50, 327, 375
cytokinin 330, 332, 334, 366
cytometry
- history of cytometry 2, 20 ff.
- image cytometry 3 ff., 9, 14 ff., 248, 418 ff.
cytosol 79, 90, 99, 170 ff., 185, 211, 251
cytotype 50 f., 71 f., 103 ff., 108, 110, 114 f., 117 ff., 125 f., 172, 283

d

2,4-D, see 2,4 dichlorophenoxyacetic acid
Dactylis 140, 191, 362
Dactylorhiza 113
DAPI, see 4′,6-diamidino-2-phenylindole
DAS-ELISA, see double antibody sandwich ELISA
DCMU, see 3-(3,4-dichlorophenyl)-1,1-dimethylurea
dead time 30
debris 34, 41, 45 f., 60, 68, 74, 78 f., 89 ff., 95 f., 98, 232, 234, 238, 282, 288, 297, 352, 384, 387, 389, 413
degenerate oligonucleotide primed PCR (DOP-PCR) 393, 395 f., 399
denaturation of DNA 177, 204, 206 ff.
detergent 68, 78 ff., 92, 272
dextran 49
DF, see dye factor
4′,6-diamidino-2-phenylindole (DAPI) 13, 44, 55 ff., 74 ff., 78, 90, 97, 107, 109, 111, 125, 148, 169, 172, 178, 179, 181, 185 ff., 203 ff., 273, 275, 278 ff., 325 ff., 351 f., 385, 387, 395, 414 f., 418, 433 f.
Diapensiaceae 124
diatoms 42, 267 f., 289, 294, 296, 307 f., 315
2,4 dichlorophenoxyacetic acid (2,4-D) 335, 365
3-(3,4-dichlorophenyl)-1,1-dimethylurea (DCMU) 258
dichroic mirrors 4, 27, 29, 295

difference of potential (DOP)   29
differential replication   114
diffraction   297, 312
3,3'dihexyloxacarbocyanine iodide (DiOC$_6$(3))   240
Dinobryon   290
dinoflagellates   267 ff., 272 f., 280, 284, 289, 301, 307 f.
Dinophyceae   273, 280, 284, 308
Dinophysis   308
DiOC$_6$(3), *see* 3,3'dihexyloxacarbocyanine iodide
Dioscorea   50
Diospyros   123
diphenylboric acid-2-aminoethyl ester (DPBA)   92, 94
diplospory   132, 141, 146
Ditrichaceae   281
Ditylum   290
DNA amounts   52, 67 f., 70 ff., 113, 118, 125, 131, 153, 160 ff., 170, 203, 271, 277, 280 ff., 337, 424, 431, 433
DNA aneuploidy   51, 105
DNA base
– composition   55, 177 ff., 184 f., 204, 212, 435
– content   55 ff., 185 f., 206
DNA content, *see also* nuclear DNA content
– analysis   10, 45, 52, 60, 141, 312, 434 ff.
– measurement   9, 10, 12, 14, 48, 67 ff., 105, 276, 311, 325 f., 339
– studies   81, 98, 279, 425 ff.
– variation   54, 95, 98, 110, 113 ff.
DNA dye   11, 13, 211
DNA fingerprinting   140 f., 146
DNA library   57, 246, 373, 378, 393, 395 ff., 408, 416
DNA ploidy   50 f., 81, 105, 109 f., 113 f., 116, 119, 121 ff., 143, 145, 149, 272, 276, 278, 282 ff., 323, 423, 430, 433 ff.
DNA staining   54, 76 ff., 149, 170 ff., 276 ff., 352, 427
DNA synthesis inhibitors   328 f., 335, 383
DNA transfection   49 ff.
DOP, *see* difference of potential
DOP-PCR, *see* degenerate oligonucleotide primed PCR   393, 395
Doritaenopsis   365 f.
dot blotting   386, 389
double antibody sandwich ELISA (DAS-ELISA)   222
double fertilization   53, 132, 136 ff., 144, 246
doublets   112 f., 354, 387, 389

DPBA, *see* diphenylboric acid-2-aminoethyl ester
droplet   8, 10, 34 ff., 48, 233, 377 f., 407, 410
Draba   117
Drosera   113
Drosophila   155
Dryopteris   124 f., 191
durum wheat   381, 389, 391, 393
dye factor (DF)   181 ff., 203, 205, 211
dyes
– aniline   2
– base-specific   55, 177, 180 f., 205, 434
– fluorescent   6, 20, 34, 37, 42, 49, 58, 178 f., 181 f., 204, 211 f., 222, 240, 272, 284, 325, 327, 378, 433
– intercalating   74 f., 169, 178, 275, 281, 433 f.
dysploidy   51

e

EB, *see* ethidium bromide
Ebenaceae   123
Ecballium   360
ecological parameters   163, 165, 284
ecology   39, 46, 51, 103 ff., 119 ff., 122, 126 f., 283 f., 303, 314
Ectocarpus   279 f.
elaiosomes   351, 362
electrostatic droplet sorter   377
Eleocharis   117
ELISA, *see* enzyme linked immuno sorbent assay
ellagic acid   91
ellagitannin   89, 91, 93
Elytrigia   117, 124
embryo
– DNA content   124, 131, 133, 135, 142 f., 149 f.
– formation   53, 131 ff., 135, 140, 142 f., 145, 150
– ploidy variation   142 ff.
– sac   131, 135, 142, 144 f., 351
Embryophyta   44
Emiliana   278, 305, 316
emission   4, 12, 14, 20, 25, 27, 29, 33, 38, 170, 219, 234, 239, 241, 245, 254 ff., 262, 264, 292, 295, 297, 311, 375, 413, 418, 434
Empetrum   115, 124
endocycle   71, 350, 366
endomitosis   105, 141, 337, 349 ff., 364
endonucleases   78
endopolyploidization   53, 141, 143, 351 ff., 360, 362 ff.
endopolyploidy   18, 41, 52 ff., 75, 80, 105, 107, 113 f., 282, 349 ff.

endoreduplication 105, 143, 237 ff., 349 f., 354, 363 ff., 368 f., 414
endosperm 53 f., 70, 74, 79, 124, 131 ff., 142 ff., 148 ff., 246, 342, 351, 356, 360, 362 ff., 365 f., 368
endosymbionts 280, 284
energy transfer 93, 212, 255, 263, 295, 418
enhancer trap lines 412
Enteromorpha 191, 277
enumeration of microorganisms 223, 227
enzyme linked immuno sorbent assay (ELISA) 217 f., 222 f., 225 ff.
ephemerals 161 f., 271
epidermal cell 48, 243, 361, 363
epidermal protoplasts 414
epigenetic factors 405
epiillumination 4
Epilobium 107 f.
epoxomycin 329 f.
Equisetum 117, 119
Ericaceae 110, 115, 140
Erwinia 219, 225 f.
Escherichia 49, 276
ESTs, *see* expressed sequence tags
ethanol 45, 78, 98, 134, 301
ethidium bromide (EB) 8, 45, 49, 74 f., 78 f., 109, 111, 169, 178, 275, 378, 433 f.
ethylene 365 f.
Euglena 268, 280
Euphorbiaceae 95, 357
Eupithecia 126
Eustigmatophytes 267
evolution 51, 56, 67, 98, 105, 117, 120, 122, 126 ff., 132, 147, 150, 263, 284, 378, 400
excitation
 – beam 26, 29, 33, 45 f., 295, 375
 – energy 43, 255
 – light 25 f., 45 f., 60, 295, 326, 375
 – fluorescence excitation 10, 219, 244
 – wavelenghts 3 f., 14, 25 f., 185, 263 f., 297, 434
expressed sequence tags (ESTs) 408 f.

f

Fabaceae 45, 48, 50, 68, 75, 78, 81, 91, 97, 117, 140, 149, 170, 178, 182, 186, 192, 194 ff., 200, 203, 207 ff., 243, 260, 276, 281, 327, 335 f., 352, 356 f., 361, 367 f., 380 ff., 425, 431
FACS, *see* fluorescence-activated cell sorter
FAD, *see* fluorescence array detector
FCM, *see* flow cytometry
FCS, *see* flow cytometry standard
FCSS, *see* flow cytometric seed screen

FDA, *see* fluorescein diacetate
FDA fluorochromasia 238, 246 ff.
Fermi-function 366
ferrichloride 94
Festuca 79, 97, 149
Feulgen (micro)densitometry 68, 86, 88, 95, 103, 169, 171, 174, 274, 282 f., 352 ff.
Feulgen staining 8, 170, 282
field bean 45, 223, 333, 382 ff., 389, 391 f., 396 f., 399
filter 4, 26 f., 29, 33, 43, 46, 221, 228, 235, 292, 377, 437
fireweed 110
FISH, *see* fluorescence *in situ* hybridization
FITC, *see* fluorescein isothiocyanate
fixation 46 f., 49, 79, 91, 109, 132, 135, 148, 252, 259 ff., 272 ff., 301, 341, 384
fixatives 45, 91, 273 f.
Flaveria 263
flavonoids 74, 84, 88 f., 91 ff.
flow chamber 22 ff., 44
flow cytogenetics 373 ff., 377 ff., 385 f., 390 ff., 395, 398 ff.
flow cytometer 6 ff., 20 ff., 29 ff., 36 f., 42 f., 48, 59 ff., 76, 86, 88, 108, 127, 139, 143, 173, 222 f., 227, 234, 244, 247, 255, 257 f., 277, 282, 284, 291 ff., 297, 299, 300, 302 f., 305, 307 ff., 313 ff., 324 f., 376, 390, 411, 415, 426, 435
 – lamp-based 83, 86 ff., 258, 433
 – multiparametric 8, 14
 – pump-during-probe 297
 – submersible 293 f., 303, 316
flow cytometric informatics 21, 30 ff.
flow cytometric seed screen (FCSS) 54, 71, 74, 131, 136 ff., 142 ff., 150
flow cytometry (FCM)
 – advantages 20, 68, 104 f., 178, 424
 – application 47 ff., 50, 59, 67, 103 ff., 111, 114 ff., 119 ff., 149, 154, 171 ff., 262, 276, 303, 315, 423, 425, 436
 – aquatic 59 ff., 284, 288, 294, 303 ff.
 – development 1 ff., 41 f., 169 ff., 232
 – fluidics 21 ff., 60, 292
 – history 2, 20 ff.
 – multiparameter 8 ff., 20, 248, 340
 – of chloroplasts 251 ff., 256, 263
 – principles 19 ff., 31, 375
 – slit-scan 376
flow cytometry standard (FCS) 30
flow karyotype 57, 373, 375 ff., 384 ff., 390 f.
flow karyotyping 57, 373, 375, 378, 385 f., 390 ff., 394 f., 399
flow orifice 233

flow rate   23 f., 233, 293, 307, 375
flow sorter   43, 60, 233, 376 f.
FLOWer database, *see* Plant DNA flow cytometry database
fluid sheath   22 ff.
fluorescein   4, 10, 12, 14, 25, 37, 44, 58, 223, 239
fluorescein diacetate (FDA)   47, 238 f., 414
fluorescein isothiocyanate (FITC)   33, 38, 48 f., 223, 239, 243 ff., 260 f., 395
fluorescence   4, 43, 123
  – activated cell sorter (FACS)   9 f., 12 ff., 435
  – array detector (FAD)   14 f.
  – channel   12
  – emission   20, 27, 38, 170, 234, 239, 241, 254 f., 375, 413
  – *in situ* hybridization (FISH)   57, 307, 351, 387 ff., 392 ff., 400
  – microscopy   3 ff., 9, 11, 15, 92, 178, 219, 222, 227, 276, 292, 352, 374
  – pulse   375 f., 389
  – quenching   48, 53, 90 f., 93, 237, 243, 257 f.
  – resonance energy transfer (FRET)   418
fluorescent antibody technique   3 ff.
fluorescent protein (FP)   49, 56, 237, 244, 251, 323, 342, 411 f., 414, 417
fluorochromatic dyes   239
fluorochromes   14, 25, 33, 37, 41, 43 ff., 49, 55 ff., 68, 74 f., 78 f., 81, 90, 92, 95, 104, 107, 169 ff., 186, 188, 190, 192, 194, 196, 198, 200, 211, 222, 224, 237, 239, 244, 272 ff., 281, 292, 352, 378, 418, 423 f., 426, 433 ff., 437
  – base-specific   107, 169, 352, 433
  – intercalating   352, 434
fluoroglucinol   247
fluorophore   33, 237, 261, 295, 411
flying spot scanner   5
Folin-Ciocalteu reagent   94
formaldehyde   47, 49, 109, 260, 273 f., 277, 301, 384
forward scatter (FS, FSC)   10, 25, 80, 92 f., 104, 123, 223, 235, 238 ff., 246 f., 259, 260, 297, 299, 315, 389, 413
FP, *see* fluorescent protein
FRET, *see* fluorescence resonance energy transfer
Fritillaria   54, 56, 159, 161, 191
FS, FSC, *see* forward scatter
Fucus   268, 279
Funariaceae   71, 281

fungal spores   58, 219 ff.
fungi   58, 218 f., 221, 228, 231, 364
furanocoumarins   91
Fusarium   221, 397

g
G_0/G_1 peak   52
galactolipids   260
galactose   260
Galanthus   118
Galax   124
gallotannins   89 f., 92–94, 276
gamete   54, 103, 108, 109, 111, 120 ff., 132 ff., 139, 140 ff., 270, 279, 350
  – unreduced   54, 75, 103, 106, 108 f., 113, 120 ff., 127, 133 f., 139 ff., 147
gametophyte   70 f., 95, 131 ff., 189, 246, 269 ff., 279, 283
garden pea   381, 391 f., 399
Gasteria   118
gating   10 ff., 50, 60, 92, 94, 262
Gaussian distribution   326
GeneChip   407 f., 416 f.
gene cloning   57, 373, 397
gene expression   18, 48 ff., 56, 232, 243 f., 333, 337 ff., 341, 405 ff., 409, 411 ff.
  – methods for analysis   41, 406 ff.
  – regulation   48, 337, 406
generative nuclei   111 ff., 123, 141
generative polyploidy   67, 70 ff., 103
Genlisea   54, 159 f.
genome
  – analysis   18, 374, 394 ff., 399
  – copy number   103
  – duplication   51
  – sequencing   57, 153 f., 204, 279, 373, 397
  – multiplication   103, 127
genome size   2, 15, 41, 52, 54 ff., 61, 67 ff., 74 f., 78 ff., 83 ff., 88, 92, 95, 97 ff., 104 ff., 108, 113, 118, 124 ff., 145, 147, 149, 153 ff., 166 ff., 170 ff., 177 f., 182, 185 ff., 202 f., 212, 267, 271 ff., 272 ff., 349, 352, 356, 358, 362 ff., 367, 369, 423 f., 430 ff.
  – algae   158, 276 f.
  – analysis   118, 157, 271 ff., 352, 424
  – angiosperms   88, 155 ff., 160, 169 ff., 202, 362
  – bryophytes   158 ff., 172 ff., 277, 280 ff.
  – gymnosperms   156 ff., 171
  – Initiative (GESI)   157
  – pteridophytes   158 ff., 172
  – studies   75, 106, 156, 169, 276 ff., 430, 435

genome size (*cont.*)
- terminology 70
- variation 52, 55, 95, 97 f., 104, 153 f., 159 ff., 166 ff., 174
genomic constitution 55, 118
genotype 48, 132, 141 f., 145, 154, 237, 382 f., 391, 405
geophyte 115
Gerbera 191 f., 221
GESI, *see* Genome Size Initiative
GFP, *see* Green Fluorescent Protein
gibberellic acid 365
gibberellins 365 f.
Gigartinaceae 189, 272
Ginkgoaceae 192, 203, 207, 358
Gliocladium 220
Glomus 364
β-glucuronidase 342, 414 f.
glutaraldehyde 273 f., 301
glutathione 247
Glycine max 81, 85, 87, 97, 170, 192, 207, 238, 281, 332, 335, 352, 432
grana 253, 259
green fluorescent protein (GFP) 41, 49, 238 f., 241, 245, 340, 342, 369, 411 ff., 420
Greya 126
GUS, *see* ß-glucuronidase
Gymnadenia 113 f.
Gymnodinium 301, 307 f.
gymnosperms 74, 142, 153, 156, 157 ff., 171 ff., 356
gyrase 58, 251, 261
Gyrodinium 307, 310

h

HAB, *see* harmful algal bloom
Haemophilus 155
hairy roots 380 f., 383 f.
haploid 3, 47, 50, 69, 70, 95, 133, 144, 269, 270, 277, 278, 282 f., 382
- parthenogenesis 109, 117, 122
Haplopappus 57, 162, 192, 379 f., 385
HAPPY mapping 393, 399
Haptophyceae 273, 278
Haptophyta 273, 277, 316
harmful algal bloom (HAB) 287, 308 ff., 317
heat-shock 242
Helianthus 95, 125, 170, 352
Helleborus 119
hematopoietic stem cells 37
HEPES 77 ff.
herbarium vouchers 46, 109 f., 426

herbicides 317, 329, 383
heterochromatin 52, 56, 86, 88, 114, 202, 275, 350, 378
heterocysts 293
heterokaryons 41, 49, 244 f.
Heteroptera 349
Heuchera 126
Hieracium 97, 118, 124, 140, 147
histogram 31 f., 38, 42, 45, 53, 68, 71 f., 78, 81, 89 f., 93 f., 96, 106 ff., 112 ff., 143 ff., 147, 163 f., 169, 172, 179, 234, 245 f., 256, 274, 278 f., 281, 326, 334, 352 ff., 354 f., 375, 412 f., 426, 428, 433 f.
Hoechst
- 33258 53, 76, 181, 278, 378
- 33342 11, 13, 178, 181, 185 ff., 192 ff., 246, 277 f.
holokinetic chromosomes 104, 113 ff.
holoploid 67, 69 ff., 274
homeostasis 339
homoploid 55, 107, 118, 123 ff.
Hordeum 87 f., 192, 207, 333, 336, 380, 431 f.
horsetail 117, 158, 425
HPLC 204
Hyacinthaceae 56, 226, 357
Hyacinthus 226
hybridization 41, 50, 75, 109, 122, 124 ff., 147, 244
Hydrangea 192 f., 203
hydrodynamic focusing 22 ff.
hydrolysis 91, 204, 219, 238 f., 247, 313
hydroxyurea 328, 330, 332 f., 335, 383
Hymenophyllaceae 172
Hypericaceae 54, 71, 139, 144
Hypericum 54, 71, 74, 134, 136, 138, 142, 144 ff.

i

IAA, *see* indolyl-acetic acid
IBA, *see* indole-3-butyric acid
ICRF 193 329, 331
IF, *see* immuno fluorescence microscopy
IFCM, *see* immuno flow cytometry
illumination 5, 9 f., 12, 14 f., 22, 34, 234 ff., 245, 292
image
- analysis 5, 7, 21, 34, 174, 267, 282 f., 352 ff., 374
- analyzer 7, 42
- cytometer 9, 15
- cytometry 3 ff., 14 ff., 248, 418 ff.
imaging 1 ff., 5, 9, 15, 21, 42, 61, 293 f., 296 ff., 300, 323, 339 ff., 342, 410

immuno flow cytometry (IFCM) 222 f., 227 f.
immuno fluorescence microscopy (IF) 218, 222, 227 f.
immunofluorescence 7, 11 f., 15, 32, 53, 243, 327
indole-3-butyric acid (IBA) 365
indolyl-acetic acid (IAA) 365
inhibitors 67, 74 f., 79, 81, 83 ff., 88 ff., 95 f., 98 f., 170, 172, 276, 328, 329, 331 f., 335 ff., 366, 383
  – cytosolic inhibitors 99, 170, 172
  – DNA staining inhibitors 276 ff.
  – DNase inhibitor 79
  – fluorescence inhibitors 67, 74, 83 ff., 89 ff., 95, 98
  – synthesis inhibitor 328 f., 335, 383
interrogation point 22, 26, 30 f., 33, 36, 255, 354
interrogation volume 23 ff.
intraspecific DNA content variation 54, 98, 113
invasion biology 119 ff.
ionic detergents 78
Isochrysis 278
isolated nuclei 56, 68, 244 f., 272, 278, 325, 353, 417
isopycnic centrifugation 232, 234

j
Juncaceae 110, 113
Juncus 110
Jungermanniaceae 159

k
Kalanchoe 363
Kappaphycus 193, 277
Karinia 308
karyotype 57, 124, 373 ff., 380, 382, 384 ff., 390 ff., 399
Katodinium 280
kinematic methods 341
kinetochores 113, 329, 396
kiwi 109
Koeleria 362
"Kranz" anatomy 254

l
lactacystin 329, 331
Lactuca 74, 119, 193 f., 202, 353
Lamiaceae 115, 193, 199 f., 203, 358
laminar flow 20, 22 ff., 34, 45, 273
Laminaria 194, 272, 277 ff.
Lamium 115, 117

lamp
  – arc lamp 3, 9 f., 12, 13, 15, 25 f., 105, 110, 255, 258
  – filament lamp 3
  – mercury lamp 25, 434
land plants 44 f., 153, 156, 158 f., 173
laser 8 ff., 22, 25 f., 29 f., 33 f., 42 f., 60, 86 ff., 114, 212, 222, 225, 227, 234 ff., 241, 246, 254 ff., 257 f., 260, 263, 275, 292 ff., 300, 314, 354, 374, 376, 411, 415
  – air-cooled 10, 12 ff., 314
  – argon 9, 10, 12 ff., 25, 275, 297, 314
  – diode 12 f., 33
  – diode-pumped YAG 13, 33
  – solid-state 12, 15, 25, 275
  – water-cooled 10 ff., 25
lectins 260
LED, *see* light-emitting diode
legumes 238, 364, 399
lens 6, 12, 14 f., 251
Lentibulariaceae 54, 159
lettuce 119
leucocyte 83, 88
life cycle 67, 69, 131, 133, 166 f., 270 ff., 278, 280, 284, 297
light
  – absorption 7, 21, 252 f.
  – coherent 25
  – emitting diode (LED) 15
  – scatter 6 ff., 10, 14, 20, 26 f., 31, 34, 38, 41 f., 45, 47, 50, 58, 61, 223 ff., 238 ff., 246, 251 ff., 259 ff., 294 f., 297, 300, 305 f., 311 f., 389 f., 413 ff., 423
  – source 3 f., 7, 15, 20, 22 f., 25 f., 37, 59 f., 295, 326
  – stress 257
Liliaceae 54, 56, 123, 141, 159, 168, 191, 194 f., 203, 358
Lilium 123, 141, 168, 194 f.
liquid sheath 20, 23 f.
liverworts 158 f., 269, 276, 281, 284
Lolium 51, 149, 363
Loranthaceae 52, 201
Lotus 140, 195
Luminex 13, 223 ff.
Lupinus 48, 56
Luzula 113
Lycopersicon 74, 79, 195, 208, 362, 380, 432
lycophytes 153, 156, 158, 425 f.
Lythrum 117, 120

m
Macrocystis 272, 278
magnetic sorting 262

maize, *see also Zea mays* 47, 56 ff., 132, 246, 256, 259, 262, 339, 363, 365 ff., 373, 382, 390 f., 397, 399, 412
malic enzyme (ME) 254, 261 ff.
Malus 108, 122
Mammillaria 365
MARs, *see* matrix attachment regions
massively parallel signature sequencing (MPSS) 409 f., 416
matrix attachment regions (MARs) 412
ME, *see* malic enzyme
mean C-level 353 ff., 364, 367
mechanical homogenization 46, 49, 57, 373, 382, 384
Medicago 50, 75, 195, 208, 332 f., 335 f., 368
megagametophyte 95, 337
meiosis 150, 160 ff., 270
Melandrium 78, 96, 195, 380
melanin 68
Melanthiaceae 113
Meloidogyne 364
Melosira 290
melting profile 204
membrane fluidity 48
mercaptoethanol 77, 79, 90, 352
MESF, *see* molecules of equivalent soluble fluorochrome
mesophyll 58, 238, 247, 253 f., 256, 261 ff., 332, 335, 379 ff., 414
 – cells 45, 48, 243, 254, 259, 379
 – protoplasts 46, 238, 379, 383, 414
metaphase
 – accumulation 336, 341, 375, 379, 383 ff.
 – chromosome 180, 329, 373, 375, 384, 394
 – indices 336, 379, 384
 – synchrony 384
methanol 273
MIA, *see* microsphere immuno assay
microarray 378, 405, 407, 410, 416, 418
Microcystis 307 f.
microdensitometry 103, 169, 171, 174, 352
microfluorometry 325
Micromonadophyceae 158
Micromonas 278, 312
microsatellite markers 154, 388, 395 f.
microscope 1–9, 11 f., 14 f., 21 f., 24 f., 68, 135, 168, 238, 245, 291, 293, 317, 352, 377, 393, 400, 407
microscopy 1 ff., 59, 61, 140, 150, 178, 219, 227, 241, 252, 254, 260, 263, 292, 297, 342, 351 ff., 396, 415

microspectrophotometer 2 f., 5
microspectrophotometry 3 ff., 7, 264, 325, 352
microsphere immuno assay (MIA) 222, 224 ff.
microspores 44, 47, 50, 123, 141, 236, 246 f., 416
microtubule 329, 331, 333
mimosine 327, 330
mithramycin 55, 68, 75, 169, 178 f., 181, 201, 385, 433
mitochondria 41, 56, 58 ff., 241 f.
mitochondrial
 – membrane potential 58, 240 ff.
 – numbers 244
mitosis 53, 71, 123, 160, 324, 329, 331, 334, 336 f., 341 f., 349, 351, 364 f., 369, 383
mitotic
 – chromosomes 18, 57 ff., 279, 373, 375, 380, 394 f.
 – index 50, 321, 333, 336, 341, 383
 – phase 53, 71, 85
 – spindle 329, 336, 383
mixoploidy 50, 368
Mnium 281
molecular
 – beacons 419
 – cytogenetics 374
 – markers 57, 126, 138, 140 f., 393, 396 ff.
molecules of equivalent soluble fluorochrome (MESF) 14, 37
monilophytes 153, 156, 158, 425 f.
monitoring applications 302, 305 ff.
monoclonal antibody 1, 11 f., 243, 326
monoploid 67, 69 ff., 274, 278
Monostroma 283
monovariate analysis 386
Moore's law 410
MOPS 77 ff., 273
mosses 158, 267, 269 ff., 277, 280 ff., 284
MPSS, *see* massively parallel signature sequencing
mRNA 341, 419
mucilage 79
multiplex assay 58
Musa, *see also* banana 50 f., 56, 118, 196
Mylia 159
Myristica 113

n

NAA, *see* naphtaleneacetic acid
NADH 240
NADPH 253, 261

naphtaleneacetic acid (NAA)   365 f.
Nicotiana   48 f., 58, 68, 196, 208, 240, 245, 260 f., 332, 337, 365 f., 380
NimbleGen   407 f.
NLS, *see* nuclear localization signal
noise   14, 25, 226, 248, 273, 419 ff.
non-ratiometric dyes   240
non-vascular plants   18, 59, 267 ff., 274 ff., 283 f.
norflurazon   414
nuclear
    – DNA breakdown   247
    – DNA content   2, 45, 50 f., 54 f., 61, 67 ff., 72, 74, 85, 98, 103, 105, 107, 110 f., 113 f., 116, 118, 122 f., 132, 139, 141, 180 f., 271 ff., 275 ff., 284, 324, 327, 334, 339 ff., 352, 355, 369, 377, 423, 426, 430 f., 434 ff.
    – localization signal (NLS)   56, 414
    – phases   67, 69 ff., 103
nuclei suspensions   46, 49, 68, 75, 79, 86, 90, 92 f., 139, 148, 277, 281 ff., 384, 427
nucleolar organizing region   351
nucleosomal fragmentation   48, 240
nucleotide   73, 408, 410
nucleotype   54, 154
nucleotypic effect   155, 160, 162, 166
numerical chromosome changes   57, 373, 390, 398

o

oat-maize chromosome addition line   382, 390
obscuration bar   235
olivomycin   75, 109, 169, 178, 201, 433
olomoucine   329, 331, 337
Onagraceae   107, 110, 112
onion   85, 328, 333, 336
Ophrys   113
optics   11, 14, 21, 25 ff., 135, 235, 287, 303, 314 ff.
Orchidaceae   114, 357, 361, 365
Orchis   113
organelle   2, 39, 41, 44, 46 ff., 56, 61, 234, 243, 251 ff., 272, 408, 419
Oryza sativa   178 f., 185, 196, 205, 208, 381, 432
oryzalin   329, 330, 383
Ostreococcus   59
Othonna   360
ovule   122, 124, 135, 142, 350 f.
Oxalis   79, 124
Oxyrrhis   272

p

palea   143
Panicoideae   135
Pap smear   4 f., 7
paraformaldehyde   273
paramagnetic microsphere immuno assay   226 ff.
Paramecium   268, 280
parasite   51, 115, 126, 364 ff.
Partec   9, 11, 13, 76, 78, 82, 105, 110, 114, 125, 149, 173, 435 f.
parthenogenesis   109, 117, 131, 132, 135, 145, 147 f., 278
Paspalum   54, 146
pathogenic   217 ff., 221 f., 225 ff., 228, 317
PCD, *see* programmed cell death
PCR   56 f., 217, 233, 275, 307, 373, 377, 387, 389, 392 ff., 396 f., 408 ff., 436 f.
pectinase   44 f., 232, 383
Pennisetum   141
PEP carboxykinase   261
pepsin   45
perennials   161 f., 166, 167, 269, 357
pericarp   143
Petunia   49, 88, 90, 119, 170, 196, 281, 326, 332, 335, 379, 381, 431 f.
Pfiesteria   269, 273, 280, 308
pH   73, 76 ff., 83, 90, 149, 204, 243, 247, 384
Phaeocystis   277 f., 311
Phaeophyceae   279, 356
Phaeophyta   153, 156, 158 f., 277 f.
Phaseolus   178, 196, 209
phenolics   56, 68, 74, 81, 84, 89 f., 98, 171, 276
phenoloxidases   68
phenotype   39, 47, 51 f., 54, 103, 115, 127, 140, 154, 160, 248, 261, 313, 355, 405 f., 417, 419
Phleum   362
Phoenix   52, 56
3-phosphoglycerate   260
phosphorylation   331, 336 f., 411
photodetector   3, 6, 21, 26 f., 29
photodiode   27
    – avalanche photodiode   29
photography   1 f., 9, 92
photoinhibition   257
photomicrodensitometry   168
photomultiplier tube (PMT)   27, 29, 295
    – multichannel   29
photosynthesis   41, 251 ff., 261, 263 f., 289, 313, 318
phycobiliproteins   12
phycoerythrin   12, 297

phylogenetic 117 ff., 167, 212, 267, 284, 308, 369, 396
Physcomitrella 71, 271, 279, 281 ff.
physical
— contig map 394, 397, 399
— mapping 57, 373, 378, 392 ff.
physiology 39, 59, 125, 166, 237, 239, 306, 311, 313, 315, 341
phytochrome 363
phytohormone 349, 362, 365 ff., 369, 417
— receptors 243
Phytophthora 220
phytoplankton 18, 42, 59 f., 173, 237, 271, 273, 275, 287 ff., 290 ff., 297, 299, 301 ff., 305, 307 ff., 310 ff., 317 f.
PI, *see* propidium iodide
Picloram 365
PicoGreen 276
pigments 12, 38, 45, 92, 245, 252 ff., 264, 271, 273, 288, 293, 295, 299, 301, 305 f., 308, 311
Pilayella 277
Pimpinella 97, 117
Pinaceae 95, 171, 186, 189, 194, 196 f., 203, 209, 358, 381, 425
pinhole aperture 3
Pinus 95, 171, 196 f., 209, 425
Pisum 75 f., 78 f., 82, 84, 87 f., 92 ff., 114, 182, 197, 205, 209, 260, 276, 333, 336, 356, 360 f., 364 f., 367, 381, 430, 432
Plant DNA C-values Database 54, 156, 158 f., 161, 166, 168 f., 171 f., 431, 433
Plant DNA Flow Cytometry Database 18, 77, 423 ff., 430, 433, 435 ff.
plant pathogens 58 ff., 217 ff., 225 ff.
plant regeneration 368 ff.
plastids 41, 238, 240, 242 f., 251 f.
ploidy 41, 50 ff., 61, 70, 72, 81, 103 ff., 126, 127, 131 ff., 137, 138, 139 ff., 145 f., 149 f., 155, 167, 272, 275 f., 278, 282 ff., 323, 349 f., 354 f., 361, 364, 366, 423, 430, 433 ff.
— analysis 104, 131, 139 ff., 150
— chimeras 51
— level 50 ff., 70, 103, 105 ff., 109, 113 f., 116, 120 ff., 124, 126, 133, 135, 138, 142 f., 145 f., 149 f., 155, 167, 272, 276, 278, 282 f., 323, 349 f., 354 f., 361, 366, 423, 430, 435 f.
— screening 41, 50 f., 81, 107, 283
— stability *in vitro* 50
— variation 50 f., 103, 115, 117, 119 ff., 126, 142 ff., 283
PMT, *see* photomultiplier tube
Poa 74, 139, 141, 147 f., 150

Poaceae 51, 54, 56, 74 f., 79, 82, 88, 97, 116 f., 123, 139 ff., 143, 146, 148 f., 178 f., 191 f., 196, 198 ff., 207 ff., 254, 256, 259, 336, 352, 356 f., 361 ff., 367, 380 ff., 425, 431
Poinsettia 95
Poisson statistics 36
pollen 44, 47 ff., 50, 111 ff., 121 ff., 126 f., 133, 135, 139 ff., 144, 149, 160, 246, 416
— unreduced 111 ff., 122 f., 133, 139 ff.
pollution 166 ff., 317
Polygonaceae 52, 92, 116, 120, 123, 197
polyhaploid 124
polyhydroxyphenols 89, 91
polymerase 328, 330, 399
polyploidization 50, 67, 69, 104, 125, 127, 147, 280, 349 f.
polyploidy 51, 69 ff., 103, 118, 120 ff., 125 ff., 279, 282, 368
— generative 70 ff., 103
— somatic 69, 71 ff.
Polypodium 124
Polysiphonia 279
polyteny 278 f., 350
polyvinylpyrrolidone 90, 352
Pooideae 135, 143
Porphyra 197, 277
Portulaca 361 f.
positional gene cloning 57, 373, 396 f., 399
Potamophila 146
pounds per square inch (psi) 23, 36, 377
Prasinophyceae 59, 278
pressure 22 ff., 36, 48, 166, 233, 282, 284, 292, 377
— differential 23 f.
primed *in situ* DNA labeling (PRINS) 57, 389, 396
primed *in situ* DNA labeling en suspension (PRINSES) 386
PRINS, *see* primed *in situ* DNA labeling
PRINSES, *see* primed *in situ* DNA labeling en suspension
Prochlorococcus 59, 289 f., 310
progeny test 131, 135, 149, 150
programmed cell death (PCD) 48, 237, 239 f., 242, 247
prokaryotes 38, 251, 275 f., 288
promoter 49, 340, 342, 412, 414 f., 417, 420
propidium iodide 43 f., 55 ff., 60, 75 f., 78 f., 82, 86 f., 90, 92 ff., 97, 106 ff., 111 f., 114, 169 ff., 178 ff., 182, 204, 219 ff., 237, 239, 246, 261, 273 ff., 281 f., 325, 395, 412 f., 415, 433 f.
proportionality factor 180, 211
propyzamide 329 f., 332 f.

Prorocentrum 280, 307
protein arrays 222 f.
proteomics 39, 234, 374, 399, 410 f.
protoplast 37, 41, 45 ff., 56 ff., 68, 75, 231 ff., 237 ff., 240 ff., 248, 272, 325, 332, 335 ff., 379 ff., 408, 411 ff., 423
– analysis 48, 231 ff., 237 ff., 416
– cell cycle 244 f., 335
– fusion 49, 244 f.
– integrity 234, 243
– lysis 46, 282, 284
– preparation 45, 231 ff., 413
– production 232, 239, 244, 246, 417
– purification 234 f.
– size 237 ff., 240
– sorting 242 f., 414 ff., 417 f.
protoporphyrin IX 240
Prunus 91
Prymnesiophyceae 273, 277 f., 311 f.
PS I/PS II ratio 253, 261 f.
pseudogamous 54, 131, 135 f., 142, 144, 147
Pseudo-nitzschia 299 f., 308
psi, *see* pounds per square inch
Psilotales 172
pteridophytes 44, 156, 158 ff., 172 ff.
pUBs1 vector 397
pulse width 234, 236, 246, 354, 389
purity in sorted fractions 377, 389 ff.
purple loosestrife 117
PVP 79, 90, 352
Pyramimonas 306
Pyrrhophyta 290

q
QC, *see* quiescent center
QTL, *see* quantitative trait locus 397
quality control 95 ff., 98, 435
quantitative trait locus (QTL) 397
quantum dots 12
quantum efficiency 43
quenching 48, 53, 91, 93, 237, 243, 257
quercetin 91
quiescent center (QC) 49, 417

r
radiation 91, 160, 167, 255
Ralstonia 220
Ranunculaceae 48, 116, 119, 124, 142, 178, 188, 203, 356, 358
Ranunculus 124, 142
Raphanus 87, 198, 355, 432
raphidophytes 267
ratiometric measurement 240
reaction center 253, 255, 257 f.
reactive oxygen 242, 247

recombinant DNA 247, 373, 396 ff.
reductant 79, 90, 253
reductase 240, 328, 330
refractive index 259, 295, 297, 312, 315
regula falsi 182 f.
repetitive DNA 57, 118, 212, 386 ff., 395
replication 53, 69, 71, 74 f., 97, 106, 114, 261, 324, 326, 328, 337, 366
replication-division phases 69 ff.
reporter 56, 223 f., 342, 419 f.
reproduction 52, 54, 74, 131 ff., 138, 140, 142, 145 ff., 150, 269 f.
– apomictic 52, 132, 146
– mode screening 74, 131 ff., 138, 143
– sexual 132, 140, 142, 147, 150, 280
reproductive pathway 53 ff., 122 ff., 136, 140, 142 f., 145 ff.
restitution 71, 146, 337, 365
retinoblastoma 337, 341
Reynolds number 23
Reynoutria 120
RFLP 141, 146, 392, 396
Rhizobiaceae 364
rhodamine 123 58
rhodamine isothiocyanate 49, 244 f.
Rhodomonas 291
Rhodophyta 153, 156, 158, 267, 269, 277, 297
Rhoeo 351, 361
rice 49, 381, 412
RNA 3, 39, 49, 53, 56, 75 f., 204, 218 f., 228, 245, 271, 275, 312, 326, 408 ff., 416 ff.
– content 3, 75, 245, 326
– transcripts 56, 417
RNase 75 f., 204, 275
root 36, 49, 71, 74, 92, 109, 113, 126, 172, 180, 182, 238, 327, 335, 341 f., 351 f., 356, 359, 361 f., 364, 383 f., 408, 414 ff.
– meristem 335 ff., 380 ff., 399
– tip 45, 49, 57, 68, 70, 85, 327 f., 332 f., 335 f., 373, 383 f., 391, 416
Rosa 123
Rosaceae 108, 117, 122 f., 191, 196, 203, 356, 358
roscovitine 329, 331, 333, 335, 337
Rubiaceae 54, 90, 116, 146, 170
Rubus 117, 120, 122
Rumex 52, 92, 94, 123
rust 397
rye 51, 82 f., 333, 381 f., 387 f., 390 f., 393 f., 398 f.

s
Saccharomyces 155, 277, 298
Saccharum 254

SAGE, *see* serial analysis of gene expression
salicylic acid   242
sample preparation   44, 46, 53, 74 ff., 81, 105, 113, 145, 204, 228, 277, 301 ff., 325, 373, 375, 379, 424
sampling frequency   291, 302 ff., 307, 317
Saxifraga   115, 198
Saxifragaceae   115, 126, 198
scanning   3 ff., 26, 86, 95, 223, 241, 254, 263, 296 f., 299, 307, 374, 396, 407
scatter, *see* forward scatter *and* side scatter
scattergram   92 ff., 220
Scrippsiella   280
Scrophulariaceae   54, 342
Scytosiphon   279
Secale   76, 82, 87, 198, 209, 333, 336, 381, 387, 394, 432
secondary metabolites   39, 44 ff., 48, 54, 68, 81, 89, 93, 98, 276 ff., 327, 427
Sedum   79
seed   44, 47, 49, 50, 52 ff., 61, 70 f., 74 f., 84 f., 99, 106, 108, 109, 120 ff., 124, 126 f., 131 ff., 135 ff., 142 ff., 158, 160, 161 ff., 167, 173 ff., 222 ff., 339, 349 ff., 358, 367, 383, 392
seedling   85, 12 f., 352, 363, 365 f., 368, 383
senescence   240
sequencing   57, 153 f., 177 f., 204 ff., 208, 212, 275, 279, 338, 373, 393, 397, 399, 407 ff.
serial analysis of gene expression (SAGE)   409 f., 416
Sesleria   117
sex chromosomes   52 ff., 274, 392
sheath
  – fluid   24, 48, 233, 292, 296, 377
  – flow   6, 23 f.
  – pressure   36
side scatter (SS, SSC)   6, 10, 14, 29, 41, 57, 60, 79 f., 92 ff., 96, 235 f., 238 ff., 247, 259, 295, 297, 413
Silene   52
silico-imaging   297 ff., 300
Sinapis   162, 199, 363
Skeletonema   293, 301
slit-scanning   26
Solanaceae   48, 49, 68, 74, 88, 119, 170, 189, 195 f., 199, 203, 208 f., 240, 260, 281, 326, 356 f., 362, 365, 367, 380 f., 425, 431
Solanum   49, 74, 199, 209, 326, 362
Solexa   410
somaclonal   146, 367 f.
somatic hybrids   47, 49 ff., 232, 244 ff.
Sorghum   363

sort purity   389 ff.
sort rate   48, 233 f., 377, 384, 390 ff.
sorter   8 ff., 21, 25, 30 f., 34, 36, 43, 60, 233, 376 f.
  – cell sorter   8 ff., 13, 21, 30 f., 34
  – fluidic sorter   12
  – fluid-switching cell sorter   34
  – high-speed sorter   11, 13, 34, 36
  – photodamage cell sorter   34
sorting
  – cell sorting   1, 7 ff., 10, 20 f., 34 ff., 39, 231, 248, 323, 342, 405 f., 408, 411, 414 ff., 419
  – electrostatic sorting   34
  – fluidic sorting   12
  – high-speed (flow) sorting   11, 34, 36
  – of nuclei   56 ff., 417 ff.
  – protoplast sorting   49, 242 f., 416 ff.
Spathoglottis   361
spectral   12, 19 f., 29, 31, 33 f., 252, 254 f., 263, 295, 411
  – band   20, 29, 31, 33
  – compensation   33 ff.
spectroscopy   2
spectrum   20, 25 ff., 61, 204, 212, 240, 254, 277, 292, 418, 434, 436
sperm cell   246
spermine   77 ff.
Sphacelaria   199, 277
Sphagnum   270, 272, 281 ff.
spinach   257, 259 f., 367
Spinacia   74, 199, 209, 257
Spirogyra   268, 276
spores   58, 132, 141, 219 ff., 270
sporogenesis   131 f., 143, 150
sporophyte   71, 131, 150, 268 ff., 278 f.
spotted arrays   407
SS, SSC, *see* side scatter
standard   83 ff., 86 ff., 430 ff.
  – animal standard   430
  – biological standard   72, 81
  – calibration standard   37, 169, 173, 424
  – curve   37 f., 185
  – external standard   81, 145
  – internal standard   83, 104, 107 f., 114 f., 169, 272, 274, 281 f., 352
  – reference standard   182, 272, 274 f., 430 ff.
  – species   67, 83 ff., 86, 88 f., 97, 274 f.
standardization   47, 67 f., 80 ff., 89, 95, 169 ff., 274 ff., 423, 426, 430 ff., 435
  – external   81, 170, 274, 430
  – internal   67 f., 81 ff., 169 ff., 170, 274, 430

– pseudo-internal 81, 426
– type of 80 ff., 430
static cytometry 8, 68
stilbenes 93
Stipa 146
stoichiometric
– errors 67, 171
– staining 78, 221, 276, 279
Stoke's shift 43, 292
stramenopiles 267
Striaria 279
stroma thylakoids 253
Strombidium 310
structural chromosome changes 373, 390, 392
subgenomic DNA libraries 57, 373
sulfofluorescein diacetate 239
supernumerary chromosomes 52, 170
surface water 288 ff., 302 f., 315
suspension 43 f., 46, 49, 57, 68, 75, 79, 86, 90, 92 f., 139, 141, 148, 231, 234, 246, 252, 263, 278, 281 ff., 292, 296, 314 f., 325 ff., 332, 334, 365 f., 373 ff., 378 ff., 399, 406, 423, 427
– cell 44, 231, 246, 292, 327, 332, 334, 365, 366, 379, 383 f., 406
– chloroplast 252, 263
– chromosome 57, 373, 378 ff., 382 ff., 387, 399
– nuclear 46, 75, 86, 90, 92, 139, 148, 277 f., 281 ff., 384, 427
– protoplast 234
suspensor 351, 362
SYBR Green 273, 275 f., 278 f.
Symbiodinium 280, 284
symbiont 51, 125, 280, 284, 364 ff.
synchronization 53, 323, 327 ff., 332 ff., 338, 375, 383 ff.
synchronized 53, 57, 326 ff., 331, 333 ff., 373, 375, 379, 383 f.
Synechococcus 290, 308, 310
syringe 8, 24
systematics 54, 98, 103 ff., 106, 114 ff., 118 ff., 126 f., 284
SYTO9 219 ff.
SYTOX Green 273, 275 f., 280, 313

t
tannic acid 92 ff.
tannin 79, 89 ff., 274
tapetum 141, 351
taxonomy 46, 61, 115, 126, 185, 283, 302, 426
TEM, *see* transverse emission mode
TEs, *see* tracheary elements

testa 143
Tetraodes 83, 88
Thalassiosira 293
thallus 269, 272, 277
thermal denaturation 177, 204, 206 ff.
thistle 117
thylakoid 57 f., 251 ff., 259 f., 262 f.
time of flight (TOF) 26, 30, 236 ff., 240, 295
tobacco 68, 237 f., 245, 272, 332 f., 335 ff., 368, 379 f., 412 f.
TOF, *see* time of flight
tomato 220, 339, 364, 368, 379 f., 396
topoisomerase 329, 331
TOTO 275
tracheary elements (TEs) 247
Tragopogon 97, 115
transcript 56, 246, 327, 405, 407 ff., 417 f.
transcription(al) 171, 218, 261, 328, 330, 337, 408, 411, 415, 417, 419 f.
transgenic 49, 231, 237 f., 240, 244, 248, 406, 411 ff., 414 ff., 420
– analysis 412 f.
– expression 237, 411
– markers 411 ff., 416 ff.
– plants 49, 238, 240, 244, 406, 412, 414 f., 417
– technologies 406
translational fusion 56, 342
translocation 57, 373, 382, 386 f., 391 ff., 397, 399
– lines 386, 392, 397
transmission 3 f., 7, 292 f., 305, 374
transplantation 37 f.
transverse emission mode (TEM) 25 f.
2,3,5-trichlorophenoxyacetic acid 365
trifluralin 383
Trifolium 149, 200, 209, 432
Trichomanes 172
trichomes 361, 363
Trilliaceae 56
Tripsacum 143
Triticum 51, 200, 210, 333, 336, 361, 363, 381, 388, 395, 432
two-step sorting 389 ff.
two-wavelength excitation 185

u
Ulex 117
Ulmaceae 79
Ulva 200, 277
Undaria 200, 277, 279
Uronema 310
Urtica 200, 363
Urticaceae 200, 203, 357, 363

v

Vaccinium 110, 115, 117, 140
Vanda 361, 365
vascular bundle 221, 254, 261, 361
VBNC, *see* viable but non culturable
vegetative nuclei 111 f., 141
viable but non culturable (VBNC) 221
Vicia 45, 68, 79, 87 f., 97, 200, 210, 223, 243, 276, 327, 333, 335, 361, 382, 385, 432
vinylpyrrolidone 90
virus 10, 12, 37 f., 42, 58, 61, 153, 155, 205, 218, 222, 225 ff., 238, 288, 291 f., 301, 310, 313, 337, 414
volumetric delivery 24, 223, 291 f.

w

walled cells 246 ff., 254, 263
weed 124, 163 f., 269
WGA, *see* whole genome amplification
wheat 55 ff., 118, 132, 140, 145, 333, 373, 379, 381 f., 388 f., 391, 393 ff., 397 ff.
  – germ agglutinin 58
wheat-barley addition line 390
wheat-rye addition line 51, 382, 390, 398
white campion 380, 383, 392
whole genome amplification (WGA) 399
whole genome sequencing 204, 211

x

Xanthomonas 222 f.
Xanthophytes 267

y

YOPRO-1 276
YOYO 275 f.

z

Zea mays, see also maiz 56, 75, 87, 92, 123, 143, 201, 210, 256, 259, 333, 336, 352, 356, 360, 362, 382, 431 f.
Zebrina 361
Zinnia 247, 332
zooplankton 42, 288, 291
zoospores 273, 280

Related Titles

Meksem, K., Kahl, G. (eds.)

The Handbook of Plant Genome Mapping
Genetic and Physical Mapping

402 pages with 77 figures and 11 tables
2005
Hardcover
ISBN: 978-3-527-31116-3

Christou, P., Klee, H. (eds.)

Handbook of Plant Biotechnology
2 Volume Set

1488 pages
2004
Hardcover
ISBN: 978-0-471-85199-8

Cullis, C. A.

Plant Genomics and Proteomics

214 pages
2004
Hardcover
ISBN: 978-0-471-37314-8

Shapiro, H. M.

Practical Flow Cytometry

approx. 736 pages
2003
Hardcover
ISBN: 978-0-471-41125-3

Givan, A. L.

Flow Cytometry
First Principles

292 pages
2001
Softcover
ISBN: 978-0-471-38224-9